A Little Corner of Freedom

The publisher gratefully acknowledges the contribution to this book provided by the Provost's Authors Support Fund of the University of Arizona.

A Little Corner of Freedom

Russian Nature Protection from Stalin to Gorbachëv

Douglas R. Weiner

University of California Press
Berkeley Los Angeles London

This book is a print-on-demand volume. It is manufactured using toner in place of ink. Type and images may be less sharp than the same material seen in traditionally printed University of California Press editions.

University of California Press
Berkeley and Los Angeles, California

University of California Press, Ltd.
London, England

Library of Congress Cataloging-in-Publication Data

Weiner, Douglas R., 1951–
A little corner of freedom : Russian nature protection from Stalin to Gorbachëv / Douglas R. Weiner.
p. cm.
Includes bibliographic references (p.) and index.
ISBN 978-0-520-23213-6 (cloth : alk. paper)
1. Environmentalism—Soviet Union—History. 2. Environmentalism—former Soviet republics—History. I. Title.
GE199.R8W45 1999
363.7'056'0947—dc21 97-40206

Printed in the United States of America

The paper used in this publication meets the minimum requirements of ANSI/NISO Z39.48-1992 (R 1997) (*Permanence of Paper*).

To my angels:
Loren and Pat, Feliks and Nadia, Nikolai and Elena,
Ol'ga, Konstantin, and Dania

CONTENTS

ILLUSTRATIONS

ACKNOWLEDGMENTS

I have been fortunate to have been able to spend considerable amounts of time living and researching in the Soviet Union and its successor states. For that I am indebted to the generous support of a number of granting agencies and foundations that have had faith that this project would one day see the light of day. A trip to the USSR in 1986 was supported by the National Academy of Sciences. IREX and the National Council for Soviet and East European Studies (contract 806–28), a Title VIII program, generously funded a key second research trip for ten months in 1990–1991 plus summer support. From June to August 1991 I had the good fortune to be a fellow at the Kennan Institute of the Woodrow Wilson Center in Washington, D.C., allowing me to reflect on the materials I had just acquired and also to obtain other materials at the Library of Congress. To its director, Dr. Blair Ruble, to the administrative assistant, Monique Principi, to my research assistant, Jason Antevil, and to the Kennan's entire staff, my enduring thanks. The Udall Center for Studies in Public Policy under its former director, Helen Ingram, and its deputy director, Robert Varady, provided a warm and stimulating atmosphere where the writing continued. Finally, the Spencer Foundation (grant no. 9500933) generously provided the opportunity for me to complete the writing of this book during my sabbatical year, even as I began yet another book project with that foundation's kind support. Each of these funding sources has my deep and sincere gratitude.

As a result of liberalization within the Soviet Union and its successor states, I was able to use a vastly larger range of sources than I could have (and did) ten years ago. Owing to the kind efforts of the then Soviet minister for environmental protection, Dr. Nikolai Nikolaevich Vorontsov, I became the first foreigner to use the archives of the Council of Ministers of the Russian Republic, housed in the former TsGA RSFSR (Central State Archives of the

RSFSR). I am glad to report that the exceptionally warm atmosphere set by Tat'iana Gennadievna Baranchenko, Natal'a Petrovna Voronova, and Liudmila Gennadievna continues to this day. Additional archival sources include: GARF (State Archives of the Russian Federation, formerly TsGAOR), RGAE (State Archives of the Russian Economy, formerly TsGANKh, with special thanks to its deputy director, Valentina Ivanovna Ponomarëva), TsKhDMO (Center for the Preservation of Documents of Youth Organizations, formerly the Komsomol Archive), ARAN (Archive of the Russian Academy of Sciences, both the Moscow and St. Petersburg branches, and its photo lab LAFOKI), RTsKhIDNI (Center for the Preservation and Study of Documents of Recent History, formerly the CPSU Archives), TsKhSD (Center for the Preservation of Contemporary Documentation, formerly the Central Committee CPSU Archives), the Archives and Library of the Moscow Society of Naturalists (MOIP), The Ukrainian Central State Archives, the Library of the Academy of Sciences in St. Petersburg, the Russian State Library (Saltykov-Shchedrin), the Russian Federation Library (formerly the Lenin Library), and the Library of the Russian Geographical Society. The archival staffs have been generous and supportive far beyond the norms of professional courtesy. I am indebted to these women and men beyond words. To Nina Vladimirovna Dem'ianenko of the MOIP Library and Archives, my special thanks.

Documentation was supplemented by numerous interviews with veterans of the movement conducted in Russia, Ukraine, and Estonia. Some of these were conducted in *zapovedniki,* or nature reserves (Prioksko-Terrasnyi, Tsentral'no-Lesnoi, Askania-Nova). To my informants, Academy of Sciences vice president Aleksandr Leonidovich Ianshin, Ksenia Avilova, Tat'iana Leonidovna Borodulina, Galina Borisovna Chernousova, Nelia Efimovna Dragobych, Iurii (Georgii) Konstantinovich Efremov, Oleg Kirillovich Gusev, Dmitrii Nikolaevich Kavtaradze, Viktor Masing, the late Andrei Aleksandrovich Nasimovich, Vitalii Feodos'evich Parfënov, Evgenii Makarovich Podol'skii, Linda Poots, Evgenii Arkad'evich Shvarts, Vladimir Vladimirovich Stanchinskii Jr., Vadim Nikolaevich Tikhomirov, the late Mikhail Aleksandrovich Zablotskii, Iurii Andreevich Zhdanov, and Sergei Vladimirovich Zonn, I owe a huge debt of gratitude. Konstantin Mikhailovich Efron has given me encouragement and friendship as well as the gift of his inestimable knowledge and wisdom.

The wisdom and friendship of my colleagues in Eurasia greatly assisted me to a more sophisticated understanding of the materials I had collected. They have given me more than I can ever hope to repay. Daniil Aleksandrovich Aleksandrov has played an immense role in encouraging me, among other things, to distinguish clearly between civic and scientific activism. This advice has been invaluable. Feliks and Nadia Shtil'mark, as always, have been the truest friends and most knowing commentators on the history of Russian nature protection. For those who seek a definitive history of the *zapovedniki* as institutions, there is only one book, and that is by Feliks Robertovich Shtil'mark. Vladimir Evgen'evich Boreiko also deserves more than mere mention. A man of big vision, he has been a creative and encouraging coexplorer in these relatively uncharted waters who has unstintingly shared his findings with me. Owing to his selfless desire to get the truth out, he has given collegiality a new dimension. I salute him. Competing with Boreiko for top prize in collegiality is Oleg Nikolaevich Ianitskii, the foremost authority on the modern environmental movements of Eurasia, who has also enriched my understanding with his. Others who have actively helped with this project are my friends and colleagues Aleksei Enverovich Karimov, who provided camaraderie during my last archival blitz, Anton Iur'evich Struchkov, Eduard Nikolaevich Mirzoian, who more than once gave me an institutional home in Moscow, Eduard Izraelovich Kolchinskii, who did the same in St. Petersburg, Viktor Kuz'mich Abalakin, Nelia Drogobych, Elena Vsevolodovna Dubinina, Nikolai Aleksandrovich Formozov, Mikhail Vladimirovich Geptner, Tat'iana Gerasimenko and Aleksandr Aleksandrovich Volkov, Elena Kriukova, Aleksei Vladimirovich Iablokov, Elena Alekseevna Liapunova, Nikolai Daniilovich Kruglov, Nina Trofimovna Nechaeva, Bernice Glatzer Rosenthal, Kirill Rossiianov, Ol'ga Leonidovna Rossolimo, Veronika Vladimirovna Stanchinskaia, Hain Tankler, Marshida Iunusovna Treus, Vladimir and Svetlana Zakharov, and countless other friends too numerous to mention. I owe special thanks to Aleksandr Sergeevich Rautian, who single-handedly saved the invaluable Viazhlinskii collection of photographs of conservation activists from the 1920s through the 1950s and who allowed me to reproduce them.

Colleagues on this side of the ocean also helped to clarify my thoughts and challenge dubious assertions. I am indebted to Valery N. Soyfer and to Stephen Kotkin for insisting as strongly as they did that the scientists' professions of Soviet patriotism might well be genuine and that it was impossible not to be, at least in small part, a "Soviet" person. The recommendations of Loren R. Graham, my lifelong friend and quondam mentor, have also found their way into this book. A conversation with Mike Urban led me to examine rhetoric as positioning, and even further incisions followed a generous critique of the introduction by my colleague Hermann Rebel. Upstairs in the sociology department, Elisabeth Clemens, truly a magician, read part of the manuscript and offered great recommendations for tightening it. Janet Rabinowitch valiantly read it twice, offering key suggestions, and Susan Solomon provided valuable guidelines for framing the story. To executive editor and publishing magician Howard Boyer go my heartfelt thanks for believing in this book and championing it. Erika Büky shepherded the manuscript through its production with a swift and sure hand, while Madeleine Adams's elegant copyediting made it infinitely more readable.

Finally, my thanks to those who have stood by me all these obsessive years—my wonderful friends, Mars (the world's leading cat), and my loving parents. Finally, I would like to thank you, the reader, who risked inguinal hernia and other bodily harm to pick up this too, too solid tome.

Introduction

In those times scientific and other publics showed their various colors—more often than not straining to approve [the decisions of the regime]; on occasion, however, even during the most difficult years, some stood up to the arbitrary use of power and to ignorance. People were expected to praise the transformation of nature under Stalin and Khrushchëv, to fetishize those programs as ones that supposedly only brought improvements. . . . But geographers cleverly devised ways to oppose these transformations even in the years of "The Great Stalin Plan." Is it possible that there were people that brave?

IURII KONSTANTINOVICH EFREMOV

When we speak about our public opinion, then it is necessary first of all to speak about scientific public opinion.

SERGEI PAVLOVICH ZALYGIN

This book is an attempt to come to grips with some very surprising archival findings. As I continued my research on the Russian nature protection movement forward in time from the years of the Cultural Revolution and the First Five-Year Plan (1928–1932), I repeatedly came across documents that testified to the unlikely survival of an independent, critical-minded, scientist-led movement for nature protection clear through the Stalin years and beyond. Through a number of societies controlled by botanists, zoologists, and geographers, preeminently the All-Russian Society for the Protection of Nature (VOOP), the Moscow Society of Naturalists (MOIP), the Moscow branch of the Geographical Society of the USSR (MGO), and the All-Union Botanical Society, alternative visions of land use, resource exploitation, habitat protection, and development were sustained and even publicly put forward. In sharp contrast to general Soviet practices, these societies prided themselves on their traditions of contested elections for officers on the basis of the secret ballot, their foreign contacts, and their prerevolutionary heritage.

To gain a sense of the boldness of these activists consider that in June 1937 the leadership of VOOP drafted a letter to Central Committee secretary A. A. Andreev seeking a meeting to upgrade the Party's commitment to nature protection and requesting authorization to travel to an international conference on conservation in Vienna set for the following year.

That very week the high command of the Red Army had been arrested, accused of working for a foreign power. Two months earlier the Soviet government had written to the International Genetics Committee postponing the Sixth International Genetics Congress, scheduled for August 1937, to some unspecified time during the next year. Stalin was shutting down the country to the outside world, and foreign contacts in one's past left Soviet citizens open to the charge of treason. An atmosphere of terror was settling on the gigantic country. And yet these scientist-activists wanted to go to Vienna! Another VOOP document dating from July 1948 reveals that the society's leaders wrote directly to the USSR Ministry of State Security to question why secret police detachments were chopping down all of the cypress trees in the Crimea. And in May 1954, barely one year after Stalin's death, the scientists' societies organized a protest meeting demanding the restoration of nature reserves eliminated three years earlier by Stalin.

Just as eye-catching were telegrams from *oblast'* (provincial) Party and government leaders "categorically opposing" those 1951 "liquidations" of the reserves at the time. Also in the archives is evidence of a number of dramatic intercessions by a succession of Russian Republic (RSFSR) premiers to protect the nature protection society itself from elimination at the hands of Kremlin authorities. Ukrainian and other archives have confirmed that such patronage by republic- and *oblast'*-level authorities of the nature protection movement was not limited to the RSFSR.

This book describes a succession of *independent* social movements for nature protection that predated and survived Stalin and all of his Soviet successors.[1] With protection from republic- and provincial-level patrons, this movement was institutionalized in a number of scientific and voluntary societies and for a time was also able operationally to control a twelve million–hectare (thirty million–acre) network of *zapovedniki* (scientific nature reserves), which acquired symbolic importance as a unique "archipelago of freedom" within the GULAG-state.

The reader may well ask how it was possible that any such movement, its institutions, and the energetic protection of them by provincial- and republic-level politicians could have existed in Stalin's terror state. Indeed, historians and Soviet specialists overwhelmingly deny that such a movement and network of patrons could have existed. For example, the social historian Geoffrey Hosking asserts:

> The Soviet Union . . . was a uniquely centralised polity, in which the Party-state apparatus governed not only the aspects of society normally associated with authority, but also the economy, culture, science, education, and the media. . . . Social interest groups had no identity separate from the nomenklatura hierarchy, so there was no question of their formulating their distinct interests, let alone of forming associations in order to defend them. That constituted the strength of the system.[2]

But Stalin and his successors did not root out all such autonomous social groups. Although we lack conclusive answers as to why the nature protection movement was not obliterated along with other institutional sources of political, cultural, and moral dissent or deviation, the following pages suggest some promising avenues of explanation. Perhaps as more archival materials become available and are examined we will find better answers. The very fact that independent social organizations continued to exist through the Stalin period and after raises fundamental questions about the Soviet system: Were there other areas of social organization besides nature protection that were able to survive as something more than naked transmission belts of regime values? Was it the regime's intention to extirpate every expression of divergent views and every manifestation of social autonomy? If so, then the persistence of various nature protection movements seems to indicate a certain lack of efficiency of Soviet rulers in the face of subjects determined to defend their autonomous selfhood. If not, we must come up with a more sophisticated picture of how Stalin and his colleagues expected to maintain effective control over society. Could the continued existence of this apparently autonomous nature protection movement actually have served the interests of the Party-state? More broadly, what changes in our picture of Soviet society and politics do these archival findings move us to consider?

* * *

In earlier works I have argued that the Soviet conservation movement in the 1920s and 1930s represented a means by which a section of educated society tried to moderate, or even halt, the juggernaut of Stalinist industrialization and social change.[3] Armed with unprovable holistic ecological doctrines that asserted that pristine nature was composed of geographically bounded closed systems ("biocenoses") that existed in states of equilibrium and harmony, conservationists warned of the dire consequences to the stability of those natural systems as a result of collectivization, industrialization, and other Stalin-era projects. They averred that only they, through their expert study of long-term ecological dynamics of pristine natural communities, could determine appropriate economic activities for specific natural regions of the USSR. They began to conduct this study in specialized protected territories—*zapovedniki*—which were off-limits to any uses except scientific research on ecological/evolutionary problems. Conservation activists, led by the foremost field biologists in the country, sought first to obtain, and then sustain, the right to a veto over unacceptable economic policies through the newly created Interagency State Committee for Nature Protection. At the same time they struggled to retain control of and to expand the network of *zapovedniki.*

As Stalin's revolution from above from 1928 to 1933 turned the country on its ear, nature protection emerged as a means of registering opposition to aspects of industrial and agricultural policy while remaining outwardly apolitical; arguments were couched in the language of scientific ecology. The picture ecologists drew of fragile self-regulating biocenoses seemed to throw cold water over the Stalinists' plans for a successful total mastery and transformation of nature.

Unlike the situation under Lenin and during the NEP (New Economic Policy),[4] scientists now found their own professional freedom under mortal threat. Moreover, on some level they perhaps understood the linkage between Stalin's plans for shackling nature and the reenserfment of society. Prominent Soviet scientists responded by marshaling ecological arguments against collectivization, acclimatization, and the great earth-moving projects. When this scientific opposition to the Five-Year Plan failed, scientists retreated to their ultimate fall-back position: a defense of the inviolable *zapovedniki* under their control.

By their charters the *zapovedniki* were absolutely inviolable. They had become the symbolic embodiment of the harmony of communities, of natural (and human) diversity, and of the free and untutored flow of life (in Anton Struchkov's eloquent phrase, the "unquenchable hearths of the freedom of Being"). As long as the "pristine" *zapovedniki* could remain independent, what was denied to human society in Stalin's Russia could be preserved in symbolic, natural form in these reserves. They formed an archipelago of freedom, a geography of hope.

Particularly revealing in this regard are the comments of Sergei Zalygin, repentant hydrologist, conservation activist, and editor in chief of *Novyi mir*, Russia's leading cultural-literary monthly: "Here is the crux of the matter: the word '*zapovednik*' means 'a parcel of land or marine territory completely and eternally taken out of economic use and placed under the protection of the state.'" But *zapovedniki* are much more than that: "A *zapovednik* is something sacred and indestructible, not only in nature but in the human being itself; it is also a commandment, a sacred vow [from its root, *zapoved'*]. And it is precisely around these meanings that the struggle over the *zapovednik* raged and indeed rages at the present time. . . . [T]he *zapovedniki* remained some kind of islands of freedom in that concentration-camp world which was later given the name 'the GULAG archipelago.'"[5]

If the defense of these inviolable institutions had become the paramount aim of Russia's elite natural scientists, such a defense required justification in biological theory. Of the two ecological paradigms available at the time—the individualistic or continuum theory of species distribution, and the paradigm of discrete, bounded, fragile, highly ordered ecological communities in a homeostatic equilibrium—only the latter fit with the research agenda of the *zapovedniki* and could provide a justification for absolute inviolability. However, we are struck by the disparity, in Gerovitch and Struchkov's words, "between the rather weak 'scientific' arguments for absolute inviolability, on the one hand, and the inspiration with which this idea was defended, on the other hand."[6] Even after a Stalinist campaign forced the *Zapovednik* Administration of the RSFSR to renounce "inviolability" in principle and to accept a new mission for the reserves—the transformation of their "pristine" nature into the more productive "Communist nature" of the future—the scientific establishment and its patrons in the *Zapovednik* Administration continued to defend the "sacred" reserves from their "profane" new tasks in practice. Indeed, the struggle over the defense and later the reestablishment of inviolable *zapovedniki* eclipsed all other environmental issues through the mid-1960s, and through the 1970s if we include Baikal, which was also a part of the geography of hope.[7]

This book confirms and develops these ideas. The early movement, which described itself as "*nauchnaia obshchestvennost'*" (scientific public opinion), a self-designation that connoted a social identity with its own values, traditions, interests, and ethical norms,[8] does not derive its sole historical importance from its accomplishments in the areas of species protection, landscape preservation, and support for multidisciplinary and unique ecological research in the *zapovedniki*. Perhaps its greatest significance for Soviet society resides in its role as an institutional "keeper of the flame" of civic involvement independent of the Party's dictates.

Activists maintained an atmosphere of internal democracy within the societies that they controlled: the All-Russian Society for the Protection of Nature (VOOP), the Moscow Society of Naturalists (MOIP), the All-Union Botanical Society, and the Moscow branch of the Geographical Society of the USSR (MGO). The epicenter of this movement was the Zoological Museum of Moscow State University, just down the block from the Manezh and Red Square, where MOIP and MGO were headquartered and where VOOP frequently held its meetings. Loren Graham, the historian of Russian science, once remarked that hundreds of Western scholars, diplomats, and journalists had long wondered whether there was any "island of freedom" in the Soviet Union, unknowingly driving right past it thousands of times!

This study underscores the resilience, courage, and determination of the nature protection activists, whose leading members were drawn mostly from the elite ranks of Soviet field biology (laboratory-based biologists and scientists from chemistry and physics were less well represented, although there was a fair contingent of geologists and soil scientists). In 1953–1955, when control over VOOP was wrested from them and placed in the hands of Party stalwarts, activists transferred their activity to MOIP, still under the control of their own people. This study also reveals their determined and creative efforts to pass on their ethos of *nauchnaia obshchestvennost'*—with its connotations of activism, service to Science, broad erudition, scientific autonomy, individual responsibility, and collective action—to succeeding generations. The most important vehicles for this were the instructional programs in field biology for children and teenagers run by the Moscow Zoo (KIuBZ), VOOP, and MOIP. Not coincidentally, today's leading zoologists and botanists, who include some of the most prominent reformist politicians such as Nikolai Vorontsov and Aleksei Iablokov, were molded in these intellectual non-Party youth groups.

In Russia, MOIP took the leading role in the creation of the first student brigade for nature protection (*druzhina po okhrane prirody*) in the Biological Faculty of Moscow State University in 1960, whose membership varied from 25 to 150 over the next thirty years.[9] Soon, almost all major universities and elite technical schools boasted *druzhiny*, which collectively reached a membership of about 5,000 by the 1980s.[10] These perpetuated at least some portion of the old prerevolutionary ethos of the "botanical-zoological-geographical intelligentsia." Members of the student brigades engaged in measuring point sources of air and water pollution, monitoring compliance with environmental laws, detaining poachers, and planning new nature reserves. Seeking idealistically at first to enforce laws that were already on the books, they soon discovered that the system was not interested in their "help"; indeed, their attempts to hold managers and hunters to the law was viewed by the authorities as oppositionist and slightly subversive. By and large, *druzhinniki* continued in the more Western- and global-oriented perspectives of their mentors.

At times, the movement served as a counterculture in that it provided for its members' most important social needs, which the larger society had failed to meet. Central were the needs felt by those who joined to be self-directed and autonomous of authority, to engage in genuinely creative work, and to serve the broader society—a variant of the old nobility's and intelligentsia's ideal of service.

"Self-sufficiency," emphasized Oleg Ianitskii, a leading social-movement sociologist, "lies at the core of the values embraced by environmentalists; they understand independence as a way of life, as a mode of everyday existence. . . . At the heart of the value-orientation described above is intelligent, creative activity." Ianitskii used the Marxian expression "unalienated labor" to describe the work of activists. Above all, this social identity was a path to a kind of "self-realization" as an individual. Activists, observed Ianitskii, "perceive independent activity as a search for meaning in life, as a testing of alternative possibilities for realizing their personal identity."[11]

The result of the efforts of the individuals examined in this study was the creation and perpetuation of significant communities that sustained their members—creative personalities who would otherwise have been ground down by the conformist, repressive Party-state system. As late as the early 1990s, Ianitskii could write:

> Environmentalists remain a united community, above all in terms of their values and psychology. Whatever the future might hold, the great advantage en-

> joyed by all those who make up the movement is that they have already found their place in life. They have discovered their goal and have linked up with fellow-thinkers. As a result, they have the inner calm and assurance of people who are aware of their path. . . . In these small circles people felt their strength, and realized that the System was not so monolithic after all. The members of the groups overcame their sense of inferiority, of their superfluity where the System was concerned. In a society without legal guarantees they acquired a real measure of social defense precisely because they became a community, that is, a genuine collective entity.[12]

Ianitskii points out the mutually reinforcing nature of the social movement and the professional background of its members. The process was iterative:

> The resistance put up by the clubs to official dictates is not only the result of the general causes cited earlier, but also has professional and ethical origins. The first of these is the feeling of members that they are participating in serious scientific work that can benefit both nature and humanity. The young biologists' clubs have really been much more than clubs in the usual sense of this word. Their key activity has not been holding meetings but getting out into the field. Many of today's environmental activists have been accustomed from childhood to taking part in expeditions and to carrying on collective work in the midst of nature. The second reason behind the resilience of these clubs is the extremely powerful ethic of group loyalty that operates within them.[13]

The British historian Geoffrey Hosking extends this self-confident moral authority to writers: "Like scientists, writers had both the moral and the social standing to make their opinions felt even in a highly repressive system. The tradition of the writer as an 'alternative government' had been established already in tsarist Russia. The Soviet government had tried to prevent the resurgence of any such 'alternative' by creating its own literary monopoly through the Writers' Union. But even this was, paradoxically, a tribute to the power of the word."[14]

This study points to a crucial difference between writers and field biologists, however. Undeniably, a number of individuals—Akhmatova, Pasternak, Kaverin, perhaps Paustovskii, and later, Pomerants, Dudintsev, Ovechkin, and others—refused to cave in to the regime's demands that literature be put at its service. But if they were an "alternative government," they had no shadow cabinet, no meeting lodge, no debates or elections, no general assemblies. And they could not. Regime surveillance of writers was so overwhelming that there was only the testimony of lone, brave individuals. Certainly, they had their networks of friends and their literary "circles," but that social site was tiny. And with the high rates of arrest, exile, or transfers "on assignment," it was also discontinuous in time and space. Writers left a written legacy, but that was not the same as an organically functioning social movement.

Equally important, the proportion of writers from the high intelligentsia or those converted to its social identity dropped sharply during the Stalin years, swamped by an influx of *vydvizhentsy*—those promoted upward owing to their more humble social origins plus Party affiliation (those prerevolutionary writers who survived, such as Aleksei N. Tolstoi, did not faithfully represent the older intelligentsia's ethos). At least for the first generation of such parvenus, gratitude to the regime far outweighed any constraints of censorship they may have felt. Only later would these writers feel that their enthusiasm and loyalty were betrayed by the political leaders, particularly in connection with the Party's promotion of policies that seemed to threaten the Russian heartland from which the writers hailed. Consequently, their protest represented not so much an affirmation of the autonomy of creative individuals as a disillusioned turn from Soviet patriotism to Russian nationalism.

Patterns of literary participation in the nature protection cause support these conjectures. Very few *littérateurs,* even during the 1950s and 1960s, aligned with the scientific intelligentsia's nature protection movement. Konstantin Paustovskii, Oleg Pisarzhevskii, Natalia Il'ina, and Boris Riabinin come to mind. Perhaps Sergei Zalygin should be placed here as well. The majority of writers who have gotten involved in environmental causes are associated with the "Village Prose" school (*pochvenniki, derevenshchiki*) and trace their genealogy back to Leonid Leonov and Vladimir Chivilikhin, themselves parvenu writers who began as grateful Soviet patriots and ended as disgruntled Russian nationalists.

"Scientific Public Opinion" in the Light of This Study

Like Russian lawyers after 1917 as depicted by Eugene Huskey,[15] "scientific public opinion" continued to view itself as a kind of professional *soslovie,* or closed guild. Owing to the corporativist, castelike, and somewhat elitist nature of "scientific public opinion," the scientists that constituted it may hardly be classed as thoroughgoing democrats, despite their observance of democratic norms within their own milieu (although the relationship of the provincial membership, particularly nonscientists, to the scientists who dominated VOOP, for example, is still unclear).

Defying simplistic understanding, the elitism of Soviet intellectuals generally and of scientific public opinion in particular derives from their claims to professional competence and moral vision. Living in a society that historically kept it in a condition of political tutelage, the intelligentsia's sense of moral superiority and purity was a form of psychological compensation for the real power and rights it lacked. While mistrusting the competence and judgment of the "dark masses," scientific public opinion was more resentful of the boorish Party leaders and bureaucratic bosses whom it regarded as having "hijacked" scientists' rightful role as arbiters of policy. As the sociologist Vladimir Shlapentokh observed, "Being well aware of their high level of education and creative capacity, intellectuals hold elitist attitudes toward others, although in most cases they try to hide them. The elitism of the intellectuals is not, however, directed so much against the masses, but rather toward the ruling class, which is quite often perceived as incompetent and selfish."[16]

The Tightrope Walk of Scientific Public Opinion

Although the complex public behavior of the nature protection movement at times resembled a guerrilla war against the regime, it can also be compared to a tightrope walk. Clearly, scientific public opinion deplored the vulgar attitudes and policy choices of the Stalinist bureaucrats at the center. Nature protection leaders wanted to be invited to assume their rightful places at the policy-making table with responsibility for environmental matters. This was not simply a self-serving wish for power and status; it was part of their professional ethical impulse and sacred duty to serve Science. Once they defined nature protection as a matter for "scientific" rather than political adjudication, a claim accepted by the scientific community at large, protecting the environment also became a sacred duty in the name of Science. Consequently, ethical norms at the very core of scientists' social identity continually impelled scientific public opinion to critique or contest official regime policies toward the environment.

It is also possible that field biologists disproportionately consisted of those who were more intensely attracted to freedom and therefore were prepared to risk more in order to defend the wild.

On the other hand, what enabled the movement to survive in the Stalinist political environment, in addition to serious patronage and protection from enlightened and/or self-interested middle-level political figures, was its perceived harmless marginality. Doubtless Nikita Khrushchëv's depiction of *zapovednik* naturalists as oddballs (*chudaki*) reflected the general views of regime leaders about these field biologists at those very rare times when they even noticed their existence. Despite occasional arrests and episodic characterizations of the movement as a hotbed of counterrevolutionary "bourgeois" professors, it was hard for the regime to perceive these ornithologists, entomologists, herpetologists, mammalogists, botanical ecologists, and biogeographers as sources of effective political speech. Marginality thus became a guarantor of the survival of scientific public opinion as a social identity.

That, however, created a dangerous contradiction for the movement, for its ethical norms demanded that it speak out when nature, and hence Science, was threatened. Too forceful a critique of policies to which the regime was heavily committed, however, could create the impression that the nature protection movement was a nest of counterrevolutionary subversion. Accordingly, scientific public opinion had to walk a tightrope, negotiating between its ethical norms and its desire to survive.

Nature protection activists had made their peace with Bolshevik rule. They were Soviet patriots and had no pretensions to supreme power. Indeed, it can be argued that they had an investment in the perpetuation of an authoritarian, centralized state regime, for they sought to use the great power of the Leviathan-state to impose their "scientific" vision of environmental quality on the country as a whole. A democratically run government might not afford such possibilities. Yet, the actual Leviathan-state in which these scientists found themselves was controlled by boorish, vulgar bureaucrats who did not recognize the eminent rationality of the scientists' alternative vision of development. Because of their wish to participate in, and not destroy, the Leviathan-state, the scientists of the nature protection movement could only hope to persuade and enlighten these bureaucrats to invite them into the circles of power. Failing that, they could only wait for "better tsars."

This dilemma also expressed itself in the movement's perennial conflict over its lack of a "mass character." A multiply determined double bind, this question concentrated many of the conflicting pressures on the movement. When the regime turned its attention to the movement and its flagship society, VOOP, it invariably leveled the criticism that VOOP had failed to become a "mass society." By this, regime arbiters meant that the Society still had an elitist, corporativist spirit and had not yet become a reliable, Soviet-type society, that is, a transmission belt to mobilize large, organized segments of the population on behalf of the regime's objectives. That was precisely the kind of organization that VOOP's leaders sought to avoid allowing it to become.

But it was possible to construe a "mass-based voluntary society" differently. Such a society, if it were truly independent, could represent a real social force in support of the goals of nature protection. At times, a platonic desire of scientific public opinion to see itself as leading such a mass movement may be fleetingly perceived in the internal conversations of movement leaders. However, the reality of VOOP becoming a truly mass society under Soviet conditions was too frightening and risky for the scientific intelligentsia. An authentically mass society, if truly democratic, might throw off the tutelage of the scientific experts. Worse yet, a truly activist mass society would certainly elicit the harshest repression from a frightened Party-state, destroying scientific public opinion and all nature protection goals in the process. Consequently, the scientific intelligentsia could not permit VOOP to become a truly popular organization.

Nevertheless, the creative leaders of the movement were able to cobble together a tolerable solution to their image problem. After World War II an aggressive effort was made to recruit teachers and schoolchildren as members. Additionally, and with somewhat greater reservations, the society began to enlist "juridical members," that is, whole ministries, factories, and other institutions that joined in the name of their workers and staffs. This, of course, made such employees' membership a formality—little more than a source of income from membership dues. However, by the early 1950s these measures boosted membership over the 100,000 mark. This solution was so inventive because it created the impression that the leaders of VOOP were building a "mass society" while ensuring that the new members—nonparticipating employees of "juridical members" and pliable schoolchildren—would not be in a position to challenge the dominance of scientific public opinion in the affairs of the Society. VOOP would only become a "mass voluntary society" in the Soviet sense after its takeover in 1955 by Communist Party bureaucrats, enforced by a decision of the Russian Republic leadership. Indeed, with twenty-nine million "members" by the 1980s, VOOP became the largest nature protection society on the planet, not to mention one of the largest nonstate businesses in the USSR.

Instrumental Shame, Protective Coloration, and Civic Honor

In January 1956 Aleksandr Formozov revealed to a large conference of Moscow conservationists his personal mortification at having to answer foreign colleagues' questions about the status of nature reserves and habitat protection in the USSR at an international gathering in Brazil the previous year. Similar statements by him and his colleagues at a variety of meetings also point to a general rhetorical strategy of using shame as an instrument to get the regime to adopt the scientists' nature protection policies.

In declaring that "we must think about all of the *zapovedniki* of the Soviet Union as we are all patriots of the Soviet Union," Formozov was not only proclaiming his authentic love of homeland. Patriotism was one thing, but being able to take pride in one's country was another, and that was at issue. Formozov and his scientist colleagues were telling the regime that they could not represent the USSR at international meetings with pride so long as the regime failed to restore the eliminated nature reserves and then move forward on that front. The high praise accorded to the new RSFSR Main Administration for Hunting and *Zapovedniki* (Glavokhota RSFSR) by Formozov ("the leadership of *zapovedniki* . . . is now in hands we can trust") and others reflected not merely their genuine pleasure and relief but also the hope that the Russian republican government would salvage the situation and remove the blot of shame where the All-Union government was mired in inaction. To point out the patriotism of scientific public opinion and of the *druzhinniki* is neither criticism nor praise; it is merely noting one more piece of evidence against a romanticized view that sees these people as antiregime dissenters.

Similarly, it would be a mistake to view the frequent premeditated professions by VOOP of loyalty to "socialist construction" and other regime goals and values (what I call "protective coloration") as simply hypocritical, tactical moves aimed at enhancing its image. A number of key leaders of the movement, most prominently V. V. Stanchinskii and V. N. Makarov, were socialists going back to 1905, albeit Mensheviks and Socialist Revolutionaries, while a good many of their colleagues shared a common mistrust of a private property–based economy. Although dissenting from the regime's specific vision of economic development and environmental policy, scientific public opinion sought to work through the state, bypassing real democratic control over resource use. The student movements also sought to change things from above by occupying responsible positions in the machinery of power. Opposition to policies does not a subjective enemy of the state make. Protective coloration was a complex and negotiated response, containing elements of both cunning and sincerity.

Scientific Activists versus Civic Activists

Of the extant speeches in the available archival record of VOOP, a handful stand out for their passion and their unapologetic, unwavering tone of conviction. Curiously, those who uttered them were some of the most prominent nonscientist members of the VOOP inner circle: Susanna Nikolaevna Fridman, Aleksandr Petrovich Protopopov, Ivan Stepanovich Krivoshapov, and I. E. Lukashevich

Through the early 1920s, conservation discourse had been not only scientific but ethical and aesthetic as well. Ethical and aesthetic positions had been frequently voiced by scientist activists alongside scientific rationales for nature protection. By the late 1920s, this integrated mixture of motives, probably shared by the bulk of scientist activists, could no longer be expressed without penalty. Over the course of the decade ethical and aesthetic arguments lost their legitimacy and were derided as nonmaterialist and sentimental. Consequently, the public redefinition of nature protection as an exclusively "scientific" problem of ecology was an adaptive response by movement leaders, who recognized that the Bolsheviks might heed those speaking in science's name but might persecute those who advanced "moral" arguments for policy. For scientists, internalizing this claim of nature protection as a scientific question additionally allowed them to fight for nature with the same sacred determination with which they would fight to defend Science. For fifteen years the ethical and aesthetic sides of the issue disappeared from view.

Accordingly, Susanna Fridman's remarks at VOOP's 1947 Congress were all the more startling. The longtime recording secretary of the Society and a nonscientist who considered herself one of the last remaining independent citizen activists in the country, Fridman dared to question the cornerstone of her scientist colleagues' authority. "Is nature protection," she asked, "or more correctly, the survival of wild nature and its capacity to blossom, compatible or incompatible with our quickly changing culture and civilization?" What was the opinion of science on that question? "Science has answered that it is compatible"; but, she wondered, what if science was wrong? In that case, she concluded, "our science is worthless, empty, and, as theory, holds no water. We know a great deal, but if we cannot [make the survival of wild nature compatible with culture], then that which we know wasn't worth knowing." This comment revealed the concealed tension between science and ethics within the nature protection cause and the parallel tension between scientific public opinion and those few citizen activists for whom "the public good" as they construed it outweighed the interests of science. Fridman, who emphasized that nature protection was a "momentous" question "not only of international but of planetary importance," was one of those few in VOOP who wanted to build the Society into a genuinely powerful independent mass society. "I must declare that in our Union we must engage in nature protection with pure and burning hearts and with passion," she emphasized, because, among the masses, "no one has any conception of the sweeping scope of this cause or of its crucial importance for the whole world. We must enter the international arena. Life itself urges us that way." As a final heresy she as much as stated that capitalist societies had "successfully tackled" a number of specific environmental problems still unsolved in the USSR.[17]

Fridman intriguingly suggested in 1957, one year before her death, that nature protection would become the universal creed that would unite humankind. "I have always believed," she wrote to her old friend Vera Varsonof'eva, "and now especially believe that the idea of nature protection will triumph, that precisely that idea will become the basis on which friendship between peoples will be built, which will give rise to common interests and to a universal common culture."[18] These were the thoughts of a civic-minded activist, not a scientist.

The difference between the VOOP scientists and the citizen activists is that scientific public opinion fought fiercely to defend its institutions—*zapovedniki*—and nature as "science," but the citizen activists defended nature as "nature," their personal civic dignity, and a larger vision of citizenship.

One could say that citizen activists provided the backbone that allowed their scientist colleagues in the movement to act more bravely. Tragically, as we see from the final letters of Susanna Fridman, the citizen activists were also the most socially isolated of all.[19]

The Students

Russia's two student nature protection movements—the *druzhiny* and the Kedrograd experiment—are treated in this study as historically and sociologically distinct from each other as well as from the citizen activists and scientific activists of the older movement. They seemed to resurrect the old spirit of the prerevolutionary *studenchestvo.* The *druzhiny,* especially the flagship brigade of the Biology Faculty of Moscow State University, were godparented by MOIP and scientific public opinion in the hope of reproducing another generation of activists in their own image. Although the students continued their teachers' commitment to *zapovedniki,* to an elitist self-image—brigades consciously limited their membership and long concentrated on the problem of poaching—and to the hope of using the monumental levers of power of the state to implement and enforce their environmental vision, they did not become yet another generation of scientific public opinion. Students sought to enter the natural resource bureaucracies rather than to stay in academe and wait vainly to be called to advise the political leadership. They sought to capture the levers of power themselves—as fishing, hunting, or water quality inspectors or nature reserve directors, administrators, and staff scientists.

The student activists of the 1960s and later raised action, not pursuit of science, to the apex of their hierarchy of values. Where scientific public opinion located its authority to speak out in scientists' reputations and erudition, the students viewed their moral authority more as a matter of course, arising out of their educational status. Historically, average citizens and Russia's various regimes (Stalin's excepted) were more inclined to indulge students' "excesses" and protests. For their part, students were impatient to engage in concrete forms of nature protection planning, practice, and enforcement, and all the more if these entailed a certain degree of adventure or even danger. This mood of action grew out of the hopes of the reform era of the 1950s and defiantly sustained itself during the long period of "stagnation" of the mid-1960s through mid-1980s. It also grew out of an entirely different education, received in the Soviet era, and out of changes in science itself; for a number of reasons it was no longer easy for young field biologists to acquire the scientific authority of the previous two generations.

During the crucial Khrushchëv years, the students revived a tradition of activism combined with the aura of moral authority that idealistic students customarily enjoyed in Russia. This produced efforts at practical resource management and nature protection that brought the students into conflict with the Soviet bureaucracy. The stifling of Kedrograd, for example, vividly illuminated the gap between broadly shared social values, highlighted by the students' idealistic efforts, and the Soviet system. It therefore constituted a critical "object lesson," which catalyzed the far larger Russian nationalist/nativist movement of the 1970s and 1980s.

Stalinism as a System

Inspired by Marxian traditions, a trio of Hungarian ex-Marxists, Ferenc Fehér, Agnes Heller, and György Márkus, argue that in Soviet-type systems all economic investments, no matter how profitable or sensible they might seem or how likely to contribute to the general well-being, are judged by their likely effect on the stability of the system in the short term. They argue that this is tantamount to generating as big a flow of resources as possible into the hands of the central bureaucrats. Moreover,

> the social usefulness of the end-product is graded according to its propensity to remain during its process of utilization under the control of the same apparatus or to fall out of it. . . . From the viewpoint of . . . a pure economic rationality, Eastern European societies are strangely and strikingly ineffective; they consistently make wasteful economic choices. This is, however, the consequence of their own objective criterion of social effectivity, of their own logic of development.[20]

Actually, there is a bit more here than a simple passion for aggrandizement. From a political standpoint, investments that seemed likely to create or enhance autonomous pockets of power irrespective of their economic and social "merit" appeared to the system as threats and were not approved. Conversely, those that manifestly propped up, reproduced, or augmented the power of the central bureaucratic apparatus were most heavily favored. Where decentralized investments seemed unavoidable, the system compensated with an increase in the capacity of the bureaucracy to monitor those potential nodes of autonomy, thus undercutting the economies achieved in the first place. This need for oversight fed the inexorable expansion of the *apparat:*

> In a society in which all exercise of power has the character of a trustee/fiduciary relation and where systematically organized control from below is at the same time excluded in principle, a constant reduplication of the systems of supervision is an inherent and irresistible tendency. Processes of decentralization dictated by demands for greater efficiency are therefore constantly counterbalanced with attempts to impose new checks (and hence new systems of control), lest any unit . . . become so effective as to be able to follow its own set of objectives. It is in this vicious circle that the apparatus as a whole continues to grow, against all (sometimes drastic) attempts at its reduction.[21]

The bottom line is that "while the increasing social costs of bureaucracy may well be considered as a specific form of exploitation inherent to this society, the numerical expansion of the whole managing-directing apparatus, which is its main cause, certainly is not in the material interests either of individual bureaucrats or of their collectivity—in fact it only enhances the competition between them. And this tendency actually prevails against the articulated will of the apparatus."[22]

Ianitskii adds, "For the System, which to serve its own interests had created an economy of extravagance and shortages, environmentally safe technology was an empty phrase, and resource-saving was actually a threat to its well-being."[23] Consequently, the commitment of all regimes from Stalin's through Chernenko's to colossal "projects of the century" becomes eminently explicable despite those projects' long-term potential to undermine the system's viability financially and environmentally; each was perceived by the regime as one of the few options for politically safe large-scale economic growth, seen as essential to the perpetuation of the system. It is likely that these projects—Stalin's "Plan for the Great Transformation of Nature," Khrushchëv's Virgin Lands campaign and opening of Siberia with the Bratsk-Angara Dam, and Brezhnev's River Diversion Project and Baikal-Amur Mainline Railroad—also were thought to contribute to the system's stability in another way, in the arena of popular legitimacy. By each promising to represent one last "great leap forward" to Communism, the various "projects of the century" propagandistically endeavored to overcome the people's ever-increasing suspicion that the regime was actually a parasitic dead-end; each Soviet ruler had a "signature" program to legitimize his claim on leadership.

The opposition by the environmental movement in Russia to these big projects and to Soviet economic development generally from the early 1930s to the 1990s could be considered a continuous record of political opposition to the regime, attacking it at its very political-economic foundations. However, such a conclusion would have to assume some understanding on the part of nature protection activists of the connection between "the great transformation of nature" and the patterns by which the regime continually reproduced itself. The evidence does not currently support such a conscious realization. For activists, what was objectionable were the visible consequences of these patterns of economic development—for "nature," for society, and for their own social identity.

The Russian nationalist-oriented activists associated with opposition to the river diversion project seem to have developed a more conscious sense of the subversive nature of their campaign. Nicolai Petro writes that the river diversion project led critics to examine some of the structural attributes of the system, and concludes: "This criticism of the bureaucracy extends far beyond Minvodkhoz to all those who will fully confuse their own narrow-minded interests with the long-term national interest. Careerism is thus disguised as social need."[24] In fact, he continues,

> It is scarcely too much to say then that many of the critics of the diversion projects, both writers and scientists, espouse an alternative worldview to the one currently inculcated by the present political system. . . . The essential components of this alternative worldview are that science is not the solution to human problems because it does not address the need for spiritual and moral values, and that proper morality should be based on patriotism as manifested in an individual's personal responsibility for his country, its history, and its culture.[25]

Still, no branch of the environmental movement was able to articulate a critique of the system rooted in political economy or to offer a clear picture of an alternative way of organizing economic life. True, there were times when the older scientists' movement and students subjected individual government officials to interrogation and even humiliation. One could even call the conferences of 1954, 1957, and 1968 (where activists demanded the resignation of USSR agriculture minister Matskevich) carnivalesque inversions of the Shakhty, Industrial Party, and other show trials of the late 1920s and early 1930s directed against the scientific and technical intelligentsia. Independent civic associations seizing such initiative would seem to constitute a prima facie case of subversion from the Party's standpoint.

Why then did the regime tolerate these implicitly subversive movements when it easily could have obliterated them just as it had so many others? Could it not see that nature protection discourse and the *zapovedniki* represented a last holdout of an alternative cultural and political resistance? Did the regime not understand the implications of the activities of VOOP, MOIP, the *kruzhki* (circles), and the *druzhiny,* that they challenged or undermined core values and policies of the leadership? Why did the activists not become object lessons regarding the transgression of the rules of permitted Soviet speech?

A conclusive, let alone unitary, answer to this problem probably will never emerge from available archival sources; we can only speculate. However, a number of explanations should be considered. First, the regime did not expect *political* speech from field biologists and geographers, whom it considered arcane eccentrics and whose economic relevance it barely acknowledged. They were at once too unimportant to worry about and too silly and strange to be perceived as serious political threats.

Second, an oppositional role was not the activists' default mode. For these individuals, nature protection was an absolute injunction or sacred duty. However, the rationality and the protective role of suppressing the political and social implications of what they were doing must be appreciated. Were their everyday agitation on behalf of nature less naively passionate and more self-conscious, their presentation of self in the public arena would have been more characterized by "bad faith." Arguably, they would have then been more vulnerable to being "unmasked" as dissidents in Soviet society. As naive "nature lovers" they presented a convincing image of harmless and somewhat ridiculous cranks and oddballs—*chudaki.* The fact that the leadership of the movement consisted of world-class scientists—botanists, zoologists, geographers—made their nature protection appear from the regime's perspective to be an eccentric and low-cost hobby. Nature protection only appeared on the regime's radar screen when those in power decided that they had other uses for the resources and lands (*zapovedniki*) used by the movement for its "hobby." At that point the activists' resistance potentially acquired a new, subversive cast . . . because someone was finally paying attention!

Third, the various movements were authentically patriotic, for the most part, and had no intention of overthrowing the system. Even when Malinovskii and Bochkarëv were publicly humiliated by activists, activists' criticisms were leveled at these men as individuals or even "bureaucratic types," not as representatives of a rotten system.

Fourth, the nature protection movement was assisted in its quest for survival and influence by high- and middle-level patronage and protection. Activists looked to institutional patrons and protectors of all kinds to give or secure them their little bit of social space, and so they were imbricated to some extent in the system. Without allies they never could have preserved and maintained their "archipelago of freedom." A series of premiers of the RSFSR and local, *oblast'*-level politicians proved to be true friends and protectors of both the reserves and the nature protection movement. Academy of Sciences president Nesmeianov had enough power to provide wiggle room for this politically aberrant group and let the Academy serve as a Noah's ark for displaced ecologists after 1951. Gosplan leaders Saburov and Zotov respectively tried to mitigate Stalin's and Khrushchëv's depredations against the *zapovedniki.* Lesser leaders had the power to give space as well. The *druzhiny* could not have existed long without the patronage of the Komsomol (the Young Communist League) and of local branches of VOOP.

The motives of the movement's patrons were not identical. Nesmeianov, it appears, was personally committed to the cause of nature protection. At the opposite end of the spectrum, the leaders of the Komsomol saw patronage of the *druzhiny* as a means of burnishing the image of the Komsomol as "liberal" while keeping tabs on the potentially disruptive *druzhinniki.* And although many of the RSFSR and *oblast'* politicians did not act out of ideologically conscious "liberal" sentiments, they did seek to protect "their own" scientists, territories, and jurisdictional portfolios from the grasp of the center. After Stalin's death the republican governments, often with the support of local authorities, took the lead in reconstituting the disbanded nature reserves. Did that patronage morally obligate the nature protection movement to restrain itself so as not to endanger their patrons? Unwritten understandings about the permissible limits of public speech doubtless played a role in framing policy dissent. This is not a story about "black hats" versus "white hats." All of these actors are located somewhere on the same spectrum. Even Aleksandr Vasil'evich Malinovskii emerges from this study not as an irredeemable "evil genius" but as a Soviet bureaucrat whose vision of a "souped up" nature was at once utopian and utterly pragmatic.

* * *

Although it might appear that my research has discovered currently sought-after seeds of "civil society," the story is not nearly that simple. Perhaps it is even a good deal more ironic than it seems, for the very conditions—tsarist and Soviet—that gave rise to the social identities explored here also made them largely self-limiting. They testify to the durability of corporativist or guildlike social identities in Russia. Such mutually uncomprehending guildlike social groups could achieve solidarity only during rare moments, such as 1905 and March 1917, but could not sustain it, allowing a Bolshevik autocracy to supplant the fallen tsarist one. The independent groups portrayed in this study do not seem to have transcended this pattern.

This study trains its sights on what James Scott has called "hidden transcripts,"[26] drawn from the worlds of documents, statements, and social practice, which testify to the existence of Soviet social sites where alternative values and visions of the world were affirmed, shared, and perpetuated. This it does not merely to find inspiring examples of resistance under dauntingly difficult circumstances but to shed light on the nature and evolution of the social identities of scientist-activists and other activists and the ways in which they experienced their world and accommodated to it. Such portraits must necessarily also point out the ways in which these individuals were *integrated* into their larger society and system; although near the margins of the officially sanctioned social order in some respects, they could not fully escape dependency on the system. Their activism never countenanced a frontal challenge to the supreme political authority of the Party-state. Nor were they equipped to join with other social groups to defend social interests alien to their own, owing to their insular, caste-based psychologies. In a word, they were not dissidents in the way we now understand that term and were not fully formed nodes of civil society. But that in no way diminishes their achievement of establishing autonomous social identities for themselves on the basis of their own internal compasses against the pressures and directives of a jealously authoritarian system.

This book is organized chronologically, although the story is not entirely genealogical. After a discussion in chapter 1 of the various social identities adopted by members of the Soviet nature protection movement and a brief synopsis of the movement's background through the early 1930s, chapters 2 through 8 trace the history of the field naturalist–dominated nature protection movement during the remaining and most repressive portion of Stalin's rule. That section culminates with an extended exploration of the circumstances surrounding Stalin's decision to "liquidate" the great bulk of the nature preserves and the concurrent investigations into the All-Russian Society for the Protection of Nature, which only narrowly escaped being shut down itself.

Chapters 9 through 13 cover the decade following the death of Stalin in 1953. They focus on the remarkable and almost single-minded efforts of this scientists' movement to pressure the authorities to restore the eliminated *zapovedniki,* efforts that included the convening of mass scientific conferences involving hundreds of participants to protest the closings. These chapters also show the adaptive flexibility of the movement, which was able to relocate its institutional site to the Moscow Society of Naturalists and to the Moscow branch of the Geographical Society of the USSR when in 1955 the RSFSR authorities quashed the independence of the All-Russian Society for the Protection of Nature.

Chapters 14 through 19 cover the period from the emergence of the university student movements of the late 1950s to early 1960s through *perestroika,* although the latter deserves more space than I have allotted here. In this last portion of the book focus on the old field naturalists' nature protection yields to a portrayal of new social actors: the student *druzhina* and "Kedrograd" groups, the emergence of a Russian cultural-patriotic nature protection movement, and the larger coalition of all of these groups in opposition to the industrial pollution of Lake Baikal and the project to divert northward-flowing Siberian and European rivers to the drier south. The book ends by noting the emergence of mass protests by ordinary Soviet citizens. The Conclusion distills new insights into Soviet history and environmental history from the story of the nature protection movements told here.

* * *

This study's focus is elusive: charting the role of organized participation in nature protection advocacy as a unique arena for affirming and perpetuating self-generated social identities, ipso facto a subversive undertaking in the eyes of the Stalinist state. Complicating this task is that nature protection provided the symbols and rhetoric around which more than one distinct, autonomous subculture was organized in Russia. Consequently, this is a study of how "nature protection" as an aesthetic, moral, and scientific concern and as a source of symbols and rhetoric was used creatively by Soviet people to forge or affirm various independent, unofficial, but defining social identities for themselves.

One who noticed this role for environmental activism is Oleg Ianitskii, who called attention to ecological protest as the formal banner under which *any* kind of political expression originally marched at the dawn of *perestroika:*

> The struggle against the dictatorship of the Center, for national autonomy and for the preservation of national culture, for civil rights and freedom, against the arbitrary rule of the local bureaucracy, for self-government and the right to participate in the taking of decisions—all of these social actions marched either partially or wholly under ecological slogans. One way or another, ecological protest during the period 1987–1989 became the USSR's first legal form of *democratic protest and of solidarity among the citizenry as a whole* . . . [although] the motivations and goals of this "ecological uprising" were quite divergent for different participants in this struggle.[27]

Was it purely accidental that emerging Soviet citizens first chose to march "under ecological slogans"? Certainly, the April 1986 Chernobyl disaster played a large role, graphically demonstrating the consequences of the system's wanton and decades-old disregard for the health and environmental safety of the population, and moving people to notice and speak out about environmental threats in their own localities.[28] But also Soviet people knew that historically, unlike political, religious, ethnonationalist, labor, or even cultural dissidents, environmental protesters were not greeted by billy clubs, water cannons, imprisonment, deportation, or exile. A host of compelling problems angered Soviet people in the early days of *glasnost'*. Any one of those could have served as the focal point of their initial public protests. People almost universally chose *environmental* issues, however, because they were aware of the low risk historically associated with speaking out in that area. This study reveals that environmental protest and activism served as a surrogate or even a vehicle for political speech continuously throughout the Soviet period. In Ianitskii's words, "Nothing arises out of a vacuum. *Perestroika* and reform measures do not represent isolated actions but an extended process, which has its prehistory."[29] That prehistory is the central subject of this book.

CHAPTER ONE

Environmental Activism and Social Identity

Some who have reflected on the prehistory of Russian environmentalism, such as the geologist Pavel Vasil'evich Florenskii, a former member of KIuBZ (the Young Biologists' Circle of the Moscow Zoo), believe that the environmentalist ethos draws its source far back in time, from the traditions of brotherhood that flourished in Pushkin's day at the Tsarskoe Selo Lycée, which then were revived in the traditions of the St. Petersburg University *studenchestvo* (radical student subculture).[1] These traditions somehow survived in the *kruzhki* (circles) that the Soviet-era nature protection movement created to ensure the perpetuation of its values and social identity:

> In the children's circle a collective was forged of like-thinking individuals with their democratic structures, independent self-governance, continuity over the generations, here were molded principles of morality, traditions of friendship, an awareness of our unity with nature and of the need for an eternal dialog with it. The free Young Naturalist life was a life-filled alternative to the dry and bureaucratized school and the decayed Pioneer and Komsomol organizations. Having been members ourselves in our childhood and adolescence of this noisy youthful community, we continue to feel to this very day that back then we swore our loyalty in friendship and our loyalty to nature. KIuBZ and its spin-off, the VOOP circle, were the nurseries where the future leaders of the nature protection organizations were lovingly cultivated and where the principles were honed that later would provide the basis for the charters of environmental organizations. . . . [S]ince [the 1950s] the nature protection movement has irrepressibly grown, realizing an "ecological niche" in all age and social groups. Its schools were the student *druzhiny* for nature protection—as well as "Kedrograd" in the Altai. Those were the milieux where the country's future "green" movement's leaders were molded.[2]

We cannot say for sure whether the continuity of the ethos of the tsarist-era *studenchestvo* was unbroken before it reemerged within the university brigades for nature protection in the 1960s. However, Florenskii and Shutova are right to point to the linkages between a decades-old nature protection movement, that movement's youth organizations (especially from the 1940s onward), and the university student nature protection brigades (*druzhiny*) of the 1960s through 1980s to which the older movement gave rise.

Tempting as it may be, however, it would be an error to conflate the distinctive groups of Russian activists into a unified "environmental movement." The earlier nature protection movement of the field naturalists and activists, the later movements of university students and of engineering and technical students, the Russian national-patriotic movement for the protection of nature, and the mass protests of the late 1980s all must be distinguished from each other sociologically despite the links between them. Although all these currents enlisted the rhetoric of nature protection, they drew inspiration from sometimes quite distinct cultural, professional, and ideological traditions. For example, the *druzhiny* echoed the hoary traditions of the Russian *studenchestvo,* whereas the "scientific public opinion" of their professor-mentors was rooted in the prerevolutionary ideology of the old academic intelligentsia. Those divergences reflected underlying social differences among the members of these various environmental activist movements: levels and kinds of professional training, professional or career status, social origin, and generational cohort. Admittedly, the distinctions made here are overschematized; nevertheless, our insight is better served in this case by splitting than by lumping.[3]

If environmental activism served as an unauthorized form of public speech, what were the "speakers" trying to say? They were not all saying the same thing. Only by understanding what anthropologist Walter Goldschmidt has called the various "human careers" of members of these distinct groups may we begin to grasp the part played by environmental activism in their struggles for self-definition and self-affirmation under evolving Soviet conditions.[4]

Scientific Public Opinion as a Social Category

To understand one of the most important social meanings of environmental activism in the Soviet period, it is first of all necessary to appreciate its connection to a Russian ideology of science and learning that emerged during the tumultuous years of the late 1850s and early 1860s.[5] In those years a "mystique of *nauka*" (science, learning), in James McClelland's phrase, gripped an entire generation of Russian educated youth. Whereas the tsar and the political system proved limited and flawed, science held out the promise of nothing less than the secular redemption of the world. Its adepts were characterized by "an enthusiasm that elevates and enthralls a person, a conviction that he is doing something that is capable of absorbing all of his intellectual inclinations and moral energies—something which . . . enters as a necessary constituent part of the much broader general movement that will guarantee the eventual elevation of the intellectual and material well-being of the public as a whole."[6] Russian scientists and academics retained this faith in the redemptive power of science up to and through the Bolshevik Revolution.

One corollary of this ideology was that a life in scholarship conferred moral superiority. A scholar not only became a knight in the army of enlightenment but also acquired through learning a superior moral vision. As a rule, liberal politics—including opposition to tsarism, support for some kind of representative democracy, belief in intellectual freedom, and commitment to civil and human rights—formed part of this vision of an enlightened future. Among professors, this "mystique of science" was colored by a shared "caste" or "corporate" sensibility.[7] They embraced a social identity that McClelland has called an "academic intelligentsia," which,

> while subscribing to the general outlook of the larger liberal intelligentsia as a whole . . . developed an additional and distinctive viewpoint of their own, which stressed the vital importance of university autonomy and the role of *nauka* in Russia's future social and cultural development. The majority of Russia's professors, in short, were more than just scholars and scientists. They formed a closely knit and articulate sociocultural group which sought to embody in its academic activities a moral commitment to progress and reform.[8]

A further component of this ideology, at least among many academics, was a high regard for basic or fundamental research, what the Russians called "pure science" (*chistaia nauka*). If science and learning were a secular religion then pure science was its most sacred precinct, undefiled by outside political, commercial, or social pressures. Pure science embodied the principle that true academics answered only to the ethical injunctions of their priestly calling.

By the first decade of the twentieth century, *nauka* became a more contested issue. Certainly not all educated Russians endorsed the ideology described above. Progressive but loyal tsarist bureaucrats, seeking to modernize the country, had an obvious stake in denying that the march of knowledge would inevitably lead to the downfall of the autocratic order. On the one hand, they lobbied for greater regime support for academic institutions; on the other, they tried to convince academics of the need to dissociate learning from antiregime politics.

At the other end of the spectrum, radicals, especially students, demanded that academics actively subordinate learning and science to the struggle against autocracy. In its later incarnation in the postrevolutionary period, this view denied the possibility of science and learning independent of socioeconomic and ideological interests and consequently came to challenge the notion of an autonomous, value-free realm of "pure science."[9]

Insensitive and repressive policies, including a pattern of disregard for academic freedom and university autonomy, characterized tsarist education policy. Although by nature basically unrevolutionary and staid, the academic professoriate gradually concluded that to defend or attain academic freedom it needed to change the political structures of the land. Motivated by its ideology of *nauka* (not by a passionate interest in politics per se) the academic intelligentsia in 1905 crossed the political Rubicon, joining the "all-nation struggle" against the autocracy. Several thousand professors even signed a declaration proclaiming that "academic freedom is incompatible with the present system of government in Russia."[10]

For Russians, therefore, science was not simply a form of employment. It was a calling, a unique form of "human career" that endowed the lives of its adepts with a transcendent moral significance. Writing in the Imperial Academy of Sciences' monthly, *Priroda* (Nature), the physicist V. A. Mikhal'son captured the precise flavor of this Russian ideology of science:

> The average German pursues *nauka* as a profitable trade—profitable not only for himself personally, but also for the people and the state. Many Englishmen and Frenchmen pursue *nauka* as an interesting and noble sport, not giving a thought to its utility. But one often finds Russians, and Slavs in general, to be motivated by a sacred enthusiasm which regards the pursuit of *nauka* as the only way to achieve a tolerable if incomplete worldview, and the search for truth as both an irresistible personal need and a moral duty before the fatherland and all of mankind.[11]

Many went beyond that to claim for members of the academic intelligentsia a generally superior vision of life, gained on the basis of their scholarly training and erudition. One of the classic expressions of this understanding of scientific public opinion was voiced by the great biogeochemist Vladimir Ivanovich Vernadskii, who in 1892 wrote:

> A society is strong to the extent that its processes are consciously determined. . . . Let us imagine a series of human societies and states. In some of them, people are given broad freedom to speak their minds, to expound and discuss opinions. In the others, this possibility is reduced to a minimum. Societies of the first kind will be much stronger and happier than those of the second. If in societies of the first category necessary collective actions are, moreover, performed on the basis of the correctly established views of the best people, and in societies of the second type these actions are performed on the basis of arbitrary decisions by chance individuals, the strength of the former societies will steadily increase. Meanwhile, the question of the existence of societies of the latter type will inevitably be placed in question, and life in them will become more squalid and difficult. . . . Russia is in just such a situation.[12]

Of course, everything hinged on who were considered "the best people" and who "chance individuals." After the Bolshevik seizure of power, leaders of the Party considered themselves to be the "best people," who, thanks to their Marxism, enjoyed a privileged view of human society and its problems. Not surprisingly, many non-Bolshevik scientists and academics continued, like Vernadskii, to view *themselves* as the "best people" and Bolsheviks as "chance individuals," and to treat the Party's claims to privileged knowledge with condescension. The Bolsheviks, like the tsarist regime before them, returned the compliment, resisting the efforts of scientists to press their claims to decision-making power. The authorities almost always prevailed; academics did not have much of an independent power base. But that did not mean that academics entirely ceded their claims to technocratic expertise.

Yet, alongside academics' pretensions to independence and power was a poignant cognizance of their ultimate dependence on the state: "Russian professors could hope for academic freedom, but they could not forget that they were state employees," the historian Samuel Kassow has observed.[13] Unlike other professionals, academics could not retreat into private practice. They preferred to cultivate the state's confidence and trust; they entered into overt political opposition only after the regime posed a threat to *nauka*.[14]

These attitudes and contingencies carried over into the Soviet period. But the Soviet drive to demolish scientific autonomy was far more thoroughgoing than anything attempted by the tsars, placing a tremendous strain on the ideology of *nauka*. Both the tsarist regime and radicals often viewed science and learning in purely utilitarian terms, as a means to achieve national sufficiency or a better material life, and disparaged what they called "science for science's sake." The Bolsheviks inherited these attitudes in double measure.

With all of the human and social sciences under the most relentless scrutiny as class-based from both the Bolshevik authorities and freelance ideological vigilantes, an overt political defense of learning in general was too dangerous. The natural sciences, however, were best able to retain a remnant of intellectual and institutional autonomy. Disciplines and approaches that could convincingly be gathered under the umbrella of the natural sciences stood better chances of survival, for Lenin and his immediate coterie still regarded the natural and exact sciences as relatively more objective than the patently ideology-ridden social sciences (although later, for a time, natural sciences too would lose their "value-free" exemption).

By the end of the 1920s astute leaders of the nature protection movement had succeeded in redefining nature protection as a branch of scientific ecology. Although the scientists who led the movement viewed nature protection as a matter of ethical and aesthetic concern on a personal level, their public discourse was almost exclusively framed in scientific terms to provide legitimacy for their cause. More important, they almost certainly believed their own contentions.

Like American Progressives such as Gifford Pinchot, the Russian field biologists who led the nature protection movement tried to make the case that questions of land use and resource exploitation were scientific and technical, best resolved through scientific study and evaluation by experts—themselves.

These scientists had convinced themselves that ecological science would sooner or later reveal to them the precise limits of permissible human incursion into natural systems. That conviction rested on their view that individual ecological communities formed the building blocks of the biosphere, earth's envelope of life. Each of these communities (biocenoses), the field biologists and ecological theorists believed, was largely self-contained and bounded, and existed in relative equilibrium. That is, within these putative natural systems, all constituent elements balanced each other; fluctuations in the numbers of one or another species would soon be followed by a return to the norm. Scientists assumed that humans were extraneous to these "natural" systems and could only harm them. In their judgment, the task of ecology and field biology was to determine for each ecological system the kinds and levels of human economic activity that could be pursued without catastrophically damaging the biocenosis.

As a vision of nature, the idea of the ecological community was static. Taken to its logical conclusion, it implied that perfect natural balance on earth could be attained only in a world without humans. It was hardly demonstrable. Nonetheless, this picture of nature—later assailed by Bolshevik critics as reactionary—retained a deep hold on Russian naturalists and ecologists for decades, largely because it served as the "scientific" justification for an entire edifice of claims and institutions connected with the role in Russian public life these scientists sought.

Thanks largely to the contributions of Grigorii Aleksandrovich Kozhevnikov, a Moscow entomologist, a strategy for allegedly determining ecologically acceptable levels of economic development had been put forward in the decade before the Revolution. Kozhevnikov envisaged a vast network of inviolable nature reserves—*zapovedniki*—dedicated exclusively to the long-term study of the ecological dynamics of the biocenoses they were supposed to incorporate. Managed and staffed by scientists, *zapovedniki*, created on tracts believed to be both pristine, intact ecological systems and representatives of even larger landscapes, would serve as *etalony*, or baseline models of "healthy" nature. Kozhevnikov proposed that these tracts be compared with areas, once similar, that had undergone human economic transformation in order to assess how much damage was caused by which kinds of economic activity. With the endorsement in the mid-1920s of this strategic vision by the leaders of the RSFSR People's Commissariat of Education and its science-management department, Glavnauka, just such a network of *zapovedniki* came into being.

Such views and strategies neatly fit within the larger ideology of the cult of Science (*nauka*), once nature protection was defined as within the realm of "science." In this spirit, scientists' opposition to elements of the First Five-

Year Plan and later Stalin-era and post-Stalin-era projects on ecological grounds was doubtless motivated by their understanding of the sacred duty of responsible scientists before *nauka.* Ditto their claims to veto power over the resource policies of the regime. However, although activists may not always have consciously understood such claims to be "political" (viewing them rather as "scientific"), they were political nonetheless, for the Party had already claimed a monopoly on all decision-making authority.[15]

Similarly, protecting and expanding the system of *zapovedniki* was not simply an instrument for assuring the adequate protection of objects of scientific interest and venues for their professional study. In the Soviet context it also represented an expansion of the realm of autonomous scientific institutions within the Soviet polity and more. Symbolically, the reserves constituted a counter-GULAG, territories that remained inviolable and hence undefiled by the kinds of social and nature transformations that characterized the social sea that surrounded them. In the words of Sergei Zalygin, they were "islands of freedom in that concentration-camp world that people would later call the GULAG archipelago."[16] That made the scientists' fight to defend and extend them a fight to extend a realm of extraterritoriality where at least nature could develop without fetters. In other words, *zapovedniki* also functioned as a geography of hope.

Caught up in these heady theories, scientists failed to ask themselves whether the theory of the biocenosis was the best model for understanding the distribution and structure of life on earth. They did not question whether they were buying into a delusional construct of "healthy" versus "pathological" nature. Indeed, it did not occur to them that they might be sacrificing science to the cult of *nauka* (Science).

With the demise of an overarching "academic intelligentsia" during the first decade and a half of Soviet rule—so-called "bourgeois" professors were thoroughly expunged from the humanities and social sciences—the survival of the ideology of *nauka* was now mainly dependent on natural scientists. However, because so many of the corporate institutions of even these academics were policed, terrorized, reorganized, or eliminated, particularly during the Cultural Revolution, after 1932 there were practically no venues for the active expression and vocal affirmation of academics' prerevolutionary social identity.[17] Some few, such as Vladimir Ivanovich Vernadskii and Pëtr Kapitsa, valued by the Soviets for their strategic importance to the economy or national security, were allowed as individual exceptions to continue to profess the old academic creed unrepressed. Most who dared to defend science or to oppose regime policies in the name of science, such as the geneticists Nikolai Konstantinovich Kol'tsov and Nikolai Ivanovich Vavilov, met tragic fates.[18]

Social identities are subject to radical mutation if they fall into desuetude—witness the Judaism of the *conversos* during the Spanish Inquisition.

They cannot survive indefinitely in the individual imagination; rather, they need a social setting with real human interactions in order to maintain a full-blooded existence. Like the *conversos,* individual academics kept their felt social identities a secret, publicly professing loyalty instead to regime values and largely acting out the public roles prescribed for them. After a while, though, many isolated and terrorized academics lost touch with their original values and perspectives. Some even came to adopt their prescribed public roles as new social identities.

Against that backdrop, then, the continued existence and independence of a few voluntary societies of field naturalists assumed a crucial importance for the survival of the old social identity. Like the *zubr* (European bison), they became a relict population. The institutions, praxis, and speech of nature protection formed the basis for the continued expression and affirmation of an unauthorized, suspect social identity.

Given Soviet conditions, the field naturalists needed to be creative in order to survive. They perfected a strategy that I call "protective coloration," which involved promoting their own aims while rhetorically professing loyalty to the regime's. Accordingly, they took up a new designation for themselves, rhetorically in keeping with new Soviet social categories. They called themselves "*nauchnaia obshchestvennost'*" (scientific public opinion). Outwardly this term had the virtue of sounding eminently "Soviet"; in the media one never ceased to hear about the support or participation of *sovetskaia obshchestvennost'* (Soviet public opinion) for one or another regime campaign. Accordingly, *nauchnaia obshchestvennost'* seemed to constitute one small subgroup of the loyal cheering section—that of scientists. Yet, for the members of the field naturalists' societies in the nature protection movement the term had another, internal meaning. *En famille* it was a self-description in which they recognized themselves as representing the last organized bastion of the old ideology of the prerevolutionary academic intelligentsia. Not that the scientists were disloyal; they were simply presumptuous and critical. For them, "scientific public opinion" was the only truly credentialed public opinion, credentialed because it was "scientific." That, they believed, entitled them to critique the policies and strategies of the regime in areas where "scientific public opinion" had determined that the regime was acting at odds with the interests of "science."

The cult of *nauka* contained important elitist elements, reflected in the ambiguous nature of the term *nauchnaia obshchestvennost'*. Despite the fact that it seemed to fit into a larger, official, choreographed group of "broad Soviet public opinion," *nauchnaia obshchestvennost'* still had at its core the prerevolutionary understanding of *obshchestvennost'*—the voice of educated, responsible public opinion. In the absence of comparably independent organizations of scientists and educated society, nature protection activists spoke as though the burden of representing scientific public opinion fell on them alone. It was unclear, however, how much of a public they really represented as the decades of Soviet power wore on. At times it seemed that they represented a constituency of historical memory, the residue of the dreams and hopes of a bygone era.

Another aspect of the nature protection activists keeps us from romanticizing them as complete democrats. Although priding themselves on their own independent initiatives, such as the VOOP-sponsored expeditions to chart and propose new *zapovedniki,* their ultimate hope was to be invited by enlightened leaders of the state to take their rightful places as the expert arbiters of resource decisions. Fearing the acquisitiveness and dark ignorance of the masses, scientific public opinion hoped to realize its nature protection programs through the mighty fiat of the Leviathan-state. Each small liberalizing shift in the political winds lofted activists' hopes that they would receive the Kremlin's call to serve. As things turned out, they spent many decades in fruitless waiting and had to content themselves with the occasional—though sometimes enthusiastic—patronage of local and republic-level politicians.

One caveat must be included here. Nestled within scientific public opinion's All-Russian Society for the Protection of Nature were a number of individuals who can be better be described as "citizen activists" representing the broader prerevolutionary ideal of *obshchestvennost'* (educated lay public opinion) rather than the narrower one of *nauchnaia obshchestvennost'.* For such truly rare relicts as VOOP secretary Susanna Fridman or longtime activist Aleksandr Petrovich Protopopov, there were few if any other independent voluntary societies where their civic concerns could find an outlet. Partly because they were not invested in defending a cult of Science, partly because they were not as dependent on the system for their perquisites and careers, and partly because they still nourished the nearly extinct ideal of the dignity of the citizen qua citizen (not scientist), Fridman and Protopopov were consistently on the front lines of the Society, holding the most militant positions. They must be regarded as admirable but tragic curiosities in the tale this book will tell.

The Tradition of the *Studenchestvo*

Another prerevolutionary social identity that reappeared in the Soviet period as environmental activism was the *studenchestvo,* or membership in the student movement. This social identity, possibly born in Pushkin's lycée as Florenskii suggests but certainly flourishing from the 1860s on, was based on students' perception that they constituted "a unique and distinct subgroup in Russian society . . . with its own history, traditions, institutions, code of ethics, and responsibilities."[19] Reflecting their intrepid, impatient psychology was the students' penchant for *skhodki,* or mass meetings, as well as street demonstrations. However, they were also capable of sustaining long-term institutions that reflected their intense in-group solidarity: mutual aid societies, independent banks, libraries, cafeterias, and even dormitories.[20]

Reaching its height during the final two decades of tsarism, to outward appearances the ideal of the *studenchestvo* seemed to be dead among Soviet college students by the 1970s; Solzhenitsyn had even unflatteringly rechristened them an *obrazovanshchina,* or "educated rabble."[21] Soviet students *had* lost their nineteenth-century corporativist traditions of solidarity; groups independent of the Komsomol (Young Communist League) were seen as potential nodes of subversion and were at times ruthlessly snuffed out.[22] From the 1960s, however, there was one exception: the *druzhiny po okhrane prirody* (student nature protection brigades), which grew out of the particular esprit de corps preserved at Moscow State University's Biology Faculty. Here again, nature protection served as a protected locus for the preservation (or resurrection) of a prerevolutionary-style group identity, in this case that of the *studenchestvo,* with its characteristic attributes of impatience, direct action, group loyalty, moral absolutism, independence, and bravado. Students were also protected by a long tradition of indulgent attitudes toward them throughout society. Without asking the permission of higher authorities, the *druzhiny* organized independent efforts to enforce environmental laws: roundups of poachers, checkpoints to ensure that New Year's trees were procured legally, and the monitoring and testing of factory discharges. Because they studied under professors who were central figures of scientific public opinion, the *druzhinniki* could not help imbibing many of their teachers' scientific ideas and liberal, internationalist, and statist ideological biases. Nevertheless, unlike their teachers, the ardent students put adventure and direct action ahead of the ideal of *nauka* in their own civic activism. They also tried to penetrate the state's bureaucracy, so as to get their hands directly on the levers of power and policy.

If *druzhiny,* spreading from Moscow University's Biological Faculty to most other important universities of Russia and the USSR, represented the *studenchestvo* tradition in its elite form, then the "Kedrograd" movement expressed the corporate student identity of those in less prestigious technical and engineering schools. By contrast with the *druzhinniki,* many of whose parents were members of the intelligentsia, these technical school students (*kedrogradtsy*) were largely first-generation college-educated and were mostly from the provinces. Arising from the vision of a group of Leningrad forestry academy students in the late 1950s, "Kedrograd" was a quixotic attempt to manage Siberian stone pine forests in the Altai so as to harvest the forest's secondary production—sables, squirrels, pine nuts, and so on—without logging the trees themselves. Fired by the optimism generated by Khrushchëv's thaw, these ardent Soviet patriots wanted to build Communism. They wanted to make the system work more efficiently by applying their expertise to the problem of resource management.

Like the *druzhinniki,* the *kedrogradtsy* used environmental activism as a means of asserting their own independent status as experts in the area of resource management, particularly forestry, and to build a special feeling of fraternity and identity. As the idealistic graduates saw their dreams dashed by cold bureaucrats, a number of the Kedrograd movement's key organizers as well as a prominent journalist who covered the story traded tarnished feelings of Soviet patriotism for a strong Russian nativism. Environmentalist images and rhetoric were as central to this shift in political/social identification as they had been to the forging of the original Kedrograd movement.

National-Patriotic Nature Protection

From such beginnings as the Kedrograd experiment, the rhetoric of nature protection mobilized a considerable number of Russians by the 1970s. They were led by prominent writers and other public figures to affirm a social identity whose highest value was ethnic Russian cultural patriotism. While individual beliefs among this group ran the gamut from a benign antiquarianism to extreme xenophobia, members shared the premise that Russian culture could not be preserved in its integrity without preserving integral Russian landscapes and the Russian village. Although members of this movement, united in such large organizations as the All-Russian Society for the Preservation of Monuments of History and Culture, frequently fought side by side with those of the scientist-led movement against the pollution of Lake Baikal or for other causes, the cultural patriots were far removed from the social identity of *nauchnaia obshchestvennost'*. Whereas scientific public opinion prided itself on its membership in an international confraternity of science, a "universal" global civilization, the cultural patriots, composed of Soviet-era writers and Soviet-trained engineers, technicians, and even scientists, did not worship at the altar of the old-style cult of Science. Rather, they claimed to speak for an equally grandiose "public," the Russian nation. The rhetoric of nature protection proved equally serviceable to both movements.

Middle-Level Officialdom

One of the curious facets of Soviet politics this study brings to light is the highly supportive role, including active patronage and protection, that Soviet republic-level and *oblast'*-level leaders accorded the nature protection movement. Unable to counter decisions of the USSR Council of Ministers and the Politburo on a whole range of more important matters, local-level leaders were able to demonstrate their independence and authority in the realm of nature protection. Precisely because of nature protection's marginality, republic-level leaders and *oblast'* first secretaries dared to oppose the center's plan to "liquidate" the *zapovedniki* in 1951, counting on the relatively low risk of such dissenting political speech. A series of Russian Republic premiers gave significant material and political support to both VOOP and the *zapovedniki,* and directly intervened to save the Society from the Central Committee's repeated attempts to eliminate it. This was one of the few policy areas where these middle-level politicians could express their independence from the center, pursue policies strictly on their own initiative, and demonstrate the "dignity" of their offices by protesting attempts by the center to confiscate or eliminate territories and organizations lodged within local bureaucratic portfolios. Nature protection, consequently, also provided the policy arena for middle-level politicians to express political identities other than simply cogs in a larger, centrally driven Party machine.

Ordinary Soviet People

Finally, beginning in 1987 when Soviet people began to test the sincerity of *glasnost',* the plazas, parks, and boulevards of Soviet cities became the locations for a remarkable series of public protests, involving hundreds of thousands of people who rallied under environmentalist—mostly public health–related—slogans. Mass environmental protest made its mark on Soviet history only to give way first to explicitly economic protests, then to overtly political protests, and finally to apathy, all largely before the official collapse of the Soviet Union. Although the period of mass environmental protests, brief as it was, is important in its own right, it marks the end of the special role of nature protection in pre-*glasnost'* Soviet society. For these ordinary people, nature protection was not an arcane exercise in identity politics (with the exception of environmental protests in the non-Russian republics, where it was often a stand-in for an expression of the local ethnonationalism).[23] Their spouses, parents, children, coworkers, and friends were slowly or quickly being poisoned by Soviet industrial and agricultural development.

By 1987 the day of scientific public opinion had nearly passed. Its successors, the student *druzhiny,* were left partly on the sidelines through their continued focus on the protection of sacred space and on the campaign against poaching, which led them to ignore the environmental public health and safety issues that troubled the population at large. Similarly, the appeal of a Russian nationalist nature protection proved limited against the backdrop of the life-and-death environmental concerns of the broader public. The closed, castelike nature of scientific public opinion and the *druzhiny* and the abstruse and self-limiting nostalgia of the cultural nationalists prevented them from assuming leadership in building a civil society.

Additionally, Gorbachëv's reforms made it possible to assert almost any kind of independent social identity openly for the first time in many decades, and people had other vehicles besides environmental advocacy to express dissatisfaction with the regime's policies and visions. Environmental advocacy had to compete for attention in a completely changed political environment. Doubtless involvement in nature protection will continue to serve as a nucleus around which groups build special identities in post-Soviet Russia and elsewhere. But it will no longer have the special—in Iurii Efremov's words, "brave"—role it had in the highly repressive Soviet polity of decades past. And its potential for the creation of an inclusive, full-blooded civil society is still to be tested.

CHAPTER TWO

Archipelago of Freedom

At a Soviet-American conference on the history of environmentalism, the historian Leo Marx voiced his surprise at the degree to which Russians, especially scientists, equated conservation in general with *zapovedniki,* inviolable nature reserves.[1] For Russians *zapovedniki* have a significance far transcending their ostensible functions as centers for ecological research and the protection of rare species and habitats, because they were central to the social identity and mission of Russia's leading field biologists, who doubled as the leaders of that country's nature protection movement. This was particularly true from the late 1920s and early 1930s, when *zapovedniki* and the All-Russian Society for the Protection of Nature were nearly the only scientific institutions that escaped elimination or stultifying Party control.

Protected territories first appeared in Russia about a century ago, the efforts of scientists attached to the Imperial Academy of Sciences, the Russian Geographical Society, and teaching institutions, as well as of private landowners such as Baron Friedrich-Eduard Falz-Fein (Fal'ts-Fein) of the southern Ukrainian estate Askania-Nova. Later, during World War I and into the Bolshevik period the state emerged as the chief patron of a growing network of nature reserves.[2]

Although some of these reserves, those established by the tsarist regime or by the People's Commissariat of Agriculture in the Soviet period, resembled American national forests in their open dedication to the propagation of commercially valuable resources, the remainder had no U.S. analogs. Organized by scientists, especially Russia's pioneers in plant and animal ecology and other field naturalists, and coming under the patronage of the RSFSR People's Commissariat of Education and its Scientific Administration (Glavnauka), the inviolable *zapovedniki* were aimed at the protection and long-term study of what were believed to be pristine, intact ecological communities. Such study, it was hoped, would reveal the ecological dynamics of "healthy" natural systems and could serve as a baseline (*etalon*) against which "degraded" communities, that is, those under economic exploitation, that had allegedly shared the same natural conditions, could be compared. The end result would be that scientists would not only gain knowledge about biological processes but would be able to use such knowledge to make expert recommendations regarding the most appropriate economic use for a given natural complex.

An array of assumptions was built into this model of the ecological community and consequently into that of the *zapovedniki* as well: that discrete natural communities existed, that they normally maintained themselves in a state of balance, that they represented "healthy" pristine nature, and, correspondingly, that humans existed outside nature as a "pathological" force. These beliefs represented unprovable assumptions about how to map nature. Although few would deny that nature is highly ramified—every life form is linked by myriad threads directly or indirectly to other life forms as well as to the inanimate environment—the idea of a tightly bounded natural community is a speculative leap. Nevertheless, Russian field naturalists heavily favored it over rival theories of how nature was put together.

Since the 1910s in America, France, and Russia, another view of nature had been advanced that, while accepting the interconnectedness of species in food chains and other relationships, denied the existence of bounded, self-regulating natural communities. However, field naturalists needed to believe in the existence of fragile, holistic ecological communities to justify their magnificent research project, based in the *zapovedniki,* that would decode those alleged pristine communities and ultimately allow the scientists, as experts, to make the key judgments about land and resource use that would prevent catastrophic injury to those systems. Although both models of nature were unprovable, most scientists opted for the one that best supported their attempt to present themselves as the expert arbiters of resource and land use.[3]

Zapovedniki were valued for a more prosaic reason as well. Those who became field naturalists were drawn to studying life forms free in their natural habitats, rather than as caged, dried, or dissected specimens in the lab. *Zapovedniki,* as undisturbed wild habitats, were indispensable to field naturalists as the last bastions where they could securely pursue their distinctive kind of scientific observations of "free," living nature. Such a role made the reserves valuable not merely to community ecologists but to a wide range of naturalists, including plant and animal taxonomists and physiologists, ethologists (experts on animal behavior), soil scientists, and geologists.

The 1920s were good years for the Russian nature protection movement. In 1924 an All-Russian Society for the Protection of Nature (VOOP) was established under the leadership of eminent field biologists and successfully initiated the expansion of the network of *zapovedniki,* based on expeditions organized by the Society. As a result of activists' lobbying and the support of a sympathetic leadership of the People's Commissariat of Education, in 1926 a unique Interagency State Committee for the Protection of Nature was established and given the power to examine all resource-related government decisions and to veto those it found excessively damaging to nature. This forerunner of the environmental impact process existed nowhere else in the world. By 1932, moreover, VOOP had 15,000 members and was supported in its program by the 60,000-plus-member Central Bureau for the Study of Local Lore (*Tsentral'noe biuro kraevedeniia*), which was led by many of the same naturalists and their supporters. A national conference on nature protection was organized in September 1929, arguably the high point of the movement's efficacy.[4]

By the end of the 1920s, though, with the onset of Stalin's triple revolution—collectivization, frenetic industrialization, and the attempted elimination of any effective form of civic autonomy—*zapovedniki* and the nature protection movement began to acquire another layer of meaning: they were among the rare physical and social spaces in the Soviet Union that had largely escaped the juggernaut of Stalin's "Great Break." As such, they came to constitute an "archipelago of freedom," unique islands in the scientific intelligentsia's geography of hope.

To understand why scientists and their allies defended the inviolability of *zapovedniki* with dogged tenacity, we must understand the cultural meaning and the political and social implications of the landscape that Stalin and his regime sought to create. The writer Maxim Gorky wrote that poets must champion "the struggle of collectively organized reason against the elemental forces of nature and against everything 'elemental' . . . in the formation of man."[5] Stalinists viewed the wild with repulsion, seeing in it the embodiment of everything outside the rational control of humans or, more correctly, of the Party's leadership. Nothing characterized the Stalinist worldview better than its unquenchable craving for total, conscious control over nature, society, and events. This phobia of spontaneity and obsession with conscious control permeated Stalin's policies toward the land and society both. Gorky's euphemistic motto for the "Belomor" White Sea Canal slave-labor project, "Man, in changing Nature, changes himself," summarizes this connection, even as the canal embodied the linkage in real life.

Faced with this overwhelming threat to their own professional freedom and perhaps comprehending on some level the linkage between Stalin's plans for a great transformation of the landscape and the reenserfment of society, Soviet scientists responded by marshaling ecological arguments against collectivization and the great earth-moving projects. When this scientific opposition to the Five-Year Plan failed, scientists retreated to a defense of the inviolable *zapovedniki* under their control.[6]

By their charters the reserves were absolutely inviolable. As Vyacheslav Gerovitch and Anton Struchkov have noted,

> the idea of the "absolute inviolability" of *zapovedniki* has been disclosed as an allegory of the age-old Russian theme of "The City of Kitezh." According to their old Russian legend, when the country had become the Kingdom of Evil and Falsehood embracing both the State and the church authorities, the Kingdom of Good and Righteousness—the City of Kitezh—sank to the bottom of a lake. Hence [it] is the idea of withdrawal from surrounding vicious life, the idea of wandering elsewhere in search of this "ideal City."[7]

Nikolai Vorontsov has corroborated that one major reason for the unique scientific milieu in the reserves stemmed from the continuing policy of repression directed at the intelligentsia, which drove leading scientists to seek physical refuge in those territories.[8] But the *zapovedniki* were more than tangible sanctuaries for the "endangered species of bourgeois scientists."[9] They had become the symbolic embodiment of the harmony of communities, of natural and human diversity, and of the free and untutored flow of life (in Struchkov's eloquent phrase, the "unquenchable hearths of the freedom of Being"). As long as the "pristine" *zapovedniki* could remain independent, what was denied to human society in Stalin's Russia could be preserved in symbolic, natural form in these reserves. Indeed, the struggle over the defense and, later, the reestablishment of inviolable *zapovedniki* eclipsed all other environmental issues through the 1960s or even 1970s, if we include Lake Baikal in the sacral geography of the academic intelligentsia.

To protect its own fragile institutions as it continued to defend the prerevolutionary ideal of "science" (*nauka*), the nature protection movement labored to present a public face of loyalty to the regime in a strategy the movement's critics termed "protective coloration." Somehow outlasting the critics, the effective leader of VOOP and of the RSFSR Main Administration for *Zapovedniki* from the early thirties, Vasilii Nikitich Makarov, raised protective coloration to an art form.[10]

Vasilii Nikitich Makarov

Born August 5 (New Style), 1887, in Lunëvo, a village not far from the provincial town of Vladimir, northeast of Moscow, Vasilii Nikitich Makarov came from peasant stock, although both his father and paternal grandfather were workers (see figure 1). After excelling in his rural school, he was recommended by his teacher for a *zemstvo* scholarship to complete his higher grades in town.[11] For two years after graduating, Makarov worked in agriculture, entering the Moscow Teachers' Institute in the fall of 1905. Soon he

Figure 1. Vasilii Nikitich Makarov (1887–1953) at age sixty.

was drawn into the vortex of protest during that revolutionary year. A member of revolutionary student circles (*kruzhki*), Makarov joined a strike committee and distributed illegal literature among workers. With the restoration of order the following year, Makarov was arrested, but he was released after three months for lack of conclusive evidence and was allowed to resume his studies, graduating in 1908.[12]

Trained as a science teacher, Makarov was posted to a school in the Volga town of Kostroma, north of Moscow, but returned to Moscow in 1911 to attend night school at the Moscow Commercial Institute to upgrade his qualifications, teaching fourth grade during the day at a school attached to the Moscow Teachers' Institute. Apparently, the punishing schedule did not diminish his effectiveness as a teacher; indeed, he seems to have had a talent for teaching In bidding him farewell, his students in both Kostroma and Moscow emphasized not only his kindness and empathy, but also his ability to inspire them to strive for a life "in science."[13]

In 1916, after meeting a physician who was a member of the Socialist Revolutionary Party, Makarov joined the SRs as well. With the overthrow of the tsar, Makarov was elected the *uezd* (county) commissar of Makar'ev *uezd* and chair of the democratic rural assembly. But his growing misgivings about the irresolute policies of the SR party led him to decline the nomination by the local Provincial Peasant Congress to stand as a deputy for the Constituent Assembly in the fall of 1917. He officially resigned from the party on January 1, 1918.

In September 1918 he was tapped to serve as the principal for a middle-grade school for workers in Moscow province and later for a number of schools in the capital itself. Rising through the educational bureaucracy, Makarov was named head of the Moscow's Bauman School District but apparently continued to teach science. This relatively placid existence ended in June 1930, when he was appointed academic specialist in the Science Sector of the RSFSR People's Commissariat of Education, almost immediately thereafter rising to deputy head and then head of the sector (which he remained until February 16, 1937). Simultaneously, he was appointed the director of the Zoological Museum of Moscow State University, to replace Grigorii Aleksandrovich Kozhevnikov, who had been forced to resign as a "bourgeois" professor. By the beginning of 1931 Makarov was also president of the All-Russian Society for the Protection of Nature, and with the reorganization of the *zapovedniki* in September 1933 became the deputy director of the Main Administration. Makarov could have achieved none of this had he not been accepted into the Communist Party in April 1928.[14]

When Makarov assumed leadership of the nature protection movement, hostile critics were already identifying the "counterrevolutionary" implications of the movement's ecologically based objections to elements of the First Five-Year Plan. To deflect these accusations, Makarov instituted a policy of "protective coloration," muting criticism of regime resource policies, pledging verbal loyalty to "socialist construction," and renouncing a commitment to the absolute inviolability of the *zapovedniki.* At the same time, however, the strategy sought to preserve the All-Russian Society for the Protection of Nature as a place where alternative visions of development could be freely discussed and to preserve the *zapovedniki* as factually inviolable, although no longer officially so.[15]

Both the movement and the regime at times revealed an awareness of discord between Soviet environmentalism and Stalinist policies and values. Recent finds in Russian archives throw dramatic new light on just how courageously "out of step" leading conservationists were with the Five-Year Plan for "socialist construction." Of course, not every ecologist always evinced courageous behavior, nor did every occasion elicit it. With the exception of ichthyologist Mikhail Nikolaevich Knipovich, the nerve of almost all prominent ecologists and zoologists withered under the ferocious attacks of Isaak Izrailovich Prezent at the All-Union Faunistics Conference in February 1932.[16] Yet, surprisingly frequently ecologists sketched out an alternative vision of land use, the use of scientists, and even civic speech. Perhaps unmatched in its time as a call for norms of decency in political discussions was a letter sent in 1931 by Makarov, now the de facto leader of the Russian conservation movement, to the Scientific Sector of People's Commissariat of Education (where he was deputy head) and to its Communist Party cell.

Only recently having become president of VOOP, Makarov in early 1931 inherited a precarious situation. VOOP had undergone a high-level audit the previous year that revealed numerous deficiencies in the work of the Society from the perspective of the regime, including "undisguised apoliticism" and ecological "alarmism."[17] Press articles ridiculed scientific societies, including VOOP, as an "All-Union *zapovednik* for the endangered species of bourgeois scientists," coming dangerously close to the truth.[18] Makarov's letter combined a surprisingly forthright objection to an excessively rough, denunciatory style of polemics with protestations of loyalty to the regime's strategies of development, "socialist construction." Because the nature protectionists' visions of development clashed with those of the regime, their averring loyalty was either conscious dissembling or self-delusion in pursuit of "protective coloration."

Makarov's letter was one of his first serious attempts to counter the ominous assaults directed at the movement he now headed. While he conceded that "Marxist-Leninist criticism" prodded "many stagnant areas of science to come alive" and succeeded in getting academics to descend from their ivory towers and to begin to meet society's "legitimate expectations" of them (*sotsial'nyi zakaz*), Makarov observed that "that was not so in all cases." Sometimes, he contended, "comrades offering critical comments have acted too hastily and made superficial judgments, not possessing the requisite erudition for a proper consideration of the problems addressed." At times, "Bolshevik" critics behaved even more irresponsibly, driven by "the preconceived aim—*whatever it takes*—to identify an enemy, reveal a [political] deviation, and to unmask sabotage and counterrevolution in science; they have 'twisted and distorted' critical material, turning healthy Bolshevik criticism into the dubious weapon of polemics and even denunciation. This unfortunate criticism, purveyed in the mass media and distracting the masses from the substance of the issue, has been harmful."[19]

Amazingly, the concrete example Makarov chose to exemplify his charges was the recent article "Sabotage in Science" published by Arnosht Kol'man in the Party's theoretical journal *Bolshevik.* Kol'man was one of the Party's key curators of science, even serving as watchdog over such illustrious figures as Nikolai Ivanovich Vavilov and Nikolai Ivanovich Bukharin during their 1931 visit to Great Britain. In strong language Makarov contested what he argued were Kol'man's false claims—that the conservation movement sought to "undermine our socialist construction and engineer a restoration of capitalism."[20] "Pointing out to Comrade Kol'man the error of his views in the given case elicited no effect." He evidently continued to remain convinced that the protection of woodlands in "sparsely wooded areas and on nonarable lands is a land mine under socialist agriculture." Similarly, Makarov accused Kol'man of failing to understand the value of the protection of unplowed steppe as a reservoir for genetic material, especially in developing drought-resistant varieties of agricultural plants, as demonstrated by Vavilov.

Additionally, the conservation leader cited an equally vicious article by two other authors also directed against his movement, and concluded:

> These [articles] also force us to consider the following questions: Is *THIS KIND OF defense* of the great cause of socialist construction of the Five-Year Plan from putative sabotage useful? Is it permissible to purvey gross distortions, as Comrade Kol'man and others have done, in full public voice? Doesn't this gladden the genuine enemies of socialist construction both here and abroad, enemies who will snatch at any opportunity to demonstrate, on the basis of isolated examples, how science is profaned in the USSR and how thoughtlessly and wantonly scientific ideas and the people selflessly serving science are trashed? . . . The Council of the [All-Russian] Society [for the Protection of Nature] insists that the Scientific Sector and the Party cell . . . rap the knuckles and head of those adepts of "leftist" witchhunting and "distortion" of the authentic character of the activity of our Society and the content of its journal. Criticism, merciless Bolshevik criticism of the entire press is an essential fact of life, but, in the opinion of the Council of the Society, the "obfuscating" tactics of [our] critics has nothing in common with that.[21]

The conservation movement's defense not only of "free" nature but of "free" science and, to an extent, of prerevolutionary norms of public communication, was fraught with risk.[22] As mentioned in my previous work and now confirmed by a host of newly available archival documents, repression did indeed strike Russian environmentalists hard during the early to mid-1930s.[23] Some few lucky ones like movement founder Grigorii Aleksandrovich Kozhevnikov were merely fired from their positions or, like A. V. Fediushin, were able to flee to distant regions. Others, like geographer V. P. Semënov-tian-shanskii, were placed on blacklists but somehow were never picked up. Many others, though, were less fortunate, and the roster of those arrested during that period abounds with important names.[24] Although not all environmentalist victims of Stalinist repression suffered because they were environmentalists, and a majority of committed activists emerged relatively unscathed from the terror, a climate of intimidation enveloped the conservation cause during the dark decades of the 1930s and 1940s.

The Purge at Askania-Nova

The episode that was most traumatic to the conservation movement was the devastating purge of Askania-Nova, engineered by Trofim Denisovich Lysenko and Isaak Izrailovich Prezent during the fall of 1933.[25] The newly available archival documents and oral testimonies do not permit us irrefutably to determine the cause or causes of the purge;[26] however, it is likely that the mass arrests of Vladimir Vladimirovich Stanchinskii and his colleagues at the Ukrainian nature reserve flowed in good measure from their resistance to Stalinists' plans for a "great transformation" of Soviet nature.

In the hope that he would bring coherence and cutting-edge research to Askania-Nova, V. V. Stanchinskii, a professor of zoology at the Smolensk State University, had been hired by the troubled reserve in 1929 as deputy reserve director for science, simultaneously joining the zoology faculty of Khar'kov State University. Losing little time, Stanchinskii organized the measurement of energy budgets by groups of organisms arranged by trophic levels and pioneered the first attempt anywhere to measure the amount of solar energy captured by plants and then subsequently passed along to herbivores, carnivores, and decomposers. While pursuing that audacious program in trophic dynamics, in 1931 Stanchinskii assumed responsibility as the principal editor of the USSR's first scientific journal of ecology, the *Zhurnal ekologii i biotsenologii,* and began to develop scientific arguments against one of the favorite nature-transforming schemes of Stalinist academics and politicians, the acclimatization of exotic plants and animals.

Acclimatization was being promoted as a means of enhancing economically exploitable biological productivity and was predicated on the idea that natural conditions were being underutilized by the existing mix of organisms. New nonnative organisms could be "inserted" into nature's "empty places" to provide new sources of ornamental plants, fruits and vegetables, game and pelts. Stanchinskii argued that there was no guarantee that the genotype (genetic make-up) of the introduced plants and animals would enable them to survive in a new habitat, making these widespread experiments a potential loss of lots of money. Even if they did have such adaptive fitness, Stanchinskii warned, their successful acclimatization could entail equally heavy costs. That was because in real life there usually were no "ecologically empty places"; introduced animals and plants, to survive, would need to outcompete endemic or native forms that subsisted on roughly the same mix of resources. Thus, the addition of a nonnative species would

Figure 2. Vladimir Vladimirovich Stanchinskii (1882–1942) in the Balitskii Penal Kolkhoz.

probably be at the cost of the elimination or even possibly the extinction of a native one. Moreover, there was the further possibility that the introduced species could serve as a vector for the introduction of new parasites and also become an unchecked pest. Underscoring the need for caution, Stanchinskii's objections threw cold water on the grandiose hopes of Stalinists for an unprecedented rearrangement or "transformation" of nature.

Following a first visit to the reserve by Lysenko and Prezent, a decision was taken in December 1932 by the Ukrainian Academy of Agricultural Sciences and supported by the Lenin All-Union Academy of Agricultural Sciences in Moscow to shut down Stanchinskii's Steppe Research Institute as "not having any real importance for socialist agriculture."[27] Although the institute had been eliminated, the *zapovednik* formally still remained, but with the arrests of Stanchinskii, Gunali, Fortunatov, and the other members of his research team in the autumn of 1933, that too fell into the hands of the sheep breeders and the "barefoot agronomists."

Sentenced to a prison-*kolkhoz* run by the People's Commissariat of Internal Affairs, where he worked as a veterinarian from 1934 through 1936, Stanchinskii was released two years before his term was up. A photograph from his incarceration shows how much he aged in the two years after his arrest (see figure 2). Yet, at the Balitskii penal *kolkhoz* (named after Ukraine's secret police chief) he enjoyed surprisingly liberal conditions of detention. He was permitted visits from his family, and he was even permitted to accept an invitation to spend New Year's Day, 1936, at the home of the great evolutionary geneticist Ivan Ivanovich Shmal'gauzen in Kiev. Under the circumstances, the invitation was a courageous act of friendship on the part of Shmal'gauzen.

After his early release, however, Stanchinskii was unable to pick up where he had left off. Forever barred from teaching, he had to rely on the loyalty and generosity of old friends to secure employment again as a research biologist. Returning to the region of his roots, Stanchinskii accepted the invitation by G. L. Grave, director of the Tsentral'no-Lesnoi (Central Forest) *zapovednik* north of Smolensk, to become deputy director for scientific research.

Olga Borisovna Lepeshinskaia's Investigation

Not long after the purge of Askania, the conservation movement lost its most devoted patron from among high-ranking members of the regime. In April 1935 Pëtr Germogenovich Smidovich died under questionable circumstances at the age of sixty-five. The year before, Smidovich had saved the *zapovedniki* from falling into the hostile hands of the economic commissariats, bringing them instead under his direct care as head of the newly established Committee for *Zapovedniki* of the Presidium of the RSFSR Central Executive Committee (VTsIK). After the trauma of the Askania purge and the conservation movement's near annihilation, the movement had slowly recovered during 1934.[28] Moderate Bolsheviks—the protectors and patrons of the conservation movement—seemed in the ascendant at the Seventeenth Party Conference; the horrors of the First Five-Year Plan, the famine, and the "left-wing" extremism of the Cultural Revolution all seemed past.

The clear skies of 1934, however, were suddenly darkened by the events of December 1; Stalin's arranged assassination of leading Bolshevik Sergei Mironovich Kirov served as the prelude to a terror that cast its pall over the whole nation. The conservation movement's horizons clouded over as well, particularly after Smidovich's death.

On April 25, nine days after Smidovich's passing, Frants Frantsevich Shillinger was fired from his position at the Committee for *Zapovedniki* without explanation, and the Science Department of the Central Committee of the Party ordered another audit of the reserves and of the conservation movement.[29]

It is unclear whether the conservation movement was aware of just how close to the edge of danger it had strayed. An ominous official report of more than one hundred pages assembled the case against the nature preserves and the conservation movement. Entitled "Report to the Science Department of the Central Committee of the All-Russian Communist Party on the Results of an Investigation of the *Zapovedniki* of the RSFSR," this document was still in the form of Ol'ga Borisovna Lepeshinskaia's notes intended for circulation among the other members of the investigation team.[30] Feliks Robertovich Shtil'mark has recently unearthed the final version of the report, countersigned by Makarov, and dated December 19, 1935.[31]

Reflecting what movement scientists told the investigators about the distinctive purposes of Soviet reserves, Lepeshinskaia noted that although foreign protected territories were directed toward tourism and other forms of profit-making activities, Soviet reserves were created to "preserve gene pools [*sokhranenie geno-fondov*], for scientific study, to enable humans to master nature, and for educational purposes." However, it would have been awkward to include this dubious claim to superiority on the basis of a largely traditional program defined by scientists—the Party sympathized more with the capitalists' approach of exploiting the reserves for revenue—and this section was crossed out.[32]

What remained in the report was an almost unrelieved portrait of the reserves as refuges for anti-Soviet politics, values, ideas, and scientists. The reserves, Lepeshinskaia held, were pervaded by "anarchy and lack of supervision and planning." It was bad enough that nature reserves were established on the basis of private or citizens' initiative; worse was that some of these activists, for example, longtime movement leader Frants Frantsevich Shillinger, "son of an emigre White Guard," deviously arranged for the creation of *zapovedniki* (e.g., Altaiskii, Crimean) on the very frontiers of the Soviet republic, the better to facilitate hostile subversion, she alleged.[33] Shillinger was additionally faulted for the alleged emigration of his son in 1924 (the family claimed that he had died) and for Shillinger's himself becoming a German citizen in 1935.[34]

Commenting first on the personnel of the *zapovedniki,* Lepeshinskaia remarked on the "absence of a firm Communist nucleus" and offered that, in general, the "SELECTION OF PERSONNEL HAS BEEN UNHEALTHY. IN THE MAJORITY OF CASES, THEY HAVE BEEN POLITICALLY UNRELIABLE TYPES, RECOMMENDED BY THE OLD-LINE PROFESSORS. AS A RESULT ALMOST IN EVERY *ZAPOVEDNIK* THERE IS A GREAT INFESTATION [*zasorënnost'*] BY ANTI-SOVIET ELEMENTS, those exiled by the Soviet regime, those arrested previously, those [class enemies] deprived of their civil rights, and even disguised bandits."[35] Partly at fault were those in the Party elite that afforded the conservation movement patronage and protection. Lepeshinskaia even fingered Lenin's science adviser N. P. Gorbunov and prosecutor-general Krylenko for their support in the hiring of class enemies (*byvshie liudi*) for the nature reserve system.

The investigation revealed the penury of the *zapovedniki* better than any scientists' petition; there were only two cars for the entire thirty-one-reserve, 8,457,436-hectare network, an absence of furniture, tableware, scientific instruments, work clothing, and shoes. There were no radios or telephones, and no electricity. One reserve staffer, Vvedenskii, even died of hunger. Security was deficient; abandoned mud or wooden shelters testified to the trespassing of poachers or even smugglers.[36] Although such conditions could well be regarded as evidence of the central government's neglect of conservation, the report implied that they were the result of negligent and deficient leadership by the Committee for *Zapovedniki* of VTsIK, an agency with no independent funding source.

More weighty were arguments against the scientific research conducted by the reserves. At once eclectic and unsystematic, overly descriptive and unrealistic, and without links to the world of practice or to outside institutions, the reserves' scientific work was also ideologically highly suspect. The crucial problem of acclimatization, for example, was largely neglected, with the exception of work on beavers and muskrats at the Laplandskii *zapovednik* and on raccoon dogs at Buzulukskii bor. Lepeshinskaia cited information provided by Pëtr Aleksandrovich Manteifel' in denouncing a number of Committee and *zapovednik* staffers as believing in "the existence of harmony and equilibrium" in nature and in the idea of "nonintervention by humans in the life of nature." Specifically named were Buturlin, Zhitkov, and Alëkhin, although these names seem to reflect more the quirks of Manteifel''s personal animus than an exhaustive list of unreformed "bourgeois professors." As Manteifel' emphasized, all was not well on the ideological front.[37]

Indeed, in addition to the ideological heresies of the nature protectors there was the implication of political unreliability, if not outright disloyalty. Lepeshinskaia saw a political cover-up in the decision of the conservationist community, including Communists, to delete the names of individuals from a conference resolution condemning erroneous "class positions."[38]

Within the Committee for *Zapovedniki*, the staff of thirty-one was named by Pëtr Smidovich on the recommendation of Vasilii Nikitich Makarov. Lepeshinskaia was clearly unimpressed. Half were "dead souls," while some, such as Alëkhin, Zhitkov, and Buturlin, were active, but were philosophical "idealists," holding "reactionary" views on nature reserves. Buturlin, a prominent ornithologist, was additionally mocked as a "walking encyclopedia" and a positivist (how that was reconciled with his idealism remains a puzzle).[39] On the ground in the *zapovedniki* themselves there seemed to be a veritable swarm of suspect "elements." I. I. Puzanov's dedication to conservation won him the label of "fanatic."[40] The deputy director of the Pechoro-Ilychskii *zapovednik*, one Pirogov, was a non–Party member of noble origin with—higher education! Another was the wife of a colonel in one of the White armies who had emigrated abroad. A third was exiled from Moscow in 1933 for his harmful ideological influence on students.[41]

The director of the Tsentral'no-Lesnoi *zapovednik,* Grigorii Leonidovich Grave, was of the landowner class and had attended classical gymnasium. Now, complained Lepeshinskaia, Grave's hereditary class instincts led him to treat the reserve as his own baronial estate. More ominously, she noted Grave's friendly association with Stanchinskii, who had been arrested as a counterrevolutionary in the autumn of 1933; indeed, it was to Stanchinskii that Grave owed his own appointment to Smolensk University's zoology department.

Foreigners visited the Committee for *Zapovedniki,* which maintained ties overseas, ipso facto a sign of unreliability. The *zapovednik* system was aswarm with "alien elements" and alien values.

For Lepeshinskaia, it was not surprising that the Committee was such a swamp. After all, Makarov was a former Socialist Revolutionary, and although "personally honest and devoted to his cause," he was also a "rotten liberal who makes a show of party loyalty" and a "man of weak character, without principles, and too mild."[42] An example of Makarov's weakness was his failure to press the accusation of extortion against the Caucasus *zapovednik* director, Krasnobryzhev, to its logical conclusion. Makarov also failed to achieve a "firm Bolshevik line" in the literature of his committee.[43] The "Bolshevik spirit was undetectable" in the training of new staff members.[44] Although Lepeshinskaia's conclusions did not call for the elimination of the nature reserves, "which give us nothing at the present time," and although she did not call for the removal of Makarov, indeed giving him credit for being selfless and informed, she did demand a strict housecleaning.[45] The bottom line was her recommendation that the reserves be turned over to the USSR People's Commissariat of Agriculture.[46]

Lepeshinskaia's conclusions contained more than a grain of truth; from the "Stalinist" Soviet standpoint the whole nature protection movement together with its institutions was an island of subversion. Why then was it spared during that terrible year? Frants Shillinger revealed in his 1937 letter to Nikolai Mikhailovich Kulagin that, in a mood of terminal despair, Shillinger had decided to take a big risk. On January 20, 1936, telling no one, he wrote a twenty-page letter to Stalin. Seemingly miraculously, the transfer of the reserves to the Commissariat of Agriculture was called off. Not only that, state allocations for the reserves were dramatically increased. Shillinger's other suggestions—to create a Main Administration for Hunting and Animal Breeding within the USSR People's Commissariat of Agriculture and a Main Administration for Forest Protection and Afforestation under the USSR Council of People's Commissars directly—were enacted as well. Ironically, Shillinger himself remained without work until his arrest on April 14, 1938, thence to be devoured in the maws of the terror machine.[47]

Why did Stalin or his immediate subordinates indulge the doomed Shillinger? Why were the serious accusations leveled by Lepeshinskaia against the movement set aside? So far the archives have failed to provide any definite answers. One plausible reason is that those at the top regarded these field biologists as too impractical, too "nerdy," and far too marginal to pose a recognizable political threat. Teachers were threatening because they shaped young minds. Historians were threatening because they could subtly undermine the Party's legitimacy by constructing other ways to explain the past and the present. Writers were threatening because they might try to smuggle into their novels, poems, and short stories encoded messages of opposition. But a zoologist who studied the effect of snow cover on the foraging habits of hoofed mammals? Or a botanist who sought to explain whether or not the steppe was a result of some sort of grazing or pasturing? Such figures, if the Party elite thought about them at all, must have been objects of gentle ridicule. Ultimately, as a "class," they were not serious enough to be worth liquidating.

Despite the unwanted scrutiny of the Central Committee, the conservation movement did not rush headlong for the cyclone shelters. Here and there VOOP managed even to expand its network. In Gor'kii (Nizhnyi Novgorod) in 1935, the energetic activist Professor Ivan Ivanovich Puzanov, a well-respected zoologist, emerged as the organizer of that city's branch of the Society. (He would play equally central roles in the Crimea and in Odessa, where he later resided, testifying to the importance of specific individuals in the life of the Society.) Nevertheless, despite isolated successes, the movement's leaders looked to the future with apprehension. Owing to a paper shortage and rising electricity costs, the monthly, and later bimonthly, journal of the Society—*Okhrana prirody* to 1931, *Priroda i sotsialisticheskoe khoziaistvo* through 1932—became an annual anthology in 1933 and ceased publication altogether in 1935.[48] Deprived of its old lodgings at 38 Sofiiskaia naberezhnaia by the Moscow Soviet, which aimed to "renovate" the building, VOOP was forced to be taken in by the Committee on *Zapovedniki,* which itself was cramped for space. Then there was the general political situation in the country.

As Stalin built his paranoid case against much of the elite of the Bolshevik party, the atmosphere of terror and suspicion took on a life of its own. Political vigilantism and denunciations were incontestable; those who tried to mitigate the terror were ipso facto guilty of protecting counterrevolutionaries, and were themselves carted off. In the deep of the night, every night, thousands were dragged from their apartments to the dungeons of Stalin's secret police. Even the army general staff was not immune from this seeming madness and was liquidated in June 1937. Recent figures suggest that it is likely that as many as two million people were arrested in 1937–1938 alone.

Against this backdrop the All-Russian Society for the Protection of Nature on June 10, 1937, drafted a letter to Andrei Andreevich Andreev, one of Stalin's colleagues among the Secretaries of the Central Committee. If the great purge then raging was madness, this letter could only be described as lunacy—or great courage. Noting that "progressive minds of all eras and peoples, alarmed at the impoverishment of natural resources . . . they have noticed, began seriously to occupy themselves with the problem of protection of the entire complex of natural treasures," the drafters of the letter then used contemporary international efforts in that area to buttress their case for more Party support for conservation. "Both in lands large and small, on the basis of weighty scientific work governmental and nongovernmental movements for conservation have expanded. Everywhere there are hundreds of scientific and citizen's mass societies, state committees, and entire departments attached to ministries, as well as special legislation, tens of nature preserves [*zapovedniki*] and a rich literature (especially in the USA)."[49] The All-Russian Society for the Protection of Nature's Executive Council, they wrote, had assembled a delegation that included the president of the USSR Academy of Sciences, Vladimir Leontievich Komarov, academician N. M. Kulagin, the deputy president of VOOP, A. P. Protopopov, Presidium member V. N. Makarov, the VOOP secretary, S. N. Fridman, and VOOP Council member V. N. Fofanov, which, they proposed, should meet with Andreev. They sought to raise four issues: (1) the expansion of the Society from all-Russian to all-Soviet status, with direct patronage from the Council of People's Commissars of the USSR; (2) the nomination of Andrei Matveevich Lezhava as president of this all-Union society; (3) permission to enter the International Center for Conservation and to attend the 1938 Vienna conference; and (4) assistance from the Party in the cause of conservation, including issuance of directives. The typed draft of the letter was then emended in pencil. Deleted was the nomination of Lezhava, a former USSR minister of domestic trade who had just been arrested and would be shot in October, but added were requests for permission for VOOP to resume publication activities and for an incremental increase in governmental subsidies to the Society. One has to know what Moscow was like in those bone-chilling days of June 1937 in order to appreciate the considerable courage involved even in drafting this letter. Regrettably, the relatively disorganized and incomplete condition of the Central Committee archives do not permit us to confirm whether this letter was sent or received. But this draft nevertheless testifies to the endurance of the conservation community in its defense of its vision of the entitlements of scientific public opinion.

In 1937 the Society also started up its second commission devoted to the study of regional environmental problems. On the model of the Crimean Commission, the Caucasus Commission convened for the first time on February 26 with veteran Society leader Aleksandr Petrovich Protopopov presiding. The growing political chill did not seem to cool the fervor of the Commission members. Still at liberty, Frants Frantsevich Shillinger, a founder of the Society, urged that the Commission not restrict its purview to the *zapovedniki* of the Caucasus. "The question [of conservation] must be posed more broadly," he continued. And while nature transformer Kh. S. Veitsman tried, circuitously, to deflect such a broad mandate as beyond the Commission's capacities, Protopopov quickly injected that "the Commission has nothing to fear by conceiving its tasks broadly. V. M. Fofanov [another Commission member] is absolutely right when he states that the Commission must not restrict itself only to collecting facts, but must evaluate the economy of the Caucasus region as well." The lionhearts carried the day, and the resolution of the meeting pledged to address conservation problems "in their entirety," although work would begin immediately on the more limited problems of forest depletion and water quality.[50]

The Conservation Congress of 1938

On April 20, 1938, the First Congress of the All-Russian Society for the Protection of Nature opened in Moscow. Although a previous all-Russian conservation congress had been held in 1929 and an all-Union one in 1933, neither had been convened under the exclusive auspices of the Society. Nor had the Society's leadership previously had to give an accounting of its activities to representatives of the membership at large. And although such stalwarts as Makarov and the Society's secretary, Susanna Fridman, still held center stage, gone were Grigorii Aleksandrovich Kozhevnikov, who had died at the previous congress, Pëtr Germogenovich Smidovich, and Shillinger, who had been arrested five days earlier.

Formally vice president but really in charge, Makarov presented the first major substantive address. One striking note repeated by Makarov (recall his letter to A. A. Andreev) was his assertion that the Soviet conservation movement remained a part of "a larger international movement" at a time when Stalin's regime was slamming shut all the windows between the Soviet arts and sciences and the outside world. He was, however, mindful enough (and probably sincere in this) to emphasize that capitalism and private property were systemically bound to plunder the environment. "If capitalists could assert a right to the air, they would," he said. "Luckily, air cannot be appropriated as private property by individual entrepreneurs" because of its ambient nature.[51] Nevertheless, great enough damage to the environment had already been done, he argued; by the late eighteenth century most of Western Europe had already become deforested, and "with every passing year the faunal web has become thinner and thinner." As recently as in the seventeenth century Eurasia experienced the extinction of the aurochs, in the eighteenth, the Steller's sea cow, and in the nineteenth, the tarpan. The

North American bison and the European bison had been driven to the edge of extinction more recently, and the situation of marine mammals had become catastrophic. Of the three groups positioned to notice this alarming turn of events—commercial hunters, sport hunters, and scientists—"only the latter could adopt a reasonably objective view rising above self-interest."[52]

Interestingly, Makarov, in his thumbnail sketch of the emergence of conservationism worldwide, reserved his strongest praise for Americans, who regarded conservation as "a national ideal."[53] Indeed, he noted, "the Americans were right when they advanced the rule of thumb that a nation's culture may be judged by its treatment of natural resources," although he was quick to add that "it must be said in advance that capitalist countries will scarcely be able to resolve their internal contradictions that flow from the nature of the capitalist system."[54]

Invoking the names of the founders of the movement in Russia—Kozhevnikov, Semënov-tian-shanskii, Borodin, Taliev, and others—Makarov chronicled the often rocky path for nature protection both before and after the Revolution. In one revealing comment, he recounted how Mikhail Petrovich Potëmkin, the onetime president of the Society, had been subjected to a long interrogation by the president of one of the Party purge commissions, who demanded of Potëmkin in consternation: "How can you, a member of the Party, have gotten involved in a cause like conservation?!"[55]

Although Makarov had little good to say about the last years of Narkompros's stewardship of VOOP after its patron, former People's Commissar of Education Anatolii Vasil'evich Lunacharskii, had been replaced by A. S. Bubnov, who was hostile to conservation, he did note that with the Society's transfer to the jurisdiction of the Presidium of VTsIK, it had experienced a revival. For the first time "juridical members," including the Academy of Sciences, the Committee for *Zapovedniki,* and the Main Administration for Forestry and Afforestation, affiliated as institutions.[56] Additionally, the number of thematic sections of the Society continued to expand, with an ornithological section formed in 1936 and a mammalogical one added in 1938, exemplifying what Makarov categorized as "academism in the good sense of the word"—linking research with practical problems.[57] Academism it was; of the 150 members of the ornithological section, forty-one were professors and an additional thirty-six were docents and senior scientific workers.[58]

Despite the purges and disruptions of the mid-1930s, VOOP refused to allow itself to be frightened or diverted from pursuing its bold goals. In conjunction with the Committee it continued to sponsor expeditions to promote the creation of new *zapovedniki* (Barents Sea, Teberda, Kazakhstan) and persisted in its studies of the ecology of endangered species such as dolphins in the Black Sea. VOOP's submission to the government of a huge amount of research data on deforestation led to a law on headwaters protection, and the Society's special study of the Crimea, long a focus of special interest among conservationists, although failing to elicit comprehensive governmental action, did result in a disbursement of 400,000 rubles for some improvements. VOOP's far eastern branch asked the State Committee on Procurements to cut target quotas on sea lions by half, which was done, remarkably, and VOOP also successfully secured the creation of a twenty-five-kilometer-wide green belt around Moscow (which was eventually built over in the 1950s).[59]

Despite the Party's refusal to allow VOOP delegates to attend the international conference in Vienna, Makarov emphasized that ties with similar foreign organizations were continuing to be maintained. With 5,000 volumes in sixteen foreign languages, all acquired through exchanges with foreign conservation societies, VOOP's library was one of the best in the world and was unique within the USSR. Sadly, the volumes were languishing in boxes; the Moscow Soviet had dispossessed VOOP of its office space, and the Society's operations were hanging by a hair, its paperwork processed on one desk in a corner of the office of the Committee for *Zapovedniki.* Komarov, the Society's president-designee, had even called on the president of the Moscow Soviet to try to straighten out the matter, but was also unsuccessful. "If the Society is acting improperly, then it must be eliminated," Makarov stoutly challenged; "if not, and it contributes to the general good, then it is to the shame of the Moscow Soviet that the Society lacks its own office space."[60] The Moscow Soviet "should think about its outrageous attitude toward social organizations," he admonished bitterly.[61]

Because a reregistration of members had not been conducted in some years, it was unknown how many of the 16,000 putative members were real and how many were "dead souls."[62] Negligent in collecting membership dues, the Society's financial situation continued to be precarious.[63]

At the evening session on April 22, 1938, the Society elected its Executive Council. V. L. Komarov was elected president, while Makarov continued as vice president and de facto leader. The inveterate secretary of the Society, Susanna Fridman, was reelected overwhelmingly as well. In addition, Konstantin Matveevich Shvedchikov, official head of the Committee for *Zapovedniki,* was confirmed in his virtually ex-officio council seat. Not surprisingly, academic biologists and biology students represented the single largest bloc on the council. Testifying to the continuing fiercely independent spirit of this Society, members rejected the candidacy of S. V. Turshu, considered more friendly to Stalinist tempos of resource exploitation, giving him only seven votes.[64] Perhaps Turshu's criticism of the Congress as too dominated by academics also had something to do with the result.[65]

Even the election of the honorary presidium, comprising prominent members of the Soviet scientific and cultural elite, became an occasion for a display of nonconformity. A number of academicians as well as Ivan Dmitrievich Papanin, whose aviatorial efforts rescued the crew of the icebreaker Cheliuskin, all received unanimous support. Otto Iul'evich Shmidt, a cosmologist and one of those whom Papanin rescued, was elected with the surprisingly large number of eight abstentions, however. Noting that the election of honorary members was "a serious political act," one member asked that those who abstained justify their positions. One who abstained, Lukashevich, then explained that his abstention was not occasioned by a lack of respect for Otto Iul'evich, but rather because he thought that others were closer to the movement's ideals: "Why was it necessary precisely for *our* society *precisely now* to advance the name of Otto Iul'evich?" Lukashevich earlier had exhibited the same fierce spirit of independence regarding the question of press access; the press had shut out issues involving conservation. "We must not view ourselves as poor relations," he thundered; "rather, we are Soviet citizens . . . imbued with passion to assist our government and people. And since that is the case, we can certainly demand space in the pages of the press and not simply timidly beg for it through intermediaries."[66] It was not always easy for VOOP to walk the fine line between political accommodation and its own robust grassroots traditions of fierce scientific and political autonomy.[67] That tradition of autonomy, however, was inextricably linked with a desire to be a fully accepted, valued, and heeded part of the power structure.

Although after mid-1938 the "Black Maria" police sedans no longer swarmed as frequently through Russia's cities in their terrifying early-morning feeding frenzies, it is inappropriate, to say the least, to speak of a return to "normalcy," let alone liberalization. Nevertheless, until the Nazi invasion of the USSR on June 22, 1941, Soviet society began a slow recovery from the trauma of the Great Purge. VOOP, too, reflected this upsurge of civic energy. It resumed its propaganda activities with a booth and lectures in Gorky Park, and successfully gained protection for polar bears from Glavsevmorput', the administration that organized expeditions, transport, and supply for the Soviet Arctic Ocean and its coastal zone.[68] In the spring of 1941 a section of the Society devoted to marine and waterway protection was inaugurated under Professor Lev Zenkevich of Moscow University.[69]

Records of the Society's activities during 1939 support this picture of heightened activity. While membership stood at only 2,553 on January 1, 1940, that figure did reflect a growth by 696 new members over the real figure for 1938.[70] Significantly, there were almost as many members of the Academy of Sciences (7) as peasants (10), and more than half as many professors and docents (55) as workers (95). Communists (127) and Komsomol members (97) were still underrepresented in this largely non-Party milieu.[71]

A new branch was organized in Astrakhan, which quickly attracted 300 new members, while in Moscow a new section on protection of the earth's crust was established by the noted geologist A. E. Fersman, a close colleague of Vernadskii. A seed bank, an herbarium for rare steppe plants, and a photo gallery of conservation figures were all established, too. The mammalogical and ornithological sections compiled lists of endangered species, which were delivered to the Main Administration for *Zapovedniki,* along with a proposal to publish a series of monographs of these interesting and threatened life forms.[72] Indeed, a special Species Commission was organized within the mammalogical section to organize this initiative.

Linked with the above efforts was an intensive lobbying campaign to stop hunting of the desman, a rare aquatic shrew. Based on field observations subsidized by the Committee for *Zapovedniki,*[73] VOOP sent the SNK RSFSR a memorandum "illustrating the real state of affairs and directly clashing with the data presented by SOIUZZAGOTPUSHNINA," the state's fur procurement agency. The result was a big victory for the conservationists; the SNK's decree No. 673 extended the ban on trapping desman to January 1, 1943, continuing a policy first set (at VOOP's initiative) in 1935. Lobbying continued for an all-Union structure for conservation as well as for the removal of responsibility for hunting matters from Narkomzem's Main Administration for Hunting and Breeding to an interministerial body.[74]

The Society cleverly called for adoption of the "newer methods" propounded by Academician Lysenko as a desirable replacement for the use of arsenic-based pesticides by the People's Commissariat of Agriculture. Such use had resulted in a massive die-off of birds, and a special trip was planned for 1940 to study the question in greater depth. Meanwhile, Iu. A. Isakov, a member of the ornithological section, conducted a study into the death of Black Sea waterfowl as a result of pollution by petroleum products.[75] Other research sponsored by VOOP included investigations of the decline of willow ptarmigan in Kalinin *oblast'* despite protection and of ways in which the Moscow-Volga canal affected avian life. The Society was heavily represented at conferences, symposia, and meetings of governmental advisory agencies.[76] Additional commissions on endangered species—walrus, sable, beaver, otter, tiger, polar bear, and others—were organized to influence public policy. New nature reserves were called for, and extensive areas were carefully surveyed and drafts were meticulously prepared.[77]

As Europe edged toward war in 1939, the Hitler-Stalin pact bought the Soviet Union a dual cushion of extra time and extra territory. The partition of Poland with Soviet absorption of its eastern half had tragic consequences for that country, not the least of which was the cold-blooded massacre by Stalin's secret police of almost 15,000 Polish Army officers in the Katyn forest of Belorussia. One zoological footnote to that terrible political drama, however, loomed large for Soviet field biologists: for the first time since 1923, wild pure-line European bison were living at liberty within the political boundaries of the USSR, in the newly acquired Belovezhskaia pushcha reserve, formerly run by the Poles.

VOOP's membership continued its slow growth right up until the war. In January 1941 it stood at 2,960, although a major setback came when the People's Commissariat of Finance prohibited financial contributions to the Society by state or economic organizations in their capacity as "juridical members."[78]

Although the Society recognized that its small membership affected its public image and effectiveness, this was not viewed as a catastrophic problem. The Society's old guard instead put a premium on the individual, the amateur, and the enthusiast. "Their role is very great," the VOOP activities report emphasized, "and it may boldly be stated that wherever there are even one or two such enthusiasts the cause of conservation successfully develops."[79] Put in less sentimentalized terms, the nature protection movement, because it was a sanctuary for individuals of a certain social type, not only was uninterested in converting VOOP into a truly mass society; its raison d'être was to preserve the Society's clublike atmosphere, which guaranteed a safe and comfortable haven for "scientific public opinion."

World War II

The World War was an unparalleled cataclysm for the USSR. War ravaged not only the Soviet Union's vegetation and wildlife, but especially its people.[80] Among the tens of millions who perished were many conservation activists; some died at the front, while others, such as professors Andrei Petrovich Semënov-tian-shanskii, his brother Veniamin, and Daniil Nikolaevich Kashkarov, were claimed by the Leningrad siege and famine. Despite the hardships, nature reserve staff scientists and other conservation workers tried to save what they could while contributing to the war effort; work on natural substitutes for rubber and on vegetation cover suitable for airfields was vigorously pursued in the *zapovedniki.*

Even wartime conditions, which relegated conservation concerns to nearly last place in the national agenda, could not stop VOOP activists from finding ways to press their programs. V. M. Bortkevich, a longtime member, wrote a brief that sought to use the commemoration of battle and massacre sites as a means of expanding protected territories. The flagship of this system would be the Central Park of the Patriotic War, to be built near Kuntsevo, just west of Moscow. Other parks would commemorate the struggles of Rus' with the steppe peoples and the like. "The cult of our forbears ancient and recent, the recognition of their valor and great heroic deeds, that is the slogan for new *zapovedniki* . . . of national honor and glory . . . and as a lesson and a warning to our enemies," he floridly concluded.[81]

Amid the mass evacuation of millions of citizens, institutes, whole factories, and ministries to the east, the war prompted another, unlikely evacuation. As Moscow fell prey to German air raids, zoologists worried about the denizens of the Moscow Zoo, some of which represented highly rare and endangered species. At the time, the RSFSR Main *Zapovednik* Administration held responsibility for the zoos of the Russian Republic in addition to the nature reserves, and so an agitated Nikolai Sergeevich Dorovatovskii, director of the Administration's Zoo Division, went hat in hand to RSFSR deputy premier Aleksei Kosygin, pleading for allocation of vehicles to transport the animals to safety. Ultimately, many were shipped out on rafts and other transport, but Kosygin could find no trucks or train cars that could accommodate the giraffes, which had to be left at the zoo. Sadly, they were soon killed by a bomb during a German air raid.

More than one hundred caretakers and scientists remained at the zoo to care for the 1,400 other, less valuable animals that were not chosen for "Dorovatovskii's Ark," and because Moscow, unlike Leningrad, still had decent food supplies, none of the animals starved. An elephant even gave birth to a calf in 1944, which prompted another visit to now-Premier Kosygin by Dorovatovskii, who suggested that the premier publicize the event to boost morale. (By contrast, the Leningrad Zoo's elephant died of starvation.)[82]

A little victory garden for the animals—mostly carrots and potatoes—was planted by the staff with an assist from prominent zoology professors V. G. Geptner, B. M. Zhitkov, Sergei Ivanovich Ognëv, and S. S. Turov, all of whom also tended to the animals themselves. Miraculously, there were no significant losses to cold, even in the monkey house, despite the zoo's low priority for fuel. Even the tropical birds survived.[83]

Other zoos were not so lucky. The Leningrad Zoo, understandably, suffered from the blockade of the city. The situation was much worse, though, in Ukraine, where in Kiev and Khar'kov the Nazis intentionally shot the zoo animals. Luckily for the European bison (*zubry*), Hermann Goering was a sometime hunter with a Gothic sensibility. He ordered twelve of the beasts shipped off to his personal estate in Bavaria, which is how they alone escaped the general bloodbath; they were sent back after the war to the Polish side of the Białowieża Puszcza (Belovezhskaia pushcha) reserve and some of them ultimately were given to the USSR by Poland as gifts (see chapter 3).[84]

Wherever the Germans came across *zapovedniki,* they inflicted sadistic carnage. Happily, the hoofed mammals of Askania-Nova were successfully herded into the steppes of Kazakhstan, but the birds of that reserve were not so lucky: they all died by firing squad. By a bit of good luck, the Germans were never able to penetrate the Kavkazskii *zapovednik* in the North Caucasus, only reaching as far as the reserve's borders.[85]

On a less lurid note, VOOP was able to print up two large print runs of posters urging care to prevent forest fires and some assorted brochures. Beekeeping courses continued to be offered by leading specialists; evidently, it had its appeal. There were sixteen on-site classes with a total of 879 students, not counting three advanced groups with 126 and five correspondence classes with 364. Next most popular was a class in poultry-raising and the breeding of small-sized stock with 315 enrollees, in third place were the orchard and vegetable-growing classes with 295 each, and pisciculture was last with 31 students. However, the Moscow Soviet—which bore serious food-supply responsibilities during the war—was dissatisfied with the low numbers in the vegetable-growing classes, and by an order of January 13, 1943, sixteen new vegetable-growing groups were organized and one new group of stock-raisers, which were completed by 484 enlistees. Altogether, 2,236 students took these victory-garden classes, learning to keep up supplies of honey, cabbage, chickens, and potatoes for the hard-pressed Soviet capital. In addition, classes to help citizens identify medicinal and other important wild-growing plants were organized in Moscow, Gor'kii (Nizhnyi Novgorod), and other cities.[86]

The Achievements of the *Zapovedniki*

Although the network of ecological *zapovedniki* prospered during the 1920s, by the mid-1930s their unique status as inviolate ecological research centers was critically impaired. Economic ministries, notably the People's Commissariat of Agriculture, derided the ecological reserves for pursuing "science for science's sake" and sought to incorporate them into their own networks of *zapovedniki,* which pursued the more narrowly utilitarian goals of maximizing the propagation of selected, economically valuable species of wild animals. Cultural revolutionaries denounced the reserves as "havens" for the despised species of "bourgeois academics." Isaak Izrailevich Prezent, a close collaborator of the notorious charlatan agronomist T. D. Lysenko, accused the *zapovedniki* of leading a counterrevolutionary resistance to such key economic programs as collectivization and acclimatization (introduction of exotic animals and plants) under the cover of scientific argument. Although we may deplore Prezent's political thuggery, we must acknowledge that there was more than a grain of truth to his charges.

Owing perhaps to Prezent's potent political connections as well as to Stalin's desire now to favor only those scientific findings that supported his economic and social policies, Prezent's attacks proved the most telling. Through his direct involvement, the highly innovative work in trophic dynamics of Stanchinskii at the Askania-Nova *zapovednik* was abruptly ended (after which the ecologist was arrested in 1934), and Prezent served notice that holistic ecological doctrines that asserted limits to humans' ability safely to transform nature were now to be regarded not only as flawed but as devised by the "class enemies" of Soviet socialism. For an unrelated reason (discomfiture with mathematics), Prezent also announced the unsuitability of attempts at the formal, mathematical description of biological phenomena. Both of these measures, in the words of two of Stanchinskii's students, had the effect that "theoretical research in biology, including ecology and biocenology, was excluded from the work plans not only of Askania-Nova but also of all scientific institutions for two decades at the very least."[87] Although this assessment seems exaggerated, especially for the period after World War II, there is no doubt that severe damage was done. With the exception of the bold interdisciplinary studies of the fir forest directed by Stanchinskii in the Central-Forest *zapovednik* until his final arrest in late June 1941, community ecology, particularly its theoretical side, was indeed stunted by Prezent's attacks.

Ironically, just before theoretical ecology's development became subject to attack, there was a growing tendency among a few ecological thinkers to emancipate themselves from a priori judgments about the nature of the ecological community. Indeed, it can be argued that such seminal thinkers as Stanchinskii and the geobotanist Leontii Grigor'evich Ramenskii were approaching the sophisticated view that the biocenosis was just another useful category *we* impose on nature to make it comprehensible and manipulable. At first, Ramenskii waged total war on what he believed to be the idealistic conception of the natural community. In his earliest view, vegetation was an unbroken continuum, whose patterns of species distribution could be explained on the basis of environmental gradients rather than on the basis of the structure of some mystical community.[88] Ramenskii admitted the conditional utility of the community concept, but only so long as its arbitrary nature was clearly recognized: "In connection with the multifaceted inexhaustibility of phenomena there does not exist nor can there be a single all-embracing classification of them, fitting for all times and situations. In fact, such a classification system is not desirable. We need taxonomies firmly linked to specific objectives, helping to solve definite scientific and economic tasks."[89] Stanchinskii, for his part, in the years just prior to his arrest, had been moving strongly away from the view of the biocenosis as "closed." Migratory animals and birds participate in multiple systems, he pointed out, precluding absolute closure. Rather, his trophic pyramids (food webs) were only "loosely ordered systems," to use R. H. Whittaker's term.

Neither ecologist's colleagues were receptive to these de-idealized views of nature, however.[90] And these views were too sophisticated for induction in the service of the regime's crude nature-transformism. Consequently, for more than a decade a theoretical vacuum surrounded a sullen and silenced camp of holists and a tiny, ignored band of antiholists. The policing of biology by Prezent and Lysenko, combined with the pervasive fear among scientists and editors, all conspired to impose an unnatural silence over ecology, so recently brimming with discussion. This meant that members of the conservation movement continued to take on faith the scientific-ecological justifications for *zapovedniki,* just as nature protection's adversaries took on faith transformist beliefs.

The *zapovedniki,* while continuing to increase in number, were in some cases unable to avoid becoming bases for the very radical transformation of nature that their establishment originally had sought to prevent. As the price of the reserves' survival, Makarov felt that he had no choice but to renounce their prior official policy of inviolability. As concessions to Prezent and other critics, acclimatization of exotic fauna and flora proceeded apace together with such other aggressive management techniques ("biotechnics") as predator and pest elimination, winter feeding of select species, and measures designed to change the mixes of tree and shrub species in some reserves to more economically advantageous ones. Incidentally, the biological concepts of conservation's critics, despite their utilitarian and commonsensical ring, were just as speculative and politically motivated as those of the movement scientists.

Despite the crippling political ravages, the 1930s and 1940s were years of solid and occasionally brilliant achievement for Soviet field biology. Much of the best work was done in the *zapovedniki.* Although that work is not the subject of this book, we may note the contributions of Nasimovich and Formozov on the role of snow cover in animal ecology, that of Kashkarov on the ways in which burrowing mammals influence soil development and, consequently, vegetational cover, and that of Stanchinskii, Rode, and their colleagues on the influence of decomposers on the specific development of soil microenvironments, to name just a few examples.

* * *

It is sometimes said that the practitioners of many trades and disciplines are self-selecting. There is a strong suggestion that those who commit themselves to biological field research have a deep and abiding love for the outdoors and an attraction to studying life in its unfettered state—"in the wild." This may be an indicator of an even broader stance toward freedom: the aversion to seeing any life forms—including humans—enchained or oppressed.[91] The entomologist Andrei Petrovich Semënov-tian-shanskii reflected this set of values, as Anton Struchkov reminds us, when he said that "freedom is necessary for nature as it is for humans."[92] Perhaps no one has made this point as strongly as Feliks Robertovich Shtil'mark, when he critiqued me for making too crisp a distinction between "scientific-ecological" and "aesthetic-ethical" camps of nature protectors:

> I am convinced that the aesthetic (ethical or emotional) approach somehow invisibly is present in all matters linked with nature protection, even if arguments of a completely different cast are uttered or written. The crux of the matter is how any given individual relates to the world around him or her and what their personal attitudes about nature are. It seems to me that Soviet (and, perhaps, international) scholarship completely underestimates the emotional-personal factor. But numerous examples may be held up to show its decisive significance—examples of how particular problems in the sphere of nature protection were solved not on a strictly scientific basis but precisely on an emotional one, conditioned by the concrete attitudes on the part of specific individuals to natural objects. There can be no doubt that the productive activity of such prominent biologists as G. A. Kozhevnikov, I. P. Borodin, V. N. Sukachëv, and many others, including N. I. Vavilov and V. I. Vernadskii, drew their inspiration from feelings of deep love for the nature of their birthplace, from that 'emotional-ethical factor' over which our author [D. Weiner] as an objective historian feels duty-bound to pronounce a harsh sentence. . . . And here too, officials of the agencies concerned with *zapovedniki* such as V. N. Makarov or F. F. Shillinger fought to create new *zapovedniki* not only owing to the requirements of their jobs, not only out of bureaucratic calculations, but from their own convictions, their purely emotional strivings to save protected [*zapovednye*] corners of Russia.[93]

I now strongly agree. Not by chance did the field biologists, in their litanies of justifications for setting aside inviolable tracts in *zapovedniki,* frequently smuggle in the "aesthetic" justification, despite the fact that such an argument was unlikely to win any points with the regime. As for Makarov, who tried mightily (and successfully) to acquire a high level of biological literacy, there is no longer any doubt about his deep emotional attachment to nature. A letter of his to the ornithologist Georgii Petrovich Dement'ev from his home village, Lunëvo, where he was vacationing in the summer of 1934, reveals some of this: "How am I relaxing? I take walks in the forest, gather strawberries, blackberries, and wild raspberries. I go fishing more rarely because the rods are not too good. And I read some in biology. . . . I'm feeling well, I'm relaxing, and I must confess that I have not fretted the absence of more serious work or other distractions of an urban variety. Twelve days have passed by like a breeze."[94] For its later inductees such as Makarov no less than for its adepts, field biology was a "calling" in the fullest sense of that word. As a corporate social identity it was tailor-made for those whose inner selves did not fully subscribe to the modernist impulse to completely control the world around us and who clung to ideals of professional mission and status in a society whose rulers sought to obliterate them.

CHAPTER THREE

The Road to "Liquidation"

Conservation in the Postwar Years

One of the few bright spots for the scientific intelligentsia in the USSR's desperate struggle against Nazism was the internal liberalization of society, owing partly to the government's need to regain popular support and partly to its need to solidify its alliance with Great Britain and the United States. VOOP tried to consolidate the authority of "scientific public opinion" after the war and eagerly looked forward to celebrating the twentieth anniversary of its founding with a slightly belated congress. In April 1947, when the congress convened, and even in late October, when the Society and the movement celebrated V. N. Makarov's sixtieth birthday, VOOP's members still held hope for the cause and for the larger world. Few suspected that these brief years would be viewed in retrospect as a golden age.

* * *

With the end of the war in view, VOOP began to redirect its energies toward its prewar concerns and jettison the activities for the war effort that had dominated the previous four years. In January 1945 the Executive Council of the Society had requested that A. S. Shcherbakov, a Stalin associate of the USSR State Committee for Defense, take off VOOP's hands the Experimental Institute for the Rehabilitation of Wounded Veterans of the Great Patriotic War.[1] Reflecting the Soviet intelligentsia's expectations that the postwar period would bring the long-desired cultural and political relaxation seemingly augured by the regime's wartime policies, VOOP's vice president Makarov in March reported on nature protection groups in Great Britain, which provoked lively discussion and a request for an analogous report about the United States. Continuing his push for a reestablishment of international links, Makarov in June opened a discussion about the invitation of foreign guests to the Society's planned twentieth anniversary jubilee.[2]

Here, VOOP's leaders were again caught between hope and caution. The most ambitious program, suggested by V. G. Geptner, included visits by foreigners to the Seven Islands (Barents Sea/Arctic Ocean) and Voronezh *zapovedniki* plus Kiëvo Lake. V. V. Alëkhin was more realistic, observing that a simpler excursion program within the Moscow area would place less strain on the Society's treasury and would not require the elaborate travel permits necessary to follow Geptner's plan.[3] Holding a congress of the Society, however, required express government permission, and bureaucratic delays frustrated the activists' hopes for a speedy convocation. Indeed, when the congress was finally held in April 1947, the invitation of foreign guests had become a moot point. The Cold War had already chilled the international atmosphere and dashed the hopes of educated Russians for the long-awaited liberalization.

A more immediate problem was filling the Society's presidency, left vacant with the death of the Academy of Sciences' president, V. L. Komarov, in December 1945. Because the Society's first two candidates—academician Nikolai A. Semashko, the People's Commissar of Health under Lenin, and academician Leon A. Orbeli, a prominent physiologist—declined for reasons still unclear, Makarov continued effectively to lead VOOP from his new position of acting president.[4]

VOOP's postwar activities were defined by continuing attempts to restore the autonomous, civilian ethos of the Society in the face of fiscal and political obstacles. When the Society's annual budget was examined by the Presidium in January 1946, it turned out that there was no budget line for "nature protection" per se. This motivated Aleksandr Petrovich Protopopov to raise the politically sensitive question of whether it was essential for the Society to continue to involve itself in tree plantings and forest management when "it should be occupied only with issues of nature protection," by which Protopopov meant protection of biota and the creation of nature reserves. Here, however, the Society was caught on the horns of a dilemma. Agreeing with Protopopov in principle, Makarov questioned where the bulk of the Society's income would come from if not from contracts with interested agencies and enterprises for landscaping and arboriculture. Was it not better, he asked, that the Society earn its funds by performing socially useful afforestation work rather than depending on "charity" from the government (which would also decrease the Society's autonomy)? Geptner added that if the Society refused contract work, its membership dues would not cover expenses, and the Society would run the risk of losing the contracting enterprises and agencies as "juridical" (institutional) members, whose dues represented significant income.[5] Maintaining the Society's viability, not to mention civic autonomy, seemed to entail a never-ending series of painful trade-offs, particularly when the choices pitted nature protection activism against the survival of the Society as an institution.

Undeterred by budgetary woes and formalities, however, the Society resumed its energetic and independent-minded defense of nature. It energetically opposed specific assaults on natural objects, such as plans to drain historic Lake Glubokoe; in that instance, the Society won a big victory, saving the lake and even securing its inclusion in the new Moskovskii *zapovednik*.[6] Remarkably, considering the aura surrounding the recent veterans of the world war, the Society won its appeals to the RSFSR Council of Ministers and the Iaroslavl' *oblast'* government to impose an 80,000-ruble fine on members of the Military Hunting Society for illegal moose hunting.[7] The Society won other, less dramatic victories as well, including thwarting the attempt by the head of a game-procurement *sovkhoz* (state farm) to wipe out an ancient colony of beavers and saving patches of forest in the Russian Soviet Federated Socialist Republic (RSFSR).[8]

Saving the *Zubr*

Also at the initiative of the conservation movement one of the USSR's most strikingly successful efforts to rescue an endangered species was begun at the close of the war. The *zubr*, or European bison (*Bison bonasus bonasus*), had been the Russian Empire's largest land mammal, its range limited to two widely separated, and geographically disparate, protected tsarist game preserves: the Belovezhskaia *pushcha* in western Belorussia and the Kubanskaia tsarskaia okhota (Kuban Imperial Hunting Preserve) in the north Caucasus. Winter forage had become inadequate for the herds in both ranges, however, and from the mid-nineteenth century special winter feeding by game wardens had become the rule. Even so, by spring the animals were thin and breeding rates were low: the animals were able to bear calves in the best case every two years, and often only once every five.[9] After birthing in May, calves nursed on mother's milk until the following June; winter feeding was critical to the survival of mother and calf.

Under this regime, the population of bison in the *pushcha* remained largely stagnant until World War I, when hostilities, a devastating epidemic, and poaching snuffed out the entire herd there. The Caucasus bison lingered until 1927, when the general climate of weak respect for the law in NEP Russia contributed to its extinction. Thus, on January 1, 1927, there were forty-eight European bison left in the world, in Swedish, Polish, German, Austrian, and Belgian zoos. The Duke of Bedford kept a small herd at Woburn Abbey, but these animals were hybrids with the North American bison. The most intensive efforts at breeding were led by the Polish professor Jan Sztolcman, who worked at the Białowieża Puszcza, which was then entirely in Polish territory, but his herd of thirty in 1929 had dwindled to nineteen in 1939, when Poland was partitioned; miraculously, only two died during the war.[10]

One of the largely unnoticed footnotes of the Hitler-Stalin pact was that it placed the largely coniferous forest in Soviet hands.[11] Perhaps at the urging of Geptner, Makarov quickly wrote to Kliment Efremovich Voroshilov, the Soviet defense minister, asking him to place the nineteen bison in a special breeding farm (*pitomnik*) under special military protection. Likely owing to his passion for hunting, Voroshilov had become a patron of the nature protection movement. Here, those interests seemed to coincide, and Voroshilov dispatched a special commandant and a cavalry guard detachment to guard the woodland giants. (By contrast, Air Marshal Vershinin did not lift a finger to save the bison in the Ruminten *pitomnik* in Kaliningrad *oblast'* from extermination by Soviet airmen-hunters. All were killed.)[12]

The *pushcha* itself, an area of almost 130,000 hectares, was almost immediately added to the *zapovednik* system in June 1940, but with war imminent even the decree of the USSR Council of Ministers was no guarantee of the reserve's inviolability. American readers will recall the attempts of loggers to cut timber in the Olympic National Park during World War II. Analogously, the legendary aircraft designer Sergei Vladimirovich Il'iushin had written to the authorities asking to cut wood in the Belovezhskaia *pushcha* for aviation veneer. Voroshilov, no supporter of mechanized warfare, refused to grant Il'iushin's request. Perhaps Stalin should have appointed Voroshilov to be head of the Main Administration instead of defense minister![13]

The vagaries of politics now worked at odds with the aims of Soviet nature protection activists and scientists. Barely had the reserve celebrated its first anniversary when the German blitzkrieg ravaged the reserve and everything else in its path. Ironically, the huge losses of the initial weeks of the war, including the loss of the reserve, were a consequence of Stalin's arrest of leading military aircraft designers during the late 1930s, including Il'iushin, and of the dictator's and Voroshilov's idiosyncratic and fatally wrongheaded prejudices in the areas of military strategy and engineering.[14]

Of the bison hardest hit were the hybrid (European-American) bison mixtures in the Askania-Nova hybridization institute and Crimean reserve; all seventy bison died during the first two years of German occupation. Although the worldwide European bison population actually increased through 1944 to 146, the brutal last year and a half of war ravaged many of the breeding areas (many maintained by the Germans), and with peacetime the bison's numbers fell back to the 1937 level of eighty-four animals.[15]

Unlike the smaller American bison, which subsisted on huge quantities of grass, the European bison required forest nutrients: bark, oak twigs, and aspen. Curiously, two generations could survive on a diet of grass, but the third would die from nutritional deficiencies. Owing to their rarity as well as to the survival requirements of these colossal bovines, captive breeding seemed the most prudent first step toward a recovery of their numbers, despite the disappointing prewar Polish experience. Eventually, Soviet zoologists hoped to restore 1,500 *zubry* to the wild. For Makarov's Main Administration, supported by VOOP, this became a mission of the highest priority, heavily charged with symbolism. Not by coincidence were "bourgeois" professors of science ridiculed as *zubry* in their scientific *zapovedniki* during the Cultural Revolution. In the same vein, Daniil Granin's biography of Nikolai Vladimirovich Timofeev-Resovskii, a giant of Russian field biology (population ecology and population genetics), was entitled *Zubr*, for that relict mammal symbolized the fate of the Russian prerevolutionary scientific intelligentsia for both its friends and its enemies.

Makarov entrusted zoologist Mikhail Aleksandrovich Zablotskii with overall leadership of the rescue operation (see figure 3). If anyone were to rescue the bison, a knowledge of their genetics would be essential. Zablotskii not only knew genetics; he had studied with the famous Nikolai Petrovich Dubinin.[16] He was also a diplomat by nature and worked well with people, although he never shirked from speaking out if the situation required it.

With Zablotskii on board, Makarov sent a note to Molotov, now foreign minister, alerting him to the urgency of restoring the European bison to viability. With the establishment of a new frontier with Poland, the USSR would recover half of the Belovezhskaia *pushcha* nature reserve. On November 12, 1946, at a meeting of the Central Council of VOOP, Zablotskii laid out his proposal to establish a breeding farm for *zubry* near Serpukhov, south of Moscow on the Oka, and gained the Society's support for a newly-created Bison Commission to approach the Russian Republic's Council of Ministers for funding and support.[17]

While trying to coordinate support from both the all-Union and the republic levels of government, Zablotskii had been conducting his own private diplomacy, involving officials of the Belorussian foreign ministry, Polish zoologists, and Makarov in his capacity as deputy director of the RSFSR Committee for *Zapovedniki*.[18] Thanks largely to his efforts, five pure-line bison—three males and two females—were donated by the Polish government to the Soviet half of the Belovezhskaia reserve in July 1946. A calf born to one of the females that September was an encouraging augury,[19] yet the experience of the war, including new epizootic infestations among the bison, and the need for genetic monitoring of such a small population both recommended a more controlled venue close to Moscow. Zablotskii continued to press for his plan for a breeding station near Serpukhov.

Finally, on April 11, 1947 the Bureau of the RSFSR Council of Ministers met to consider the bison situation. Both Shvedchikov, the official head of the Main Administration, and Makarov attended. Geptner, who was unable to attend, sent a written argument against keeping rare species in border reserves as too risky, and he supported Zablotskii's plan for a Moscow-area reserve with captive-breeding facilities. These arguments had swayed Makarov, who was unusually literate scientifically for a layman, and he asked that all

Figure 3. Mikhail Aleksandrovich Zablotskii (1912–1996).

five pedigree bison constituting an additional new gift from the Polish government be transferred to the Prioksko-Terrasnyi *zapovednik,* site of the proposed breeding station. With the usual strong support of Deputy Premier Aleksandr Vasil'evich Gritsenko this package was approved by the Russian Republic leaders, and Zablotskii was soon on his way to Poland to see Żabiński, president of the Polish Bison Society, to nail down Polish approval for the transfer of these animals to the RSFSR.

Not one to simply take without giving, despite the USSR's overwhelming ability to dictate terms to its "allies," Makarov proposed a shipment of beavers, Bactrian camels, moose, and other animals, including polar bears, to Poland in exchange. The Poles were very gracious and the new bison—three females and two males—were sent to their new home in 1948. Indeed, there seemed no end to Polish generosity: three pedigreed descendants of the Belovezhskaia line were delivered to Prioksko-Terrasnyi in 1951. In all the USSR received nineteen pedigreed European bison from Poland between 1946 and 1951, which represented no small contribution; on January 1, 1947, there were only forty-four such bison in Poland and ninety-three throughout the entire world.[20] By the 1970s, Zablotskii's labors had borne fruit for all to see: through 1980, 170 animals had been reintroduced into appropriate environments to start independent herds of their own.[21] Significant for us is the unusual initiative demonstrated by Makarov and the scientist-cum-diplomat Zablotskii as well as the independent diplomatic support they received from the leaders of the Russian Federation.

* * *

Conservation activists also had their share of disappointments. A Moscow region *zapovednik* first proposed in 1940 to be composed of eight separate tracts with an area of 50,000 hectares had been approved by the Moscow *oblast'* authorities that same year. The war delayed confirmation by the republic-level authorities, however, and only a letter to Kosygin signed, among others, by VOOP's president, Komarov, got matters moving again in mid-May 1944. In 1945 an "Oka River Interdisciplinary Commission" headed by Sukachëv delineated the exact territories to be protected. However, a last-minute objection by the Main Administration of Forest Protection and Afforestation to the inclusion of Pogono-Losinii ostrov, a patch of undeveloped forest to the northeast of the capital whose tongue extended into Moscow proper, impelled RSFSR deputy premier Aleksandr Pavlovich Starotorzhskii to remove the tract from the proposal; Losinii ostrov remained under the control of the objecting agency. Makarov and Protopopov wrote a memo to the RSFSR government on February 24, 1945, calling the forestry agency's objections "unfathomable and unacceptable" and arguing that only *zapovednik* status could guarantee the integrity of the largest forest in the vicinity of Moscow; if the tract were not included, the value of the *zapovednik* would be significantly reduced, they added.[22] Nevertheless, the exclusion of Losinii ostrov was not reversed.

Contrasting with the indifferent attitude of the central authorities in the Kremlin toward nature protection was the active support and patronage accorded conservation by the government of the Russian Federation and by individual *oblasts*. A local initiative, supported by a resolution of the Primorskii *krai* Executive Committee on October 23, 1945, sought to expand the Sudzukhinskii *zapovednik* from 138,000 to 325,000 hectares, allowing the inclusion of the southern portion of the Sikhote-Alin mountains. This request was forwarded by Konstantin Shvedchikov to Russian premier Aleksei Kosygin, who speedily signed the change into law on January 4, 1946. Instructively, the letter two days earlier to Shvedchikov from Starotorzhskii announcing the decision revealed that Russia's leaders apparently endorsed the "*etalon*" rationale for *zapovedniki;* the expansion was approved "in light of the fact that the existing territory does not represent *fully* the natural conditions of the southern part of the Primorskii region."[23] Kosygin also signed on to the creation of two completely new reserves, the Visim and Denezhkin kamen' *zapovedniki*.[24]

More dramatic support by the Russian Federation for its Main Administration for *Zapovedniki* came three months later, following a letter of April 30, 1946, from Ivan P. Bardin, a prominent metallurgical engineer and leading member of the Academy of Sciences, to Lavrentii P. Beria. In his note, Bardin sought the return of the Il'menskii *zapovednik,* hard by the crucial Cheliabinsk-Kyshtym facilities of Beria's nuclear empire, to the Academy, even though it had been under that institution's aegis for all of one year, 1935. The request for transfer worked its way through the USSR Council of Ministers and back to the Russian Republic, where Mikhail I. Rodionov had just succeeded Kosygin as premier. In his response to A. A. Andreev, Rodionov agreeably offered any assistance to the Academy in its research in the *zapovednik,* but firmly refused to approve the transfer. Perhaps because the dispute was between a scientific institution, albeit an all-USSR one, on the one hand, and a lower level of administration, on the other, the central authorities did not feel particularly invested in the outcome, and did not seek to overturn Rodionov's decision.[25]

Behind the modest gains of the postwar period was the cultivation of personal ties between the leadership of the conservation movement and that of the RSFSR and other republics, and more local politicians. Links between the political patrons of the movement and VOOP had been especially strong when Lunacharskii and Smidovich were in office, but had weakened with Lunacharskii's retirement and Smidovich's death, the subsequent great purge, and the war. Rebuilding them was a top priority for VOOP.

An important opportunity for conservationists to present their case in person to the RSFSR leadership arose with the decision by the Russian Republic to issue a decree on nature protection. A draft was solicited from VOOP and was discussed at a meeting of the Operativnoe biuro (Operations Desk) of the RSFSR Council of Ministers on August 9, 1946.[26] The draft, incorporated into a longer brief, was another attempt by activists to educate the political leadership. A short history of the conservation movement in Russia underscored the great promise of the late tsarist and Lenin periods left unfulfilled in the Stalin era. Rodionov and his colleagues were reminded that tsarist Russia was a participant in the first international conservation congress in 1913 and that the cause had been endorsed by such luminaries of Bolshevism as N. K. Krupskaia (Lenin's wife) and Lunacharskii in their day. But "not only among the population at large but among the leaders of the economic *apparat* as well attitudes toward nature here at best are based on primitive utilitarian positions." VOOP had succeeded in uniting within its ranks "all of the leading naturalists of the USSR," but their scientific understanding of the issues had not penetrated the general public and political leadership, whose "superficial and untutored observation" continued to regard natural resources as limitless.[27]

Three days later Makarov addressed a session of the Council of Ministers of the Russian Republic, making a strong case for increased political and financial support for VOOP, the *zapovedniki,* and conservation generally.[28] Claiming only 5,183 members in VOOP, Makarov urged the Council of Ministers to require local governments to provide support to VOOP branches. Additionally, he requested that conservation matters be raised routinely at all levels of government in the RSFSR as well as in the press, that an institute for the study of problems of conservation and nature protection be established, that a new RSFSR statute on nature protection be enacted, and that conservation be included at all educational levels in all programs of study. More specifically, Makarov requested permission to resume publication of *Okhrana prirody,* the release by Gosplan RSFSR of two tons of paper, a subvention of 150,000 rubles from the RSFSR Council of Ministers' reserve fund, a dependable printing plant from the system controlled by the State Publishers (OGIZ), and the convocation of a congress of the Society in December 1946. Perhaps Makarov's most controversial proposal, which, along with that to resume publication of VOOP's journal elicited a question mark in the margins by the premier's aide, was to replace the Main Administration for *Zapovedniki* with one that would also be responsible for broader questions of nature protection generally.[29]

One result of Makarov's appearance was the promulgation of a decree on September 25, that among other things, called for more scrupulous observance of conservation laws and principles, awarded VOOP a 100,000 ruble subsidy for expenses, and delegated to the Society together with the Main Administration for *Zapovedniki* the responsibility of drafting a law for nature protection in the RSFSR to be submitted to the Council of Ministers in January 1947.[30] Despite this recognition, militant members of the Society were dissatisfied with the decree. Geptner wanted to know why the decree made no mention of reviving the Society's journal, *Okhrana prirody,* which had already been approved in principle, while Giller deplored the decree's silence on the issue of paper supply, which was controlled by Gosplan of the RSFSR, concerns that had been included in the draft resolutions submitted by Makarov to the Council of Ministers at their meeting.[31]

Pursuant to these instructions the RSFSR Council of Ministers was soon presented with two draft decrees on conservation, one from VOOP and the other from the Main Administration for *Zapovedniki.* Where the VOOP draft was broad, reflecting the scientific intelligentsia's pretensions to veto power over resource use, Shvedchikov's Main Administration's draft sought to limit the scope of the law to the territory of the *zapovedniki,* considering a broader bill unrealistic.[32]

This narrower bill, however, was rejected as well by the Bureau of the RSFSR Council of Ministers on June 3, 1947, and VOOP was asked to report back in a month with another draft law text. But as the slow wheels of the Soviet bureaucracy turned, changes in high politics raced on ahead, rendering moot the hopes of the immediate postwar period.[33] The RSFSR law was enacted only in 1960.

Among the greatest sources of frustration for the activists was the absence of a monthly journal. A journal was confirmation of the Society's social importance as well as an important vehicle for communication and bonding both domestically and with nature defenders abroad. In late 1946, Makarov and VOOP secretary Zaretskii wrote to the Press Department of the Central Committee about this. Coming after Winston Churchill's Fulton, Missouri address and a general souring of the international atmosphere, Makarov's letter could have been a serious tactical error, stressing as it did the importance of foreign models and foreign ties. In the absence of overt repression, however, those were concerns that Makarov and his colleagues were unlikely to disavow; international standing and a feeling of belonging to the world's scientific community were important mainstays of the self-image of these Russian scientists.[34]

Because it went so heavily against their hopes and their own sense of self, the nature protection activists were among the last to come to terms with the new isolationism of the Stalin regime. Although their request for a journal with a circulation of 50,000 to appear four times a year beginning April 1947 did not elicit an immediate response, by the following year, miraculously, *Okhrana prirody* (Protection of Nature) was off and running, albeit as an irregular anthology with a more modest print run—3,000 copies. Although any kind of approval might strike us as surprising, the Press Department of the Central Committee's decision reveals a capacity on the part of high Party bureaucrats to avoid "surplus repression"; perhaps, the Party censors thought, there was no harm in allowing these quaint activists a minor concession or two.

During these years of transition VOOP continued to be dogged by the pull of contrary aims: retaining its ethos as perhaps the country's sole remaining, intact defender of *nauchnaia obshchestvennost',* or corporate scientific autonomy, and becoming an influential mass organization that would be well connected within the system. The ideal of *nauchnaia obshchestvennost',* it must be emphasized, was not a fully democratic one, for it regarded the educated—especially scientific—elite as the truly authentic and enlightened representative of all of society. Nor was it an all-out oppositional stance toward the regime; the scientist-leaders of VOOP tried to stay within the bounds of acceptable dissent, a threshold in continual flux. Dissent, then, became transformed into an elaborate theater of identity, where scientists could defend the ideal of *nauchnaia obshchestvennost'* symbolically while avoiding the GULAG.

Yet, the scientist-leaders of VOOP also cared about nature, saw nature protection as a preeminently *scientific* problem within the realm of their expertise, and wanted to be effective. A draft of a letter from Makarov and Zaretskii to Andrei A. Zhdanov in 1947 illuminates this quandary. In it, Makarov and Zaretskii announced that VOOP was embarking on a new stage in its history, becoming a truly mass organization "while conserving its scientific base." For this, they sought a candidate for the Society's presidency who would enjoy broad public authority. Three political figures had been proposed as candidates: I. A. Vlasov, chairman of the Supreme Soviet of the RSFSR; S. V. Kaftanov, RSFSR minister of higher education and prominent supporter of Lysenko; and A. V. Gritsenko, a deputy premier of the RSFSR and movement patron. However, all three candidates refused to allow their names to be considered without express agreement from the Central Committee of the Party. Hence the letter to Central Committee secretary Zhdanov.[35]

The choice for president ultimately settled on academician Nikolai V. Tsitsin, an early apparent supporter of T. D. Lysenko and sometime personal acquaintance of Stalin whose political reliability gained him membership in the Academy of Sciences and the directorship of the Academy's Main Botanical Garden in Moscow. Like Lysenko, Tsitsin was a botanist specializing in hybrids; on the surface he sometimes appeared to be more charlatan than scientist, a regime toady, and a Lysenkoist. Unlike Lysenko, however, Tsitsin was not a hopeless biological illiterate; consequently as Valery Soyfer has noted, his "attitude toward Lysenko shifted with the political winds."[36]

The choice of Tsitsin, which seemed to go against the overwhelmingly anti-Lysenkoist sentiments of the VOOP board and executive council, in fact followed both logic and tradition. The scientist-activists of the conservation movement had always turned to prominent and loyal public figures to serve as the nominal leaders of VOOP and the Main Administration for *Zapovedniki,* posts Makarov was unable to assume personally, in part because of his Socialist Revolutionary past.

Tsitsin turned out to have a few surprises up his sleeve. Competing with Lysenko for leadership in biology, Tsitsin attempted to exploit what in 1947 and early 1948 seemed to be a withdrawal of regime support for his rival. In a letter of February 5, 1948 to Stalin, Tsitsin daringly revealed his opposition to Trofim Lysenko, writing apparently in response to a note from Lysenko that Stalin had asked to be sent to Tsitsin:

> Many representatives of biological science in recent years have been living in an atmosphere in which they feel shut in, an atmosphere of fear of one-sided criticism and of biased exposition of many questions that have already been solved by them.
>
> An organized discussion could identify much that is new and of value both for theory and for practice and could enable us to assess what is of value, to discard that which is not or which is actively harmful. A discussion might allow us to find a common line in tackling the problems of our agriculture.
>
> That is why I consider Lysenko's formulation incorrect in principle as it has been presented in his letter. To demand acceptance by all scientists of our country of his scientific point of view as if it were a command to be obeyed, I would even believe, from the point of view of Comrade Lysenko's interests, is simply awkward and tactless.
>
> I ask you, Comrade Stalin, to permit us to hold such a discussion at the nearest future time.
>
> Tsitsin[37]

The same day, Tsitsin sent an extended scientific critique of Lysenko to A. A. Zhdanov, with a copy to A. A. Kuznetsov, in which he assumed an even more urgent tone:[38] "I may boldly state not fearing to exaggerate matters that the situation now is such that the normal development of biological and agricultural sciences . . . has become impossible without the intervention and serious assistance of our Party and government."[39] Tsitsin rejected Lysenko's characterization of his opponents as "reactionary scientists" and added dramatically: "If they agree with Comrade Lysenko, the Soviet people should then . . . deport to the camps such reactionary individuals, whose hands created our best . . . plant varieties, and, on the other hand, recognize as 'progressive' . . . the theory of Lysenko, on the basis of which, by the way, during the course of twenty years, not one acceptable variety has been produced, notwithstanding the numerous promises and loud assurances."[40] Wondering how Lysenko could be president of the Lenin All-Union Academy of Agricultural Sciences, the Academy of Sciences' Institute of Genetics, the Siberian Research Institute for Cereals in Omsk, and the Institute of Genetics and Selection in Odessa, as well as editor of *Agrobiologiia,* all while holding grossly erroneous views, Tsitsin accused his rival of "surrounding himself with a claque of unscrupulous individuals" and charged that Lysenko had transformed Vaskhnil into "a vacuous bureaucracy, excluding all scientists except his own loyalists."[41]

Figure 4. Delegates to the VOOP Congress, 1947.

Tsitsin then debunked Lysenko's vague and primitive Lamarckism and stated: "Such 'dialectical materialism' as exemplified by Comrade Lysenko, in philosophical language, has to be called metaphysics."[42] From the standpoint of scientific argument, Tsitsin's extended critique of Lysenko was passably coherent, reflecting a far greater ability to cope with questions of genetics and developmental biology than his intellectually stunted rival, who excelled only in deviousness, theatrics, and imagination. To his credit, Tsitsin also protected a limited number of plant geneticists at the Botanical Garden. The VOOP scientists hoped, no doubt, that he would play the same role for them. All in all, the *chudaki* of the nature protection movement proved surprisingly adept at this high-stakes chess game of survival and social identity.

The VOOP Congress of 1947

The long-awaited delegate congress—nine years after the previous one—was brought to order by Makarov on April 26, 1947 (see figure 4). Delegates elected a working presidium and an honorary one (the Politburo) and, after a greeting by Old Bolshevik F. N. Petrov, commenced its real work. One of the more memorable addresses was that of Susanna N. Fridman, the longtime secretary of VOOP from its founding through the war, who voiced the feelings of the founders' generation: "We are the generation already exiting from life." She had not come to the tribune, however, simply to pass the baton. A bigger question was on her mind: "Is nature protection, or, more correctly, the survival of wild nature and its blossoming, compatible or incompatible with our quickly changing culture and civilization?" "Science," she continued, "has answered that it is compatible, and, I would go further, that if that is not the case our science is worthless, empty, and, as theory, holds no water. We know a great deal, but if we cannot [make the survival of wild nature compatible with culture], then that which we know wasn't worth knowing."[43] With these remarks Fridman had exposed—for a remarkable instant—the submerged tension between ethics and science within the nature protection cause. Because nature was held to have a normative, healthy state that was identifiable by scientific experts, the scientists who led the nature protection movement promoted the view that nature protection was a fundamentally *scientific* problem. But Fridman was suggesting that nature protection was fundamentally a problem of values and ethics; science as a system of social knowledge and organization could be fundamentally flawed in its ethical vision, in which case it was important openly to defend more compelling alternative ethical positions. Taken to its logical conclusion, Fridman's talk raised the question of whether the Russian nature protection movement wished to represent *scientific* opinion or a broader *public* opinion in the spirit of Russian moral and political activism from Radishchev to Tolstoi to the Marxists and other socialists. Although outweighed in the leadership by scientists, nonscientists such as Fridman, Krivoshapov, Protopopov, and, to a certain closeted extent, Makarov always represented a minority within the movement who viewed nature protection as a civic and ethical imperative rather than as a defense of part of the empire of science, however sacred. It was rare, though, to hear this explicitly; the scientists' hegemony within the movement was nearly total. That was not surprising: openly ethical speech was far more dangerous under the Soviets than scientific speech, which tended to submerge its ethical positions.

In line with her wider conception of nature protection as the problem of protecting life itself, Fridman—as Makarov had done earlier—raised the call for replacing the Main Administration for *Zapovedniki,* which she characterized as just another economic agency, with something broader and more authoritative to handle conservation policy questions. "Nature protection is a momentous question," she averred, "not only of international but of planetary importance," but it has become "not only unpopular, but, in fact, odious. And that is our failure." Challenging all sorts of narrow orthodoxies and emphasizing the moral poignancy of the issue, Fridman called for a new educational offensive by activists fueled by an independent *moral* vision. "I must declare that in our Union we must engage in nature protection with pure and burning hearts and with passion," she proclaimed, for, among the broad masses, "no one has any conception of the sweeping scope of this cause or its crucial importance for the whole world. We must enter the international arena. Life itself urges us that way." In perhaps the ultimate heresy, she concluded that "it is not necessary for us to wage a struggle with the world of private property over those specific problems which those societies have already successfully tackled."[44]

Perhaps inspired by Fridman, Krivoshapov was equally blunt in his critique of Soviet economic and ideological rigidity: "We have a planned economic system, but there is no sense to it. We write laws, focus our attention on delineated issues, but things never get further than producing a document." The only way out, he said, was to raise VOOP's status to the all-Union level and generally to elevate the level of culture of young adults, focusing on the middle schools.[45]

On the morning of April 28, the congress held its final session to hear the concluding remarks of the Society's acting president. Remarkably, Makarov tentatively engaged the difficult questions raised by Fridman and Krivoshapov. Addressing the questions of education and youth, he urged the adoption of a prewar Estonian statute that required those seeking certification as teachers to pass a special exam in problems of nature protection and natural history. "We, of course, have had nothing like this in memory," he lamented, courageously holding up a "bourgeois" legal precedent as a model. Perhaps his courage, like Fridman's, was stimulated by the realization that "the old guard is little by little leaving its posts . . . and our ranks are thinning."[46] Would there be a new generation to which the founders could pass the torch? The Society's demographics were far from encouraging, for there had been no appreciable influx of young people into the Society in the two years following the war.

Last, Makarov touched on aesthetic questions of nature protection, which were ideologically among the most sensitive for Soviet conservation. "I here would like to fully associate myself with the comments of Comrade Bogdanov of the Bashkirian ASSR and consider that the aesthetic importance of nature protection must not be sidelined from VOOP's field of action. We must care for and protect not only the paintings of Kuindzhi, Shishkin, and Levitan, which we treasure as works of great aesthetic value, but those natural scapes that inspired Kuindzhi, Shishkin, and Levitan."[47] "I have always been amazed," he continued, "that people are conscious of the value of these products of human creativity but find it impossible to perceive the beauty of nature and protect the actual nature [that inspired these paintings]."[48]

Makarov then shared a personal recollection:

> I sometimes recall a particular time in my life when I was in the Crimea; there I used to be terribly struck and upset by the following picture: a few lonely pines standing on a high precipice. That scene had always upset me, and I was traveling once with a friend with whom I would frequently talk about things, and he was perplexed by the power of a devastated forest to upset me. "Why does that scene touch you so?" he asked. "It would be nice to build a beautiful palace where those pines now stand." I answered him that the palace

might indeed be beautiful and that it might captivate me for the moment, but that I might not pay it any attention the next time. But I could see pines ten times and they would still stir me, because they tell much . . . because they are more valuable to me than a palace built in their place. It seems to me that we love nature through its specific examples, and, loving nature, we also love our homeland. For that reason, it is in the interests of the homeland and of cultivating love for it that we must care for the preservation of the most ancient examples of our own land's nature.[49]

During a final question and answer session, a number of delegates asked about past and future press coverage of the movement. Makarov and other organizers assured the delegates that the entire domestic press, as well as overseas press representatives, had been informed about the congress. Makarov admitted that there were no articles in *Pravda* or *Izvestiia,* and promised to find out the reason for that. "Perhaps they are covering more important questions now than the work of our congress," he wryly observed.[50]

Tsitsin chaired the first meeting of the new Central Executive Committee, which met on May 15 to elect a presidium.[51] The scholarly secretary, Zaretskii, offered a list of eleven, which was immediately amended by M. A. Zablotskii to include V. G. Geptner, and by N. A. Gladkov to include S. N. Fridman and K. N. Blagosklonov. A proposal by Tsitsin to limit the nominees to the original eleven with an option to expand later was put to a vote, and passed over surprisingly strong opposition, sixteen to eight, with one abstention. Ratification of the eleven as a group then proceeded smoothly, with twenty-three in favor and only two abstentions.[52]

Under the new president, Tsitsin, and his first deputy, Makarov, the new Presidium of VOOP included a founder of the Society, F. N. Petrov, and long-time activists A. P. Protopopov, a retired agronomist, and ornithologist G. P. Dement'ev, who became second deputy president. D. V. Zaretskii was elected scholarly secretary, and the remaining members included I. S. Krivoshapov of the Main Spa Administration of the USSR Ministry of Public Health; the USSR minister of higher education, S. V. Kaftanov; the USSR minister of the timber industry, G. P. Motovilov; G. A. Avetisian, an expert on bees at the Academy of Science's Institute of Evolutionary Morphology; and the Moscow University geology professor Vera Aleksandrovna Varsonof'eva.[53]

Although VOOP was not given the opportunity by the regime to celebrate either its twentieth or twenty-fifth anniversaries, one anniversary was warmly marked: V. N. Makarov's sixtieth birthday on October 20, 1947.[54] The presence of 285 people at the Executive Council meeting, a record crowd, testified to the genuine affection Makarov inspired. More than fifty greetings from agencies and societies were read, with an additional hundred messages from individuals and private groups. Makarov was truly in his prime, bathed in the appreciation and devotion of his colleagues and followers. A motion was presented to make Vasilii Nikitich an honorary member of VOOP, a high honor in Russian academic culture. It passed unanimously.[55] And a handsome photograph of Makarov was included in the second fascicle of *Okhrana prirody,* which appeared in 1948.

Finally, 1947 was remarkable for the appearance of the first major popular work on nature protection in the USSR, Makarov's sixty-page soft-covered book, *Okhrana prirody v SSSR.* With its attractive cover showing bison peaceably grazing in an alpine meadow of the Caucasus *zapovednik,* where they were being reintroduced, the message of *Nature Protection in the USSR* was serious : if we continue destroying habitat we could put an end to evolution itself. To make this point as strongly as possible, Makarov cited a letter from Russia's great paleontologist V. O. Kovalevskii to his brother, dated December 27, 1871, in which Kovalevskii wrote: "The vertebrate kingdom, especially Ungulata [hoofed mammals] now is simply in flight, seeking refuge anywhere they may find it. There will be no place for them to develop and to evolve into new forms; for this they will need thousands of years of a free and unfettered existence."[56]

From the end of the war until the summer of 1948 was a transitional period, in which the dying embers of hope for a postwar liberalization could still occasionally be fanned. The new realities of the Cold War and of an almost airtight and militantly anti-intellectual isolationism began to be felt with the onset of the Zhdanovshchina (the Party's new campaign for cultural orthodoxy, led by Central Committee secretary for ideology Andrei A. Zhdanov) in 1947 and, in the natural sciences, with the final battle over genetics that took shape during 1948.

Leonid Leonov and the Green Plantings Society

During the 1930s, under the banner of the "Green Cities" movement, groups of citizens focused their efforts on the cosmetic improvement of factories and urban neighborhoods through the planting of trees, shrubs, and flowers.[57] This was a far cry from the radical antiurbanism of Leonid Sabsovich and the original utopian theorists of "Green Cities" during the late 1920s and very early 1930s, but with monstrously dehumanizing industrial landscapes emerging out of the Russian mud, it was better than nothing.

The constituency for such a tame meliorist movement had grown by the end of the war. More important, the movement had found a spokesperson in Leonid Maksimovich Leonov, the author of some of the more memorable epic novels of the First Five-Year Plan. In 1947 Leonov published a long article in *Izvestiia,* "In Defense of a Friend." The friend was nature, specifically the Russian forest. He took to task municipal administrations that invested billions of rubles in urban greening but had nothing to show for it. "It would be nice . . . more often to grab [these officials] by the buttonhole and take them on foot on a tour of their imaginary groves," he wrote sarcastically. "Let them admire . . . the pitiful remnants of their courageous armchair leadership." He issued a challenge to Moscow's authorities: "We must start this great crusade in defense of our Green Friend in Moscow."[58] Evidently, it worked.

In 1947, with the Moscow city government's support, the Main Botanical Garden, the Timiriazev Agricultural Academy, and other prominent institutions had joined together as the All-Russian Society for the Promotion and Protection of Urban Green Plantings, which, Leonov noted, had also gained the crucial support of the leadership of the Russian Federation. To help "our green friend" was not simply a matter of aesthetics or utilitarian self-interest, Leonov emphasized. It was a question of patriotism: "We took a collective vow in '17 making it our duty to transform our fatherland into a place more beautiful than all the Floridas and other capitalist Edens." Now, he wrote, "we have placed the issue before an all-national *veche,*" using the archaic term for the town meeting of medieval Rus'. Leonov's article is interesting on a number of levels, perhaps most because it foreshadows the attraction of Soviet patriots—later to become Russian nationalists—to that unparalleled symbol of the *Russian* land, the Russian forest.

By 1948 the first stage of Leonov's vision was a material fact. The chairman of the RSFSR Council of Ministers, Mikhail Rodionov, approved the charter of the urban greening society on June 23, 1948 and sent the materials to A. A. Kuznetsov of the Central Committee for final approval.[59] The president of the organizational bureau was Nikolai Aleksandrovich Maksimov, director of the Academy's Institute of Plant Physiology, and Leonov, who was also a deputy to the USSR Supreme Soviet, became vice president.[60] A local Moscow society, DOSOM (the Voluntary Society for the Greening of the City of Moscow), was formed as well.

DOSOM and the All-Russian Society for the Promotion and Protection of Urban Green Plantings (VOSSOGZN) would be curious footnotes to the great sweep of Russian and Soviet conservation history were it not for just those qualities that made them seem platitudinous and banal. When the heavy hand of state repression once again was raised against VOOP, Makarov's strategy of protective coloration called for a merger with those conformist societies: the subversive VOOP core would be shielded and disguised by the patriotic and trivial veneer of urban greening.

The Alma-Atinskii *Zapovednik* Problem

On June 12, 1948, the Main Administration for *Zapovedniki* of the Kazakh SSR sent a plea to the Expediter of the USSR Council of Ministers to stop the Kazakh Ministry of Forestry from seizing a majority of the woodland belonging to the Alma-Atinskii *zapovednik.* A similar threat faced the Borovoe reserve, whose forests served to sustain natural climate and bathing water conditions at the nearby elite spa, Borovoe, which had housed much of the Academy of Sciences during World War II. To make matters worse, the forest ministry's plan, hatched in late 1947, redefined an additional large tract of steppe as "forest" to justify the seizure of that parcel as well.[61] In a letter supporting the plea of their Kazakh counterparts, the Presidium of VOOP concluded that the forestry's plan would constitute a "liquidation" of the two *zapovedniki.*[62]

Although it was unclear whether this was simply another of the numerous opportunistic depredations by "economic commissariats" on the reserves that were so common in the 1920s and early 1930s, or reflected the postwar crisis in fuel supplies, or represented something even larger and more ominous, the Kazakh forestry plan was the conservation movement's first serious challenge from a regime agency in more than a decade. An ominous sign was that the plan, an addendum to a Union-wide forestry decree, was passed by the USSR Council of Ministers on May 17. The last-minute protests by the head of the Kazakh Main Administration for *Zapovedniki* and of VOOP hinged on being able to engineer a rare emendation of existing legislation. They were not successful.

Another disruptive development was a decision by the Department of Agitation and Propaganda of the Central Committee to close down all small publishing houses attached to agencies and voluntary societies. VOOP's own facilities were closed on August 2, 1948 on order of the Moscow *oblast'* publishing authorities. In an impassioned letter to the Agitprop Department head Suntsov, Makarov protested that such a move "is tantamount to liquidating the activity of the All-Russian Society for Nature Protection, which has been in existence since 1924." Makarov argued that the Society was not receiving any publication subsidies and that the publication plan for 1948 had already been approved by the Central Committee's Press Department.[63]

A further disappointment came in December when Central Committee secretary and USSR deputy premier Georgii Malenkov turned down a request by Makarov and Aleksei Vasil'evich Mikheev, secretary of the Communist Party organization among the staff of the Main Administration for *Zapovedniki,* seeking to establish a scientific research institute for problems of *zapovedniki* and nature protection.[64] The proposal for such an institute had been kicking around since 1940, and had gained the important support of the Academy president, Sergei I. Vavilov, in November 1945.[65] Academician V. N. Sukachëv, Professor A. N. Formozov, and others moved it along in May 1948 in a letter to the chairman of the RSFSR Supreme Soviet, Ivan Andreevich Vlasov.[66] They even included as one of the research aims the study of the inheritance of characters acquired by wild animals and plants as a result of "the actions of their conditions of existence (of the external environment)," which could be construed as at least rhetorical acceptance of a Lysenkoist view of heredity.[67] Such an institute, however, was clearly not a top priority for the Soviet regime, whose cities were still piles of rubble three years after the war. A deputy chairman of Gosplan of the USSR, A. Lavrishchev, one of those delegated by Malenkov to sort through this question and recommend a course of action, provided a terse justification for his negative conclusions: the *zapovedniki* already have perhaps 180 permanent scientists among them, with an additional eight attached to the Moscow-based Main Administration. Any scientific problems that they could not handle should be passed to the Academy of Sciences or institutes of the Ministry of Agriculture.[68] D. Shepilov and F. Golovchenko of the Central Committee's Department of Agitation and Propaganda, similarly asked by Malenkov to perform an analysis, contacted two Academy institute directors and the secretary of the Academy's Biology Division, Aleksandr Ivanovich Oparin. Their conclusion was equally blunt: "The creation . . . of a special institute will only lead to parallelism in the organization of the study of living and nonliving nature and will encumber an unnecessary expenditure of state funds."[69]

Amid these disappointments were a few hopeful signs. By November 1948, after persistent lobbying, VOOP obtained a larger, permanent suite of offices, a move that required official approval of the USSR Council of Ministers.[70] Happily, the approval for the issuance of an official VOOP insignia pin (*nagrudnyi znak, znachok*) was somewhat simpler, resting only with the more sympathetic Council of Ministers of the RSFSR.[71] Membership, too, seemed to be on the rise. The Sverdlovsk branch had about 2,000 members, the Kabardinian ASSR branch 500 plus another 1,500 in the youth section; there were 500 members in Saratov, more than 2,000 in Kazan', and 10,000 schoolchildren in Gor'kii.[72] And in 1948 the Society was finally able to restart publication of its journal, *Okhrana prirody*.

Yet the Society was suffering from an unmistakable malaise. In his report on the affairs of the Society from the 1947 congress to September 1949, Makarov struck an uncharacteristically doleful note. The number of branches had stagnated, Presidium attendance had fallen on average to only three or four, and the December 1948 Plenum of the Central Executive Council drew a meager eighteen on the first day and sixteen on the second. Only half of those council members present in Moscow bothered to come to hear Makarov's report.[73] Was it inertia, fatigue, or something else? To answer this, we turn to events unfolding in the larger society.

CHAPTER FOUR

Zapovedniki in Peril, 1948–1950

We cannot wait for kindnesses from nature; our task is to wrest them from her.
IVAN VLADIMIROVICH MICHURIN

One of the first whiffs of trouble for the nature protection cause appeared in the East, in the Alatau of Kazakhstan, with an attempt by that republic's forestry authorities to wrest control of the Alma-Atinskii *zapovednik* in order to open that area to logging. The ideological and political climate of the Soviet Union, set by Lysenko's triumph at the notorious August 1948 session of the Lenin Academy of Agricultural Sciences and by the deepening of the Cold War, gave a new edge to the resource-motivated attacks on the conservation movement by economic ministries.

The Lysenko victory and the outlawing of classical genetics that followed did not have much immediate effect on the *zapovedniki* aside from requiring each reserve director to conduct discussions about "the situation in the biological sciences."[1] Lysenko and Prezent's purges of the major university biology departments and academies, however, cut deeply into the ranks of supporters of the movement. The purges extended beyond the confines of genetics and cytology to field biologists—botanists and zoologists; they even became a vehicle for settling scores with such convinced Lamarckians as Boris Evgen'evich Raikov, who still held aloft the banner of scientific autonomy and the spirit of *nauchnaia obshchestvennost'*.[2] Although the nature-transformation enthusiast Pëtr Aleksandrovich Manteifel' managed to get Aleksandr Nikolaevich Formozov fired from the Scientific Council of the All-Union Institute of Game Management and from the editorial board of the *Zoologicheskii zhurnal* (Zoological Journal) after Formozov, D. A. Sabinin, I. I. Shmal'gauzen, and M. M. Zavadovskii came out against Lysenko at Moscow State University, that was small potatoes compared to the damage inflicted by Isaak Izrailevich Prezent alone. In a dual calamity for professional biologists, in 1948 Prezent was named dean of the biology faculties of Moscow State and Leningrad State universities concurrently. As Lysenko's ideologue,

Prezent had no peers in his ability to "unmask" bearers of the ethos of scientific public opinion.[3] In short order Formozov, Shmal'gauzen, Zavadovskii, and Sabinin were expelled from the university (Sabinin shot himself as a result). When the tidal wave of firings had subsided, 3,000 instructors and professors had lost their jobs. Nor was the USSR Academy of Sciences system spared.[4] The human costs incurred, not to speak of the pedagogical ones, were incalculable.

With the university biology departments captured by Lysenko's minions and the USSR Academy of Sciences on a very short leash, the conservation movement was under greater pressure to defend the last institutions still under the control of the scientific intelligentsia, the *zapovedniki* plus a handful of scientific societies including VOOP, MOIP, and the Botanical and the Geographical Societies of the USSR.

Now the *zapovedniki* unexpectedly came under siege as well. Had the Kazakh Forestry Ministry's claim on a nature reserve in that republic been an isolated incident, perhaps the threats to the integrity of the reserve system would have stopped there. Ominously, though, there were parallel developments at the center.

As early as April 19, 1947, RSFSR premier Rodionov was warned by the ailing Konstantin Shvedchikov, head of the RSFSR Main Administration for *Zapovedniki,* that a draft law on the management of forests inside the *zapovedniki* was being prepared at the request of the USSR Council of Ministers, which had met on April 4.[5] Republican authorities were given three months to provide their input into the draft; responsibility for the major portion of it rested with German Petrovich Motovilov, who was named USSR minister of forestry at that meeting. Shvedchikov insisted that "it was indispensable to include basic provisions that would guarantee the inviolability of the *zapovedniki,* strict observance of the regime of inviolability which allows for the successful fulfillment of [their] tasks and goals." This paragraph was underscored by Rodionov when he read Shvedchikov's note, and the Russian Republic leader penned in the margins the significant phrase: "the *zapovedniki* are not to be handed over to the USSR Ministry of Forestry."[6]

Shvedchikov kept up the pressure on Rodionov, insisting on the need to defend the RSFSR Main *Zapovednik* Administration "as an independent agency, autonomous of the USSR Ministry of Forestry," as well as to resist the forestry ministry's newly announced claim on the Buzulukskii bor *zapovednik* near Samara.[7] He need not have worried. Tellingly, when Rodionov's deputy premier, A. V. Gritsenko, and Motovilov jointly sent USSR deputy premier Georgii Malenkov a memorandum outlining elements of the new draft law, they pledged their continuing commitment to the inviolability of the reserves.[8] Doubtless this was the result of the strong position taken by Rodionov and Gritsenko, who had the backing of Andrei Zhdanov, Malenkov's major rival for Stalin's favor.

No doubt hidden political maneuvering, perhaps by Malenkov, soon allowed Motovilov's ministry to backtrack and to advance claims to manage *zapovednik* forests independent of the republic-level *zapovednik* administrations. The new text of the draft law reflected this changed position; this was a direct rebuke to the Russian leaders. Responding to this development, one hot-blooded *referent* (junior advisor) of the RSFSR Council of Ministers in a memo to Rodionov called for strong resistance by the Russian Republic. "It is not difficult to notice," he wrote, "that the Ministry [of Forestry] does not understand the role of the state reserves and their basic tasks. . . . For that reason their proposals either partially or completely ignore the need for a regime of inviolability in the *zapovedniki.* Moreover," he continued, "the Ministry of Forestry feels that forestry . . . should be under their control despite the fact that the Main Administration for *Zapovedniki* is subordinated to the RSFSR Council of Ministers, which sets its tasks and provides its funding. In light of the above," he concluded, "I consider it essential to support the proposals of the Main Administration . . . and make them the bases for the draft decree of the USSR Council of Ministers."[9] The *referent*'s memo, which made the republic/center conflict explicit, apparently was seen by Deputy Premier Gritsenko first, for only on September 6 did the latter send Rodionov his own note repeating the advisor's positions and calling for Rodionov to write to Malenkov personally.[10]

Rodionov did indeed write to Malenkov on September 18, noting that

> the regime of inviolability is ignored in the draft law of the Ministry . . . and the chief role of the state *zapovedniki* is likewise not taken into account there as well. . . . In light of the fact that the previous experience of ministerial control over nature reserves has been shown to be a failure and is in contradiction with the Fundamental Law on *Zapovedniki,* it is necessary to transfer all [remaining] nature reserves on the territory of the RSFSR belonging to ministries to the control of the RSFSR Main *Zapovednik* Administration.[11]

The next month, on October 23, 1947, Aleksandr Vasil'evich Romanetskii, a functionary from the RSFSR Ministry of State Control's Working Group on Forests (but sympathizing if not in league with the central USSR forestry authorities), wrote to Gritsenko, outlining a "compromise" text for the decree. "Taking into consideration that the *zapovedniki* include more than 5 million hectares of woodlands as well as the necessity to establish forestry policies for each reserve, plus the insignificant number of qualified forestry specialists in the *zapovedniki,* we must agree to the indispensable oversight of forestry measures in the reserves on the part of the Ministry of Forestry. . . . In light of the above, it is necessary to reject the positions of Comrade Shvedchikov."[12]

Trying to find a solution that would satisfy both sides, Rodionov responded to A. A. Andreev, a secretary of the Central Committee, who had asked Rodionov to review the "USSR Council of Ministers' draft" once again. Although he now granted the USSR Ministry of Forestry the right to overfly the reserves, Rodionov held firm on the question of republican sovereignty.[13] In particular, Rodionov rejected any condition enjoining the Main *Zapovednik* Administration to submit annual reports to the USSR Ministry of Forestry. One concession Rodionov did make was to agree to turn over the entire 10,500-hectare *zapovednik* Buzulukskii bor to the USSR Ministry of Forestry's Borovaia experimental forestry station.[14] However, that was a small territorial price to pay in order to secure the integrity of the system as a whole.

The archives yielded a final note from Rodionov to Andreev, dated November 4, 1947. Over the previous week or so Rodionov apparently decided to stiffen his resistance and now rejected the center's latest draft outright. Although that draft explicitly prohibited only mowing and pasturing in the reserves, he explained, his RSFSR Council of Ministers insisted on a much broader ban on economic activities consistent with the traditional regime within *zapovedniki*. Rodionov also backtracked on his earlier agreement to hand over the entire Buzulukskii bor reserve to the USSR Ministry of Forestry. "The RSFSR Council of Ministers," he explained, "likewise cannot assent to the transfer of 3,500 hectares of the Buzulukskii *zapovednik* to the Borovaia . . . station, because this would deprive [the reserve] of a valuable forested tract rich in fauna as well as a significant number of residential buildings, although the remaining territory is of little value to the reserve."[15] All in all, it was a remarkable and plucky political display.

One is tempted to ask why men as busy and highly placed as Gritsenko and Rodionov sank as much time and political capital into defending nature preserves as they did. Certainly most important was the fact that the reserves fell under the direct administrative responsibility of the Russian Republic; the Russian leaders were protecting *their* turf, *their* portfolio of responsibilities, and *their* scientists from encroachment by outsiders, in this case the Kremlin. Second, the nature protection activists had established personal contacts with the republic-level leadership and had the opportunity to explain their scientific program to that leadership face to face; to a certain extent, as is evident from the archival correspondence, the republican leaders even came to identify the scientific program of the activists as *their* program as well. Third, it is possible that Russia's leaders felt patriotic pride in their protected territories, much as they may have felt for the European bison whose restoration they supported. Finally, there was the sense that nature preserves were an issue over which dissent or even resistance would *not* lead to a final tour of the Lubianka's basement. The political marginality of the *zapovedniki* in the eyes of the Kremlin made it possible for their local patrons to defend them—and their own bureaucratic "honor" and "turf" besides—without becoming, ipso facto, dangerous reactionaries.

Matters remained quiescent through the late spring of 1948, during which time the staff of the Main *Zapovednik* Administration compiled its annual report on the status of the reserves and their work.[16] Now expanded to thirty-one reserves with a total area of 9.2 million hectares, the Main Administration saw its nonsalary budget, particularly for scientific research, decline in absolute terms from 1947 levels. Correspondingly, the amount of funds earmarked for forestry measures continued to rise.

Notably, a special section of the report was devoted to the forests of the reserves, including an update on the status of the inventory of the reserves' total forested area, as well as a table showing the progress of such measures as sanitary logging, hiring of forest fire fighters, afforestation, and prophylactic clearing.[17] This was almost certainly a reaction to the pressures put on the reserves by the USSR Ministry of Forestry. Specifically rejected, however, was the aerial spraying or broadcast application of pesticides and other chemicals; biological methods were the only permissible means of controlling pests. So far the reserves and their patrons had kept the faith.[18]

Some of the more interesting scientific research themes listed in the report included biological pest control methods, including the control of fungi; causes of pest outbreaks, especially insects; the process of change of quaternary landscapes; the "forest vegetation and the soil" system and its role in the cycling of materials in the layer subject to wind erosion; the natural restoration of disturbed biocenoses of *zapovedniki*—for example, at charred sites—and the formation of new ones on newly emerging sand bars, islands, and so on; the protective role of various types of vegetation relative to the hydrological regime and erosion; the most important species of protected animals and plants; the natural change of biocenoses in *zapovedniki* and its causes; and changes in biocenoses as a result of human activity.[19] The report admitted that despite the successes and the great amount of factual material gleaned through direct year-round observation in nature over various geographical zones, "the *zapovedniki* have still not been able to solve one of their most important scientific problems, namely studying the regularities that . . . determine their natural productivity."[20]

Although a serious admission of mission failure, this was not as immediately worrying as the observations made in a June 1948 report about the Main *Zapovednik* Administration's personnel and research staff, addressed to Premier Rodionov.[21] The report, probably compiled with the participation of Romanetskii, pointed to the frequent replacements of chief bookkeepers of the reserves as signs of poor management. Worse, it impugned the Main Administration for its laxity in selecting staff; in effect, Shvedchikov, who was described as "too old, often ill," and a "weak" administrator, and Makarov were operating refuges for politically unwholesome elements in addition to refuges for plants and animals: "In a number of *zapovedniki*," charged the report, "there are ethnic German senior researchers (Griumer, Knorre) who were seized and exiled to distant parts of the USSR by the MVD [the USSR Ministry of Internal Affairs, which controlled the camps and exile regime]; when the war ended the Main Administration went and invited these same Germans to conduct scientific work in the *zapovedniki.*" Other staff members were identified as relatives of those repressed for anti-Soviet activity or for coming from suspect social backgrounds.[22]

The efforts of Rodionov and Gritsenko to deflect the center's attention from the forests of the *zapovedniki* were themselves sidelined by the continuing urgent demand for lumber. Pressed by a decree of the USSR Council of Ministers of May 17, 1948, the RSFSR cabinet eight days later issued its own directive under the signature of Gritsenko. Under apparent political duress, the Russian republic's government surrendered to the USSR Ministry of Forestry the right to scrutinize and approve the forestry management plans of the *zapovedniki,* to oversee their execution, and to subpoena any materials from *zapovedniki* in connection with any investigations it might conduct, and obliged the Main Administration to provide the ministry with annual reports of plan fulfillment. The conservationists and their patrons had lost across the board.[23]

With this first battle lost, Makarov again resorted to protective coloration. In a major article in the Main *Zapovednik* Administration's *Nauchno-metodicheskie zapiski* (Scientific and Methodological Notes), he noted that the reserves from 1940 had already been engaged in forest management, including fire control and prevention, and maintained that "all of this shows that the idea of absolute human noninterference in nature is alien to Soviet *zapovedniki.*"[24]

This was partly true. One of the early forest management schemes was in the Voronezh *zapovednik,* where the forester Mitrofan Petrovich Skriabin thought that management could hasten succession from aspen, ash, and birch to oak, where he then hoped to freeze succession. The forest had been cut down under Peter I, and second growth had slowly taken hold. Skriabin wanted to quicken the arrival of the oak stage. His efforts were pursued during the 1930s and 1940s. But this was the rhetoric of political coloration; nothing would have made Makarov and his scientist allies happier than being able to abandon all these forestry, predator-control, and pest-control diversions in favor of fundamental research.

The Stalin Plan for the Great Transformation of Nature

To the misfortune of the conservation movement, however, the wood procurement question, which had driven events thus far, was now supplemented by a vast ideological campaign that evoked the heroic rhetoric of the Great Break and the First Five-Year Plan. Hard on the heels of Stalin's endorsement of Lysenko's monopoly in biology, on October 20, 1948, the party and state jointly announced a "Plan for Shelter Belt Plantings, Grass Crop Rotation, and the Construction of Ponds and Reservoirs to Secure High Yields and Stable Harvests in Steppe and Forest-Steppe Regions of the European Part of the USSR."[25] In conjunction with this massive program of afforestation of the southern steppes, a Main Shelter Belt Administration was created under the USSR Council of Ministers, with a Main Expedition led by Vladimir Nikolaevich Sukachëv as its operational arm. Authorities on all levels had to provide progress reports before the year was out.[26]

Although the practical goals of increasing cereal crops in those famine years were salient, the symbolic importance of the plan was immense. Here was the renewed offensive on "counterrevolutionary," anarchic first nature and its replacement by a "planned" second nature. The iconic power of the image of thousands of kilometers of sturdy oaks breaking the strength of the parched eastern winds (*sukhovei*) was endlessly exploited in films and news clips of the era and resonated with the revived image of the "Fortress USSR" withstanding capitalist encirclement.

Hundreds of authors tried to outdo each other in celebrating the power of Soviet science to transform the planet. In the words of Zhores Medvedev, "Lysenko's cult in these years was blown up to fabulous proportions. . . . His portrait hung in all scientific institutions. Art stores sold busts and bas-reliefs of Lysenko. . . . In some cities monuments were erected to him."[27] Only the adulatory Stalin cult, then at its apogee, overshadowed the cult of "Michurinist biology" and of Lysenko himself. At least in the realm of propaganda the Stalin Plan for the Great Transformation of Nature represented the triumphal fusion of Stalin's great political and social vision with the unsurpassed biological "know-how" of Lysenko and Michurinist biology.

Amid the torrent of schlock that found its way into print was Prezent's authoritative article "The Refashioning of Living Nature," which better than most captured the spirit of the campaign. Although "bourgeois professors assure us that nature will not tolerate human interference and will avenge itself with natural disasters for intrusion into its regularities," Prezent began, "[miraculous] possibilities are opening up before Soviet Michurinist biologists!" Now, Lysenko has proposed to defend the grain fields with squadrons of trees, he continued. But Lysenko has also noticed that the grain protects the trees against their common enemies; so wheat should be planted among stands of trees in the forest-steppe, just as trees should be planted among the wheat:

> Field- and forest-protecting plantings—trees and bread—what a wonderful idea about cooperation and struggle in the green kingdom of plants. And what about the construction of ponds and reservoirs! Truly, never in the history of the world was there ever and could there ever be such a huge scale of hydroconstruction! . . . Bourgeois degenerates assert the idea that it is impossible to

> create new natural forms of animals and plants through training under new conditions. . . . Western philosophers and biologists dejectedly repeat their refrains of the "decline of culture," "nature's revenge," and "the protection of nature from humans."

However, he concluded, "Soviet biologists are joyously creating new kinds of life, are renewing and enriching living nature, and together with our entire people are building Communism."[28]

Among the key goals of the "Stalin Plan" were "overcoming the lethal influence of *sukhoveis* on agricultural crops" as well as implementing soil conservation in the Povolzh'e, North Caucasus, and Black Earth regions. Over a fifteen-year period ending in 1965, it was intended to establish seven large-scale shelter belts, some of which were almost 1,000 kilometers long.[29] Responsibility rested with the USSR Ministry of Forestry. By 1965 the area under shelter belts was projected to be 5,709,000 hectares.

Sukachëv's Main Expedition was divided into three smaller ones, each responsible for a major geographical region; Sukachëv's close colleague Sergei Vladimirovich Zonn was named to head one of them, rescuing him after he lost his academic position following the Lysenko victory. In fact, Sukachëv turned the whole Expedition into a "refuge for Weismannist-Morganists" and tried to save as many persecuted biologists as circumstances would allow.[30] Academy president Sergei Ivanovich Vavilov, who was secretly intervening on all fronts to save genetics, himself gave recommendations to Sukachëv about whom to hire for the project. Tellingly, the first was the country's most outspoken defender of genetics and Lysenko's sworn enemy, Nikolai Petrovich Dubinin.[31]

After two years of work, the Expedition was investigated by a commission from the Central Committee. Zonn told Sukachëv that he would serve as the latter's front man; the academician should stay in the background and plot strategy. Much was at stake, for the Expedition's importance now transcended the practical question of the shelter belt; it had become a kind of Noah's ark for geneticists and field biologists, including a large contingent of ecologists–conservation activists. The first commission's report was generally positive, but the shadow of Lysenko's animus toward Sukachëv and the genetics and conservation communities continued to hover over the Expedition.[32]

Luckily, there were intelligent individuals even in the bowels of Stalin's Kremlin. One was Iurii Andreevich Zhdanov, son of the late Party secretary, who had studied organic chemistry at Moscow State University and had done a "brief internship in genetics under V. V. Sakharov, [becoming] convinced of the validity of Mendel's laws." Later, he studied philosophy of science with the highly regarded Bonifatii Kedrov.[33] Not long after the first commission, a second was named by the Central Committee, this time by Iurii Zhdanov's Science Department. The Commission had to investigate, among other things, all the work of the Expedition on the ground (the actual plantings). Zhdanov, who was married to Stalin's daughter, Svetlana, and therefore had some degree of political protection, called Zonn to his office before the expedition set out again and told him to give him only the facts; Zhdanov would handle the political defense of the selection of cadres and of the scientific approach to the plantings.

Vastly complicating the situation was the Lysenko-style "scientific" framework in which the Expedition was supposed to work. "Oak" (*Quercus*) was the only genus of tree permitted to be planted. That already tied one hand behind Sukachëv's back. To this day, no one knows who was the author of that order. The Stalin Plan, it is thought, was prepared at the former Kamennaia Steppe Experimental Station, which later became the Agricultural Institute for the Central Black Earth Belt. At that time at the institute, it seems, were foresters of a pro-Lysenko cast. Ironically, the experimental station was established by Georgii Fëdorovich Morozov, the founder of Russian holistic forest ecology and Sukachëv's mentor. Sukachëv understood, first, that not all local ecological conditions were suitable for the planting of oak trees. Second, and more important, the primitive conception that created the oak tree fetish was erroneous. The Lysenkoists believed that the shelter belts would function by physically impeding the dry winds from the east. For that reason they favored a physically bulky tree such as the oak. The scientific leaders of the Expedition, however, understood that the shelter belts would aid agriculture not so much as windbreaks but by holding soil moisture in the ground after the melting of the snow, preventing soil erosion and making for a much more gradual release of moisture through the soil to the growing crops. Not the sheer size of the tree but rather its root structure and its ability to thrive within the climatic and ecological conditions of the region were the most important variables. Zhdanov helped the scientists to fight against the "dictatorship of the oak monoculture."[34]

Aside from the selection of tree species, another technical question soon posed ideological difficulties for Sukachëv and his colleagues. Over centuries of trial and error, farmers in various parts of the world have developed a reasonable sense of how densely crops can be sown or seedlings planted. Of course, desperation has sometimes overruled common sense, but in the main farmers have come to understand that each individual plant of whatever species needed to have a certain minimum area from which it could extract water and nutrients without debilitating competition from other members of its species. During the nineteenth and twentieth centuries plant physiologists and ecologists have recast this agronomic folk wisdom in scientific terms: the physiological requirements, mechanisms, and structures of each species. The appearance of Darwin's *Origin of Species* placed these physiological investigations within a framework of intraspecific competition. Those having traits better adapted to a given environment were better able to extract resources than their less fit cousins and therefore survived in greater numbers to reproduce their own kind. Although there were dissenters from this model, such as Prince Pëtr Kropotkin, who countered that animals, at least, were just as prone to cooperate to get food and resources as to compete for them, by the 1940s (except in official Soviet biology after the August 1948 session) Darwin's competition-oriented theory, now united with genetics, generally carried the day.

Sukachëv's entire scientific opus was based on an acceptance of natural selection, although it coexisted uneasily with his concept of the relatively harmonious, static biogeocenosis—in which intraspecific competition only increased the efficiency of resource extraction within species, never rocking the foundations of the community, all of whose species components were more or less coadapted to one another. This Darwinian commitment to intraspecific competition led Sukachëv vehemently to reject Lysenko's new idea that "overpopulation has never existed, does not now exist, and never will exist in nature," which was accompanied by Lysenko's denial of intraspecific competition.[35]

Even Zhdanov, however, was powerless to silence the Michurinist chorus led by Lysenko that criticized the ecological approach of the Expedition. Professor N. P. Anuchin, Leonid Leonov's "tutor" in forestry, was particularly vehement in asserting that "Afforestation in the Steppe Does Not Need Scientific-Sounding Teachings and the Biogeocenosis," as one of his articles was titled.[36] Once again, as in the 1930s, ecology, genetics, and the nature protection movement found themselves on the same front lines, under attack for daring to assert that there were natural barriers to rearranging nature according to political whims. From the standpoint of science proper, the spearhead of resistance to Lysenko was a triad of journals, most notably Sukachëv's *Bulletin of the Moscow Society of Naturalists* (the other two were the *Zoological Journal* and the *Botanical Journal*).

The Stalin Plan cannot be characterized simply as "good" or "bad." The project did much for erosion control in localities where the scientists were allowed to use their best judgment. Many influential scientists and conservation activists, such as ex-USSR minister for environmental protection Nikolai Nikolaevich Vorontsov, consider its abandonment to be one of Khrushchëv's bigger mistakes. Moreover, under Sukachëv the Expedition was an institutional *zapovednik,* sheltering those who had been persecuted in the wake of the "August session."

On the other hand, it amplified the campaign for the great transformation of nature, which worked heavily against nature protection and community ecology and reinvigorated the pharaonic GULAG-managed canal, hydroelectric dam, and reservoir construction projects. An offshoot was a renewed attack on the scientific bases of community ecology (or biogeocenology, to use Sukachëv's widely accepted locution).

1949–1950: A Darkening Sky

Attempts by the RSFSR leadership to provide political cover for the Main Administration and the conservation movement ran up against an increasingly charged political atmosphere, which put that leadership itself in mortal peril. In this climate, regime vigilantes rediscovered that the conservation movement and *zapovedniki* were out of step with the regime's ethos. On February 21, 1949 Romanetskii wrote to warn Gritsenko that Makarov was planning to hold a conference on zoological research in the *zapovedniki* with more than 100 participants from February 22 through 26, even though permission had not been obtained from the Central Committee of the Party. Fadeev of the RSFSR Ministry of Finances had evidently provided 10,000 rubles for the event, but Romanetskii recommended last-minute cancellation, noting that the theses had not been politically reviewed and that the meeting was to be held in an inappropriate venue, a basement area without natural light. It was unclear whether Romanetskii was more worried about electricity bills or about the "underground" nature of the gathering.[37]

Romanetskii's second complaint to Gritsenko, on March 29, 1949, came after the fact and placed the spotlight on the subversive nature of the movement's ideology of *nauchnaia obshchestvennost'*. By going ahead with the full-scale conference, Romanetskii charged, Makarov had overstepped his bounds; Gritsenko, Romanetskii reminded the deputy premier, had only authorized an "expanded plenum of the Executive Council." Makarov needed to account for this personally in Romanetskii's and Gritsenko's presence.[38]

As it happened, Makarov had written to Gritsenko on January 26 asking permission to hold two meetings. He pointed out that no similar conferences had been held since 1933.[39] A program was outlined that included mostly technical talks by A. N. Formozov, Vsevolod Borisovich Dubinin (the director of the Academy of Science's Zoological Museum and a leading parasitologist), L. L. Rossolimo, S. V. Kirikov, and E. M. Vorontsov (Stanchinskii's relative and student). His own speech, "Tasks of Zoological Research in Light of Michurinist Biology," written later, on February 10, was intended to provide maximum political cover for the gathering. In it Makarov made the requisite perfunctory bow to the doctrine of the inheritance of acquired characteristics and the staged theory of development, and praised regime philosopher V. M. Iudin and Michurinist acolytes M. F. Ivanov and L. K. Greben', who had played such regressive roles in Askania-Nova. He also included attacks on M. M. Zavadovskii, N. K. Kol'tsov, A. S. Serebrovskii, and I. I. Shmal'gauzen, old supporters of the movement, by name. Finally, he noted that the first page of the latest issue of *Nauchno-metodicheskie zapiski* included a quotation from Stalin, which, incidentally, was the only time that the great leader appeared in such an honored spot in a Soviet conservation publication. For the greater good, much like the legendary princes of Rus', Makarov sinned against his own civic conscience and his personal values. By contrast with others who denounced to save their own skins, Makarov was motivated by a painful burden of responsibility to the movement as he understood it. He had reached the limits of protective coloration, pushed there by an extremist politics. If anything else could be said in his defense, it is worth noting that all those he denounced had already been singled out in the violent, orchestrated campaign and had already lost their jobs.[40]

Another blow landed almost immediately afterward. On March 13, 1949 Rodionov was removed from office and arrested, and he was later executed in connection with the so-called Leningrad Affair. (Gritsenko was replaced slightly later.) Coincident with Rodionov's fall was an April 25 report by A. Safronov, state councillor of finance, first rank, to the new Russian premier, Boris Nikolaevich Chernousov (see figure 5), and his deputy, Mikhail Mikhailovich Bessonov.[41] In it, Safronov charged that "the Main Administration had not exercised the necessary leadership over the financial activity of the *zapovedniki*" and that in a number of cases even "abetted the violation of financial-budgetary discipline."[42] The agency was accused of overspending to the amount of 285,000 rubles, including 40,000 for salaries of personnel above the number officially permitted for the system and 196,000 on purchases of equipment "from private persons and in stores on personal account." These budget overruns, Safronov alleged, "were illegally concealed" through credits from the salary accounts of scientific workers who did not exist; the Main Administration claimed that there were 174, although on January 1, 1949 there were in fact only 142 in place.[43] Thus, Pechoro-Ilychskii *zapovednik* received salary credits of 228,000 rubles despite an actual need for only 180,000.[44] Particularly galling was the cost overrun on research even when the thematic plan was "significantly underfulfilled."[45]

Each new development brought the conservation movement closer to the brink. By late 1949, the Secretariat of the Central Committee of the Party had become involved. Its Agricultural Department had requested an investigation of the Main Administration, to be jointly conducted with the RSFSR Council of Ministers. The findings, not surprisingly, revealed a festering alien colony in the Soviet body politic. Only four of twenty-six scientific directors of reserves were Party members or candidates, and only twenty of 111 scientific researchers; most of the 207 party members and candidates in the 890-person system were reserve directors, workers, or bookkeeping personnel.[46] Five scientific workers were found to be moonlighting. Decrees of the center, it was further charged, never reached the grass roots.[47] The Main Administration was derelict in its financial manage-

Figure 5. Boris Nikolaevich Chernousov (1908–1978).

ment, scientific research activities, and in its "selection, appointment, . . . and training of cadres." Research "did not respond to the demands of the economy for the quickest possible expansion of economically valuable wild animals and plants." The bottom line seemed unsparing: "The methods of leadership over the reserves have fallen behind the times."[48]

Yet the appendix to the report that contained the inquest committee's recommendations reveals the protecting hand of the RSFSR government. The highest priority was to find a new head to replace Shvedchikov, who had finally succumbed to a long and debilitating illness. Second, the report recommended *increasing* the numbers of scientific staff and forest wardens, reviewing the possibility of salary increases for the scientific staff of the reserves, allotting the Main Administration another, more convenient set of offices, and reviewing the statute on the Main *Zapovednik* Administration and its structure.[49]

In response to the impending report, on November 14, 1949 Makarov sent to Bessonov a seventy-three-page memorandum "On Essential Measures for Improving the Work of the State *Zapovedniki* of the RSFSR," requesting a budget of 20 million rubles as well as motor vehicles of various types.[50]

However, the official appointment of Aleksandr Vasil'evich Malinovskii (see figure 6) as the new head of the Main Administration on December 28, 1949 effectively ended Makarov's de facto leadership of the system during the interregnum.[51] Among Russian scientists, conservationists, foresters, and game specialists, Malinovskii remains enigmatic and controversial, and the degree to which he exercised initiative in the drama of the next two years is still a riddle.

Born in 1900 in the town of Kirzhach, Vladimirskaia guberniia, Malinovskii's roots were modest; his father was an engraver. As did many children of workers who sought upward mobility, upon graduation from the Kirzhach gymnasium in 1918 Malinovskii began his working career as a teacher in a rural school. He spent the years 1919–1920 in the Red Army. Upon graduating from the Petrograd Forestry Institute in 1923 with a specialty in forest engineering he began work as a stumpage assessor, later directing forest management teams and procurement assessment expeditions to Moscow and Gor'kii *oblasts* and to the Udmurt ASSR and the Transbaikal and Far Eastern regions for the People's Commissariats of Agriculture, Transport, and Communication, and VSNKh. From 1934 through 1942 he worked in the USSR People's Commissariat of Forests and then the Department of Forest Management of the Main Forest Protection Administration of the USSR Council of People's Commissars. From 1942 to March 1944 he directed the Briansk Technical Forestry Institute, which had been evacuated to Kirovsk *oblast'*, joining the Party in June 1943. In March 1944 he was named chief inspector of the State Forest Inspection Service under Motovilov; in three years the agency would be elevated to the USSR Ministry of Forestry.[52]

His official biographies do not mention his activities between 1945 and his appointment as head of the Main *Zapovednik* Administration. However, Malinovskii served as director of forests of the Soviet Military Administra-

Figure 6. Aleksandr Vasil'evich Malinovskii (1900–1981).

tion in Germany (SVAG) during that period, a post that may have recommended him to the attention of Beria and others in the Politburo.[53] Described in his *nomenklatura* dossier as "an energetic and experienced worker," Malinovskii's record was clear of administrative or Party reprimands except for one rebuke from the Agricultural and Forestry Administration of SVAG for publishing an article in a German forestry journal about the theme of shelter belts, evidently in late 1948 or 1949. With no relatives abroad or fallen victim to Stalin's repression, and with two medals for Valorous Labor (one for his wartime services), Malinovskii had little personal cause to doubt the progressive and constructive nature of Stalin's revolution. For someone like him, the Stalinist view of the world corresponded to his personal one; it echoed his own ideas about "common sense."[54]

The circumstances of Malinovskii's appointment are unclear. Officially, a commission of three representatives of republic ministries was involved in the formal ceremony of leadership transition.[55] However, archival sources imply that the whole operation was overseen by the Central Committee, whose Agricultural Department had initiated the investigation of the Main Administration in December 1949.[56] On January 23, 1950 Chernousov received a memo from one of his aides, A. Prokof'ev, who argued that insofar as the Agricultural Department of the Central Committee had investigated the Main Administration itself and had come up with measures for the improvement of the agency's work, it was wiser for the RSFSR government to confine itself simply to formally issuing the decree certifying the new appointment (rather than acting on the proposals submitted in November by Makarov).[57] We know that in the postwar period the Central Committee's Cadres Department increasingly bypassed Shvedchikov with its appointments of retired military officers and security police as directors of *zapovedniki*.[58] Moreover, the head of the Main *Zapovednik* Administration was a *nomenklatura* position, subject to Party approval. Consequently, it is entirely likely that the Agricultural Department of the Central Committee had played a guiding role in Malinovskii's selection as well.

The first glimpse the old guard activists got of their new head was at a special meeting on January 10.[59] The activists, still behaving as if it were old times, presented Malinovskii with a list of issues they wanted him to raise with the RSFSR Council of Ministers. They included reviewing the statute on the Main Administration, extending to administrators and scientists holding academic degrees the same salary scale and benefits as all other degree holders,[60] increasing funding for work-related travel between the reserves and the center, and finally, providing the Main Administration with offices that met "hygienic norms."

At the meeting, before Malinovskii presented his own vision of affairs, Makarov alluded to the new, less secure political environment in which the movement and the Main Administration now found themselves. "I believe," he said diplomatically, referring to the RSFSR/Central Committee investigatory commission, "that the commission, owing to an insufficient amount of time, was truly unable to examine all of our work, our manuscripts, published works, etc., the work of our colleagues who actively conduct their research in the *zapovedniki*. But the facts are that around the *zapovedniki* through mighty efforts we have succeeded in creating a large circle of active, committed researchers—among the most prominent scientists of our country."[61] Shtil'mark and Geptner's commentary on this seems right: "the commission was interested least of all in scientific work, manuscripts, and even the true state of affairs. The question of the radical reorganization of the *zapovedniki* had been decided ahead of time, and the new head's task was to put it into effect concretely."[62] But was Malinovskii's idea of "reform" identical with that of the Kremlin bosses? Meanwhile, an embattled Makarov signed over control of the agency at the meeting in the presence of the three RSFSR transition commissioners, an act that inaugurated a period of unprecedented turmoil, loss, and confusion for the agency and the movement.

* * *

An early preview of things to come involved the fate of the Seven Islands reserve's branch outposts on the southern island of Novaia Zemlia in the Arctic Ocean. Since 1947, protection of these territories had dramatically helped the recovery of eider duck populations, and their scientific director, S. Uspenskii, wrote to Makarov in early December 1949 to lobby for an expansion of that territory from the narrow littoral of Novaia Zemlia inland to include all representative landscapes of the island.[63] Most crucially, the Novaia Zemlia lands needed to be incorporated as a separate *zapovednik,* argued Uspenskii, as their financing, administration, and supplies were complicated by the great distance separating them from the main reserve, located in the Barents Sea. Complicating management tasks further was the existence of another branch of the reserve near Murmansk on the northern coast of the Kola Peninsula. Important populations of seals, walruses, reindeer, arctic fox, and other life forms would benefit from the creation of a separate reserve.[64] Uspenskii also appealed to Makarov to persuade the RSFSR Main Administration for Hunting to stop the "rapacious destruction of reindeer, polar bears, walruses, geese, swans," and colonial birds, and to impose a complete ban on killing fauna and taking birds' eggs.[65] Responding to this letter constituted Malinovskii's first official act as system director.

Malinovskii noted that the birds are present on the islands of Seven Islands reserve only four or five months out of the year, from late April to the end of August, and researchers visit the islands only during that time. The remainder of the year they work in labs and offices in Leningrad and Moscow, he wrote. The absence of suitable winter domiciles, equipment, and a library pose a barrier to year-round work there. Because the Kandalaksha *zapovednik,* also in Murmansk province, shares the same general research profile as the Seven Islands reserve, Malinovskii proposed integrating the administration and scientific research with headquarters at Kandalaksha. The merged reserve would be called the "State *Zapovednik* for Eider and Colonial Birds." Research expeditions to the Barents Sea islands would take place only between April and September and would be based in Kandalaksha as well. Malinovskii also endorsed the creation of an independent *zapovednik* on Novaia Zemlia, generally according to the boundaries proposed by Uspenskii and supported by the Arkhangel'sk *oblast'* Soviet in its letter of January 19 to the RSFSR Council of Ministers supporting the move. Finally, Malinovskii proposed transferring the existing staff of Seven Islands to Novaia Zemlia.[66]

We can see from this episode that, first, Malinovskii's inclination from the start was to streamline the system, eliminating all units that seemed to duplicate others' functions. Second, he was pragmatic. He could be persuaded by images and arguments that appealed to his common sense, such as the need to protect the breeding stocks of commercially valuable birds and sea mammals, but he could hardly be expected to support the protection of fuzzily defined biogeocenoses, particularly if they seemed to be a research indulgence of terminally impractical field biologists. Yet his support for a new, independent Novaia Zemlia *zapovednik* should deter us from accusing him of seeking to *obliterate* the system from the start, much less of having initiated those plans.

Meanwhile, the forestry situation as viewed from the Kremlin had become urgent if not critical.[67] A worried Presidium of the USSR Ministry for the Forest and Paper Industry met on March 16, 1950 to discuss huge shortfalls in planned production even as Stalin was preparing a draft decree "On the Unsatisfactory Underfulfilment by the USSR Ministry of the Forest and Paper Industry of a Plan for Timber Cuts and Delivery of Wood-Based Products to the Economy during the First Quarter of 1950."[68] This decree followed on the heels of an earlier one of January 11. With production of commercially usable timber at only 57.6 percent of targeted quantities for the quarter, hundreds of key ministry staffers, including the minister G. I. Orlov and his deputy, were out in the key lumber-supply regions, trying to ensure that on-site machinery breakdowns were promptly fixed.[69]

Orlov promised to make up the shortfall by intensifying summer logging and increasing efficiency. However, he noted, targets had grown so great that additional help from the state would be essential to meet them: a minimum of 90,000 more workers, more housing and supplies for workers on site, and additional tracts provided by the USSR Ministry of Forestry on which coniferous trees could be harvested, particularly if they were located near railroad lines.[70] Although Orlov did not mention the forested areas of *zapovedniki* specifically, his pressure on the USSR Ministry of Forestry doubtless stimulated forestry minister Bovin's efforts to gain control of those 8 million additional hectares of forests with already existing, if rudimentary, infrastructure.

About that time a draft law was sent by Malinovskii to the USSR Council of Ministers.[71] "State reserves fulfill an important economic function in preserving, restoring, and increasing supplies of game and other commercially valuable plants and animals," as well as pursuing the great work of studying the natural conditions of the Soviet Union, the opening paragraph granted. However, the text alleged, the organization of *zapovedniki* did not always take account of whether a given area required protection or was appropriate for the purposes pursued by reserves. "The experience of the *zapovedniki* has shown," the text continued, "that the imposition of a regime of inviolability in some cases did not permit us to use the accumulated reserves of game and in other cases hindered the solution of tasks . . . involving the restoration of basic types of vegetation and the implementation of active measures that would promote the increase and improvement of protected life forms."[72] This echoed Malinovskii's remarks to the January 10 staff meeting, where he announced that "the *zapovednik* must be that laboratory that will actually demonstrate, in nature, what can be accomplished under human influence, under the influence of goal-directedness." He went on to challenge the old-line activists' fixation on inviolability: "I am interested to know why, if a forest *zapovednik,* say, occupied an area of 15,000 hectares, all 15,000 must be inviolable. Why can't we organize it so that 5,000 are inviolable, 5,000 are used for other purposes, and 5,000 are dedicated to the transformation of nature? Nobody has spoken along these lines. . . . We must obtain results that are of interest to the economy."[73]

It is no exaggeration, though, to describe Malinovskii's draft decree as a radical, even epic departure from even the most "protectively colored" rhetoric and recommendations of Makarov. In the name of the USSR Council of Ministers and its chairman, Stalin, the draft called upon the RSFSR government to eliminate an entire slew of reserves: Verkhne-Kliazminskii, Gluboko-Istrinskii, Privolzhsko-Dubninskii, all in Moscow *oblast';* Visim in Sverdlovsk *oblast';* Kliaz'minskii in Vladimir *oblast';* Kuibyshevskii in Kuibyshevskaia *oblast';* Sredne-Sakhalinskii in Sakhalinskaia *oblast';* and the Tsentral'no-Lesnoi in Velikoluzhskaia *oblast'.* Seven Islands in Murmansk and Arkhangel'sk *oblasts* was included as well, although Malinovskii envisaged its partial reincarnation as a colonial sea birds reserve. The liquidation of the *zapovedniki* was to be completed by October 1, 1950 and their territory distributed to ministries and agencies according to an appendix included with the draft legislation.[74] Another list of *zapovedniki* were identified for reduction in area: Altaiskii, Barguzinskii, Caucasus, Kondo-Sosvinskii, Kronotskii, Pechero-Ilychskii, Saianskii, Sikhote-Alinskii, Sudzukhinskii, and Chitinskii. The remainder were to retain their current boundaries.

Second, the RSFSR Council of Ministers was urged to revise its statute on *zapovedniki* and their Main Administration, recognizing the value of exploiting the stock of commercial and game animals inside the reserves and the need for active measures to restore and "improve" the condition of typical vegetation and other objects of protection.[75]

Interestingly, the draft supported extending the wage scales and benefits of those with academic degrees working in agriculture and covered by the legislation of 1946 and 1947 to degree holders in the reserves.[76] Finally, it recommended that the State Staffing Commission increase the staff allotments for both the Main Administration and for the reserves themselves, especially for scientists, forest experts, and wardens, and that the RSFSR guarantee the publication of the scientific proceedings of the reserves as well as popular scientific literature about them.[77]

Was this a preemptive strike by Malinovskii to offset what he believed were potentially worse initiatives from the Kremlin? That is, was this the ultimate in protective coloration? Or was Malinovskii fulfilling a decision already taken at a much higher level? Was it the first stage of a more far-reaching plan to dismember the reserves, or did policy in this area develop on an ad hoc basis? Or was Malinovskii a true believer in utilitarian values, whose own quite specific understanding of the appropriate role for nature reserves (as laboratories for the transformation of nature and the increase of commercially valuable species) was manipulated by officials high above to carry out a massive land grab? The available archival record does not permit us definitively to answer these questions.[78] Although the policy conclusions are in keeping with Malinovskii's own management philosophy, that in itself is no proof that he initiated the proposal to eliminate or radically reduce more than half the system; had Malinovskii written the proposal on his own, he would have had to conduct an in-depth analysis of the reserve system in a scant two and a half months.[79] This would have been possible, but we also know that the Central Committee Agricultural Department had just concluded an investigation of its own.

One clue to the riddle is contained in the appendixes 1 and 2, entitled "List of *Zapovedniki* to Be Liquidated" and "List of *Zapovedniki* of the RSFSR Subject to Reduction of Area." Of the nine reserves with a total of 332,800 hectares, 2,600 hectares were to be turned over to the Kandalaksha *zapovednik,* and the remainder (with the exception of fewer than 10,000 hectares that were to be handed over to local governments) was designated to be ceded to the USSR Ministry of Forestry.[80] Of the ten reserves slated to be reduced in size, their aggregate area was to go from 8,784,800 to 5,591,200 hectares , a reduction of 3,183,600. Taken together, the RSFSR reserve system would decline from 9,117,600 to 5,591,200 hectares, or by 39 percent. Of the total, all but 115,000 hectares was allocated to the USSR Ministry of Forestry.[81] In the light of later developments, such a plan would appear positively liberal.

It therefore seems plausible, even likely, that Malinovskii had been given powerful cues, if not explicit instructions, to produce such a draft at the behest of his erstwhile colleagues in the USSR Ministry of Forestry, who from 1947 had been coveting the forests of the nature preserves, with their roads, domiciles, and other infrastructure. Nevertheless, Malinovskii's vision of a reformed system was colored by his own genuine concern not to vitiate the practical scientific research in the reserves. As it turned out, however, that did not go far enough. To what extent all this had already come to the direct attention of the great barons of the Politburo in mid to late 1950 is still far from clear.

CHAPTER FIVE

Liquidation

The Second Phase, 1950

In all likelihood Malinovskii was not acting particularly cynically when he promoted reduction of the reserves of his system; in his opinion, his scientists did not need all of those good woodlands for their research, and they did not have the right to lock up the resources of the Soviet state. Only if we understand Malinovskii's vision of his new post in this way can we make sense of his support of both the reduction of the system and the upgrading of the salaries and conditions for the scientific researchers in it. Malinovskii was not a villain but a Soviet bureaucrat whose visionary plans for a reconstructed, "souped-up" nature were utterly pragmatic. He had a hard time comprehending the abstruse and murky doctrine of the biogeocenosis, and could hardly be expected to agree that this doctrine, which seemed to him an unproven scientific fetish, should constitute the justification for the whole regime of inviolability in the *zapovedniki.* Nor could he be expected to sympathize with the symbolic meaning of that inviolability for the scientific intelligentsia: that an "archipelago of freedom," a tangible geography of hope saved from the profane clutches of Stalinist transformation, still persisted in the Soviet state.

* * *

Although Malinovskii apparently prepared the draft legislation in secret and consulted with none of the old guard, by early February 1950 they were acquainted with some of its elements and by late spring the veteran activists had premonitions of a crisis.[1] One confirmation came in April, when provisional approval for a projected *zapovednik* near Nal'chik in the Kabardinian ASSR of the North Caucasus by the RSFSR Council of Ministers March 1, 1949 was abruptly rescinded by that same body on April 6, 1950.

The reason given was that the alpine landscapes to be protected in Karbardinia too much resembled those already protected in the Caucasus *zapovednik.*[2]

Only on March 4 did Malinovskii deliver his plan for "eliminating the shortcomings" of the Main Administration to the RSFSR government itself, which was where he should have sent the draft in the first place. The note to Bessonov advised the Russian deputy premier that "for a radical improvement of the work of the *zapovedniki* we need a certain change in the principles that govern the management [*khoziaistvo*] of the *zapovedniki,* improvement of the material conditions and amenities of their scientific workers, and an increase of funding for scientific work."[3]

The plan was divided into four sections, addressing scientific research, the territorial extent of the system, finances, and general measures. Part 1 proposed a more rigorous selection process for research projects and deadlines for their completion, as well as a review of the quality of the scientific research staff, which was to be conducted before July. The plan also provided for a reorganization of the membership on the Scientific Advisory Council and the Scientific-Methodological Bureau of the Main Administration. These measures were delegated to the deputy head, Makarov, and to A. V. Mikheev, the head of the Scientific Department of the Main Administration. On the crucial point about an alteration of the principles of reserve management (as well as on the territorial issue) Malinovskii informed Bessonov that he had already reached an agreement with the USSR Ministry of Forestry, before whose governing collegium he was to personally present his plan in three short weeks.[4]

By this time, RSFSR premier Chernousov must have wondered whether Malinovskii was a stalking horse, at least inadvertently, for the Kremlin, particularly the USSR Ministry of Forestry; it appeared that Malinovskii's primarily loyalties lay there and not with the RSFSR government. In the byzantine intrigue that superficially controverted the politics of bureaucratic institutions, the Russian cabinet chief now became the agency's chief defender while the agency head continued to act as the sometimes willing agent of its executioners.

Chernousov fired the first shot of resistance on May 26, 1950 in a long letter to the USSR Council of Ministers protesting a decree by that body of April 28 inspired by Aleksandr Ivanovich Bovin, USSR minister of forestry (since November 20, 1948), that mandated the geographical relocation of the Central Sakhalin *zapovednik* to the northern portion of the island. Justification for the relocation was found in the USSR Council of Ministers decree of June 16, 1948, which permitted the USSR Ministry of Forestry to assert authority over the forests of that reserve, founded one month before.[5]

Chernousov based his arguments on a memo of May 13, 1950 from his deputy, Bessonov, who wrote, "I consider it essential to petition the USSR Council of Ministers for the preservation of the Central Sakhalin *zapovednik* within its current boundaries." Bessonov asked his chief to write a letter to Stalin personally once Chernousov was able to forge a common position with the Main Administration. Notably, in his protest Chernousov overturned the conclusions of his "own" bureaucrat, Malinovskii, who had included the Central Sakhalin *zapovednik* on his list for elimination. Malinovskii, still obedient to his nominal chief, sent a memo on May 9, 1950 supporting the retention of the reserve.[6]

Arguing that the "territory of the *zapovednik* had great value for science because it was representative of Sakhalin's natural conditions as a whole," Chernousov pointed out that the tundra where the Kremlin sought to relocate the reserves had "no scientific value . . . whatsoever." He added that the RSFSR had invested one million rubles on the reserve's organization, money that would now be thrown away. Finally, he observed that the forested areas coveted by the ministry were located in scarcely accessible alpine areas far from any rivers along which cut timber could be floated. "It is not expedient" (*netselosoobrazno*) to move the reserve, stated Chernousov in the accepted formula, concluding his note with an appeal to preserve it in its current boundaries.[7] For the time being Chernousov managed to hold back the tide.[8]

To mitigate the larger threat to the system as a whole, namely, the audit of the system ordered and conducted (jointly with the RSFSR) by the Central Committee's Agricultural Department, Chernousov held a meeting of his cabinet's Bureau on May 24 with Malinovskii present, and prepared an official decree to address the revealed deficiencies of the Main Administration. Published on June 8, its most stringent provision was that the Main Administration repay its debts and remain debt-free. Its thrust was to address the damaging charges against the reserves and thus disarm them.[9]

The assault on the institutions of *nauchnaia obshchestvennost'* in the area of nature protection emanated not only from behind the forbidding Kremlin walls but also from inside. Malinovskii was rapidly remaking the Main Administration along the lines of his pragmatic, even anti-intellectual inclinations. Scientists turned to Bessonov and Chernousov, their political defenders. In one agitated letter of June 16, 1950, the academician A. A. Grigor'ev, director of the Academy's Institute of Geography, claimed to have information that the Main Administration had eliminated its position of director of publications and had virtually done in (*svernulo*) its publications activity. After arguing that the kind of interdisciplinary field research done in *zapovedniki* was unique, Grigor'ev wound up with an appeal that was at once pragmatic, patriotic, and based on a defense of *science* as an unquestionable good:

> In the interests of the further development of geography here in our homeland we consider it essential to continue regular publication of the works submitted by the scientific researchers of the *zapovedniki,* who work in difficult, often dangerous conditions in sparsely settled arctic, taiga, and high mountain regions far removed from the cultural centers of the country. . . . The liquidation of publication activity by the *Zapovednik* Administration will elicit unfavorable conditions for the development of the detailed geographical study of our country. It is urgent that this question be reconsidered.[10]

After Bessonov sent on the letter to the Main Administration to ascertain the veracity of the charges, Makarov, who reviewed the letter, cleverly tried to turn the issue from intellectual norms to finances. Presumably before sending it on up to Malinovskii, the lame-duck deputy head penned in the margins that the Main Administration should raise the question of a subvention of 500,000 rubles with the RSFSR Council of Ministers and also be permitted to restructure publishing activity on a self-financing basis. Such a request from the Main Administration, however, is nowhere to be found in the archival record. Malinovskii sat on his hands, digging in for a long siege against Makarov, the entrenched field biologists of his agency, their allies, and their alien culture. Chernousov, evidently, had little operational control over Malinovskii's management of the Main Administration itself and was certainly powerless to remove him; presumably Malinovskii answered to higher authorities.

Sensing the limitations of Chernousov's political reach, Makarov now sought new patrons, this time at the all-Union level. In June 1950 he composed a letter to the head of Gosplan USSR, Maksim Z. Saburov, explaining why *zapovedniki* had become "a fully equal and essential link in the system of scientific research institutes of the Union." He emphasized their role as "a marvelous school for the training of young new researchers of nature" and mentioned that in the Il'menskii *zapovednik* alone three hundred university students did their summer practice. Makarov repeated the old arguments about the need for undisturbed *etalony* (baselines of natural processes) and, while admitting the existence of many deficiencies, argued that they resulted more from a lack of adequate support for the reserves system and official limitations on its freedom of action than from any shortcomings of the system itself. High turnover of staff, he explained, was the almost inevitable result of miserably low salaries and indescribably primitive living conditions. Moreover, the administrative fragmentation of the reserves among republican systems was "abnormal, as they all pursue common goals and share the same methods of work, . . . require a single set of goals and conditions, . . . and they interact more with ministries and agencies on an all-Union level than they do with those on the republican level."[11]

Makarov requested that the reserves be designated scientific research institutes and be placed under the Department of Education and Culture of the USSR Council of Ministers. Additionally, he petitioned for the creation of a State All-Union Committee for *Zapovedniki* and Protection of Nature with a Central Research Institute for *Zapovedniki* and Protection of Nature subordinated to it. "Carrying out these measures," he concluded, "will make our *zapovedniki* institutions worthy of the great Stalin epoch." Malinovskii signed the draft letter, although it is impossible to say with what degree of enthusiasm.[12]

Sensing an opening, USSR forestry minister Bovin in July sent a detailed letter to the USSR Council of Ministers as a whole, requesting a full review of the principle of inviolability of the reserves across the Soviet Union.[13] Vladimir Boreiko, who has also investigated this episode, explains that Georgii Malenkov was then in charge of forestry in the USSR and it was "apparently at his initiative that as early as July 20, 1950 the USSR Council of Ministers asked the republican councils of ministers and Gosplan of the USSR to submit proposals on . . . measures to improve the activity of the *zapovedniki.*"[14]

As a result of Bovin's lobbying, an all-Union committee to investigate the reserves was created, headed by Gosplan USSR chairman Saburov.[15] Saburov attempted to provide a fair hearing for a wide range of constituencies, in particular the republics and *oblast'* levels of government. In six months the plan to truncate the reserve system progressed from the draft decree of Malinovskii to the constitution of a Union-wide committee.

Disturbed by the turn of events, RSFSR deputy premier Bessonov asked Malinovskii to convene a meeting of leading staff members of the Main Administration who were also Party members to discuss the fate of the reserves. This meeting took place in early August and was attended by the shadowy figure of A. V. Romanetskii, a functionary of the RSFSR Ministry of State Control, who was almost certainly also colluding with the Kremlin authorities.[16] At first glance, Malinovskii's report to Bessonov that the leading staff had no objections to the reduction of territory of a number of reserves seems incredible. However, here again we see the hand of "protective coloration" and Aesopian language at work. Instead of voicing overt opposition to the plan, particularly with Romanetskii present, the majority offered the opinion that final territorial boundaries of the reserves should be set by the local *oblast'* governments. They counted, probably correctly, on the sympathies of local Party and government machines; these were people to whom they had ties and who felt proud to have these scientific research bases in their bailiwicks. Makarov, Bel'skii, and Mikheev did voice their opposition to the transfer of *zapovednik* territory to game farms, however, and once again raised the question of creating an all-Union administration for the reserves. Doubtless the old-line activists were convinced that to save the *zapovedniki* they would have to go over Malinovskii's head.[17]

With trembling hand, on August 3, 1950 a horrified Makarov scribbled out a note to the VOOP scholarly secretary Sergei Vasil'evich Kuznetsov. "The Society [VOOP] cannot stand on the sidelines on this question," he wrote, "for many of these *zapovedniki* (Moskovskii, Tsentral'no-Lesnoi) were established at the initiative of the Society. I ask you urgently to retype my rough draft, to collect signatures, and to send them off to the addressees."[18]

Makarov's letter, officially signed by Kuznetsov and G. P. Dement'ev, Makarov's co–vice president, was sent to Saburov the following day in the name of the entire VOOP Presidium.[19] "VOOP has received information," the letter opened, "that under your leadership a commission to review the network of *zapovedniki* . . . has begun work, and that, in particular, the question of the complete liquidation of the following *zapovedniki* . . . has been posed."

The marked *zapovedniki* were defended case by case. Arguments were drawn from history as well as from science. The authors reminded Saburov that many of the reserves were deeply connected with the general history of Russian science, such as the Verkhne-Kliazminskii reserve, where the Academy of Sciences' Hydrobiological Station, Russia's first, was established in 1891. The Visim reserve, created on the initiative of Sverdlovsk University and approved by the USSR Council of Ministers, contained the west-slope Urals landscapes depicted by Mamin-Sibiriak. The Tsentral'no-Lesnoi *zapovednik*'s forests "served as a secure haven for partisans" during World War II. The letter concluded with a plea to spare the reserves: "To destroy them is easy, to resurrect them will be impossible."[20]

Makarov's mobilization of scientific public opinion was successful. First to speak up was the Far Eastern branch of the Academy of Sciences in distant Vladivostok, with a telegram from the acting chair of its Presidium addressed to VOOP: "The Far Eastern Branch of the Academy . . . considers the closing of the Sikhote-Alinskii and Sudzukhinskii *zapovedniki* inexpedient and impermissible. We insistently ask you to take all measures in your power to block this liquidation. We are sending a detailed justification in a longer official letter."[21] The telegram was promptly sent to Saburov with a cover letter in the name of the VOOP Presidium.[22] Across the country scientists and their allies were closing ranks to defend these scientific institutions.

Meanwhile the all-Union authorities were speaking through their deeds. A decree of the USSR Council of Ministers of July 24 ordered the Main Administration to have the Caucasus *zapovednik* make available alpine pastures to local *kolkhozy* (collective farms) of the Adler *raion*.[23] Even when their actions seemed outwardly beneficent, as when the USSR Council of Ministers ordered the Main Administration and the Council of Ministers of the Iakut ASSR to create three special reserves in that huge region by 1953, they portended deep changes in the functions, management, and meaning of the reserve system. Significantly, the three Iakut reserves bore the strange designation "*zapovedniki/rezervaty*," reminiscent of the old "*okhotnich'i zapovedniki*" of the People's Commissariat of Agriculture system in the 1920s. Their chief function was to serve as preserves for valuable commercial fur-bearing mammals, notably sable and arctic fox.[24] This shift infuriated the nature protection activists, who had spent thirty years battling against any kind of utilitarian profile for *zapovedniki.* (Interestingly, Malinovskii felt that the overtly commercial purposes of these Iakut reserves warranted their subordination to the RSFSR Main Hunting Affairs Administration rather than to his unit, since for him, "*zapovedniki*" connoted bases for scientific research, even if they were not inviolable.)[25]

Taking advantage of Malinovskii's temporary absence from Moscow, Makarov, as acting director of the Main Administration, sent a letter to Bessonov on October 24 requesting a general meeting of the various staff of the *zapovedniki* in Moscow for February 1951. Makarov anticipated an attendance of fifty and requested 36,500 rubles for expenses. All the pressing questions were to be on the table, and eight directors were lined up to speak.[26] Three days later, the deputy expediting secretary of the RSFSR Council sent a terse reply: "the Council . . . deems it inadvisable to convene the active staffers of the *zapovedniki* at this time."[27] Evidently Bessonov and Chernousov believed that speaking out could only worsen matters at this point.

Saburov and his committee completed their work on November 18, when the report "On Rectifying the Work of *Zapovedniki*" was sent to the Presidium of the USSR Council of Ministers. Boreiko has justifiably characterized Saburov's recommendations as "quite liberal," particularly when they are compared with the decree Stalin eventually signed. In Boreiko's judgment, Saburov had "attentively studied the opinions of the republics, the USSR Academy of Sciences, and the major *oblast'* executive committees," most of which rallied to the defense of the reserves with greater or lesser forcefulness.[28] The Gosplan systems had historically been havens for all sorts of specialists and as a result they exuded more liberalism and *intelligentnost'* (intellectual gentility) than most of the other bureaucracies. This relative liberalism was also made possible by Gosplan's relatively low importance in the Kremlin hierarchy.

Concretely, Saburov proposed the elimination of only three *zapovedniki* in the RSFSR (instead of the twenty-six that ultimately were abolished) and in Ukraine only the reduction of the area of the Chernomorskii *zapovednik* (as against the elimination of nineteen). Although Saburov proposed to eliminate eleven reserves, they represented a minuscule portion of the systems' total area: only 341,000 hectares out of a total area of 11,596,100 hectares for all reserves subordinated to the various republics' councils of ministers.[29] However, fourteen of the surviving *zapovedniki,* stood to lose 3,860,000 hectares, leaving all USSR *zapovedniki* with a total area of 7,395,000 hectares. (Saburov did not mention how many of the 1,864 employees of the various republican *zapovednik* systems would be dismissed.)[30] Saburov's ax cut most sharply into the giant Siberian reserves, but even here no reserve was fatally debilitated.

Saburov's plan was friendly to conservationists in other respects. First, Saburov unequivocally recommended supplanting the republic-level agencies with an all-Union Main Administration for *zapovedniki,* directly attached to the USSR Council of Ministers. Second, he supported activists' insistent demands for their own publishing house to disseminate the scientific findings of research in the reserves. Finally, sensitive to the conservationists' concern for the autonomy of science, Saburov recommended that the USSR Academy of Sciences maintain supervision over the reserves' research programs; as Saburov doubtless understood, the alternative was direct political oversight or technical supervision by the USSR Ministry of Forestry.[31]

Unhappily for the nature protection activists, Saburov's proposals fell victim to new political developments in the Kremlin. For reasons still obscure, on November 24, 1950 the matter was abruptly transferred by the Bureau of the USSR Council of Ministers from Saburov, a Malenkov ally, to the newly appointed minister of state control V. N. Merkulov, Beria's right-hand man, who was charged with developing final recommendations for the reserves.[32] Merkulov, formerly Beria's deputy in the NKVD, was, to put it mildly, a man with a past. On October 27, he finally gained command in what had evolved into a rough Soviet equivalent of the FBI. As Boreiko has written, one month into his tenure the "Case of the *Zapovedniki*" landed in his lap, a made-to-order vehicle for proving himself in his new job.[33]

Serving on the committee alongside Merkulov were Nikita Khrushchëv, then head of the Moscow Committee of the Party, I. A. Benediktov, the USSR minister of agriculture; A. I. Kozlov, deputy chair of the USSR Supreme Soviet, head of the Agriculture Department of the Central Committee, and Benediktov's successor in Agriculture after Stalin's death; N. Skvortsov, who became first deputy minister of state farms; as well as USSR deputy minister of forestry V. Ia. Koldanov and A. Safronov, who had both served on an earlier commission investigating the reserves. After close examination of the archival documents, Boreiko identified Merkulov's deputy, Pavel'ev, the ministry's chief state investigator, A. Kalashnikov, and his aide Fetisov as the ones who ran the case day to day. Fetisov organized a working group that included various *apparatchiki,* including Malinovskii.[34] Malinovskii may not have known about this ahead of time; he sent a note to Bessonov on November 27 abruptly canceling his month-long vacation until the following year on account of the new investigation.[35]

Merkulov wasted little time. Two hundred investigators (*kontrollëry*) from the USSR and republican ministries of State Control were assigned to the campaign. By the end of the investigation in mid-December, more than sixty *zapovedniki,* including some of the most remote ones, had been visited by State Control agents.[36] Nature protection had finally made it onto the radar screen of the Kremlin's most powerful politicians.

An eleven-point guide to the investigation had already been generated by Kalashnikov on November 24 for Merkulov's final approval.[37] In addition to compiling histories and profiles of all *zapovedniki,* the "Program of Investigation" prodded the State Control agents to learn whether *kolkhoz* land had ever been transferred to the reserves and taken out of use, whether *zapovedniki* were illegally renting out land or resources for private exploitation, and whether there was excessive turnover or "infiltration" (presumably by anti-Soviet elements) of reserve staff; every employee's background was to be checked. Agents were to try to clarify the absolute minimum amount of territory necessary for the *zapovednik* to fulfill its function. Investigators were to find out how subjects of research were chosen and who approved them. They were to scrutinize finances and look for wasteful spending. They were also to determine whether the reserves had fulfilled the logging and forest management goals set since 1948 by the USSR Ministry of Forestry. Finally, investigators were to elicit and record the opinions of local, *oblast'*, and republic-level leaders about the reserves.[38]

A number of individual documents survive from this campaign. One extant "*akt*" or bill of findings and accusations shows how one reserve, the Verkhne-Kliaz'minskii *zapovednik* of Moscow *oblast',* was investigated. A team of three agents from the RSFSR Ministry of State Control, two relatively senior, began work on November 28, the day after their ministry was mobilized.[39] Over the next ten days, they learned that while the reserve exceeded its 1949 quotas for sanitary cutting, it fell more than 50 percent short of its quota of construction-grade timber.[40] Even more serious, reserve director G. P. Kornilov and his bookkeeping and forestry staff were accused of illegal sales of cut timber to local *kolkhozy* and *lespromkhozy* (logging enterprises) and other actions to the detriment of the state's coffers. Finally, the inspectors seemed to hold the reserve responsible for the seventeen recorded incidents of poaching and other injury to *zapovednik* property; more troubling yet was that the perpetrators in only ten of the incidents had been caught and brought to justice.[41]

On December 9, the day after the inspectors had written up their findings, Director Kornilov responded with his own letter. He pleaded ignorance of the admittedly complicated fee schedules for sales of timber (he sold the reserve's timber to *kolkhozy* at a lower rate than he should have, based on the stumpage values) and took immediate steps to restitute the state for the lost income. As for the poaching incidents, Kornilov explained that because his reserve was located in a densely populated region, the overall number of such incidents was in fact relatively small. It was ironic, he implied, to be accused of indifference when he had written an article about this very problem in a Moscow newspaper six months earlier. The exemplary record of his reserve in preventing forest fires for three years running had been ignored, he charged. Mistakes connected with the sales of wood were largely the work of the *zapovednik*'s undertrained forester, who had been dismissed in March. Last, Kornilov complained that the investigators said not a word about the scientific, cultural, or educational work being done by the reserve; "evidently," he noted with cynicism, "[those things] did not warrant mention in your evaluation."[42]

Resistance to the investigation also was voiced by political patrons of the nature protection movement. Responding to a Central Committee request to familiarize themselves with the results of a damaging series of charges against deputy director Basalaev of the Altaiskii *zapovednik,* the RSFSR premier's office agreed with some of the charges but refuted others. Most significant, the RSFSR letter flatly contested Basalaev's countercharges (an attempt by the reserve deputy director to exculpate himself) that the scientific workers of the reserve were "anti-Michurinists." "The director of scientific research . . . Comrade Dul'keit/candidate of science/himself works on the problem of 'Animal Ecology . . . in Connection with Snow Cover.' The very title of his research theme is already proof positive that Dul'keit does not detach life from its environment," the letter argued. Interestingly, it then appealed to the scientific authority of A. A. Nasimovich, "who assessed [Dul'keit's] work as satisfactory" and who "did not identify any Weismannist orientation" in it. The letter's authors accepted the charge that the scientific findings of the reserves were not leading to practical applications but pointed to a sharp shift in research themes in the RSFSR *zapovednik* system since 1950 to address this objection. By refuting accusations, strategically acknowledging some shortcomings, and promising improvement, the RSFSR government hoped to soften and blur the image of failure painted by the State Control investigation and thereby save the system.[43] Even more notable is that the RSFSR leaders turned to leading scientist-activists such as A. A. Nasimovich to muster scientific arguments in defense of the reserves.[44]

By far the most dramatic attempt to halt the evisceration of the reserves system was made by the core group of activists themselves. Led by Moscow zoologists A. N. Formozov, S. I. Ognëv, G. P. Dement'ev, G. V. Nikol'skii, and others, activists requested a meeting with Merkulov to present their case. Astonishingly, Merkulov agreed to the encounter, which took place in the early afternoon of December 28. However, he was careful to balance the presence of activists Formozov, Dement'ev, Nikol'skii, E. S. Smirnov,

A. A. Rode, N. E. Kabanov, and P. A. Manteifel' with foes of the reserves: academician A. I. Oparin, USSR minister of forestry A. I. Bovin, and USSR deputy minister of agriculture S. V. Potapov.[45]

No record of the meeting has survived except some brief notes of Formozov's. Nonetheless, these convey the activists' realization of how little they could influence events. "I know for a fact," wrote Formozov, "having participated in that meeting personally and having personally argued with Merkulov (Prof. Nikol'skii can substantiate this), that our conclusions were not even considered. We know precisely and can find witnesses to the fact that the decision was taken before the findings of the 200 investigators were received. . . . What role the Main Administration played in this is unclear. History will sort it out, and each will receive according to his just deserts."[46]

Precipitating the meeting with Merkulov was a convocation of an expanded plenary session of the Main Administration's Scientific Council three days earlier, almost exactly one year from the day Malinovskii took the helm of the agency. Malinovskii, as chair, opened with the understated observation that "the Main *Zapovednik* Administration is living through a rather interesting moment." According to him, the first historical phase of the Main Administration's activity was devoted to the "preservation of parcels of land, of fauna, and of flora," which he characterized as a "passive stage" whose time had already come and gone. "The *zapovedniki* may no longer continue along that path." The very survival of the Main Administration, he explained, depended on joining the movement for the "active intervention in nature" now sweeping the land. In particular the direction of scientific work would have to change. In fact, Malinovskii himself had prepared a new scientific work plan for 1951. This was a direct challenge to the old intelligentsia's ideal of scientific autonomy, to which Smidovich, Makarov, and Shvedchikov had always deferred.[47]

Until now, Malinovskii acknowledged, scientific plans had been developed by the reserves themselves:

> The directors of *zapovedniki* in most cases, devoted perhaps little or, at the very least, insufficient attention to science. . . . Each scientific worker . . . drew up a plan for him/herself in accordance with his/her wishes or inclinations, and sometimes the sum of themes pursued . . . did not accord with the profile of the *zapovednik*. . . . In this . . . lurked a basic and fundamental mistake, namely, that plans drawn up by the *zapovedniki* were thought to reflect local needs. . . . [T]his, regrettably, was not always the case and for that reason this year we have tried to compose the plan from the top down.[48]

Malinkovskii also criticized the overly "descriptive" and insufficiently applicable character of previous research. In some cases, he charged, researchers had not even developed reliable instructions for counting some commercially valuable species of mammals. True, this information might be contained in articles, but practical folks in the economy needed accessible instructions *as such*.[49]

Malinovskii raised another drawback to allowing scientists to draw up their research plans independently of a controlling, coordinating center: "harmful parallelism" of research conducted by other institutions.[50] And if the *zapovedniki* are distinguished from all other classes of institutions by the multi- or even interdisciplinary nature of their study of nature, should that quality not be assured by central planning?[51] From Malinovskii's perspective, a logical corollary of eliminating anarchic independence in the development of research themes would be the elimination of the scourge of "*mnogotemnost'*" (too many different and uncoordinated themes); he had already reduced that number from 192 themes in 1949 to 85 in his first year on the job.[52]

Malinovskii, an outsider who had not adopted the values and perspectives of the old-line conservation activists, subjected many of the scientific claims of the activists to stringent, unsentimental scrutiny and saw what he considered fuzzy concepts, self-indulgence, and internal contradictions. He applied his own standard of "common sense" to the reserves and their work and found them wanting. But to the activists he could appear only as a scheming, evil hangman of their cause; Makarov was reported to have described Malinovskii as "the evil genius of the *zapovedniki*."[53]

So far, the question on everybody's mind had not yet been uttered or addressed. Sergei Ivanovich Ognëv, a doyen of the old zoologists, finally broke the silence by asking whether the system would expand, contract, or stay the same. Malinovskii did not shrink from responding. "I don't have any connection with this particular question, but fully share the alarm of many of you present about the future development of this cause."[54] He explained that he had no information about which reserves would be made smaller, but added that the question of "whether it makes sense to retain all the *zapovedniki* that have been organized" in the past was indeed on the table.[55] Malinovskii argued that the regime regularly approved new proposals for *zapovedniki* on the basis of individual cases; at no time, however, did the principled questions emerge of whether the network as a whole made sense or whether the principle of inviolability was valid and justifiable. Now, however, "Life has gone forward and everything has changed."[56]

As an example of inviolability as a *dysfunctional* regime Malinovskii pointed to the experience of the Tul'skie zaseki and the Tsentral'no-Lesnoi *zapovednik*. In both, lack of forest renewal led to steep decline in the moose population. "Tul'skie zaseki must be characterized by high-productivity oak stands, and only then will the reserve fulfill its designated task, if all of our wishes were bent on restoring these stands. But a regime that is established for the good of forestry would not permit a rapid accompaniment of aspen and lime trees along with the oak stands."[57] Malinovskii was right; everything depended on one's definition of the ultimate goal of the reserves and on one's time horizon.

For Malinovskii the Tsentral'no-Lesnoi reserve was an example of redundancy. When in 1936 the State Forest Protection Service was created, protected forests of the first category were created in all of the surrounding woodlands. The question naturally arose: Is the Tsentral'no-Lesnoi *zapovednik* still necessary? This was not a cause for despair:

> The question of the *zapovedniki* is being resolved in a positive way. Everything will be preserved, but the network of *zapovedniki* will be reexamined to make sure that there is no parallelism, there are no superfluous units, and . . . the issue of the improvement of the future work of the *zapovedniki* will be addressed. . . . [A] certain portion of the *zapovedniki* . . . will be liquidated. Which ones, I simply am not able to say, but I must state directly that, as director, I personally believe that a number of *zapovedniki* are indeed superfluous.[58]

Malinovskii implied that the decision was not entirely his.

CHAPTER SIX

The Deluge, 1951

The research agenda promoted by Soviet ecologists was not readily comprehensible to ordinary folk or to Soviet bureaucrats. In an anthropological sense, we may speak of the biologist-activists and the bureaucrats as belonging to two separate cultures, trying to communicate across a wide gulf of language and values.

The conservation movement's marginal social position created two distinct political problems and dilemmas. In "normal" times the movement's obscurity had helped to save it from destruction, but that status remained a challenge to the scientists' own sense of their social identity and mission; their civic conscience and sense of entitlement to help shape public policy drove them to speak out, to try to become visible, like moths attracted to a flame. Yet their marginality left these scientist activists open to the charge of "irrelevance" and of being "cut off from life" at times of political and rhetorical-ideological mobilization. The movement activists, however, had developed an exquisite dance—flying close enough to the flame to feel the heat, yet being able to sense that threshold beyond which they would be incinerated and thus to turn back just in time. Flying close enabled them to feel as though they had risked, they had dared, they had satisfied the demands of their scientific-civic ethos and had preserved their professional dignity. Turning back was the triumph of common sense.

* * *

In the Kremlin, Merkulov and his team were busy sifting through the thousands of pages of descriptions, financial information, and denunciations regarding the Main Administrations of the various republics as well as the sixty-odd *zapovedniki* personally visited by the agents of State Control. "Absurd, dreamt-up facts were collected by [the agents]," writes Boreiko, "designed to besmirch the *zapovedniki*. Thus, in the Khomutovskaia steppe *zapovednik* the administration, it turns out, did not take 'adequate steps to combat agricultural pests and weeds that represent a great threat both to the *zapovednik* as well as to the fields of [neighboring] *kolkhozy*.'"[1] Askania-Nova had an excess of "unreliable" workers. The Moscow *oblast'* reserves were assessed as "superfluous" because they were located in the existing green-belt around Moscow.[2]

Some of the most lurid incriminations were contained in an eight-page report, "Notes on the Work of the State *Zapovedniki* for Comrade Stalin," which Merkulov sent off to the dictator immediately after everyone had recovered from the New Year's holiday.[3] Testifying to Stalin's personal involvement in the matter by the fall of 1950 is the notation from Merkulov at the head of the document of the copy preserved in the Party Archives: "In fulfillment of your order to investigate the work of the state *zapovedniki*." (A second copy was sent to Malenkov.)[4] In Georgia, Merkulov charged, local authorities had improperly transferred to the Khevskii and Telavo-Kvarel'skii reserves 9,800 hectares of farmland and pasture that had been "eternally granted to *kolkhozy*."[5] In fact, explained Merkulov, this was indicative of a more serious violation of Soviet law. According to the 1939 statute on *zapovedniki*, only Union republics had the right to organize reserves. However, noted Merkulov, "many *zapovedniki* have been organized by decisions of *oblast'* executive committees." One extreme example was the Dargan-Atinskii State *zapovednik* in Turkmenia, which had not even been organized by the *oblast'* authorities but by those in the *raion!*[6] This was local political autonomy out of control.

Another of Merkulov's arguments was that some *zapovedniki* were too large for their staffs to manage effectively. In the giant Sikhote-Alinskii reserve in the Far East, for example, each ranger had to patrol an area of 1,800 square kilometers. For Merkulov it followed that the area of the reserve should be slashed.[7]

Other reserves, such as the Alma-Atinskii *zapovednik*, should be abolished because they had "lost their value for science" through illegal grazing of flocks.[8] Still others supported frivolous or "accidental" research topics "that flowed from the personal whims of the scientific researchers." "Contrived" (*nadumannaia*) and "useless" themes included the Denezhkin kamen' reserve's study of "the feeding strategies of quail in Ural alpine-taiga habitats in a year of complete harvest failure for berries" and the Caucasus reserve's study of "the feeding habits of the lynx as a means of understanding its role of predator in the *zapovednik*." One research project of the Tul'skie zaseki reserve tried to incorporate Lysenko's theory of staged plant development and was included by Merkulov presumably as an example of a "contrived" theme.[9] Additional arguments included the alleged lack of economic application of research done in the reserves ("since 1945 the *zapovedniki* of the RSFSR have spent 20.9 million rubles on science but have not come up with one practical recommendation for the economy"), forest fires (207 in 1949 and 1950 with losses of 675,000 rubles of timber), poaching of timber by individuals (515,000 rubles stolen in Georgia from 1948–1950 alone), and timber simply going to waste because the reserves permitted no logging.[10]

Finally, reviving the charges heard during the First Five-Year Plan period and in Lepeshinskaia's report, Merkulov charged that the *zapovedniki* were havens for politically unreliable elements. This was particularly true of the deputy directors of the reserves for scientific research: those of the Altaiskii and Il'menskii *zapovedniki* served in institutions of the White regime during the Civil War, while the head of the Darvinskii (Darwin) reserve was a former noble had who served time for "counterrevolutionary agitation." The deputy director of the Main Administration itself, Makarov, the report charged, "is a former SR [Socialist Revolutionary]," while a senior scientific staff member, Georgii Gustavovich Bosse, was not only an SR but a member of the government of Kaledin (a White general in the Don Cossack Region). Of ten workers of the Main Administration and twenty leading administrators and scientists of the Ukrainian system, no fewer than twelve had been either prisoners of war or living in territory occupied by the Nazis.[11] Given the existing obligation of the Ministry of Forestry to protect Soviet forests and of hunting administration authorities to protect wildlife, Merkulov concluded that there was little reason to maintain the current elaborate network of nature reserves.[12]

Accompanying these notes in the archive is an extended memo, written partly in pencil, recording the results of consultations with the leaders of the Union republics about the proposed changes in the reserve system. On December 30, Kalashnikov met with Chernousov, who saw no choice but to go along with the recommendations.[13] On January 2 he spoke by phone with Secretary Mel'nikov of the Ukrainian Central Committee, who responded to Kalashnikov by hot line ("VCh") the next day. The Ukrainians wanted to save Askania-Nova, proposing to turn it over to the Ukrainian Academy of Sciences, but otherwise went along. The Belorussian premier, Kleshchev, also on the hot line, tried to trade the life of one *zapovednik* for another. Betting all his chips on an effort to save the large Berezina *zapovednik,* which, he argued, held "great importance for our republic," Kleshchev only weakly defended the Vialovskii reserve, although he explained that it was one of the few places in the USSR where the rare, acclimatized Père Daniel's deer (originally from Manchuria) bred in the wild. In Georgia, Party secretary

Charkviani requested that the Lagodekhskii and Teberdinskii reserves be transferred to the Agricultural Division of Georgia's Academy of Sciences, the same stratagem used by the Ukrainian Party leader.[14]

Although hobbled by the presence of Lysenko and his allies in the Presidium of the USSR Academy of Sciences, President Sergei Ivanovich Vavilov and his scholarly secretary, A. V. Topchiev, were the first to protest officially, writing on January 15, 1951 to Malenkov in his capacity as a secretary of the Central Committee. Commenting on Merkulov's conclusions, the Academy leadership reported that its Presidium "considers the question of the reduction of the network of state *zapovedniki* to be insufficiently studied at the present time," an acceptable way of stating that it was a poor idea.[15]

Science leaders were more outspoken in Ukraine, where forces quickly mobilized to repel Merkulov's attack. On January 13, the Ukrainian republic's Council of Ministers convened a conference on the problem of *zapovedniki.* Ukrainian Academy vice president P. S. Pogrebniak spoke out sharply against the plans to cut the system, particularly in Ukraine. The conference had made such an impression on local leaders that premier D. Korotchenko sent a request to the Kremlin to leave the majority of Ukrainian reserves in place. As Boreiko demonstrates, despite the pressure exerted by Merkulov in response, Korotchenko and First Secretary Mel'nikov dragged their feet, haggled, and tried to save even small parcels of protected land.[16]

In desperation, conservation activists played their last card—their personal connections to Politburo of the Party. When Ivan Dmitrievich Papanin (see figure 7), hero of the Cheliuskin rescue, retired from the directorship of the Main Administration for the Northern Sea Route (Glavsevmorput') at the end of the war, the Central Committee department of cadres was challenged to find a suitable sinecure for the amiable but not intellectually stellar aviator. Someone hit on the idea of placing him at the head of the Moscow branch of the Geographical Society of the USSR. After all, Arctic aviation and exploration were only a step apart.

Papanin settled in to his new position with his customary conviviality and quickly became part of the circle of activists. The Moscow Society of Naturalists rented half its suite of four rooms in the Moscow State University Zoological Museum on Herzen Street to the Geographical Society's Moscow branch. The Zoological Museum (see figure 8) was one of the Moscow headquarters of the conservation movement, and the mainstays of MOIP and of VOOP were mostly the same people. Geptner, Formozov, and a whole series of others were even on the staff of the museum. Until Papanin and his secretary found a new building for the Geographical Society in the early 1960s opposite the Historical-Archival Institute, by force of geographical proximity Papanin became a member of the activists' social network.

Thus, when news about the Merkulov plan was received, Andrei Alek-

Figure 7. Ivan Dmitrievich Papanin (1894–1986).

sandrovich Nasimovich and Eduard Makarovich Murzaev, nature protection activists in the Academy's Geographical Institute (having been fired by Malinovskii), went to Papanin in alarm. Papanin, who had come to understand some of their perspectives, agreed to use his political capital in a last attempt to prevent this donnybrook for the *zapovedniki*. He called his friend Kliment E. Voroshilov, a member of the Politburo. According to Nasimovich, Papanin "drew a vivid picture of the alarm experienced by scientific

Figure 8. The Zoological Museum of Moscow State University.

public opinion." A longtime ally of nature protection as well, Voroshilov promised to help and indeed actually tried to intercede with Stalin. However, he was unable to achieve more than a reprieve of a few months. The scientists' last hopes evaporated.

The fate of the reserves was almost certainly sealed. One of Stalin's aides, Sukhanov, had written on the cover sheet of Merkulov's report: "The *zapovednik* question has been decided." The note was dated January 24, 1951.[17] Voroshilov, who, though a nominal member, had not been invited to a Politburo meeting in years, was not the person to stop this avalanche.

The spring of 1951 was a time of deceptive quiet. True, the Korean War was raging and tensions in Europe were still high following the end of the Berlin blockade. However, Beria's fall from favor, the Mingrelian Case, and the arrest of the Czech Communist leadership would not take place until the fall, and the Doctors' Plot and Stalin's final orgy of paranoia were still to come. Malinovskii sent Premier Chernousov a progress report in early May, setting out in reasonable detail his new, more practical initiatives in the *zapovedniki.* In the Astrakhan reserve, for instance, scientists determined the seasonal feeding patterns of predatory fish and then informed the Northern Caspian Fisheries Administration of the best time to release commercial stocks of fish fry. To the northeast, in the Il'menskii *zapovednik,* scientists designed a wind-powered aerator to mix oxygen in frozen lakes to prevent massive fish-kills from anoxia. In the European Russian Arctic, at the Pechoro-Ilych reserve, an *instruktor* of Military Unit no. 74390 attempted to train moose for use in military transport.

The number of European bison, Malinovskii reported, had risen to forty-one from thirty the year before, and six were transferred to zoos. The Main Administration's debt had been nearly eliminated; it was only 21,000 rubles, 18 percent of what it had been the year before. Finally, Malinovskii reported that he had addressed "a major shortcoming . . . the infiltration of our cadre by [previously] repressed individuals and by insufficiently qualified workers." No fewer than forty-four employees of the Main Administration had been replaced. Malinovskii's report was decidedly upbeat.[18]

The director chose not to make explicit his own view that the *zapovednik* system also needed to be pared drastically to become an optimally useful part of Soviet economy and society. Ironically, this was at odds with the position of his nominal superiors in the RSFSR government. Indeed, an independent audit conducted by the RSFSR Gosplan and Ministry of Finance revealed that Malinovskii had failed to spend 5.7 percent of the monies allocated to the Main Administration for capital construction in 1950. Reporting this information to Premier Chernousov on May 25, Deputy Premier Bessonov editorialized that this occurred "at the same time that many *zapovedniki* are experiencing an acute lack of residential housing" and other needs. However, as Bessonov pointed out, the whole question "is under review in the Presidium of the USSR Council of Ministers."[19] Actually, the "*zapovednik* affair" had made its way to the desk of Stalin's much-feared secretary, Poskrëbyshev.[20] Neither the RSFSR leaders nor Malinovskii would be able to exercise much influence on the final decision. Bessonov recommended that Chernousov not approve Malinovskii's report, as the situation was still not fully clear. However, he did suggest that the findings of the RSFSR Gosplan and Ministry of Finance report be sent to Malinovskii "so that he might take action to eliminate the existing deficiencies in the work of the *zapovedniki*" (in other words, that he should spend the monies allocated by his superiors to the reserves and not make policy on his own).[21] Up to the last, the RSFSR leaders were determined to try to act as much as possible like masters in their own house.

In the shadow of the months-long silence of the Presidium of the USSR Council of Ministers, which now had to act on the conclusions of the Merkulov report, the Main Administration's Scientific Council reconvened on May 21, 1951. Once again, Malinovskii attempted to tackle the big question first. "Before discussing [other] questions," the director began, "I wanted to remind those present that during 1950 the Main Administration went through a not exactly everyday experience," referring to the Ministry of State Control investigation.[22] "I consider it essential to briefly present the recommendations reached by the State Control commissions," he continued.

In the name of political realism, he argued, "the Main Administration, analyzing the past, considered it necessary to amend the statute on *zapovedniki,*" readily conceding that this was "an incredibly ticklish subject." With some justification, Malinovskii alluded to "a whole series of contradictions between separate classes of protected entities in the *zapovedniki.*" In the Crimean reserve, for example, there was the conflict between the ungulate population (mainly deer) and the goal of forest regeneration. In the Voronezh reserve, forests needed to be actively managed if they were to continue to support a growing beaver population.

"All of this taken together," he suggested, "demands a decisive shift from any fixation on inviolability to a concept of *zapovednik* management [*zapovednoe khoziaistvo*]— . . . the rational activity of humans aimed at the attainment of the basic goals set for the *zapovedniki.*"[23] Although failing to inform the activists and council members of his own active role in truncating the system, Malinovskii displayed an unexpected forthrightness in setting out his positions and the logic behind them. In light of his closing appeal—"I ask you to pose your questions as sharply as possible, since by bringing them out into the open we enable ourselves to resolve them"[24]—we are left wondering about this Soviet bureaucrat: was he a consummate cynic or a straight-shooter who genuinely believed that his vision brought greater benefit to society? I suspect that the latter is closer to the truth.

In closing the conference Malinovskii made another revelation about the current crisis and his role in it. There were some, he recounted, who made the following argument during the sessions of the Ministry of State Control investigatory commission: "Let the Ministry of Forestry handle forest administration and the Ministry of the Fishing Industry handle fish stocking and breeding; let's turn all the *zapovedniki* over to them and there will be greater benefit that way." However, "After long debates I was able to demonstrate that no, that is not the case. . . . [T]he *zapovedniki* have their own job to do. And I must say that it was really shown to be the case and it is within this framework that we continue to do our work."[25]

Indicative of Malinovskii's pragmatic, task-based understanding of the functions of *zapovedniki,* so alien to the visions of the old field biologists, was his enthusiasm about the number of wolves shot in the reserves during his first year as director.[26] Other projects that reflected his personal sense of purpose included game censuses, the provision of salt licks and winter feeding stations for game animals, and the construction of artificial bird houses and refuges for wildlife during periods of river flooding. He considered bird banding and the nature log (*letopis' prirody*) useful as well.[27] His was a rather commonplace conception of the general good; tangible, material, and attainable in a short period. In fact, it was much more a Soviet philistine (*meshchanskii*) outlook than a heroic Stalinist vision of the massive transformation and transfiguration of the world. However, because it required the intrusion of the "profane" world of Soviet economics and power relations into the "sacral" realm of the *zapovedniki,* no meeting of minds was possible between Malinovskii and the old activist scientific intelligentsia.

Like Makarov twenty years earlier, Malinovskii pleaded for the chance to let the *zapovedniki* become bases for the transformation of nature in their own way, as research institutes for acclimatization, predator and pest control, and managed forest succession. The only difference between the two men was that Makarov was engaging in protective coloration. Malinovskii meant every word.

1951: Summer and Fall

As a decision about the fate of the system seemed to approach, malcontents in the reserves sensed opportunities to engage in denunciations against resident scientists while reserves' defenders made last-ditch appeals to political patrons and potential intercessors. In June, one disgruntled worker of the Lapland *zapovednik* sent up a *donos* (denunciation) of the reserve's director, Ivan Osipovich Chernenko, its scientific director, Oleg Izmailovich Semënov-tian-shanskii, and his wife, Maria Ivanovna Vladimirskaia.[28] Chernenko, it was alleged, was so "panicked by the State Control investigation" that he overworked the reindeer hauling the "guests" from Moscow and Leningrad. The Semënov-tian-shanskiis, after the reserve was cut off from the outside world for two months in late spring owing to the floods, used government nets to fish daily for food for their table, where they also fed the director and the bookkeeper and her husband. Oleg Izmailovich was further impugned with shooting "an unlimited amount of game birds" for the same purpose.[29]

There was no mistaking the class antagonism that pervaded this sullen letter. "Instead of sharing some of the fish with the hungry workers, the scientific director fed his three dogs until they were full," the writer complained. "It's time to put an end to this extended family [*semeistvo*] of scientific idlers who receive government monies and live at the expense of the workers of the *zapovednik,*" the author protested. "Chernenko fires all those who try to introduce the Soviet way of doing things to the *zapovednik,*" concluded the writer, "and the Main Administration doesn't take any action."[30]

As the letter was sent to the RSFSR Council of Ministers for disposition, Malinovskii was asked to respond. That he was not opposed to scientists per se is reflected in his response of September 14 to Bessonov, following an investigation of the reserve in the interim. The charges, declared Malinovskii, were "utterly baseless." Indeed, he countered, there was no food deficit in the reserve, only "the absence of a full assortment of products," and the right to fish for personal need extended not only to the scientific staff but to all workers of the reserve, including the complainant.[31] The new Main Administration head was not interested in inflicting surplus repression, especially at the instigation of freelance informers and vigilantes. But he was certainly not going to stand in the way of Stalin and Merkulov. Increasingly, his association with the liquidation of the reserves completely overshadowed his acts of decency toward individual scientists in the minds of the vast majority of conservation activists and field biologists of the USSR. Probably by fall 1951 a demonized perception of him had taken hold, which would persist in the scientific community until his death.

"Scientific public opinion" was expressed in unusually strong terms. For example, in the name of the Presidium of the Moscow Society of Naturalists (MOIP), president Nikolai Dmitrievich Zelinskii, a leading Academy chemist, and vice president Vera Aleksandrovna Varsonof'eva, professor of geology at MGU, sent an angry letter to the Main Administration regarding the threat to eliminate the Visim *zapovednik.* "The draft plan for liquidation . . . is eliciting the righteous protest of scientific public opinion in Sverdlovsk," they wrote, "which the Presidium of MOIP shares."[32]

Government authorities on various levels protested as well. One of the most interesting and emphatic protests was sent to Chernousov from the Executive Committee of Velikie Luki *oblast'*.[33] Pointing out the great importance of the scientific research done in the Tsentral'no-Lesnoi *zapovednik* of his *oblast'*, G. Kharin, chairman of the *oblast'* Executive Committee, added that there were a great many other scientific institutions that had a big stake in its continued existence, including the Academy of Sciences and Moscow University. "For some unknown reason the head of the Main Administration . . . Malinovskii has introduced a recommendation to the Council of Ministers about the liquidation of the *zapovednik,*" wrote Kharin, who may not have been aware of *which* Council of Ministers (the USSR's) was really making this decision. "Meanwhile," he indignantly complained, "this recommendation was not cleared" either with the relevant *oblast'* organizations or with other *oblast'* organizations that had an interest in the continued existence of the *zapovednik.* "In light of the above," Kharin concluded, "the Velikie Luki *oblast'* Executive Committee . . . decisively voices its opposition to the recommendation of the Main Administration to liquidate the *zapovednik* and asks that it be rejected, for the Tsentral'no-Lesnoi *zapovednik* is the only scientific research institution not only in Velikie Luki *oblast'*, but in a whole number of *oblasts* of the northwestern region of the RSFSR, not to speak of its importance in regulating water flow."[34]

Another *oblast'* chief who weighed in was V. Ivanov, chair of the Khabarovsk *krai* Executive Committee, whose letter spoke on behalf of the Kamchatka *oblast'* committee as well. "[Our committees] categorically oppose the liquidation of the Kronotskii *zapovednik,* the only one in the region, . . . and call upon the RSFSR Council of Ministers to obligate the Main *Zapovednik* Administration to restore appropriate scientific research staff levels there."[35]

In Belorussia, the Party and government leadership collectively sent an urgent letter to no less than Malenkov, Stalin's first deputy on the USSR Council of Ministers. While agreeing to the elimination of the Vialovskii reserve, the Belorussian premier, A. Kleshchev, and that republic's first secretary, N. Patolichev, deemed it "beneficial to preserve the Berezinskii State *zapovednik.*" Not surprisingly, they opened with technical rather than cultural arguments to bolster their case, noting that the surrounding *kolkhozy* (to whom the forests would presumably be transferred) were already well enough endowed with woodlands and arguing that plowing the former woodlands would cause a catastrophic drop in ground-water levels over a large area, while the sandy-crumbly soils would not support crop cultivation. Navigable rivers would be placed in jeopardy, and the water regime of the fragile overworked soils of the farms surrounding the *zapovednik* would be disrupted.[36] Further, the elimination of the reserve would "deprive the republic of the possibility of the field station–based study of natural flora and fauna" in an area representative of most of the republic's natural features.[37]

Finally, a letter to the RSFSR Council of Ministers from M. Gorbunov, deputy chair of the Sverdlovsk *oblast'* Executive Committee, illuminates the patron-client relationship that the scientist activists succeeded in cementing over a period of years, if not decades. Championing the continued existence of the Denezhkin kamen' reserve, Gorbunov wrote:

> There are no *kolkhozy* or *sovkhozy* [collective farms or state farms] that have any interest in obtaining lands of . . . Denezhkin kamen'. There are also no logging organizations with claims on the forests of the *zapovednik.* Even if such were the case, the status of the forests as watershed forests for the Volga basin would prohibit exploitation. If any part of the *zapovednik* should be turned over to the Ministry of Forestry, no savings or economic gains will result. . . . The entire territory is valuable for scientific research and the scientific societies and scientists of our *oblast'* have spoken out against any violation of the integrity of the territory of this *zapovednik.*[38]

The same arguments applied, continued Gorbunov, to preserving the Visim reserve. Having defended his dignity as a local political leader and the interests of his clients, Gorbunov was also realistic enough to know that his protest would carry little weight with those making the decisions Thus he concluded that "if, despite our opinion, the liquidation goes through," the Executive Committee requests that the Ministry of Forestry convert the entire area into a game preserve (*zakaznik*) for beaver, which would at least minimize the disruption to the natural complex.[39]

It would be a mistake, though, to conclude that the opinions of the republics and the localities counted for nothing in this substantial land transfer. Elaborate tables detailing the responses and reactions to all of the individual "liquidations" and truncations of reserves were compiled for Merkulov (and, ultimately, for the Presidium of the USSR Council of Ministers) on the eve of the issuance of the official decree.[40] Those dissenting were in the minority at the *oblast'* level, but the above examples were joined by the Central Committee of the Azerbaijan SSR (which sought to save both of its marked *zapovedniki*), Turkmenia (which sought to spare most of the Dargan-Atin reserve), and Ukraine. The leaders of the RSFSR were silent this time.[41]

Had the regime not additionally employed strategies of deception, the number of protesting *oblasts* might have been greater yet. Five years later, at a conference of *zapovednik* directors and staff, Ivan Osipovich Chernenko related how a potential protest about the liquidation of the Laplandskii *zapovednik* was derailed. In August 1951, Chernenko recalled, he and Oleg Izmailovich Semënov-tian-shanskii, who were on the scientific staff of that reserve, made a trip to Murmansk, the *oblast'* seat, to settle some questions about future projects and to raise the issue of the decline of fur-bearing game animals in the *oblast'*. When they arrived at the offices of the *oblast'* Executive Committee—the regional government—they were unexpectedly told that their *zapovednik* was slated to be liquidated. Immediately they sent a telegram to Malinovskii and quickly received a reply: "No one is intending to liquidate your *zapovednik.*" Chernenko and Semënov-tian-shanskii showed the telegram to the local authorities, correcting the rumors that they had heard, and, their anxiety put to rest, returned to the reserve. "A month had not passed," Chernenko continued, "when we received another telegram, in September, decreeing the liquidation of the *zapovednik.* . . . If it weren't for the [first] telegram, we could have raised this issue at the *obkom* [the regional Party headquarters] and in the *oblispolkom* [the regional government], presented our point of view, and without a doubt the Laplandskii *zapovednik* would have been saved. But we received notification of the liquidation when we already had no opportunity for appealing it."[42]

Stalin personally reviewed Merkulov's recommendations on July 25 and signed order no. 12535-r, which authorized a final drafting commission consisting of Khrushchëv, A. I. Kozlov, Benediktov, Skvortsov, Bovin, and Safronov (a deputy premier of the RSFSR) to prepare the final draft legislation. Stalin's order had two stipulations: that the materials be ready for presentation to the USSR Council of Ministers in two weeks, and that the rump *zapovednik* system be left with an aggregate area of no greater than 1.5 million hectares.[43] Cognizant, no doubt, of Malinovskii's sincere support for moderate cuts in the system, Stalin's secretary Poskrëbyshev called the Main Administration chief to remind him that Stalin, not he, called the shots.[44] The Kremlin was covering all possible contingencies.

The decree was published on August 29, 1951, one of the darkest days for nature protection in Soviet history. Simply called "O zapovednikakh" ("On *Zapovedniki*"), decree no. 3192 obliterated 88 of the 128 extant reserves, while the aggregate area of the reserves fell from 12.6 million hectares, or 0.6 percent of the overall area of the USSR, to 1.384 million hectares, or 0.06 percent. Of the forty reserves that survived, most were smaller—in some cases, unrecognizably smaller—versions of their former selves. Two provisions of the decree were on the activists' wish list: the unification, at long last, of all of the disparate republican systems into a centralized all-Union one, the USSR State Committee for *Zapovedniki,* with the status of a minor ministry; and the extension of pay scales of scientific researchers in agriculture holding degrees to research scientists in the *zapovedniki.* The Academy of Sciences was assigned methodological and scientific oversight and leadership of the new reserves system.[45]

Some local politicians such as P. I. Titov, secretary of the Crimean *obkom,* saw the decree as authorizing an open season on his *oblast*'s Crimean *zapovednik.*[46] Writing to Malenkov, Titov posed as a defender of the reserve's forests, which were critical to erosion control on the southern slopes of the Crimean uplands. That worthy environmental goal, however, was being undermined by the protection of some 2,500 European red deer and 1,500 roe deer in the *zapovednik* itself, not to mention 6,500 of the assorted ungulates in the surrounding forests, which, he claimed, were eating the new forest growth. Although Titov had repeatedly raised the problem with the Main Administration, it was always deflected. Now, he sought Malenkov's sanction for a thinning of the herds as well as a provision for continual culling. He also asked Malenkov to initiate a review of the staff breakdowns in the reserve to insure that the reserve fulfilled its responsibilities in this area.[47] Further examination of the documents points to Titov's concern over *surrounding* forests' economic losses as the primary motivation for his letter.[48]

Malinovskii responded to Titov on September 20, reassuring him that the culling of the herd would begin shortly, subject to approval of the USSR Council of Ministers. A note of November 3, 1951 from Malinovskii to A. I. Kozlov, head of the Central Committee's agricultural department, confirmed that Malinovskii had given the *zapovednik*'s director orders to organize the deer hunt. The final document of this episode was a note to Malenkov from the agricultural department informing him that a representative of the Main Administration was on his way to the Crimea to make sure Malinovskii's order was carried out and to determine the main direction for the scientific research of the *zapovednik.*[49] Malinovskii's solicitude for the Crimean forests was a harbinger of things to come.

With the elevation of the RSFSR Main *Zapovednik* Administration to all-Union status came a parting of the ways with the RSFSR administration. This was marked by a letter of Malinovskii to Chernousov in which, among other things, Malinovskii informed the premier that the entire staff of the previous Main Administration, with the exception of two bookkeepers, had been let go. Makarov, who had served officially as deputy head of the system for seventeen years and had factually led it for twenty, was another casualty of the "restructuring."[50] Of the twenty-eight *zapovednik* directors of Malinovskii's new all-Union system, seven were fired as well.[51]

Like Russia, all of the other Union republics were forced to issue independent decrees abolishing their former systems of reserves and turning those territories and property over to the new Main Administration and the USSR Ministry of Forestry. Another decree of October 29, 1951 regulated the relatively small transfers of land to *sovkhozy* and *kolkhozy*.[52] Ukraine lost nineteen reserves, Georgia sixteen, Lithuania thirteen, Turkmenia four, and the other republics ten altogether. In Askania-Nova, where even the Lysenkoists had not touched 24,000 hectares of relatively undisturbed feathergrass steppe, 2,800 hectares were immediately sown to crops and an additional 20,600 passed to local farms.[53]

Protests were of no use at this point, but the republics began to look for ways to mitigate the damage. Lithuania cleverly managed to organize game preserves within its thirteen liquidated *zapovedniki*, with the former reserves' forests declared category 1, that is, exempt from commercial cutting.[54]

To understand the emotional impact of these decrees on the activist biologists of the Soviet nature protection movement, it is necessary to recall how bound up *zapovedniki* were with the scientists' sense of identity and mission. They were the priests, the interpreters, and ultimately the keepers of these sacred territories, which they thought they had saved from the profane Stalinist mire. *Zapovedniki* were the last tangible remains of prerevolutionary civil society, the ideal of *obshchestvennost'*. Now, that mire had burst through the invisible gates of the reserves and would cover them too. The activists' own "archipelago of freedom" was being wiped off the map. "In scientific and educated circles this truncation [of the reserves] is regarded as a catastrophe," wrote longtime VOOP Presidium member A. P. Protopopov to I. I. Puzanov, zoologist activist and friend. "Personally, I cannot reconcile myself to this state of affairs," he averred.[55]

Protopopov called attention to the exquisite irony that the "Geografgiz" Publishing House had just issued the "marvelous two-volume *Zapovedniki SSSR*" (*Zapovedniki of the USSR*), a beautifully illustrated and richly detailed guidebook to the reserve system as it existed on the eve of its destruction. Its poetic and inspiring introductory chapter was written by the tragic standard-bearer of the reserves' cause, V. N. Makarov; it was to be his last publication. "In Moscow," Protopopov continued, "they are already impossible to get a hold of; they sold out in three days. I look upon this publication as a literary monument on the grave of the *zapovedniki* described in its pages. But I do not intend to weep over this grave because the monument upon it is calling us to battle."[56]

Death and Rebirth

Now equivalent to an all-Union minister, Malinovskii spent the autumn months of 1951 preparing the new statute that would govern the operations of the reserves. It was approved by the USSR Council of Ministers on October 27.[57] On March 19, 1952, it was finally published in a brochure whose copies were numbered in a restricted print run to insure security.[58] What was new was the emphasis: the first priority of scientific research in the *zapovedniki* was "the solution of practical tasks of agriculture and forestry, the fishing industry, and commercial hunting" (3.b). A number of separate articles reiterated the control over logging and the oversight of forest management by the USSR Ministry of Forestry.

Also by October 1951 Malinovskii had prepared for the forestry ministry totals of the amount of forested land area projected to be under active management in 1952 in the reserves. Of a total area of 1,307,750 hectares, managed lands already accounted for 853,600 hectares; only 257,400 hectares, mostly in steppe or arid regions, were exempted, but of this territory, 163,800 hectares were to be under other kinds of management. In other words, fewer than 100,000 hectares were to be left "wild."[59]

If the *zapovedniki* were stripped of almost all of the economically attractive lands and therefore were no longer objects of predation by the economic ministries, Malinovskii's presence did little to stop the almost constant stream of security checks and investigations that the "organs" continued to carry out. On February 12, 1952, Beria himself ordered Deputy Minister Pavel'ev of the USSR Ministry of State Control to investigate a denunciation made on New Year's Day by Darvinskii *zapovednik* director P. A. Petrov and to report back to him and to the Bureau of the Presidium of the USSR Council of Ministers no later than February 25. For emphasis, perhaps, Beria signed his instructions in red ink.[60]

Petrov's denunciation opened with a general critique of the *zapovednik* system, using much the same language as Merkulov and even Malinovskii had done. Even after Malinovskii's cuts, claimed Petrov, the system was still grossly overstaffed; in his own reserve the deputy director, head of warden patrol, three heads of laboratories, and one of the two hydrologist-technicians should all be fired. The two chauffeurs should also be let go, since there were no roads to drive on; the cars should go to neighboring collective farms. By far the most sensational accusations were directed against Malinovskii. The new chief had prevented Petrov from firing four crew members of the reserve's motorized launches whom Petrov described as "malingerers." "For your information," Petrov wrote, "the majority of these people are under the little roof of the head of the Main Administration, Comrade Malinovskii." Malinovskii was protecting those who had served time for violations of Article 58, those who had allegedly been translators for the Germans, and other politically reprehensible types. "As strange as it seems," Petrov continued, "such people are in Malinovskii's immediate working entourage." Petrov singled out Iu. A. Isakov, who had served from 1934 to 1937 on the Baltic–White Sea Canal complex. Not only did Malinovskii keep him on, he even wrote in support of Isakov's petition to have his conviction reversed. Similarly, Malinovskii repeatedly attempted to have senior researcher E. N. Preobrazhenskaia released from her eight-year sentence and was ultimately successful. Malinovskii gave her work in another *zapovednik.* Other former political prisoners and ethnic Germans were named in this ugly irruption of envy and resentment.[61]

"Knowing Comrade Malinovskii well as someone who is politically illiterate, who has never worked in positions of responsibility, uninformed in his field, a careerist and a vengeful person (irrespective of the fact that he himself brought me over with him to work)," concluded Petrov, "I with all my Bolshevik vigilance and straightforwardness decided to inform you and the Vologda *oblast'* executive committee . . . .The improper attitude of Comrade Malinovskii toward our *zapovednik* as well as his firing of . . . administrative workers who were members of the VKP(b) [the Communist Party], while at the same time surrounding himself with people who do not inspire trust, all this raises questions about his future tenure as head of the Administration."[62]

By February 13 an on-site investigation was carried out by the USSR Ministry of State Control. The team of three, the now familiar Kalashnikov plus K. P. Ivanov and V. A. Vinogradov, found Malinovskii's letter on Isakov to be "objective; we cannot find anything reprehensible."[63] More than that, they were able to turn the tables on Petrov, learning that "in his official forms Comrade Petrov hid the fact that he was excluded from the VKP(b) in 1938. With an education of only two grades of primary school, he also wrote on his official forms that he graduated from the Leningrad Technical Forestry Institute. . . . In his attempts to explain all this before the bureau of the *raikom* of the Party, Comrade Petrov announced that he had done these things by mistake."[64] Malinovskii retained his position, and the matter apparently ended there.

The Question of Causality

According to Aleksandr Leonidovich Ianshin, who was an active member of MOIP at the time, Lysenko was central to the process of the liquidation. First, he emphasizes, the actual liquidation was planned after the August 1948 Session. Many of those whom Lysenko persecuted had been extended a welcome to work in the *zapovedniki,* which gave him one reason to exact revenge on those institutions. Second, the slogan Lysenko loved to repeat, allegedly drawn from the writings of the plant breeder Ivan Michurin ("We cannot wait for kindnesses from nature; our task is to wrest them from her"), ran exactly counter to the continued existence of *zapovedniki,* with their regime of inviolability and their celebration of "pristine" nature. Third, Lysenko and I. I. Prezent, his close associate, had had their start in the early 1930s with attacks on the *zapovedniki* and their "contemplative" approach to transforming nature. The fight against genetics had distracted Lysenko and Prezent before they could wipe out this nest of enemies of socialist construction, so the argument goes, and unfortunately no one else noticed it until Lysenko returned to finish the task after the mop-up of the geneticists in late 1948. Ianshin believes that the initiative rested with Stalin himself, probably as a result of discussions with Lysenko, who then had relatively easy access to the dictator.[65]

Nasimovich, who was more of an insider in the Main Administration than Ianshin, believed that Beria's hand guided the whole process of liquidation. Reputedly the secret police chief saw the heavily wooded *zapovedniki* near the borders of the USSR as hideouts for foreign spies. Moreover, he allegedly served with Khrushchëv on a Party commission charged with increasing the area of arable land and commercial logging.[66] Nasimovich regards the appointment of Malinovskii as heavily influenced by Beria, who employed the forester as his hatchet man.[67]

Shtil'mark and Heptner, although they remain formally agnostic about the ultimate initiator, find no difference between the positions of Malinovskii and those of Bovin or Stalin. There is considerable merit in their judgment that, although "we have no knowledge about the details of the appointment of A. V. Malinovskii to his new post, there is no doubt that the recommendation came specifically from the forestry organs where he formerly worked and was well known." Contacts between Malinovskii and Bovin and other forestry leaders, they speculate, could have been strengthened in the realm of leisure, where interest in hunting was an attribute of being socialized into forestry.[68] Taking issue with Boreiko's reticence to characterize Malinovskii as a full supporter of the evisceration of the system at the outset, Shtil'mark and Heptner argue that

> he was nevertheless hardly an accidental and unwilling executor. . . . A professional economic-oriented forester and lumberman, . . . he had already also been "anointed" with an academic degree and a previous directorship of an institute. . . . This gave him the opportunity to present himself not as a primitive liquidator of the *zapovednik* system, but as a kind of "theoretician,"

> which carried great weight with his superiors, often men with only average education. For that reason the opinions held by Malinovskii, "*our own* scientist," carried particular weight for them in disputes with professors, members of the Scientific Council of the Main Administration, and scientific public opinion. Malinovskii came out against these groups with the slogan of "*zapovednik* economic management" [*zapovednoe khoziaistvo*], an "alternative" of sorts, but in actual fact a prospectus for a *pogrom* against the *zapovedniki,* which suited the administrative-party elite just fine, disguising its chiefly consumerist goals. Not by chance during the next phase of his career, already in the USSR Ministry of Agriculture, . . . did Malinovskii arrange the transformation of the best of the remaining *zapovedniki* . . . into so-called "*zapovednik*–game management economies" [*zapovedno-okhotnich'ia khoziaistva*]—factually, into imperial hunting preserves.[69]

Suggesting at least some complicity on Malinovskii's part in the evolving plans for the liquidation is evidence that Merkulov entrusted him, together with Koz'iakov, with the task of composing the memorandum outlining the proposed future use of the territories of those *zapovedniki* slated to be liquidated. (Curiously, Malinovskii sought the permission of his nominal superiors in the Russian Republic premier's office to release his memorandum back to Merkulov for use the next day, August 18, 1951, at the meeting of the Presidium of the USSR Council of Ministers, which was to ratify the liquidation decree.)[70]

On the other hand, Shtil'mark and Heptner do acknowledge features of Malinovskii's public persona—his "drive" and his "complete conviction in the rightness of his own positions"—that accord with a less cynical assessment of his activity, an assessment that Vladimir (Volodymyr) Boreiko seems to prefer. It is to this more nuanced view of Malinovskii as a sincere believer in voluntarist, experimental forestry schemes to enhance and augment nature's own productivity and utility—as contrasted with the cynicism of the Party apex by the late 1940s—that I now lean. At this point nothing can be proved. However, one piece of indirect evidence from the archives makes this view more plausible.

A little over one year after the truncation of the reserve network, Party secretary Malenkov passed along to A. I. Bovin, still USSR forestry minister, a very long complaint against Bovin he had received from . . . A. V. Malinovskii. At issue were Malinovskii's objections to the forest management instructions issued by Bovin's ministry, which Malinovskii had set out in a thirty-six-page treatise with a twelve-page addendum, entitled "On the Basic Principles of Forest Management." Malenkov additionally signaled A. I. Kozlov, head of the Central Committee's agricultural department, and asked him to figure out what the disputation was about and take appropriate measures.[71]

Malinovskii castigated Bovin's 1952 forest management instructions precisely for their "conservative character," for acting "as a brake on the development of forest management." Specifically, Malinovskii had two major objections to Bovin's approach. First, Bovin continued to employ the old system of *bonitet,* whereby trees were classified by quality and features of the wood as a basis for managing forested tracts. Thus, underwatered and overwatered pine groves would be managed similarly because they were in the same *bonitet,* and therefore management, category. Malinovskii sought instead a transition to Georgii F. Morozov's idea of "forest types" (*tipy lesa*).

Second, Malinovskii reproached Bovin with using an incorrect—capitalist—model in the practice of the engineering of forest plantations:

> The history of forest engineering tells us that when they composed their plans, forest planners took as a point of departure the idea of a normal forest, that is, they directed the entire force of their activities to bringing the forest to a normal condition. . . . They understood the term *normal forest* to mean that condition in which all planting must, as a rule, be of purely the same [species] composition, meet a certain density of plantings, and maintain a uniform area for specified age groups of trees. On this idea of the normal forest the capitalist theory of continual and uninterrupted income [from uninterrupted cropping] was based.[72]

However, argued Malinovskii, the idea of the "normal forest" completely failed to correspond to conditions of Soviet forestry and, "quite naturally, was thrown overboard." Regrettably, though, "neither forestry engineers nor workers in forestry have determined the character and structure of the forest that best suits the corresponding purposes and utilities that should determine each category of forest. In the new instructions . . . there are no indications of how to solve this problem." Malinovskii himself had hoped that the regeneration and/or engineering of commercially desirable "forest types" would be one of the central new objectives of research in his *zapovednik* system, but he was dismayed to see that there was no echo of support of this from Bovin and the ministry.

Additionally, Malinovskii sought to base logging strategy on waiting for trees to achieve their maximum growth rates before being logged, which would be adjusted by taking into account information about size requirements of logs in demand.[73] Malinovskii asked that "forest types" be officially instituted as a framework of thinking about forests, and that the Institute of Forests of the Academy be involved along with institutes of Bovin's ministry in developing these profiles for different regions.[74] He believed that science could optimize forestry: "for each forest type establish the desired composition, optimal density, and age distribution of the plantation depending on the purpose of the forest and the forest group to which it belongs."[75]

Bovin sent a reply to Malenkov on January 5, 1953, defending his instructions as approved at a conference of 150 experts on February 14–17,

1951, and later approved by the Collegium of his ministry. However, he promised to call a conference in two to three months to discuss, again before experts, Malinovskii's proposals. Thwarting later historians, the conference was postponed, with Malinovskii's concurrence, to December 1953 owing to the reorganization of the USSR Ministry of Forestry in early April 1953, and thereafter the issue apparently faded.[76]

This not terribly coherent exchange on forest structure and exploitation, although by no means conclusive, strongly suggests that Malinovskii held strong ideas and convictions about forest management—strong enough to compel him to take on a higher-ranking minister and even to take his case to Malenkov, the second most powerful politician in the country. If he had been Bovin's creature completely in 1950–1951, why were his original proposals rejected as insufficiently far-reaching and the matter turned over to Merkulov in State Control? If he had been Lysenko's or Stalin's creature entirely, then how do we explain his protection of "formerly repressed" scientists, such as Iu. A. Isakov, or his support of the RSFSR Council of Ministers' defense of the Sredne-Sakhalinskii *zapovednik* in April 1950, then individually threatened with elimination?[77] Indeed, Malinovskii took science seriously, and he is credited by none other than his bitter political foe Nasimovich with developing an important method of conducting censuses of fauna under snow cover.[78] We must take him seriously as well. Following Boreiko's line of thinking, it makes more sense to regard Malinovskii as an authentic forestry visionary, an idealistic utilitarian, particularly when viewed against cynical politicians such as the land-grabber Bovin, the faux visionary Lysenko, the evidence-manufacturer Merkulov, or the supreme boss, Stalin. This picture is strengthened by Malinovskii's decent behavior toward such scientists as Isakov, Preobrazhenskaia, and others.

Amid conjecture and disagreement about the personal responsibility borne by various individuals in this episode, Shtil'mark and Heptner have advanced an indisputable conclusion:

> The tragedy of Soviet *zapovedniki* depended not only on who was the hangman, who held the ax; it is more important to understand . . . who handed down the sentence and why such punishment was inevitable. There is a paradox in the fact that the *zapovedniki* were the creations and pride of Soviet power and were sentenced by that very power to their demise. The inevitability of their purge was sealed by the vulgar materialist principles that inescapably shaped the destructive consumerist attitudes toward nature [of the regime], sugar-coated in a demagogic ideology about its transformation in the interests of people (in its next phase, they became slogans about the enrichment and improvement of nature).[79]

It is this Soviet-style production orientation (it is difficult to call it consumerist except with respect to natural resources) that conservation activists rightly understood was at the core of the system's values and vision.

CHAPTER SEVEN

In the Throes of Crisis

VOOP in Stalin's Last Years

Nature protection's high-water mark in the postwar period was 1947 and the first few months of 1948, when the network of reserves was still expanding, republic and local governments were reasonably supportive, and the "center" had not yet begun to covet the forests of the reserve network or to view nature protection societies as political centers of infection.

A letter written just days before the August Session of 1948 from Makarov and VOOP secretary Zaretskii to the RSFSR Council of Ministers reflects VOOP's valor during its heyday.[1] Not long before, activists of the Crimean branch of the Society had reported with alarm to the VOOP Presidium in Moscow that units of the Ministry of State Security's Defense Administration (*Upravlenie okhrany*) were leveling the distinctive, spirelike cypress trees on the southern coast of the Crimea. Although this paramilitary "campaign" is one of the most inscrutable expressions of Stalin's arbitrary rule, an anecdotal explanation soon made the rounds. Apparently—perhaps after a sojourn in the south—Stalin was said to have complained that "the cypress tree is the tree of death; it belongs only in a cemetery."[2] The dictator's offhand remark was construed by his entourage as a policy injunction, and soon detachments of blue-epauletted state security troops were scouring the resort towns of the Crimea with chain saws, on the lookout for the dendrological threat.

"To try to clarify the cause for this mass logging, we sent inquiries to the Defense Administration of the Ministry of State Security and to the Main Resort Administration," wrote the VOOP officers. However, as they informed the Russian Republic leadership, their letter was not dignified with a response, which is why the Society now sought the help of the RSFSR government:

> The Presidium of the Central Council of the Society asks you to issue instructions to the Crimean *oblispolkom* [regional government] to create a special Commission to find out the causes and extent of the logging of cypress trees. Included in the work of this Commission should be the president of the Crimean branch of the All-Russian Society for the Protection of Nature, Comrade Studenkov, the director of the Crimean *zapovednik,* Comrade Rybal'chik, and the director of the Nikitskii Botanical Garden, Comrade Kaverg.[3]

The archival paper trail vanishes at this point, making it impossible to trace what happened afterward. Nevertheless, even this short document lights up, if only dimly, the mental and social worlds of the nature protection activists. We see them challenging the prerogatives of the dreaded secret police, acting on the dictates of personal and collective honor and of civic duty. Perhaps on some level they realized that their social marginality, the authorities' view of them as *chudaki,* provided some modicum of political protection. It even seems reasonable that their presentation of self, when it rose to the level of consciousness, sought to accentuate this harmlessly eccentric public image. Yet writing the letter above, not to mention lodging a direct "inquiry" with the secret police, took unimaginable courage in 1948. For the activists, saving those trees was not a trivial question, but precisely the kind of public-policy question they felt entitled, even obligated, to address as scientific experts and citizen activists.

* * *

Previous support by the RSFSR Council of Ministers and the fading pre-August 1948 hopes for a renewed cultural liberalization also promoted an attempt by Makarov in early April 1948 to gain approval for a merger of VOOP with the newly reconstituted *kraeved* societies (voluntary societies for the study of local lore, crafts, folkways, and nature), its old and "natural" allies. The Bureau of the Council of Ministers met April 2 but decided to put off their decision for a week. Although Deputy Premier Gritsenko supported Makarov's proposal for the merger, his recommendation, presented as a report, was turned down. Makarov was urged to re-edit the draft. The activists were reminded that they were still operating within the constraints of the decree on *kraevedenie* of June 10, 1937, which forbade any unification of their forces.[4]

The rebuffs to the merger proved to be precursors to even worse news for the conservation society. On July 2, 1949, the first of a series of investigations into VOOP's activities was launched by the RSFSR Ministry of State Control following reports that VOOP had exceeded its statutory authority to conclude contracts.[5] Twenty days later Makarov, Dement'ev, Nikolai Borisovich Golovenkov (the senior editor of VOOP publications), Mikhail Petrovich Beliaev (the senior bookkeeper), and Sergei Vasil'evich Kuznetsov (the scholarly secretary) were all brought in to meet with N. V. Savitskii, the deputy minister.[6] What most disturbed the investigators was the alleged "overstepping by VOOP of its publishing rights and the payment of significant sums of money to nonmembers for the completion of work not sanctioned by the charter of the Society."[7] The recommendation to RSFSR premier Chernousov was mild by Soviet standards: that "the Bureau of the RSFSR Council of Ministers hear out the leaders of VOOP and give them appropriate directives on putting their publishing activities in order."[8] The conclusions of the republic-level authorities reached the desk of acting USSR minister of state control A. S. Pavel'ev on August 10, 1949: "VOOP factually overstepped . . . its rights and, with the aim of increasing its own funds, completed work that was not sanctioned by its Charter and did not have any direct link with the goals and tasks of the Society."[9]

Seeking to explain the circumstances under which the Society was driven to violate its charter, Makarov argued that "experiencing great difficulties with funds and having weighed its options, the Society for the Protection of Nature agreed to the proposal by the All-Union Military-Hunting Society to do a rush publication of brochures. . . . The Society was materially interested in taking this order because the All-Union Military-Hunting Society promised to pay the bill immediately; the Society was experiencing extreme fiscal difficulties."[10] Makarov further admitted that the Society had identified "the possibility of a quick turnaround on printing postcards." This led the Presidium to "approve the printing of postcards with flowers, even though that had no direct connection with the Society's tasks."[11] Regarding the payment of outside jobbers, Makarov argued that the "diversity of publications of the Society, the large print-runs, the absence of our own printing facilities, . . . the urgent deadlines for some of the publications (nature calendars, literature for Tree Day, for Bird Day, etc.) of necessity forced us to use some workers . . . and to pay them, in the opinion of the Society itself, higher fees than would have been the case during the normal course of publishing."[12]

After receiving "an appropriate oral directive," the Society reported to Savitskii on December 30 that all extraneous publishing activities by the Society had ended, with the exception of work contracted for before the July investigation.[13] In August 1950 the VOOP Presidium decided to stop the printing of sixteen postcards of the series "Michurinist fruits" with an attempt to recover the paper purchased by VOOP for that purpose.[14]

In a letter to Premier Chernousov of September 16, Bessonov reported that he had reexamined the question of VOOP's publishing activities at Chernousov's request. After meeting with Makarov, Dement'ev, and Kuznetsov, he decided to permit the Society to publish materials already prepared, some of which were already at the printers, on condition that the work would be done by staff released by the Society. Simultaneously, he ordered the Main Administration for the Printing Industry, Publishing, and the Book Trade to examine the question of allowing the Society to continue publication. His final request was that Chernousov remove this issue from the scrutiny of the Ministry of State Control.[15]

With this reprieve VOOP leaders intended to use 1950 to put their house in order. According to a report sent by Makarov and his colleagues to the Central Committee Department of Propaganda and Agitation in April 1950, VOOP listed sixty-five regional organizations (thirty-six chartered and twenty-nine organizational bureaus) with a total membership of more than 30,000, with an additional 40,000 in its youth section. Active members, as identified by Makarov, numbered 2,265, mostly in Moscow and Sverdlovsk.[16] Attached was a list of all Presidium and Central Council members, their places of work, and their Party membership. Significantly, six of twelve members of the Presidium were non-Party; indeed, they had now become a majority with the exit of VOOP secretary Kuznetsov from the Presidium.[17]

At the plenary meeting of the Central Council, Dement'ev noted "a positive shift" in the general operations of the Society. In particular, he welcomed the appearance of so many provincial branches and expressed the belief that "the periphery has been created."[18] Nevertheless, the larger Soviet reality could not but intrude on this hopeful assessment. The incremental expansion of the Society was a tactical victory, but the Society was still sustaining strategic losses.[19]

Makarov told the group that he had been trying to convene another congress of the Society for April 1950 but that the question required permission from the Central Committee of the Party. "When I was called by Central Committee secretary Comrade [P. K.] Ponomarenko and I described to him the tasks of the Society and its work, he recommended not to rush but to postpone the congress to a more propitious time. In March, he noted, were the elections, and in April was already the spring sowing campaign. However, he offered help with the organization, providing the Society could supply him with the agenda, written summaries of the Society's activities, etc."[20] Makarov resolved to try for September.[21] Protopopov, who, like others, was doubtless surprised to hear any reports of Central Committee support for the Society, had this to say: "Of course this news is to be welcomed, . . . but if they want to really be helpful to us, [let them recognize the creation of] an All-Union Society."[22] It was not the attitude of grateful, cowering subjects.

A letter of July 1950 to Kliment Efremovich Voroshilov, one of the deputy chairs of the USSR Council of Ministers, appealed to him to help with a curious and infuriating problem. Although local governments, such as the Kalinin *oblast'* Executive Committee and the Kabardinskii ASSR Council of Ministers, had agreed to help the Society by funding staff positions for the

Society's new branches, "for purely formal reasons" the State Staff Commission repeatedly rejected their requests on VOOP's behalf.[23] Where adequate staffing had been approved, as in Moscow and Sverdlovsk, the Society prospered, argued the VOOP leaders; where permission for a permanent staff had been denied, as in Stalingrad *oblast'*, work went poorly. Even the *oblast'* Executive Committee's deputy chairman, who served as the president of the branch's organizational bureau (in existence since December 1946) did not have an effect. It was calculated that each *oblast'* branch would require about two or three paid staffers. If it did not receive permission to maintain such salaried staff, VOOP warned that it might have to close down its local initiatives. "The Society is not asking for any money from the state budget nor from local budgets," they underscored. "All this will be constructed on the basis of the funds of the Society itself."[24] If the Society chafed under the Party-state's tutorial strictures that hemmed in citizen activism, it used every available opportunity to mount a challenge to them.

VOKS

One of the most interesting aspects of VOOP's increasingly besieged institutional existence was the fate of its ties with similar foreign organizations. Ties with the Poles had been maintained since 1930, when the Łódż Center for Natural Science had publicized Soviet nature protection activities in its *Czasopismo Przyrodnicze.* During World War II, much of Poland was transformed into a killing field, and, as far as can be judged from the archives, all contacts with the Russians were severed for several years. With the conclusion of the war, though, Polish civil society began slowly to rebuild. Again Łódż became a center for the nature protection movement, and it was there that the Polish League for Nature Protection was founded.

In June 1950 this league now sought updated information about the Russian society. Its president, E. M. Potęga, was particularly keen to know about the organizational character of the society and what its relationships were to the *kraeved* and natural science societies of Russia as well as to the trade unions. The Poles also requested information on whether conservation was being taught or promoted in educational institutions, and asked for a package of available literature.[25] VOOP did not directly receive Potęga's letter; rather, it was sent to Makarov from the East European countries division of VOKS, the Soviet agency that supervised all cultural contacts with foreigners.[26] Maintaining the impersonal remove that the times demanded, Makarov obligingly prepared a package of recent numbers of the VOOP journal, the Society's charter, and other materials for the Poles, but sent them all back through VOKS without a personal reply.

Such was the climate of suspicion that even contacts with a like-minded organization of a fraternal ally needed to be kept at arm's length. It goes without saying that contacts with nature protection organizations of neutrals, let alone those of the "Western" bloc were fraught with even greater dangers and complexities. These were well illustrated by the example of the tortuous attempts by Austrian nature protection activists to make contact with the Russians. On January 23, 1951, a package arrived for Makarov from L. Kislova, a member of the board of VOKS. In addition to some German-language journals there was a Russian translation, dated December 11, 1950, of a letter originally in German from the Austro-Soviet Friendship Society. This letter, addressed to VOKS, was itself a reworking of an antecedent letter from the Austrian Nature Protection Movement, which apparently was filed in the Vienna offices of the Austro-Soviet Friendship Society. "In Austria," the Friendship Society's letter began,

> there exists a large and popular movement of a nonpartisan character . . . supported by a significant fraction of teachers, the so-called "Austrian Nature Protection Movement."
>
> They forwarded to us four copies of their journal *Thierpost* for re-forwarding to you and expressed the desire to receive information . . . on wildlife in the Soviet Union for publication in this journal.
>
> We assume that a similar nature protection movement does not exist in the USSR; however, your representative has made mention of such organizations as "Friends of Birds," embracing mostly schoolchildren.
>
> On their advice we ask you, at the first opportunity, to send us photographs . . . and other materials. This could have great significance, not least because American and English propaganda targeted at the Austrian schools has quite diligently and skillfully used such innocent and guileless formats as information about protection of wildlife in their countries.

For this reason, wrote Dr. Otto Langbein, secretary of the Austro-Soviet Friendship League, it would be useful to obtain analogous information about the Soviet Union, "which would be a desirable counterweight to the lying propaganda of the capitalist countries."[27] Aside from herself urging Makarov to send Langbein the desired materials, VOKS's Kislova included the following admonition: the materials were to be sent via VOKS and should be of the kind "not to raise objections."[28]

In contrast with VOOP's response to the Poles, in which the Society—via VOKS—provided (and continued to provide) their Polish counterparts with information and materials though eschewing the political risks of attempting or encouraging direct communication,[29] VOOP sat on its hands respecting VOKS's request that VOOP lend its efforts to Cold War propaganda efforts in Austria. This dawdling—or noncooperation—did not go unnoticed. In a note to Makarov dated June 30, 1951, the acting director of the Central European division of VOKS, T. Solov'eva, reminded Makarov that six months had gone by since the initial request.[30]

Revealingly, VOOP leaders in October 1951 evinced a completely different response to the Austro-Soviet Friendship Society upon receipt of an entirely different kind of communication from them. Through VOKS, the Society received a letter from Helmut Gams, a botanist who was also on the board of the friendship society. Active in the newly created International Union for the Protection of Nature (IUPN), Gams sought Soviet participation in the society, even hinting that there was sentiment for electing the Soviet representative a vice president of the international organization.[31] VOOP had sent in a membership application in July 1948, but lack of permission from the Party had stalled its attempt to join. Now the VOOP leaders hoped that they could turn the Cold War to their advantage. After all, didn't the renewed invitation come from a leader of the Austro-Soviet Friendship Society?[32]

Hobbled by the treacherous international atmosphere and the requirement that all contacts with foreign scientific and activist organizations be handled through VOKS, the VOOP leaders adroitly pursued a nuanced strategy of keeping international links open while remaining aloof from Cold War campaigns to demonize the West. Nature protection, as they tried to present it, was an ideal that transcended class struggle and the rivalry of international blocs; for Soviet scientists it was another instance of their credo that the "International" of world science had moral injunctions that transcended everyday politics.

"On the Sidelines Where Important Tasks Are Concerned"

On August 31, 1950, an article about VOOP, "On the Sidelines Where Important Tasks Are Concerned," appeared in *Kul'tura i zhizn'*. The main charges leveled at the Society were that it had been slow to provide practical help for the "Stalin Plan for the Great Transformation of Nature," that it lacked a "mass character," and that its scientific sections suffered from "academism." The Society's Presidium decided to acknowledge the essential validity of the charges. However, they cunningly blamed these shortcomings on lack of funds, lack of space, their continued frustration in gaining permission to merge with the Green Plantings Society (discussed below), and the refusal of the State Staff Commission of the USSR Council of Ministers to allow VOOP to fund central and local staff positions using its own funds, a right possessed by other voluntary societies.[33] Gauging the real attitudes of the core scientist activists to the prospect of converting VOOP into an authentically "mass" society is difficult. Doubtless the activists were torn between actually becoming effective, which necessitated transforming the Society into a broad-based mass movement, and maintaining the comfortable, clubby haven for the "lost tribe" of the prerevolutionary intelligentsia. Each choice had drawbacks. Even were the scientific activist elite to retain control over a mass organization, should it become *too* effective, that effectiveness could change the regime's perceptions, dislodging activists from the "safe" category of *chudaki* into the more dangerous one of political malcontents. On the other hand, maintaining a smaller membership continually laid the Society open to charges that it had lost its links with "the masses" and with "life itself." True, those charges tended to be raised only at annual review time (*otchët*), but no one knew when such an accusation might be used against the organization. VOOP's solution was to try to keep membership high but heavily stacked with schoolchildren and "juridical members," who could be counted on not to involve themselves in the Society and therefore would not alter the organizational culture.

* * *

Even in this time of crisis for nature protection, some of the old spunk of the Society was evident. Unfortunately, the old activism did not issue in "activity" that could endear the Society to its critics. Recalling the similar letter of 1948, the Society sent a letter to the RSFSR Council of Ministers in February 1951, under the signatures of Dement'ev and Kuznetsov, informing the republic's leaders about reports coming from the Society's members in the Kolyma GULAG of northeast Iakutiia, the northeast part of Siberia. There, in a huge part of the USSR under virtual direct rule by the USSR Ministry of Internal Affairs (the secret police), suppliers of the Dal'stroi and Kolymtorg GULAG empires had been destroying huge numbers of waterfowl annually every year since the end of the war.

> These hunters, without any kind of controls on them, are taking tens of thousands of *molting* ducks on contract with *kolkhozy*. [However], the shooting of molting wildfowl is prohibited by law. . . . The organs of the Ministry of Supplies and the MVD [the Ministry of Internal Affairs] pay no attention to such rapacious and barbaric slaughter of wildfowl. As the citizen hunting inspectors—members of VOOP—tell us, the representatives of the MVD, on business trips to places where game is procured, failed to bring a single charge against anyone for poaching and failed to bring the malefactors to justice.

The Society demanded a ban on the hunting of geese in the Kolyma River delta, where the birds nest and gather for migrations, calling the practice both "predatory" and "illegal." Supplementing the charges of barbarism and poaching, the Society's argument came down to people: if these practices continue, they warned, they could undermine the food base for some of the small peoples of the far northeast, precipitating an ethno-demographic catastrophe.[34]

A Question of Merger

Despite the Society's risky critiques of the environmental practices of the state security system, it had no death wish. Especially after finding itself the target of investigation in July 1949, VOOP's leaders turned to the tried and true strategy of protective coloration. In particular, VOOP leaders intensified their efforts to merge with the All-Russian Society for the Promotion and Protection of Urban Green Plantings (VOSSOGZN), founded in 1947 under the auspices of the RSFSR Ministry of Municipal Services. At least twice in 1950, in January and in September, VOOP tried to revive the merger plans.[35] Deputy Premier Bessonov, who seems to have been given responsibility for VOOP as well as the Main *Zapovednik* Administration, turned out to be maximally supportive. Urging Premier Chernousov to write to the Central Committee in support of VOOP's request, Bessonov in an October 5, 1950 note reminded his chief that in January a decision had been taken by the RSFSR government to do so and that the appearance of the critical newspaper article had made the situation more exigent.[36] Bessonov even prepared a draft of such a letter to the Central Committee for Chernousov's approval and sent a follow-up note with the encouraging data that VOOP's membership as of November 1 had grown to 54,000 adults and 72,000 in the youth section, while the Green Plantings Society included 60,000 members.[37]

In the meantime Chernousov had, through his aides in the "Forestry Group" attached to the RSFSR Council of Ministers, solicited additional opinions and information on the question. From the president of the Green Plantings Society, G. I. Lebedev, who was also director of the horticulture pavilion of the All-Union Agricultural Exposition, he received a brief history and status report on the Green Plantings Society.

The critical juncture in the emergence of that society was Leonid M. Leonov's article in *Izvestiia* in December 1947 "*V zashchitu druga*" (In defense of a friend), which elicited "an unprecedented lively response both in the local press as well as in letters to the editor of *Izvestiia*."[38] Local organizational committees of the Green Plantings Society sprouted everywhere: Kuibyshev, Novosibirsk, Sverdlovsk, Ivanovo, Tula, Kislovodsk, and other places.[39] Like VOOP, the Green Plantings Society ran up against the unyielding barrier of the State Staff Commission, which allowed the society to retain just two paid full-time employees, a deputy president of the organizational bureau and a secretary-typist.[40]

Such a paltry staff could not cope with the massive enrollment into the ranks of the society. In Leningrad, tens of thousands of adults and children joined up, and more than seventy enterprises enrolled as "juridical members" in a three-month span. The dues potential alone was enormous.

However, not having a single full-time paid staff member, the Leningrad organizing committee was forced to cut off the processing of new members. After a "categorical" denial by the State Staff Commission of its request for at least one such worker, the Leningrad organizing committee shut down. In Kuibyshev, the local branch of the State Bank refused to pay out deposited membership dues to the organizing committee on the legal technicality that the committee had no bookkeeper, because the State Staff Commission refused to allow the organizing committee to create such a position. On similar technical grounds financial authorities in Kislovodsk shut down gardening classes run and paid for by the local branch of the society.[41] Only where powerful local politicians ran interference, such as in Northern Ossetia, where the president of the organizing committee was the chair of the Presidium of the Autonomous Republic's Supreme Soviet, was there even a chance of successful operations.[42]

Where allied organizations and affiliates had staff, Lebedev argued, they blossomed. The Society for the Promotion of the Greening of Moscow counted more than 100,000 adult and junior members; it also had twenty-six full-time administrative units. The All-Georgian Society "Friend of the Forest," led by that republic's chair of its Supreme Soviet, V. B. Gogua, and with eleven full-time staff, had more than 800,000 individual members and 4,000 "juridical members."[43] Although Lebedev did not comment on the proposed merger, his report could only strengthen the argument that the mass character of the membership of the Green Plantings Society would be an excellent tonic for that chronic VOOP deficiency, and that the intellectual resources embodied in VOOP's membership would provide a splendid complement to the enthusiasm and numbers of the Green Plantings Society.

A radically different note was struck by A. V. Malinovskii, who was asked by Bessonov to provide an evaluation of VOOP. Not only was he against the merger, he was against the very existence of the Society: "My negative attitude to this Society flows from the fact that, judging by the draft charter, it will not have real possibilities to participate actively in carrying out measures for transforming nature." Conservation, he argued, was already built into the Soviet system, and therefore no special societies or institutions were needed. In the best case, the Society on the initiative of individual members will send up plans for the transformation and protection of nature to agencies and ministries. However, individual Soviet citizens already had that right, and so there was no need for a middleman.[44]

Worse, VOOP's charter "awards many far-fetched and unrealizable functions to the Society." Singled out among these by Malinovskii was VOOP's self-appointed mission to attract broad sectors of the population in support of organizing *zapovedniki* "when that is the prerogative of the Main *Zapovednik* Administration."[45] Another was the Society's declared aim of organizing scientific expeditions studying natural resources in order to identify rational uses for them.[46] A third was VOOP's intention to enlist citizens to check permits for culling of populations of protected species; again, argued Malinovskii, such goals presumptuously usurped the functions of existing state institutions.[47] Malinovskii's bottom line was that VOOP was "distracting a number of specialists away from active participation in measures for the transformation of nature in line with fulfilling the state's economic plan."[48] He opposed the continuation of the Society in any form, merged or not. In that opposition Malinovskii perceptively identified VOOP's insubordinate role in defending and expanding the domain of scientific autonomy from the Party-state. Meanwhile the merger plans remained on hold.

The Nature Almanac of 1951

The continuing crisis of VOOP's publishing operations, however, overshadowed all other matters. On December 23, 1950, at a small meeting in Makarov's apartment, it was revealed that Kuznetsov's story that the almanac had been delayed by Dement'ev (who, Kuznetsov alleged, still had the manuscript), was a brazen lie. Dement'ev stunned the small group by saying he had never laid eyes on the manuscript until now, in Makarov's apartment, where a carbon copy was brought. "I was forced to bother Comrade V. N. Makarov, who was dangerously ill, three times during this whole period," Bel'skii said in despair. "It is my opinion," he concluded, "that the question of this deliberate . . . undermining [of the will of the Society] must be investigated."[49]

S. M. Preobrazhenskii added that "as soon as . . . Makarov got sick, one arbitrary act followed another by S. V. Kuznetsov. . . . [He] refused to take anyone else into account. Such shenanigans are impermissible in a civic organization!"[50] This sentiment was strongly held in the democratically oriented society, as evidenced also by Krivoshapov's admonition to Kuznetsov that "the question of publication . . . is one to be decided by the Presidium and not by a single individual."[51]

The tension between the two ideological camps had become so intense that personal relations even among Presidium members were strained to the breaking point. Kuznetsov was detested by a good percentage of the Presidium. And Molodchikov, whose politics had already alienated him from the Society's mainstream, felt the need to correct a misimpression that he was involved in the sabotage while conceding that his review of the prospective issue was indeed negative. After his review, Molodchikov complained, "Comrade S. M. Preobrazhenskii pretended not to know me and turned away when we ran into each other." This was doubtless only a small portion of the human fallout of the Society's protracted crisis.

In addition to demonstrating that VOOP could keep its commitments to a production plan—thereby deserving to retain and gain expanded publishing rights—publishing was also a matter of honor for the old-guard activists. However, the Society's bookkeeper reminded his colleagues that publishing entailed risks that were at least equally weighty: 46,000 rubles had already been spent on the almanac, and printing and paper costs would run another 125,000 rubles. Moreover, given its tardiness, its distribution and sales might founder; poor sales of the 1948 almanac caused massive losses and served as grist for the investigative mills of the Ministry of State Control.[52]

Using some imagination, Krivoshapov proposed a limited almanac, to cover a half year beginning in June, but the Presidium voted to nullify all editorial changes and print the volume in full as it had been constituted in September. Kuznetsov was isolated with his lone abstention against eight ayes (although Krivoshapov's approval was qualified, calling for a reduced print run).[53] The Presidium had voted to uphold the honor of scientific public opinion.

A meeting of the Presidium in May reopened the thorny question of renaming the Society. A veteran master of "protective coloration," Makarov himself initiated the discussion, suggesting that "the Society may not stand on the sidelines regarding the tasks of transforming nature," and that consequently the Society should be renamed the All-Russian Society for the Promotion of the Transformation and Protection of Nature. (Makarov neglected to say that he had promoted almost the same name change in 1930, but permitted a reversion to "the All-Russian Society for the Protection of Nature" in the late 1930s once scrutiny of the Society's affairs diminished.)

Two old veterans, Protopopov and Preobrazhenskii, had no objections, as they were aware of political and rhetorical expediencies. Others, like the younger Gladkov, objected that whereas many organizations were involved with transforming nature, only VOOP was dedicated to protecting it. "'Protection of nature' should occupy the leading position in the Society's name," he said, supported in this by Krivoshapov.[54]

Sincere transformers of nature, such as Molodchikov, also spoke out in favor of the name change, but not as an exercise in protective coloration. If the Society "cannot move in step with the new demands [of the times]," he warned, "it should be liquidated." P. A. Manteifel' had a slightly different understanding of the matter: "Transformation and preservation are one and the same thing." That is, the only "nature" that will be preserved under Communism is that which has been transformed. Makarov tried to remind Manteifel' that to transform nature intelligently one needed to have a broad base of protected raw materials. If all the vegetation were destroyed, we would use up our original raw materials.[55]

Ultimately, two variants were considered by the Presidium: the All-Russian Society for the Promotion of the Protection and Transformation of Nature, and the All-Russian Society for the Promotion of the Transformation and Protection of Nature. Predictably, the first—putting "protection" ahead of "transformation"—received seven votes to the latter's four.[56] It was around such symbolic and semantic questions that the old-guard activists, frustrated in the real world of politics and public affairs, reaffirmed their values and social identities. In these minor battles they could experience surrogate victories for the success that eluded them in the fights for the *zapovedniki* and the protection of the Moscow green belt. Their rhetorical concessions were always reluctant and ultimately revocable, measured out word by word.

Events, though, continued to outstrip the anxious activists' fears and responses. Perhaps driven beyond the threshold of decency in part by the violent polemics, Kuznetsov, who was also secretary of the Party organization within VOOP, and P. V. Ostashevskii, his deputy, wrote what can only be described as an out-and-out denunciation of Makarov and the old-timers to RSFSR deputy premier Arsenii Mikhailovich Safronov, who was also then occupied in the investigation of the *zapovedniki*. In four typewritten pages, the denunciation raised the most serious political accusations against the guardians of scientific public opinion. Singled out as "undeserving of trust and meriting dismissal" were longtime VOOP recording secretary Susanna Fridman and a former director of publications.[57]

The literature published by the Society was "apolitical," charged Kuznetsov and Ostashevskii, "particularly the nature almanacs." Despite the severe criticism of the almanacs in the press since 1948, Makarov not only "failed to understand the essence of the criticism" but "even defended the . . . almanacs."[58]

From the perspective of Party spirit (*partiinost'*), Makarov, although formally a Party member, was glaringly deficient as well. Although the Party organization numerous times asked him to present a report on the Society's activities to a meeting of the Party organization, he declined to do so, using every excuse in the book and considered such reports . . . "unnecessary."[59] Similarly, despite repeated entreaties both oral and written to call meetings of the Party members of the Presidium in advance of full Presidium meetings so that key issues could be discussed and positions set, Makarov likewise refused to accommodate the Party group.[60]

Kuznetsov exploited Makarov's difficult decision in the late 1940s to have the Society design and produce postcards as an example of how Makarov had turned VOOP into a commercial operation.[61] The age of the leaders of the Presidium and Central Council also testified to Makarov's preference for those "who had fallen behind contemporary realities" and who were unfit for hard work. All of this boiled down to one conclusion: "V. N. Makarov has irrevocably lost his political face. . . . All of the activity of the Society is contained within the narrow confines of nature protection measures, but nothing is done to promote the practical realization of the Stalin Plan for the Great Transformation of Nature."[62] The letter writers sought Safronov's help to reorient the goals of the Society and to remove their opponents.[63]

The political pressures began to mount. That summer in closed session the Party organization of the RSFSR Main *Zapovednik* Administration examined charges against Makarov for his introductory article to the anthology *Zapovedniki Sovetskogo Soiuza*.[64] Makarov's article, it was alleged, failed to outline clearly the basic tasks of the *zapovedniki*—economic and scientific—while it raised esthetic issues, which should be "secondary," to the same level of significance. Other accusations asserted that Makarov went on at "too great a length about the history of the question of nature protection . . . in Tsarist Russia and in capitalist countries" and that his article was pernicious "from the perspective of the Marxist-Leninist worldview"; by highlighting some of the environmental achievements of prerevolutionary and capitalist societies it would "mislead the public."

Those charges were sent to the Party organization by Malinovskii in the form of a written denunciation. This was particularly bad faith in light of Malinovskii's fulsome public praise of Makarov just two months later, in September 1951, when he wrote on Makarov's official employee evaluation (*kharakteristika*):

> Comrade Makarov is a genuine enthusiast in the cause of *zapovedniki*. He enjoys great authority among employees of the system . . . and in scientific public opinion. A number of published works of Comrade Makarov illuminate the tasks and content of the scientific work of *zapovedniki*. . . . Comrade Makarov is politically literate, ideologically solid, and works on raising his intellectual and political level. He takes an active part in the work of the Party organization and for a number of years has served as an unpaid propagandist, leading a discussion circle on the history of the VKP(b) [Communist Party] and is currently leading a circle on the study of dialectical and historical materialism.[65]

Bravely, the Party organization, while slapping Makarov on the wrist, rejected Malinovskii's harshest and most damaging accusations.[66]

Most ominous of all, Romanetskii, who formally was attached to the RSFSR Council of Ministers as head of the group attached to the Expediters' Desk, and Svetlakov, another aide, in late spring had provided a nineteen-page report on the history and current status of VOOP for the RSFSR leadership.[67] The report could not have been more ruinous in its charges and implications of political unreliability. An analysis of the nature almanac for 1950 yielded a verdict that "extreme apoliticism characterized all the articles. . . . Nowhere do we see mentioned the fact that capitalism is incapable not only of organizing planned activity in the transformation of nature but of preventing the rapacious abuse of its resources."[68] The report also discerned a lack of Soviet patriotism.[69]

The nature almanac for 1951 was no better. Authors' attempts to be politically correct landed them in trouble anyway. Thus, when one contributor placed Stalinist science charlatan Ol'ga Borisovna Lepeshinskaia at the head of a list of scientists that also included Lomonosov, Mendeleev, and Miklukho-Maklai, the authors of the report described such a ranking as "incomprehensible."[70]

Each journal/anthology (*sbornik*) issued by the Society was scrutinized for stylistic and political lapses and errors of emphasis. Makarov was savaged for his article in the first anthology, "Nature Protection in the USSR and the Tasks of the Society." "From this article," wrote the investigators, "we may come to the conclusion that [international] priority in the field of nature protection belongs not to Russian scientists and to the Soviet state but to scientific figures of Western Europe and, in the first instance, America"; they concluded that there was a "need immediately to prohibit this society from engaging in the publication of such 'scientific' works."[71]

Capping the charges was their observation that Makarov and VOOP cited the work of I. I. Shmal'gauzen (whom Makarov was forced publicly to denounce in late 1948), the great genetics theorist whose career was virtually ended by Lysenko.[72] It is hardly astonishing that Romanetskii and Svetlakov saw no future for the Society. Their conclusion, to recommend "liquidation" of VOOP, flowed seamlessly from their litany of the Society's political errors and transgressions.

VOOP's leadership was informed about the Romanetskii report by Makarov himself, who had attended the recent meeting chaired by Bessonov at which Romanetskii presented his conclusions. It was not an easy task for the VOOP chief, whose health was beginning seriously to decline. Makarov was, as usual, diplomatic. "The speaker did not mention the positive aspects of our society's activities," he said, "only noting [its] shortcomings and mistakes. The report was tendentious and [I] alone was blamed for everything."[73]

Makarov explained that he did not speak in rebuttal because he was caught unprepared by the vehemence of the attacks and could not speak without supporting materials. Kuznetsov, who was also present, likewise declined to respond. Bessonov as chair was left no choice but to pass along Romanetskii's conclusions: "The Society was not capable of real work [and] did not contribute anything of benefit. . . . The need for the Society for the Protection of Nature has passed and for those reasons the Society must be liquidated."[74] By sometime in July, RSFSR deputy premier Safronov had already drafted a memorandum to Malenkov entitled "On the Liquidation of VOOP."[75]

With the fate of the Society hanging by a thread, Makarov wrote a long memorandum to Premier Chernousov, defending the Society and asking for an opportunity to argue its case before the RSFSR Council of Ministers.[76] At a meeting of the Bureau of the RSFSR Council of Ministers on July 27, with

Bessonov chairing, two decisions were taken: to accommodate Makarov's request for a large meeting and to initiate an investigation into VOOP's finances by the RSFSR Ministry of State Control.[77]

At 1:00 P.M. on August 2, 1951, twenty-five individuals gathered in the offices of Deputy Premier Bessonov. Thirteen represented VOOP, although Kuznetsov, Molodchikov, and Manteifel' could hardly be called "friendly witnesses." The remaining twelve included not only Bessonov, but also P. S. Melikhov of the USSR Ministry of Forestry, Malinovskii, and Romanetskii, no friends of VOOP.[78]

Manteifel' was only partly damaging, on the one hand conceding that Makarov was "not a bad person" and that "the Society was needed," but on the other, repeating his accusation that the Society was out of step with the times. "There is a respectable number of people whom I would call preservationists [*konservatory*]," he said. "But it is time to replace these people, and it is not necessary to close down the Society on their account. That reconstruction . . . the Stalin Plan . . . can never succeed without input from the Society."[79] *Zapovedniki,* analogously, were also needed, but they had to be transformed from "passive" institutions to "laboratories of living nature, in which we study ways of reconstructing nature," he insisted.[80]

Romanetskii again presented his arguments for shutting down the Society: apoliticism, obsolescence, uselessness. This time, however, the activists were not caught off guard. The feisty Krivoshapov immediately responded with a lecture on the importance of an activist public: "In our country the decrees of the Party and government are carried out by attracting the participation of the broad mass of the people. To ignore the existing . . . voluntary societies would be a mistake."[81] Boldly Krivoshapov interrogated the hostile functionary: "You say, Comrade Romanetskii, that the *Nature Almanac* is apolitical. I do not comprehend [your accusation]. In what does its apoliticism consist? Could it only be in the fact that the names of naturalists are set on a par with those of political figures? . . . The question, in the form that Comrade Romanetskii has posed it," he concluded, "should be swept aside."[82]

Dement'ev mounted a surprisingly strong defense of the word *protection.* "This term shouldn't frighten us," he explained, "because there are things in nature that we are unable to produce. We need to protect them."[83] Reminding the government leaders of the practical benefits contributed by voluntary scientific activism, Dement'ev pointedly recalled that the commercial viability of the beaver, moose, and sable was attained only with the central participation of VOOP. In a decree of 1946 there were instructions to carry out systematic censuses of the basic commercial species in the republic. However, the decree did not specify *who* was to carry them out. VOOP processed the data and provided the crucial recommendations for the Main Hunting Administration. If anything, concluded Dement'ev, the Society should be upgraded to all-Union status in appreciation.[84]

Perhaps the most dramatic expression of civic conscience among the defenders of VOOP was uttered by the seventy-year-old Aleksandr Petrovich Protopopov. A manifesto of the ethos of *obshchestvennost'*, its particular power was that it was delivered in the halls of power. "I am a member of the Presidium of the Society," Protopopov began. "This civic position [*eto obshchestvennoe polozhenie*] compels me to speak out with candor before the [government] leadership to which we have been called to give an accounting. I . . . listened to the talk of Comrade Romanetskii with a great feeling of pain. . . . It was written with an intention to smear Comrade Makarov, who is the founder of this cause, and to present . . . fundamental accusations."[85] Protopopov had only derision for the accusation that personal material interests motivated leaders of the Society:

> There are interests, but not of that kind. I have worked for more than twenty-five years. I have an interest only in organizing the public for constructive tasks [*obshchestvennoe stroitel'stvo*], in helping our state to create a well-run economy. . . . You have cut to the quick of the honor of a public activist [*obshchestvennik*]. And I will not permit anyone to smear either myself personally or other members of the Presidium, who have been working honestly for many years, by alleging that we have been bearing all of the burdens of work simply in order to see our names in print. All of our publishing activity . . . is a great cause and our backs are straining under the heavy burden we carry.[86]

For the first time, noted Protopopov, VOOP was being investigated by the Council of Ministers. However, he offered, this was not "an in-depth" effort but a superficial one, "unworthy of the Council of Ministers *apparat*."[87] Protopopov finally tackled the issue of political unreliability among the activists. "There are no conservative elements among us," he declared, perhaps intentionally "misunderstanding" the epithet *konservator*, which was used by VOOP's accusers to denote preservationists. "All of us are people of Soviet ideology, with Marxist training. . . . We do not accept conservatives in our midst and if we discover them, then we remove them ourselves."[88] This was more than a diversionary half-truth. Of course there were *konservatory*—in the sense of nature preservationists—in VOOP, and Protopopov himself was among their leaders. Yet all of these scientists and activists were more or less loyal Soviet citizens (or at least reconciled to Soviet power) and many were even patriots. None overtly promoted political ideologies antagonistic to "Soviet ideology" and a few, including Makarov, were perhaps even sincere Marxists. Nevertheless, Protopopov's statement was *objectively* subversive, for it claimed for citizen activists a sphere of honor, dignity, and autonomy of action that transgressed all boundaries set by the regime.

The deputy premier had the last word. Although a stenogram was recorded, it was Bessonov's prerogative to interpret the consensus of the meeting. Stretching the truth somewhat, he asserted that "all of those comrades who spoke expressed support for the preservation of the Society, proposing a reorganization of [its] work . . . and a change of its name, in order to orient the activity of the Society to serving the interests of the state, and to serving the interests of those measures pursued by the Party and the state with respect to the transformation of nature."[89]

Although most of the comments were "correct and incontestable," Bessonov, particularly in the presence of such operatives as Romanetskii, Shcherbakov, Koz'iakov, Malinovskii, and Melikhov, who either worked directly for Kremlin agencies or whose ultimate loyalties might lie with the Kremlin rather than with the RSFSR government, needed to single out Protopopov's intervention as "incorrect" and even "having an insulting quality." Addressing the aging activist, Bessonov prodded:

> You consider all of the work of the Society to be completely faultless, crystal pure. That is not so. . . . There were scoundrels in the Society who pilfered money. . . . You heaped praise on Comrade Makarov; you said that he was worthy, businesslike. But don't you see? Comrade Makarov also suffers from a whole series of fundamental errors. Comrade Makarov knows what they are and must correct them. I personally regard your observation in connection with this as not completely correct.[90]

Bessonov admonished the Society to take seriously the errors revealed at the meeting and "take very severe steps to avoid repeating them, particularly in the publishing sphere." Closing with self-criticism, Bessonov admitted that he had given the Society too loose a tether and did not press the Presidium to confront some of the "fundamental questions." The next move, he indicated, was the Presidium's, for it would have to develop a plan for the reorganization of the Society and its work.[91] That would give Bessonov political room to help to save the Society.

Five days later, Bessonov summarized the meeting for his colleague A. M. Safronov, another deputy premier. On the archival copy, the key sentence of this document was underlined in red pen, presumably by Safronov: "All who spoke at the meeting . . . spoke out against the liquidation of the Society." Further, Bessonov reported on his instructions to the Presidium of VOOP to submit a reorganization plan to the RSFSR Council of Ministers by August 20, 1951. Finally, he noted that on orders of the Bureau of the RSFSR Council of Ministers, the financial activities of VOOP were still under investigation by the RSFSR Ministry of State Control.[92]

Many of VOOP's fiscal travails had their roots in the unreliable commercial environment in which the Society was forced to operate. As of October 1, 1951, thirty-two prepared manuscripts were awaiting printing, twenty-two of which were physically at the typographers, some of them since 1948. The system provided no consumer protection from the lethally irresponsible sluggishness of the printing houses or the corruption of the State Arbitration Bureau, which ruled against the Society in an important case.[93] Precisely 494,027 rubles and 39 kopecks had been spent on the publications, most of which was unrecoverable. To revive attempts to publish these works, which included two *sborniki* (the journal-anthology *Okhrana prirody*), R. Gekker and V. Varsonof'eva's work on the protection of inanimate nature, works on forest and garden insect pests, gardening tips, and the notorious postcard series, would involve additional expenditures of 200,000–300,000 rubles. Altogether, the works currently stalled could have produced an income of 2,000,000 rubles; even discounted 25 percent, they could have netted the Society 500,000 rubles after taxes. However, the Society had current bank assets of only 110 rubles, with further expected expenditures through the end of the year of 345,000 rubles.[94] P. S. Bel'skii summed up everyone's gloom: "The collapse of the Society is at hand."[95]

Unanticipated and legally dubious taxes levied by the state also contributed to the debacle. Reinforcing the Alice-in-Wonderland nature of the situation, the state had imposed the taxes precisely *because* the literature of the Society was not for sale: because of printing delays, VOOP had lost its tax exemption.

Desperation pushed VOOP's leaders to take some decisive steps. They would petition the USSR Ministry of Finances to return the 25 percent of all profits seized as "taxes," which would net 151,000 rubles. They would temporarily stop all new publication activity. They would ask the branches to eliminate their debt to the Central Executive Council. They would temporarily end payments to lecturers for the organization of exhibits, conferences, meetings, and scientific expeditions. And they would call a plenary meeting to report to the Society's activists in mid-November.[96]

VOOP was reviewed again by the Bureau of the RSFSR Council of Ministers on September 5, this time with the RSFSR minister of state control, N. Vasil'ev, and Premier Chernousov present along with Bessonov. Bessonov and Vasil'ev were charged with preparing a draft resolution for putting VOOP's affairs in order," as Chernousov's order had it.[97]

Soon afterward, Chernousov sent Georgii Malenkov a substantial letter on the situation with VOOP.[98] Although VOOP had failed to attract the masses to its society and had neglected to reorient its work to the transformation of nature, wrote Chernousov, "these shortcomings . . . do not serve as a basis for labeling VOOP a useless and obsolete organization." Indeed, the Russian premier continued, "the RSFSR Council of Ministers considers that the existence of a mass citizens' organization that could render assistance in solving the imposing tasks of the transformation and protection of nature is exceedingly valuable and desirable." Chernousov closed with a request that the Central Committee permit the RSFSR Council of Ministers to continue to oversee the reorganization of VOOP and to allow a congress of the Society to be held in Moscow in October 1951 for that purpose.[99]

Chernousov was again occupied by VOOP's problems when the Bureau of the RSFSR Council of Ministers held its fifty-fourth session of the year on October 10. With Makarov present as well as Deputy Premier Safronov, N. Savitskii from the RSFSR Ministry of State Control presented the report on VOOP's fiscal troubles. State Control, working with Safronov, was to prepare a draft decree for action by the government within a five-day period.[100] By October 13, the draft, "On the Illegal and Fiscally Improvident Disbursements of Funds in VOOP," was prepared, undergoing a slight change to replace the Main *Zapovednik* Administration, now upgraded to all-Union status, with the RSFSR Ministry of Forestry as the agency charged with supervising the Society.[101]

On October 26, the Bureau of the RSFSR Council of Ministers considered the draft decree. Still running interference for the Society, Chernousov, it appears, made the final changes himself. Stricken from the text of the decree at that meeting were clauses that called for Makarov's removal as the acting president of the Society (although he was blamed for its plight) and assigning the future supervision of VOOP to the RSFSR Ministry of Forestry. A last instruction to the Society was to identify those culpable for mismanagement and to take appropriate measures against them.[102]

Published on October 31,[103] the decree was discussed at a meeting of VOOP's Presidium on November 13 and again on December 19, as members sought to gird the Society for another year of trial. Gurgen Artashesovich Avetisian, appointed to chair the commission to investigate the "failings" of the Society, was now again thrust into the limelight. Makarov, in declining health, urgently appealed to the Presidium to select a second deputy president of the Society. Nominated by Dement'ev, seconded by Makarov, and supported by the unlikely duo of Protopopov and Kuznetsov, Avetisian was elected unanimously.[104]

A litany of the Society's mistakes and oversights with a generous dollop of contrived self-accusations, Avetisian's report was a long one. Its length assured that it would satisfy the recondite requirements of the Soviet political game of self-criticism. All the Society's dirty linen was aired: the nonparticipation of Presidium members; the feud between Makarov and Kuznetsov over the past year, which interfered with normal activity; the choice of some authors and artists on the basis of nepotism; an oblique reference to an overpayment to one of Makarov's sons for some artwork; and a problem with the formal office of president (Tsitsin temporarily resigned in 1950–1951 owing to illness, and had not been active before and after his temporary resignation). The main problem, though, was the failure of the Society to involve itself with the great transformation of nature.

The showdown came at the Presidium meeting of December 19, chaired by Varsonof'eva. First to speak was Avetisian, who presented the VOOP commission's report. Although Makarov had seconded Avetisian's nomination for deputy president at the previous meeting, the entomologist hardly returned the favor. His recommendations included suing Makarov's son, among others, for the overpayment made to him, and removing Makarov as deputy president. They also included changing the name of the Society, asking Tsitsin to return to an active role as president, appointing Dement'ev acting president, assigning greater personal responsibility to individual Presidium members for specific functions, and petitioning the USSR and RSFSR governments for renewal of the Society's tax exemption, a change in staffing rules, and other pressing needs. Miraculously, Kuznetsov emerged almost unscathed.[105]

Vera Aleksandrovna Varsonof'eva was among those stunned by the singling out of Makarov. "How did it happen," she asked, "that only Makarov has ended up bearing the full responsibility?" She called for the entire Presidium to shoulder the burden of responsibility. However, someone else immediately rejoined that "the entire Presidium cannot be permitted to step down, because that would be seen as a [political] demonstration against the Council of Ministers."[106]

The Society was caught on the horns of a dilemma. To remove Makarov from the leadership would be a surrender of the treasured autonomy of their citizens' movement. Worse, it constituted an affront to the central values of the activists' fiercely defended independent social identity: honor and loyalty to friends and colleagues even under the pressure of regime threats or blandishments. Not to remove Makarov, conversely, seemed to the activists to imperil the very survival of the Society, one of the handful of institutions remaining in the Soviet Union in which these values of civic activism and autonomy could be affirmed, expressed, nurtured, and propagated.

The ever-cautious Dement'ev counseled that "the Council of Ministers is not interested in subjective reasons [for what happened]. We must furnish a solution for putting our future work in order." P. P. Smolin added that "to acknowledge that we were not up to the job and to resign en masse is not an option. . . . We have to go the path of sacrificing individual members."[107]

Clearly upset by these options, Krivoshapov, who was on the commission, tried to find a way to reconcile the moral imperatives of *obshchestvennost'* with the Society's survival: "The government has entrusted [our] organization itself with finding the people responsible for allowing the violations and to put forward its own recommendations." Agreeing with Makarov, he noted that "the Society cannot rely on membership dues alone for its survival." For that reason, Makarov's emphasis on developing sales of publications as a source of income was fully understandable. Moreover, Kuznetsov kept the Presidium in the dark as to the actual mechanics of the contracts that were concluded with the artists and compositors, so that even if Makarov's son had been overpaid it was the responsibility of the staff to place the issue before the Presidium in a timely way. "The *apparat* let us down. They undermined the whole Presidium." Therefore, he announced,

> It is impermissible to single out only one person to walk the plank. The great work done by V. N. Makarov for the Society is a matter of common knowledge. Yet, we are bound to carry out the decree of the Government. V. N. Makarov should not remain as president; we need to name a fresh face to that office. But the Presidium and the *apparat* are also responsible [for the Society's problems]. V. N. Makarov's reputation remains untarnished. The entire Presidium should accept full responsibility for everything and S. V. Kuznetsov . . . bears no less responsibility.

Now recommending that the Society attempt to salvage its sense of honor by spreading the responsibility broadly, even while removing Makarov as acting president, Krivoshapov more than anyone personally embodied the tortured contradictions experienced by the old guard.

If removing Makarov was a political human sacrifice to save the Society, removing Kuznetsov was a fully deserved punishment for betrayal and ethical malfeasance. Noting that "the moral responsibility of a civic activist is different from that of a white-collar worker receiving a salary for his work," contrasting the cases of Makarov and Kuznetsov, respectively, Krivoshapov had no trouble concluding that the paid secretary, Kuznetsov, was guilty at least of gross negligence and ought to be removed without regret.[108] Where Makarov's blunders were committed out of his sense of dedication to the Society, Kuznetsov's were the result of *khalatnost'* (total irresponsibility) and ignorance, if not active ill will toward the movement that paid his salary.

Another old-timer who was distraught at the choices confronting the Society was Protopopov. He proposed what he believed was another, marginally adequate moral compromise. Because three officers handled the finances of the Society, those three—Makarov, Kuznetsov, and Dement'ev—should pay the political price demanded by the authorities: removal from their positions as officers. Clearly pained, Protopopov apologetically concluded: "We have all said our piece and we must come to some resolution. We feel extremely awkward regarding Vasilii Nikitich Makarov. But I believe that the three . . . should resign their offices." Seeking to retain a shred of honor, Protopopov proposed retaining Makarov as deputy president, even while removing him as acting president.[109] Curiously, Avetisian was able to impose his own resolution of the problem over the attempts by Protopopov and Krivoshapov to spread the blame more broadly. This was likely a consequence of his recent election as the Society's deputy president.

Betraying his completely different mentality, Kuznetsov bluntly offered that "our society is not a parliament. We are not required to put in our resignations. We need to identify concretely those responsible. . . . I do not reject my responsibility. You propose to fire Kuznetsov as a person ignorant of nature. But I am really unsuitable in your eyes because I helped to uncover deficiencies in the work of the Society."[110] Both views were correct, and they overlapped more than Kuznetsov realized. Kuznetsov's inability to empathize with the ethos of the Society, which included a love of nature, led him to uncover "deficiencies" in the Society's work. These "deficiencies" stemmed from VOOP's desperate attempts to protect and promote its members' own vision of human society and of environmental responsibility. And that represented an active divergence from the Stalinist vision of utilitarian transformation of nature and society, a vision that Kuznetsov fully shared, and which led him to blow the whistle on this tribe of academics and activists so out of step with official values.

A. V. Mikheev brought in one more discordant element of Kuznetsov's style: in contrast with the Society's openness, Kuznetsov created a climate of fear among the salaried workers. "They fear Kuznetsov like the fire," charged the old activist, "and are afraid to speak out against him."[111] Kuznetsov proved the point by charging that Mikheev was not officially a member in good standing of VOOP any longer, having failed to pay his three-ruble membership dues.[112]

Long silent, Makarov now confronted the awkward and ethically difficult task of responding to his friends and colleagues. "I know," he began, "that I permitted a number of mistakes to be made, but I do not intend to be the scapegoat. We have mixed up the principles of an official state institution with those of a citizens' organization," he continued, referring to what he believed was the unfair demand by the government that he take "political" responsibility for contracts that were concluded by the entire Presidium, and not by him alone. "It is terribly sad that, after fifty years of service, I have come to this end. Morally it is not right to heap all the blame onto one person. The Council of Ministers instructed the Presidium to find others responsible as well, and on this account I am not in agreement with the conclusions of the commission."[113]

Rising to support Makarov was Varsonof'eva who, inter alia, criticized "the formulation of P. P. Smolin that there had to be [human] sacrifices." "We, the entire Presidium, all of us, must bear the responsibility for the work of the Society," she insisted. Acknowledging that Varsonof'eva was right, Smolin quickly qualified his prior statement by adding that he had not intended his words to mean that Makarov alone should shoulder the blame. Indeed, Smolin supported the idea that he should remain deputy president, even while giving up the acting presidency of the Society. But neither should the whole Presidium resign. Smolin felt that the difficult choice of naming a few names was essential in order to save the Society.[114]

Sensing moral censure from the old guard, Dement'ev unexpectedly now offered to resign as first deputy (and currently acting) president. Protopopov wanted to summon a plenary meeting of the Central Council to resolve these thorny problems. The meeting threatened to stall in the quicksand of moral confusion and helpless paralysis.

At this juncture Avetisian revealed his tough and pragmatic political will: "We cannot elect a new Presidium. . . . Our tasks are great and our resources and opportunities are meager. The Council of Ministers insists on the removal of V. N. Makarov and we must submit."[115]

The commission's recommendations were adopted with only one change, but it was a significant one: the additional recommendation that S. V. Kuznetsov be removed as the Society's secretary. It was also decided to call a full meeting of the Central Council and to confirm Dement'ev as acting president. Makarov, who was removed as deputy president, remained on the Presidium and in the Central Council. Nevertheless, with Makarov's release from the post of acting president, the Society's sense of honor had been wounded. Was sacrificing its leader of twenty-two years really the price the Society had to pay to stay in business? Was there a price that might be too great? An authentic crisis of the spirit was gripping VOOP.

CHAPTER EIGHT

Death and Purgatory

When the plenary session of the Central Council of VOOP met on January 24, 1952, no one could remember when the last one was held. Inescapably the *aktiv* (active membership) had to be brought up to date on the recent developments and brought along on the cosmetic and other changes that VOOP was now forced to face. Many were struck by one thing above all else: V. N. Makarov was no longer at the helm. In his place as acting president was Georgii Petrovich Dement'ev (see figure 9), under whose name the invitations to the meeting were sent out. (Kuznetsov was removed as secretary as well but retained his seat on the Presidium, as did Makarov).

Even before any of the reports were read, the meeting agenda became an object of controversy. Protopopov, in his usual feisty style, proposed putting the recall of the entire Presidium up for a vote, while he and A. V. Mikheev both urged another vote to draft uncompromised authoritative figures for the Central Council as that body prepared the Society's general Congress. Geptner, also displaying early initiative and seeking to keep the meeting focused on the biggest strategic questions, succeeded in eliminating an unofficial report on the work of the Auditing Committee.[1]

First on the agenda was Dement'ev's report on VOOP's activities during 1950 and 1951, which included a public reading of the new decree. This was followed by a report from the Society's bookkeeper, who tried to explain VOOP's muddled finances. In view of the tense uncertainty among VOOP stalwarts, the conclusion of the formal presentations opened the floor to an unparalleled and passionate inquiry into the movement's body and soul.

Ushering in the debate was a flurry of hard questions for the presenters, particularly regarding the leadership's disregard for participatory democracy and grassroots opinion. Gladkov led off with the unstated concern on everybody's minds: "How did a change of leadership [suddenly] take place?

Figure 9. Georgii Petrovich Dement'ev (1898–1969).

We are all used to seeing the signature of V. N. Makarov," he said, referring to the meeting announcement letters. Another guardian of internal democracy, Susanna Fridman, wondered why the members of the Central Council were being "ignored" and why that body had not been convened in so long. Someone else pressed for an explanation of why today's meeting was declared a "closed" one; V. P. Galitskii of the Moscow *oblast'* branch even questioned whether there was a provision in the bylaws to hold a closed meeting, adding that the statutory authority of the officers elected by the

1947 Congress had already run out, leaving the actions of *any* official body of the Society legally dubitable. A Congress had to be called immediately.[2]

After offering some clipped and guarded responses, Dement'ev threw open the floor for general comments. First to jump in, Varsonof'eva called for a recall of the entire Presidium. She was followed by Fridman, whose unaffected eloquence was often married to the most independent sentiments. "I have been a member of the Society for twenty-six years," she began. "In the recent past I have pulled away from the Society. The entire history of this Society has flowed through my very person," but now the Society's mood was unrecognizable to her. Fridman was particularly wounded by some of the leadership's unseemly readiness to jettison Makarov. "Why have only two members out of the whole leadership paid for the debacle?" she asked. "In the work of . . . Makarov there were some negative moments," but these were "owing to the gentleness of his character." She admonished the assembled group to remember "that he carried the entire cause of the Society on his shoulders and fought for the cause of nature protection. He is a historical figure. . . . [H]is health is ruined, and if there have been mistakes they were not 'Makarov's' but the entire Presidium's and Central Council's collectively." She reflected on the rights and obligations of civic activists:

> Unquestionably, the Central Council was ignored; they didn't want to summon us. However, we ought to have summoned ourselves. Now, [only] at the insistence of a group of members of the Council we have been convened. The older members of the Presidium and the . . . Council ought to remember that the Society for the Protection of Nature has always held itself to the highest standards. . . . Now the Society is called a "pork barrel" [*kormushka*] and that is true, as there are people who have attached themselves like leeches onto the Society and are helping themselves to things. We have lived to see this picture of shame; it is essential that we replace the entire Presidium.[3]

Relaying comments she had recently heard from former members of the Council, Fridman tried to move the discussion from the technical matters of finances or political expedience to the fundamental question of values: could VOOP survive as the institutional guarantor of the demanding ethos of the scientific intelligentsia? One value Fridman held dear was integrity; another was loyalty. "We failed to shield Makarov," she charged; "they dragged his name in the mud." For that alone the Presidium should have resigned. Now, it was up to the Central Council "to proclaim that the character of V. N. Makarov is unblemished, and the Government will support us on that. We all must be held responsible, and not V. N. Makarov alone."[4]

From the other side of the spectrum, members more in tune with green plantings and pragmatic, if not Stalinist, ideals of transformation of nature were not moved by Fridman's concerns. Well represented in the Moscow *oblast'* branch organization, they had a very different set of complaints, in part directed exactly at the elitism of *nauchnaia obshchestvennost'*. Lakoshchёnkov, a member of the Presidium of the Perovskoe branch, blamed VOOP's current troubles on its failure to become a mass organization. It could have done so in 1947, but it chose "to close in on itself."[5]

Even the *apparat,* bane of the scientist activists, had its defenders. Galitskii, a leader of the Moscow *oblast'* branch and firm member of the "transformist" camp, complained, "There is one line that is being promulgated here, to defend Makarov and to accuse the *apparat.* . . . I recall how Makarov treated the Moscow branch completely improperly. In my view the reason there are no staffs or activity in the branches is that the Presidium and Makarov are at fault. . . . You can't pin the blame on the *apparat.*"[6] Galitskii's intervention immediately provoked an angry commotion.[7]

One of the most poignant rejoinders to Galitskii came from Bel'skii, editor of the nature almanac until secretly dismissed by Dement'ev and Kuznetsov in December. Kuznetsov, Bel'skii insisted, had not only threatened him on the phone and accused him of obstruction; Kuznetsov was leading a faction in the Society, concentrated among its paid staff. "Kuznetsov ran the entire Presidium," charged Bel'skii, calling to mind an analogous secretary and another collective leadership; "we must immediately replace the working *apparat.*"[8]

Another relatively old timer, V. G. Geptner, returned to the question of Makarov, which for many was a point of honor. Echoing Fridman, Geptner averred that "the name of V. N. Makarov will be inscribed in the annals of the history of nature protection," while that of Galitskii, "whose remarks were tactless and impermissible," will nowhere appear in that history. Makarov, who worked "selflessly," though gravely ill, was let down by the Presidium, "while some paid staffer in the Society [Kuznetsov], a member of the Presidium, did not understand his duties. The support staff of the Society took on the contours of one-man rule. S. V. Kuznetsov came [to us] from the army [in which he served for twenty-eight years] and could not refit himself for civic work. . . . The *apparat* of the Society turned out not to have high standards; this applies equally to Kuznetsov himself and to his assistants. The Presidium stands guilty," charged Geptner, who repeated his call for the mass resignation of that body's members at that very meeting.[9]

Geptner's remarks throw an interesting light on the social attitudes of the old-line scientific intelligentsia. It was prepared to work with people of different social backgrounds so long as those others adopted the stringent behavioral and moral codes of the elite. It seems that Geptner and his colleagues always stood ready to suspect vulgar values and amorality in people from plainer social backgrounds. Sadly, the dishonest and vulgar Stalinist workers and retired military folk among the VOOP staff only confirmed these prejudices.

Most sensed that the Society was in the midst of a cultural-ideological crisis. Who would win out—the old guard, the accomodators (like Dement'ev), the small minority of committed transformers of nature (like Molodchikov and Manteifel'), or the cynical *apparatchiki* who hid behind regime rhetoric to create cozy "cash cows" for themselves in the machinery of the Society? Would the Society survive this test?

At the end of the long session a compromise resolution was finally hammered out. Topping the list of items was the resolve to "devote particular attention to the need in every way to broaden the work of the Society on questions of the transformation of nature and rendering assistance to the great construction projects [of Communism]." Accordingly, the Society's official name was changed to the All-Russian Society for the Promotion of the Transformation and Protection of Nature.

On the sensitive question of leadership, the resolution endorsed the removal of Makarov as acting president while retaining him on the Presidium and in the office of alternate deputy president. With admirable tact, especially in light of Dement'ev's less than total popularity (particularly among backers of Makarov), the resolution asked the academician N. V. Tsitsin to return to active leadership of the Society as president. Finally, Kuznetsov was to be replaced as scholarly secretary "by someone more qualified."[10] The resolution expressed readiness to use a legal suit to recover the difference on the "overpayment" Makarov's son and other publishing-related subcontractors received in 1949. Although politically there was probably no way to save Makarov's acting presidency, the revolt of the old guard *aktiv* had let the infirm old veteran retain a shred of his dignity. It had also sent a powerful message to the (largely Communist) *apparat*.

The final order of business was to revamp the leadership of the Society. After all nominations to the Central Council were approved unanimously, the meeting turned to reorganizing the Presidium. As an acting president was still needed, Krivoshapov proposed that Avetisian be named. And after that recommendation and the nomination to elect Chernenko to the Presidium were approved unanimously, the meeting finally ended. Even as Soviet meetings went, this one had been a marathon.[11]

Even as the activist core met to debate its future, the Society continued to preoccupy those in power. In the offices of the *referenty* of the Russian Republic's Council of Ministers, aides conducted analyses of VOOP's current situation and prospects. The membership total of 131,686 as of January 1, 1952 was superficially impressive until the number of adult members—22,718—was isolated. Some branches were unusually successful, such as the Voronezh branch with its 33,199 members or little Kabardinia with 5,825, but these were anomalies that reflected the presence of one or a number of particularly passionate patrons or organizers, often schoolteachers.

But giant Stalingradskaia *oblast'* had only seventy members and Krasnoiarsk only thirty-two. "From this portrait," concluded *referent* N. I. Koz'iakov, "we may conclude that VOOP as a mass organization, resting on the support of a broad network of local chapters and branches, in fact does not exist."[12]

Koz'iakov's recommendations seemed to open the door once again to the Society's liquidation. He urged, first, rejecting VOOP's own proposals for reform; second, leaving open the question of convening a general Congress of the Society (basically a recommendation against holding one); and third, preparing a draft letter to the Central Committee and a draft decree of the RSFSR on the liquidation of VOOP and on the reorganization of its local chapters and branches into local societies for the transformation and protection of nature along the lines of the *kraeved* societies, which were not united into any central organization.[13]

Yet that same day Bessonov wrote in pencil on the bottom of Koz'iakov's memo: "The Society must solve one chief problem: how to create strong organizations at the grass roots and how best to structure them organizationally." Perhaps the RSFSR Council of Minister junior aides such as Koz'iakov believed that their superiors would accuse them of soft-headed liberalism unless they sought the "toughest" recommendations. However, the two most powerful leaders of the Russian Republic, Chernousov and Bessonov, were now protecting VOOP even from their own aides and *referenty*. Determining the motives of Chernousov and Bessonov would require an entire archival exploration of its own. However, one chance archival record indicates that their more solicitous attitudes toward the Society than those of the "center" extended to areas outside of the nature protection movement. On January 25, 1952, coterminous with many of the events just related, Georgii Malenkov received a letter from the Party secretary of the Novgorod *obkom*, A. Fëdorov. Fëdorov was describing a demographic upheaval in his province that was, we know with hindsight, only in its opening phase:

> In connection with the difficult economic situation in the *kolkhozy* of our *oblast'*, a percentage of the *kolkhozniki* . . . have left . . . and now are living in the cities and workers' settlements of the *oblast'*. According to approximate data collected by the Central Statistical Administration of the *oblast'*, in the cities . . . there live more than 35,000 who do not work. This constitutes up to 30 percent of the population of working age living in cities and settlements.

These ex-*kolkhozniki*, noted the Novgorod Party chief, once they receive passports, were no longer subject to mobilization for obligatory agricultural duty, were exempted from the tax on agriculturalists, and enjoyed benefits on the same level as workers, such as rations of meat and other agricultural products. "We consider the situation as it has evolved to be abnormal," wrote Fëdorov, "when this blabbering do-nothing part of the able-bodied population cannot be sent off to cut timber, mine peat, or work in the collective farms. . . . Instead, we often have to siphon off our best workers to join the logging brigades," he complained. Fëdorov sought Malenkov's help in changing the law so that *oblast'* and *raion* authorities would be empowered to mobilize not only agriculturalists but also the urban nonworking population.[14]

The next day Malenkov appointed a committee made up of P. K. Ponomarëv, M. A. Suslov, and N. S. Khrushchëv to examine the question together, a sign of the problem's importance. In the meantime, he assigned the head of the Central Committee's Agricultural Section, A. I. Kozlov, to independently analyze the issues. Finally, he sent copies of the letter to Central Committee secretaries Shvernik, who routinely handled labor matters, and Gorkin, and to Premier Chernousov.[15]

We do not know the opinions of the others, but we do have Chernousov's direct response to Malenkov, sent on February 20, 1952: "The RSFSR Council of Ministers believes that at the present time it is inadvisable to extend the law on labor duties to the nonworking population of cities and workers' settlements. It is possible to attract the population of cities and settlements to logging projects and . . . other work without resorting to changing the existing legislation," the premier concluded. This pronouncement seems to have influenced Kozlov, who on April 4 wrote to Malenkov to express his agreement.[16] On the face of it Chernousov appeared to favor a contractual framework for dealing with ordinary citizens and with organized social groups such as VOOP. Fëdorov's proposal may be regarded as an attempt to extend Stalin's reenserfment of the peasantry according to the old Russian model enshrined in the Law Code of 1649: that there was no statute of limitations for runaway serfs. Accordingly, we may view Chernousov's position as an attempt to halt enserfment of the entire Soviet population, championing instead the old German medieval principle: "City air makes you a free person."

That same week, Malenkov and Kozlov were sorting out a new problem. A group of disgruntled leaders of the Moscow *oblast'* branch of VOOP, led by Central Council members V. Galitskii, N. Podlesnykh, and V. Lakoshchënkov, had sent a denunciatory letter to Malenkov in his capacity as Party secretary. In the usual pattern, Malenkov typed in the margin: "Comrade Kozlov: Please figure this all out and report to me." Although not unusual to Kozlov, such letters were rare within the Russian nature protection movement. Such a thing had happened only once before, when Kuznetsov wrote his notorious letter about Makarov. This latest letter was written by provincial Party members who were not part of the intellectual tradition and social world of the old guard.

At the heart of the letter was a political attack on the old guard. "Over a long period of time," the six signatories wrote,

> the Central Executive Council of VOOP has consisted of one and the same faces. V. N. Makarov has been acting president now for twenty-eight years. . . . There are no representatives of the *raion*-level organizations of Moscow *blast'*. . . . The Central Executive Committee is a closed group of scientists and those whom they approve. They consider themselves specialists on questions of nature protection. However, over the existence of the Society, they have failed to publish one article . . . in *Izvestiia, Pravda, Moskovskaia Pravda, Komsomol'skaia pravda,* etc. . . . Until 1947 only a small circle of people, numbering in the hundreds, knew about the existence of VOOP.[17]

In a twist, the old guard was accused of turning VOOP into its own private "cash cow" (*kormushka*); "under Makarov's wing there emerged a group of lovers of the good life." The letter writers resurrected the charges of "gross violations in disbursements" stated, but not specified, in the report of the RSFSR Ministry of State Control of October 31, 1951.

In a series of vignettes attacking individual members of the Central Council, the authors revived the charges of nepotism against Makarov in connection with his son's contract with the Society for art work. Protopopov was accused of spending 13,500 rubles on the Crimean Commission of the Society, which he headed, without the commission having once ever discussed shelter belts in the Crimea or the Northern Crimean Canal. "Pursuing vain self-promoting aims to make himself personally more popular, he spent 3,000 rubles of common monies on making a newsreel in which he is the central figure."[18]

Referring to the meeting of January 25 just past, the complainants described it as "an unsightly picture of a cliquish family circle. . . . Instead of severe Bolshevik condemnation . . . we heard how members of the Central Executive Council tried to blur the essence of the case and to vindicate . . . Makarov," they charged. The authors of the letter had particularly harsh words for Susanna Fridman, who had had the temerity to assert that whereas in the USSR "nature protection was not on the high level [it should be]," in America, conservation had been led by a president, Theodore Roosevelt. "This slavish groveling before America seemed to us to be, at the least, out of place and strange," they proclaimed. Capping Fridman's errors was her reproach of the Presidium and her defense of Makarov's honor and his position in the Society. Her line of exculpating Makarov was followed by the majority: Geptner, Varsonof'eva, Protopopov, Avetisian, Krivoshapov, Molodchikov, Gladkov, Bel'skii, Mikheev, and others. For that reason, although the Central Council "in words supports the transformation of nature, in actual fact, as before, it clings to nature protection, which has outlived its time and enjoys no popularity among the broad masses of toilers."[19]

"We activists appeal to you with a request," the letter concluded, "to order the Central Executive Council to convene a Congress without delay." At such a Congress, the petitioners hoped, the Society could be cleansed of its "obsolete" leadership and name, to become the All-Russian Society for the Promotion of the Transformation of Nature.[20]

Placed with this letter in the Party archives was a copy of Romanetskii's report of August 1951 to Chernousov recommending liquidation of the Society. The report's presence strongly implies that Malenkov and Kozlov were beginning to follow the misadventures of VOOP with greater interest.[21] Apropos of the denunciation itself, Malenkov received a report from one of Kozlov's aides in the Agricultural Department of the Central Committee on March 22, 1952. Recommending no new action, the aide's memo simply relayed RSFSR deputy premier Bessonov's communication to the aide that Makarov had been removed for unsatisfactory leadership of the Society and noted that the letter writers had been duly informed.[22] Once again, Bessonov to all appearances had massaged the Party secretariat with assurances that VOOP was on the road to institutional and ideological recovery and that extraordinary steps, especially by the center, were not needed.

As official attention to VOOP waned in the spring and summer of 1952, the resignation of Makarov from the Presidium seemed to mark the end of an era for the Society and the movement. Makarov's resignation letter of March 14, read to the entire Presidium on June 3, was formally motivated by reasons of failing health. Additionally, Makarov was named to the newly organized Committee on *Zapovedniki* of the USSR Academy of Sciences and was also serving as scholarly secretary of the Main Expedition for the Establishment of Shelter Belts, even leading a philosophy seminar organized by the Expedition's Party chapter. Here we see the compassionate hand of the academician Vladimir Nikolaevich Sukachëv, who headed the Expedition and provided many a scientist and activist with employment and a safe haven in those years. However, weariness, disillusionment, and a need to vindicate himself could be discerned in Makarov's diplomatic leave-taking:

> The course of my life has been lived within that of the Society, and so this announcement for me was not easy to make. However, there was no other way. Over the course of more than a quarter century in my activity within the Society I have been governed by its interests alone. No one can produce, unless one stoops to conscious distortion, a single fact to substantiate that during my entire period of service to the Society I ever used it to promote any kind of personal interests, much less material ones. I never counted on receiving recognition or encouragement, for I was compensated with a feeling of moral satisfaction, convinced as I was of the usefulness of a rational approach to the natural resources of the Motherland and of the protection of its nature for the people.[23]

It is probably fitting that at that same Presidium meeting, the tradition of "protective coloration" designed by Makarov was reaffirmed with the creation in VOOP of a Section for the Transformation of Nature.[24] Even

Geptner, Protopopov, and Fridman did not stand in the way, but rather greeted the new unit as a necessary evil.[25] To the Society's credit, a warm message to Makarov was composed in the form of an official resolution: "deep regret" was expressed regarding his illness and overworked condition, but there was also a request—that "Comrade Makarov take part in the resolution of particularly important questions and make himself available for consultations . . . in light of his profound and encyclopedic knowledge of questions on the protection and transformation of nature." And there was also "the hope that with the improvement of Comrade Makarov's health, he will again take active part in the activities of the Society."[26] The old sense of honor had not yet completely given way to politics.

Yet the Society knew that it had to take care of politics. By late June, Avetisian and Varsonof'eva had delivered the Society's draft resolution on its internal reform to the Council of Ministers, where it was examined by Koz'iakov and then sent on to Deputy Premier A. M. Safronov. Its main points paralleled the Central Council resolution of January 25:

1. The Society was to be renamed the All-Russian Society for the Promotion of the Transformation and Protection of Nature.
2. Accordingly, a new charter was to be drafted.
3. A Congress was to be set for September 1952.
4. The Society's publishing house and operations were to be restored.
5. VOOP should be exempt from taxation as a nonprofit organization.
6. The Society claimed that its debt had been reduced to 114,000 rubles from 406,000.
7. A suit was brought against Z. A. Fridman, B. V. Makarov, and a third artist for recovery of overpayments.
8. V. N. Makarov and S. V. Kuznetsov were both removed from their positions.
9. A Section on Green Plantings was organized.[27]

Because VOOP had in effect only been given a reprieve on its survival, Koz'iakov's memo also reflected his charge to solicit the opinions of other important agencies in this matter. Chadaev of RSFSR Gosplan expressed the assent of his agency to the change of the Society's name but opposed any independent publishing rights as contradicting the resolution passed on October 31, 1951. Chadaev also demurred on tax exemption and increased staff levels in the central apparatus.[28]

Koz'iakov's own recommendations included postponing any Congress until an investigation could be conducted on the status of the branches and chapters of VOOP. "In order to bring clarity to this matter," he recommended that the commission be headed up by V. Liudinovskii of RSFSR Gosplan, with Zhukov, Shinev, Kutuzov, Dubrovina, Leont'eva, Denisov, and

Avetisian as members, to report to the RSFSR Council of Ministers after three months.[29] The RSFSR Council of Ministers enacted these recommendations the same week.[30]

By September 6, the Liudinovskii Commission had completed most of its investigation, and Avetisian was invited to respond or to send other relevant materials within the week.[31] The report, sent to Chernousov, had been based on a study of nineteen *oblast'* and *krai* branches of VOOP.[32] The conclusions it reached were seriously damaging:

1. The Society "to date has not transformed itself into a mass organization." Of the claimed 150,000 members, 110,000 were in the youth section, and many from the first group failed to pay dues. One glaring but not atypical example was the Moscow *oblast'* organization, with 20,400 youths and 7,900 adult members. Yet dues collected for the first half of 1952 amounted to only 570 rubles.[33]
2. The Central Council, with forty-one members, was elected in 1947 but had only met three times since, with poor attendance. The Presidium, distracted by publishing activities, completely neglected organizational activities in the *oblasts* and *krais*. Moreover, "in its publishing activities the Presidium . . . allowed major errors of an ideological nature" and produced published materials of "low quality," all of which, along with "the grossest financial violations," led the RSFSR Council of Ministers in the decree of October 31, 1951, to prohibit the Society from publishing anything independently.
3. The provincial branches had been run out of provincial capitals or cities by a small group of individuals. What activities they did organize, such as planting a fruit orchard and nursery in Moscow *oblast'*, "do not have a mass character." Mostly, activities boiled down to episodic lectures in natural science.
4. The commission did acknowledge that the provincial branches were inadequately staffed. Yet the Society's functions had increasingly been subsumed under official governmental activities. "As concerns the inculcation of properly understood love for nature in children and youth, that, without a doubt, must rest fully and totally in the hands of the organs of public education and such social organizations as the Komsomol and the Pioneers, with their linked system of measures for socializing the new person into Communist society. Protection of monuments of nature and history must be concentrated in *kraeved* organizations and in state *zapovedniki*."

Like Romanetskii's report of the previous year, Liudinovskii's conclusions were unsparing. "Basing its judgment on all of the above," began the last paragraph, "the commission has been unable to identify functions that may be assigned to the All-Russian Society for the Protection of Nature and for that reason does not consider the Society's continued existence into the future to be advisable. The liquidation of this society, according to the charter now in effect, may be achieved either through a resolution of a Congress of the Society or by a decision of the government."[34]

Only Avetisian dissented from this death penalty. In his dissenting opinion he countered with facts of his own about the participation of almost the entire memberships of the provincial branches in "Day of Forests" and "Bird Day." His most powerful argument, however, spoke to the contention by the Liudinovskii commission that the Society's functions were already being performed by existing government, administrative, and scientific research institutions and agencies. In fact, he said,

> it's exactly the other way around. The unprecedented scope of projects on the transformation of nature carried out by governmental organizations requires that we attract to their side broad groups of the population (scientists, specialists, *kolkhozniki,* workers, and youth) performing *volunteer civilian work* [emphasis in the original] in the cause of the promotion of the transformation of nature and a protective attitude toward existing natural wealth and that which we create. And no state or scientific organization can fulfill the functions of civic activism and public opinion [*obshchestvennost'*].[35]

Avetisian cleverly cited the directives of the Nineteenth Party Congress, held earlier in the year, which "obliged [the Party] to mobilize the broad masses of toilers to fulfill and overfulfill the Five-Year Plan." In light of the drastic reduction in area of the *zapovedniki,* argued Avetisian, VOOP was essential to take up the slack in efforts to protect the flora and fauna of the country. "The Society for the Protection of Nature over twenty-eight years of its activity has, without any expenditure of money by the state, conducted work that was beneficial and needed by the country," he concluded. "It is necessary to preserve the Society and to assist it in reorganizing its work in connection with the new tasks that flow from the Plan for the Great Transformation of Nature."[36]

More bad news emerged, however, in mid-September, when a parallel investigation by the State Trading Inspection of the RSFSR Ministry of Trade into the stores run by VOOP uncovered numerous infractions and even outright criminal activity. "In individual stores, as a result of the absence of oversight on the part of the Presidium of . . . the Society, dishonest people wormed their way in, forging links with speculators and using the Society's outlets for personal gain. Many goods were sold at prices higher than normal, . . . with the proceeds of the sale bypassing the coffers of the Society."[37]

By November 3, when the RSFSR Council of Ministers once again prepared to assess VOOP's viability, there were ample grounds available to shut down the Society if that seemed politically inescapable. Inside the Soviet Union, meanwhile, the political climate had chilled. A new memo of November 3 sent by Koz'iakov to Deputy Premier Maslov summarized both Liudinovskii's conclusions and those of Avetisian, adding that in light of VOOP's failure to become a mass organization Koz'iakov considered "the conclusions of the commission of Comrade Liudinovskii in the main to be correct." However, he recognized that the question was of a "fundamental" nature and therefore asked Maslov to review the materials personally and to summon the members of the commission together with representatives of VOOP to a meeting at the Council of Ministers.[38]

In the meantime, at the VOOP Presidium meeting of October 7, 1952, there was yet another attempt to discover who prevented the publication of the ill-fated nature almanac. An extended report by the chair of the Society's Auditing Committee, Mikhail Aleksandrovich Zablotskii, on the Society's affairs, especially finances, from 1950 to 1952, failed to answer that question, leading Vera Varsonof'eva incredulously to repeat, "Who, contrary to the decision of the Presidium and contrary to common sense, acted . . . to squelch the [almanac]?" Although Zablotskii at first responded that there were still "no clear causes," under questioning from Varsonof'eva, Kuznetsov's explanation, which Zablotskii had dismissed as self-exonerating, emerged as the most compelling. The former secretary had averred that Tsyriul'nikov, a censor, had intervened.[39]

Support for Kuznetsov's story came from an unexpected quarter. Bel'skii, the almanac's original editor, confirmed that the censor had in fact called and approved publication on April 30, 1951, after what he described as a long series of obstacles strewn in the path of publication by Kuznetsov within VOOP. Accordingly, after making some last-minute changes, Bel'skii received official permission to publish and submitted the almanac to the typesetter. However, on May 12 there was another call from Tsyriul'nikov with the instruction: "Temporarily delay the printing of the almanac." On learning of the delay Makarov, according to Bel'skii, went personally to Tsyriul'nikov and then wrote a special letter to the Moscow *oblast'* and City Censors Board (*oblgorlit*) protesting the action of the censor and noting that such an action placed VOOP in critical financial condition.[40]

As far as Bel'skii knew, Makarov never received a reply to his letter. However, the issue was discussed in the Presidium. "Why they banned [the almanac], I don't know in detail," stated Bel'skii. "However, I assume that there was meddling here by the *apparat* of the USSR Council of Ministers. . . . Why they were interested up there in this case" Bel'skii did not know; he only recalled that at the time the authorities were reviewing a good deal of the Society's literature, not just the almanac.[41]

Kuznetsov noted that the *oblgorlit* responded to his inquiries with the abrupt reply: "You are not going to receive any explanation." And with that the investigation of the affair came to another dead end.[42] One thing was clear. At the highest levels of power, the Society had finally appeared on the radar screen.

What the Auditing Committee's report (and ensuing discussion) did bring to light was a whole series of infractions, losses, and malfeasance committed by Kuznetsov and his associates over the previous two-year period: a 141,000-ruble honorarium paid to an author without the essential preliminary review process; the illegal sale of more than seventeen tons of paper to a typographer; overpayments to others, including Dement'ev; the systematic deception of the Presidium and the Central Council regarding the true financial status of the Society; and, despite the presence of a librarian, a severe deterioration of the holdings.[43]

On this last issue Susanna Fridman grew passionate. She recalled a unique giant colored map of the world with all of the protected territories of all countries marked. There was also Ruzskii's enormous prerevolutionary illustrated history of the Belovezhskaia *pushcha,* which had been published in lavish style at the turn of the century, and Zablotskii's works on the reestablishment of the European bison population. Ten boxes of slides—support materials for lectures—had been meticulously collected, as had rare publications of the Chinese Nature Protection Society and the Italian society, of which last VOOP's library had 144 publications. Now, all were missing. "They reflected the activities of the Society," Fridman said, "historical junctures in the work of the Society. Take, for example, the reestablishment of the European bison in the Soviet Union. This accomplishment is unique in the entire world! Where could [Zablotskii's works] have absconded?"[44]

Krivoshapov was unsparing:

> The attempt of Comrade Kuznetsov once again today to present himself as innocent of all these dealings and matters at the very least seems like a ruse to avoid responsibility. . . . This could have happened only because the Presidium and first of all the Auditing Committee did not check up on the work of the *apparat,* entrusting supervision of all matters of the Society to Comrade Kuznetsov. I propose in the interests of bring the work of VOOP back to health to remove . . . Kuznetsov from his position and to hand over the findings of the Auditing Committee with all appropriate materials to the investigative authorities so that they may bring the guilty to justice.[45]

Varsonof'eva was also incensed that the Presidium's trust in Kuznetsov had been betrayed, and, as a result, the existence of the organization was at stake. "I always have a hard time believing in the bad acts committed by people," she said. "One tries in every way to find some [saving] justification for the person, but we cannot do that now because a great cause is going under." With that she called for an abandonment of the "liberal, sentimental" approach that she and the old guard historically professed toward human foibles and seconded Krivoshapov's call to turn over the evidence of wrongdoing to prosecutorial authority.[46] Protopopov strongly endorsed this move, advocating expunging Kuznetsov from VOOP. So did Geptner, who turned to Kuznetsov and said point-blank: "Sergei Vasil'evich, you must remove yourself from the Society. Your doings have finally caught up with you today." However, Geptner also had some harsh words for the Presidium members, whose absenteeism often left important matters to be settled by a few persons or, worse, by the *apparat*.[47]

"Howsoever strange it may seem," interjected Susanna Fridman, "I will now speak as S. V. Kuznetsov's defense lawyer." For her the fault lay in the Presidium's appointment of him—"a military quartermaster or a line officer, I don't know"—in the first place. "Wasn't it at all possible to find someone else to run the affairs of the Society who was even slightly acquainted with the complicated idea of nature protection?" she asked. Kuznetsov simply wanted a cushy position. But the Presidium knew beforehand that he was "illiterate" on the protection of nature, she charged: "What he knows about nature he sees through a window." And now the Society was on the ropes. "That is the way it always is, when the shoemaker bakes pies and the baker stitches shoes." Almost as an afterthought she asked, rhetorically:

> And are the individuals heading up the Main *Zapovednik* Administration really any better? We prided ourselves on the successes in the cause of nature protection, we developed Soviet methods as well as a whole series of special directions [of research], our successes have been noted in the press of many nations and even in the press of our adversaries. But in spite of all that, with one sweep of the pen, without the participation of scientific public opinion [*bez uchastiia nauchnoi obshchestvennosti*], the network of *zapovedniki* was destroyed, and what is left amounts to crumbs.[48]

Fridman was interrupted by Avetisian, who felt that her sudden digression on the *zapovedniki* had sidetracked the meeting from the Auditing Committee's report. But Fridman returned to her main point, which was about appointments, propriety—and the ethos of scientific public opinion:

> I am not speaking as a heartsick woman but am logically assessing matters at hand. A crime was committed when people appointed a person foreign to the idea of the protection of nature and to any form of scientific activity, a simple economic bureaucrat, and now *he* [emphasis added] must answer for that. In my opinion, however, that is not just. I know that I won't achieve anything with my remarks and that I will not rehabilitate Kuznetsov. Even as an administrator he was shoddy; the office was in a bad state and the archive even worse and the library even worse than that![49]

As Fridman saw it, the scientists and activists of VOOP had relinquished control over their own movement, and that was the crime. Nevertheless, it was not completely clear how much freedom VOOP or the Scientific Advisory Council of the Main Administration had really had in naming their secretary or director in late 1949 and early 1950.

Following Avetisian's endorsement of the Auditing Committee report, a unanimous vote approved Kuznetsov's exclusion from the Society, and it was agreed that the audit was to be turned over to the state's investigative bodies.[50] This matter was put to rest in January 1953 when Acting President Avetisian compiled a devastating four-page bill of charges against Kuznetsov.[51]

Only one more immediate threat to the Society remained: the newly constituted state investigative commission on the activities of VOOP chapters and branches in the *oblasts* and *krais.* Avetisian proposed electing a committee of himself, Chaianov, Krivoshapov, Motovilov, Varsonof'eva, Geptner, and Protopopov to draft a detailed memo to the office of the chairman of the RSFSR Council of Ministers, Premier Chernousov. That motion, too, was approved, and the meeting adjourned.[52]

Contrary to appearances, the flurry of investigations was driven not by any desire of the *RSFSR* government to harass VOOP, but by something far more sinister: the unwanted attentions of the Party's Central Committee. As may be seen in a memo from Deputy Premier Vasilii Alekseevich Maslov to the new Russian premier, Aleksandr Mikhailovich Puzanov, who replaced Chernousov on October 20, 1952,[53] the investigations of VOOP and a parallel one of the Green Plantings Society conducted by the RSFSR Ministry of Municipal Services, which similarly called for the liquidation of that society, were conducted on the orders of the Central Committee, stimulated by the denunciations received from Kuznetsov and G. I. Lebedev of the Green Plantings Society (VOSSOGZN).[54] Although the Ministry of Municipal Services report had already called for the liquidation of the Green Plantings Society, Chernousov at an August 27 meeting of the RSFSR Council of Ministers Bureau urged postponing any conclusive action. As a patron of his republic's own voluntary societies, Chernousov's only weapon against the Central Committee was delay. And he (and his successors) used this weapon with consummate skill.

With the Liudinovskii and Ministry of Municipal Services reports in hand and Avetisian's minority report strongly urging reconsideration, Chernousov once again convened the leaders of VOOP, their investigators, and other interested parties. With the Russian premier's guidance a clever compromise was found that fit with VOOP's long-term intentions: both societies would be liquidated but then reincarnated in a merged, new All-Russian Society for the Promotion of the Transformation and Protection of Nature.[55] In a memo from Koz'iakov to Maslov of December 20, the aide now not only urged the deputy premier to accelerate the formal process of merger but to ensure that the local branches of the new society would be provided with adequate staff levels, "since the absence of staff was one of the major reasons for the weak organizational activity" of VOOP and the Green Plantings Society. To assist this process, Koz'iakov attached his draft of a letter to be sent from the new premier Puzanov to Georgii Malenkov explaining the changes to be undertaken.[56]

On the basis of the preparatory work of his aide Koz'iakov, Maslov now sent Puzanov the draft for an official RSFSR Council of Ministers resolution creating the new society,[57] and at a meeting of that body's Bureau on March 4, 1953, Puzanov authorized Avetisian and Lebedev of the two societies to quickly deliver reviews of their societies' activities. At the same time he officially designated Maslov to head a committee (which included Liudinovskii, Chadroshvili, and Avetisian plus three others) to propose an appropriate timetable and course of action.[58]

A merger between VOOP and the Green Plantings Society had been advocated by the leadership of the former because such a merger seemed to provide the endangered nature protection community with a cloak of "protective coloration"; the Green Plantings Society from the first had a more pronounced ideology of transformation of nature. Besides, the Green Plantings Society would bring in, at least on paper, hundreds of thousands of new members—representatives of the "masses"—which would lift the stigma of elitism from the conservation movement. Yet the VOOP leadership counted on remaining the brains of the new organization, whereas the former Green Plantings Society would provide the strong back, muscled arms, and padded pockets, in a relationship not unlike that of the Menshevik Party to the Socialist Revolutionaries in 1917. However, that was a political gamble.

For their part, convinced nature transformers, Stalinists, and opportunists in both organizations were also supporting a merger, betting that they would be able to oust the old guard and inherit the movement's infrastructure. Writing to Maslov, Vasilii Pavlovich Galitskii of the leadership of the conservative Moscow *oblast'* branch of VOOP already on December 30, 1952 presented his wish list for the Presidium of the new organization.[59] Not surprisingly, he called for the retirement of Makarov, Protopopov, Avetisian, Geptner, Krivoshapov, Molodchikov, Varsonof'eva, Fridman, Preobrazhenskii, and others of the old guard.

Although Maslov was receiving input both from the Stalinists and from the old guard,[60] the most important input derived from the Kremlin, for on November 29, 1952, at the behest of the Agricultural Section of the Central Committee, the Party's Central Committee ordered yet another investigation into the societies. This time the Party announced that it would conduct it jointly with the RSFSR Council of Ministers. In other words, there would be oversight from the center.[61]

A new investigation of VOOP was ordered by Maslov on April 16, 1953. The results, presented by his aide Kostoglodov the next day, were highly negative. The Society's Central Council had not yet met once since January 24, 1952, and the decisions taken concerning reregistration of members and convening a conference of representatives of the branches of the Society were not followed up with action. Presidium meetings continued to attract only a disappointing percentage of its members. Of sixteen staff members who were supposed to be at work on April 16, only seven were present. According to the chief bookkeeper, two staffers made appearances only to claim their paychecks.[62]

Kostoglodov also provided Maslov with thumbnail biographies of some of the key members of the Presidium, turning up, in the case of scholarly secretary Ivan Osipovich Chernenko, both a *kulak* background and a Party reprimand in 1935 for hiding it. Repeating the notion that the Society's emphasis on publishing was motivated by financial rewards to individuals for their written output, Kostoglodov saw no justification for the Society's continued operation. Instead, he recommended that the Central Council meet in plenary session to abolish the Society—"self-liquidation."[63]

The other side continued to lobby Maslov, with Avetisian informing the deputy premier that "an organized scientific-citizen-based movement for nature protection has a long history and enjoys widespread participation both here and abroad. At the same moment that the Council of Ministers is deliberating on whether or not our society should continue to exist, in Kiev the Congress of the Ukrainian Society for Nature Protection is opening." Not only were there analogous societies in even more republics, but also in fifty-four nations around the globe, united in the International Union for the Protection of Nature, "which, through VOKS, has invited our society to become a member as well."[64]

For the RSFSR leadership, getting the Central Committee off their backs was the top priority. This they did by soliciting the opinions of the *oblast', krai,* and autonomous republic leaderships within the RSFSR as to whether they supported the continued existence of the two voluntary societies, albeit in merged form. The majority voiced such support.[65] Representative of the responses received was the letter sent by the chairman of the Council of Ministers of the Kabardinian ASSR, who wrote, "In light of the fruitful quality of the work of the Kabardinian branch of VOOP, [our] Council of Ministers considers that under any circumstances it must be preserved."[66]

With the shadow of Stalin's Central Committee looming, the RSFSR leadership's designation of a largely Stalinist, "safe" organizing committee for the new, merged society becomes more readily understandable. The organizing committee for the new society looked more like Galitskii's wish list than the old guard's. Tsitsin, Avetisian, and Krivoshapov represented what was left of the old Presidium, but the strongest and most articulate champions of the intelligentsia's ethos—Varsonof'eva, Fridman, Protopopov, Geptner, and Makarov—were absent. Ranged against them was a potent lineup of Stalinist bureaucrats and convinced nature-transformers: Galitskii, Chadroshvili, Lotsmanov, Gusev, Melekhov, Malinovskii, G. I. Lebedev, Mel'nikov, Manteifel', and three more government figures.[67] With Chernousov gone, VOOP's personal ties to the Russian Republic leadership were again disrupted, and Puzanov took the politically safer route of naming figures who were ideologically compatible with the Stalin Plan for the Great Transformation of Nature. Even including Avetisian in the *orgburo* had become risky: a letter denouncing him and his supporters to the RSFSR Council of Ministers, written by Galitskii in late August 1953, reminded the Russian Republic leaders of accusations against Avetisian in the press in late 1948 for "distorting" the Lysenko line in Soviet biology and the satire directed against him in *Krokodil* in June 1952.[68]

Perhaps delayed by the monumental political events of the spring and summer as well as by the desire to allow the Central Committee's attention to stray from its preoccupation with the fate of such "marginal" organizations, Puzanov took official action on the merger only on July 15, with a decision to postpone any Congress of the new society at least until June 1954.[69] The decree was published only on September 5, 1953, under Puzanov's signature, one day after receiving a letter from Avetisian asking about the cause of the delay and urging haste.[70] Maslov and Puzanov were still writing letters to Khrushchëv and Pospelov in the Central Committee asking permission for the new society to hold its opening congress as late as September 1954. These letters were rebuffed as well. With good reason the Russian Republic leaders did not wish to pester the authorities excessively on this account.[71]

At one of the last meetings of the Presidium of VOOP qua VOOP, on June 2, 1953, an announcement, shocking yet not totally unexpected, opened the gathering: Vasilii Nikitich Makarov had died that very day. Perhaps no other figure better embodied the tragic predicament of the nature protection movement. For a long time, against the odds, Makarov had succeeded in creating, preserving, and sometimes even expanding the institutional home base for Russia's lost tribe of prerevolutionary scientific intellectuals. Though not of that caste by birth or education, he made himself into a member through his contact with "better credentialed" scientific activists in the Soviet period, particularly since he took the helm of VOOP and the *zapovednik* administration in 1930. Though gentle by nature, he also displayed great courage and acumen in making space in Stalin's hostile world for his little countercultural social *zapovednik*. A master of protective coloration, he knew when to engage in it and understood which principles could be temporarily sacrificed for the sake of even more important ones. He compromised in order to protect the last symbolic islands of "purity." His death was tragic because he died thinking that his achievements were being "liquidated." Even his own institutional "children," the Main *Zapovednik* Administration and VOOP, either actively cast him aside or were forced to do so at the risk of their own survival.

No doubt deeply ashamed of its recent betrayal of its longtime chief, the Presidium of VOOP now paid its respects to Makarov's wife, Klavdiia Arsen'evna, and decided to pay for the funeral and corporately to take part in it.[72] In September, at VOOP's last Presidium meeting, the Society made plans to honor Makarov's memory in a more lasting way: a portrait of him was commissioned to hang in the Society's headquarters, and a committee was formed to posthumously publish another edition of his 1947 book *Okhrana prirody v SSSR* (*Nature Protection in the USSR*).[73]

In a letter to Vera Varsonof'eva shortly after Makarov's death, Susanna Fridman struggled to provide a final assessment of her boss and colleague of so many years: "I saw that his trusting nature, his soft character, . . . and even his personal modesty got in his own way and that of the cause. He always was in the shadows, and dragged the Society into the shadows as well, when what was needed was to create a big hubbub."[74] With this observation she came close to identifying the fatal contradiction of the nature protection movement as a voice of scientific public opinion in Stalinist Russia: it wanted to stand for an alternative vision of development but wanted support and acceptance from the system at the same time. Yet even the astute Fridman could not openly say that the problem was at base structural, not a result of less capable movement leaders. "Where are the Borodins, the Talievs, Kozhevnikovs, Komarovs, Smidoviches, Fersmans, and Makarovs [now]?" she wrote to Varsonof'eva in the same letter, complaining of an ongoing degeneration in the quality of movement activists since the days of the founders.[75] Perhaps what isolated Fridman was that for the majority of her fellow movement activists—representatives of scientific public opinion—the modus vivendi worked out by Makarov was "good enough," even if it fell short of her high standards of civic activism.

The All-Russian Society for the Promotion of the Protection of Nature and the Greening of Population Centers

With a new name, the All-Russian Society for the Promotion of the Protection of Nature and the Greening of Population Centers began to order its affairs by the late autumn of 1953. A Presidium of the Organizing Com mittee was elected with a majority drawn from the old guard—by some miraculous coup—with Avetisian assuming leadership of both the plenary Organizing Committee and its Presidium.[76] Of note was the exclusion from the Organizing Committee and the Presidium of G. I. Lebedev, one of the former officers of the Green Plantings Society, probably because of his letter denouncing a number of his opponents in that society and VOOP.[77]

With Galitskii removed, Manteifel' rebuked, and the transformist-oriented Moscow *oblast'* branch of the newly merged society placed in a kind of receivership, the new hybrid society started exploring the political opportunities that were slowly opening up in the Soviet Union.[78] Just as the violent storms of Stalin's last years seemed on the verge of washing all the achievements and "safe houses" of the movement to sea, the tyrant died, and a new weather system, not without its own dangerous irruptions of turbulence, blew in. Some pockets of blue could now be spied amidst the thunderheads.

CHAPTER NINE

VOOP after Stalin

Survival and Decay

In their efforts to engineer a merger with the Green Plantings Society, the new leaders of VOOP were still employing the politics of protective coloration. Stalin's "Plan for the Great Transformation of Nature" was still a watchword in June 1953, and the Society sought to link itself to a political agenda endorsed by the state power. However, protective coloration was a strategy fraught with peril for the integrity of a social movement. It required a convincing outward display of loyalty in some key areas so that a certain internal freedom as well as political freedom of action could be maintained in other areas. It required that the movement project the appearance of a group of quaint, even slightly irrelevant (from a utilitarian Soviet perspective) old-line scientists, more interested in discussing questions of faunal distribution than challenging economic or political decisions, while it quietly defended and expanded its "state within a state"—the *zapovedniki*—or took aim at select individual policies. Such a strategy was effective enough during times of "normal" Stalinism, but even its greatest practitioner, Makarov, had been powerless in the climate of terror of Stalin's last years. If it took a mastermind such as the late Makarov to make protective coloration work in the best of times, what could be expected of his far less gifted successors? Under Avetisian, the strategy inexorably began to overwhelm what it was supposed to protect.

* * *

Gurgen Artashesovich Avetisian's (see figure 10) finest hour was bracketed by his valiant defense of VOOP as the lone dissenting voice on the RSFSR Gosplan Commission of 1952, on the one hand, and the triumphant convocation of the "three societies" *zapovednik* conference of May 1954, on the

Figure 10. Gurgen Artashesovich Avetisian (1905–1984).

other. The influx of pragmatic planters, foresters, and horticulturists into the reorganized VOOP through its merger with the Green Plantings Society, combined with closer monitoring of the Society's activities by the RSFSR Council of Ministers, however, created the preconditions for major shifts in the Society's direction and operations. In an ominous departure from the Society's traditions, even Avetisian himself began to behave in a high-handed manner.[1]

The period 1953–1955 was an interregnum for VOOP as well as for Soviet society as a whole. In contrast to the thrust of the liberalizing changes in Soviet society, however, the interregnum in VOOP ultimately led to the suppression of the autonomous ethos of scientific public opinion within the Society and to its takeover by corrupt Communist time-servers. This new period posed an even greater challenge to the old-timers. Indications mounted that the old traditions were being supplanted by a new approach to doing business. Lush, secretive bureaucratization quickly created a barrier between the new bosses and the old stalwarts. Emblematic of these developments was the way that the All-Union Congress of VOOP was convened in August 1955.

After long delays, a decree of September 5, 1954 of the RSFSR Council of Ministers marked the official inauguration of the new All-Russian Society for the Promotion of the Protection of Nature and the Greening of Population Centers, VOSOPiONP (although I will continue to refer to the Society as VOOP, to which name it reverted in 1959). The old Organizing Committee, composed of members elected to the Presidium of VOOP in 1947 and to the Presidium of the Green Plantings Society in 1951, was replaced by a lean new committee of seven: Avetisian, Dement'ev, Motovilov, Krivoshapov, V. I. Egorov, an agronomist, A. N. Volkov, and N. B. Golovenkov.[2] A new charter was prepared under the guidance of Avetisian, who remained chair of the Society's Organizing Committee pending the convocation of the founding Congress, set for 1955. In the meantime the Society hobbled along in a state of organizational limbo. Eight years had passed since the previous Congress, and VOOP was in severe violation of its charter. The Society had repeatedly petitioned the Party for permission to hold a Congress, which was repeatedly denied. This then served as fodder for Party accusations that the Society was delinquent in upholding its charter provisions.

On June 20, 1955, the Organizing Committee met at last to set the agenda for the Congress, which needed to be submitted not only to the RSFSR Council of Ministers but to the Central Committee of the Party as well. Although activists had proposed focusing on two questions, the ratification of the charter and elections of an official leadership, the Central Committee, which had been consulted beforehand this time, recommended adding speeches concerning fundamental principles and positions that would guide the Society's work. This delicate task was entrusted to the Party group within the Organizing Committee. Reports and talks were provisionally scheduled from Tsitsin, Formozov, Kozhin, the botanist Bazilevskaia, and Egorov.[3]

On July 29 the Organizing Committee finally got word that six days earlier, the RSFSR Council of Ministers had been given permission by the Party to allow VOOP to convene its first conference. At the new campus of Moscow State University on Lenin Hills the government had set aside 100 dorm rooms for visiting delegates. The main event was to be held in the university club. After waiting for the better part of a decade to hold such a meeting, the nature protection activists were given a scant two weeks to get the word out and make their preparations.

Almost furtively, without publicity, the VOOP Congress was convened on August 15. The old-timers were not invited. One remarkable document illuminating the episode is a pained letter from Susanna Fridman, VOOP secretary from the Society's inception until 1947, to interim VOOP president Avetisian: "I was completely shocked by your totally accidental mention of the convocation of the Congress," she opened, charging that she most likely never would have heard about it at all were it not for her unrelated request for other information from Avetisian. Fridman was as much saddened as she was outraged by the slight: "We old veterans of the conservation cause have been waiting for some years now impatiently for just this Congress. We dreamed of meeting one more time, discussing many issues of concern to us, summing things up, and perhaps clasping each other's hands for one last time. Most important, we hoped to pass on our passionate commitment to conservation to the young generation."[4]

However, the Congress was called in mid-August, observed Fridman, exactly at the time when "all scientific researchers are on vacation or on expeditions." To Avetisian's excuse (in his letter to her) that it had been decided not to invite many activists so as to keep costs down, Fridman replied that it was wrong to have slighted veterans and even founders of the Society, many of whom would have paid their own way in any case. Fridman reminded Avetisian of her own decision to turn down a 1,000-ruble award from the Presidium for her work organizing the Society's archive, a decision motivated by her concern for the Society's rickety finances.

Now, five founding members found themselves "thrown overboard" after thirty years of passionate service to the cause. Fridman was especially concerned that her exclusion from the Society's Executive Council would deprive her of an indispensable credential in her continuing efforts to propagandize on behalf of conservation in the media and in society; understandably, she feared that she would be called on to explain why she was no longer a member of the Society's governing body. "Of course," she reproached Avetisian, "had the old guard been present at the Congress, none of this would have happened. Neither Smidovich nor Komarov nor Makarov would have allowed anything like this." On the contrary, they would have proposed that the five living founding members be granted lifetime honorary membership on the Executive Council.

The snubbing of Fridman and the old guard was cause for yet another disappointment. Urged on the previous year by Professor G. G. Bosse, Fridman was at work on a major history of conservation, support of which she had hoped the Congress would provide. She had counted on the official endorsement of VOOP, but now, "to her great sorrow and humiliation," she felt abandoned. Unbowed, she vowed that she and a group of veterans would continue the project, and implicitly raised the prospect of unflattering portrayals of such recent scandals as the Kuznetsov affair and the 1955

Congress. Further, Fridman promised to write to others returning from field trips to let them know of these developments.[5]

The 1955 Congress and Its Aftermath

Russian conservation history is awash in first Congresses. In 1929 there was the First All-Russian Congress of Nature Protection Activists and in 1933 the First All-Union Congress for the Protection of Nature in the USSR. In 1938 the First Congress of the All-Russian Society for Nature Protection (VOOP) was held. And on August 15–17, 1955, the First Congress of the All-Russian Society for the Promotion of Nature Protection and the Greening of Population Centers convened in Moscow. (In June 1995, I might add, the First Russian Congress for Nature Protection was held.)

With its 130 voting delegates and 201 guests, it was a decent-sized affair, overshadowing the intimate 1947 Congress with its fifty-three delegates. Yet, confirming Susanna Fridman's worst fears, its atmosphere was alien to all previous Congresses. Only three of the eleven members of the Presidium—Avetisian, Dement'ev, and Professor P. A. Polozhentsev of Voronezh—could be called old-timers; the recognizable hearts and souls of the movement—Varsonof'eva, Formozov, Geptner, Fridman, Nasimovich, Zablotskii, Protopopov, Gekker—were all absent. Their places were taken by gardeners, selectioners, and presidents of provincial chapters.

Making the keynote address was Nikolai Nikolaevich Bespalov, one of Puzanov's deputy premiers who now, it seemed, bore principal responsibility for the fate of the movement. Bespalov's remarks seemed to be a continuation of the generally supportive attitude of the Russian Federation leadership toward nature protection: "Our people rightfully demand not only comfortable and attractive housing but beautifully laid out parks and gardens, and residential quarters, streets, and courtyards luxuriating in greenery and flowers. . . . Thus far we have only a handful of cities and population centers that meet these demands," and despite the annual investment of 500,000,000 rubles in urban landscaping and greening, ultimate success would depend on mobilizing the army of citizen amateur gardeners and nature lovers.

> That is why the Congress . . . is so important. We must hope that it will facilitate the transformation of nature protection and . . . greening into a truly mass movement. In this cause we must not limit ourselves to government decrees although they are, of course, necessary. Agitation and propaganda are of paramount importance, as is upbringing and explanatory work, particularly among youths and schoolchildren. This is one of the principal tasks of the Society. . . . [Despite the fact] that nature protection and greening have been engaged in here for a long time, we cannot describe the existing situation as

> favorable. Rapacious attitudes toward nature and green plantings are not rarities at the present time. . . . We must say bluntly that the local soviets until now have paid little attention to nature protection and green plantings. Especially here is where citizens' organizations must mobilize the attention of the population. Citizens' oversight over the proper use of natural resources must occupy a great place in the work of the Society and its branches. . . . We should also recall that nature protection and greening pursue a variety of aims, not only economic but also cultural and esthetic.

Bespalov concluded by noting that eight years had passed since the 1947 Congress, during which time "the economic and cultural needs of the country had greatly increased." That, in turn, demanded "a decisive mobilization in the area of nature protection and greening, and, in particular, a mobilization of the work of the Society."[6]

The most arresting and disquieting moment came near the close of the gathering, when one of the few old-timers present, Professor Pëtr Artem'evich Polozhentsev of Voronezh, took the floor. "It is awkward to express [my] feelings and impressions," he reflected, at first talking around the subject. "I have in mind the absence at this Congress of the distinguished activist for the protection of nature Comrade Makarov, who has left us forever."[7] Echoing Susanna Fridman's letter to Avetisian, Polozhentsev now alluded to the other "absent presence" at the Congress—the *living* activists from among the generation of founders who were not in attendance. Polozhentsev's remarks revealed an incipient perception that an era in the life of the Society had ended and that VOOP, now VOSOPiONP, had fallen into the hands of "new people":

> I wanted to recall the enthusiasts of nature protection, by force of whose efforts our society not only managed to survive but also to have the opportunity to convene this present Congress. Are those present here aware that our Society was on the brink of obliteration? It is with a feeling of gratitude that I now recognize the following comrades: [F. N.] Petrov, Avetisian, Dement'ev, Motovilov, Varsonof'eva, Bazilevskaia, Krivoshapov, Protopopov, and others. It is also necessary to name those comrades who, working in the [Society's] paid staff, also maintained their support, such as Golovenkov and others.

Polozhentsev made a point of thanking some of the "newer" defenders of VOOP—Avetisian, Dement'ev, Golovenkov—in recognition that political realities would never again permit the control of VOOP by the Society's founders. With those thanks came the tenuous hope that the Avetisians and Dement'evs would be able to hold the line against the Volkovs, Egorovs, Manteifel's, Malinovskiis, and other convinced or cynical transformers of nature. "There are those who are trying to accuse the Organizing Committee of poor preparation for the Congress," Polozhentsev concluded, in a final attempt to lend support to Avetisian. Implicitly he seemed to recognize that the convocation on short notice, the unpropitious selection of the season in which to hold it, and the disturbing omissions of the founders may well have been out of Avetisian's hands, the decisions of a higher authority:

> Many of the biggest defects of the Congress, though, were scarcely in the competence of the Organizing Committee. But [even so] our Congress is taking place in a marvelous building and those of us who traveled here [from afar] were able to find housing with no difficulty. And that is all the work of the Organizing Committee, . . . [work performed] particularly under those conditions, when our voice is barely heard by those who should be encouraging us in our work. [Instead], they should say "Thank you, comrades, for your love of nature, for your efforts to enrich and to beautify our Motherland."[8]

An indication of the new order within the Society was quickly revealed in the report of the Charter Editing Commission, headed by Vasilii Vasil'evich Prokof'ev of the "Znanie" society. Noting that there had been a number of suggestions for the best possible name of the new society, including "Society of Friends of Nature" and "Society for the Transformation of Nature," Prokof'ev explained that the name was already a moot point insofar as the RSFSR Council of Ministers insisted on the existing cumbersome formulation "because it believes that the Society must chiefly orient itself toward population centers." Their view, he continued, was that "the Society must not take upon itself broad responsibility for the fulfillment of governmental measures," perhaps an allusion to the former VOOP's energetic and autonomous initiatives in the creation of *zapovedniki* and in the enforcement of anti-poaching laws.[9]

With dues set at three rubles for full adult members and fifty kopecks for youths, the Congress completed its work by electing a new Central Council of thirty members through secret ballot. Of the core group of old-timers, only Krivoshapov, Polozhentsev, and Formozov, elected in his absence, were now represented, along with second-generation members Avetisian and Dement'ev. With a clear majority, the "new people" were in the driver's seat.[10]

The first session of the Society's newly elected Executive Committee, which met on August 19 at the conclusion of the Congress, is one of the defining moments in the history of the merged society. The presiding officer of the August 19 session was not a member of the movement at all, but the same N. N. Bespalov, a deputy prime minister of the RSFSR, who had given the keynote address at the Congress. That in itself was highly unusual at a meeting of a voluntary society with VOOP's traditions. Then, announcing the order of business, which was the election of the Society's new president and vice presidents, Bespalov let it be known that the Society's pretensions to autonomy were a thing of the past. "Having weighed the various possible candidates," Bespalov Solomonically pronounced, "we have inescapably decided to recommend as president of the Central Executive

Council of the Society G. P. Motovilov," the former USSR minister of forestry. The vote was unanimous.

Nikolai Vasil'evich Eliseev, a veterinarian and head of the Russian Federation's new Main Administration for Hunting and *Zapovedniki,* was unanimously elected first vice president. Aleksandr Nikolaevich Volkov, head of the Moscow Plant Protection Station and president of the Moscow *oblast'* branch of the Society, was elected as the other vice president. In this coronation of bureaucrats there was one small jarring note when Ivan Stepanovich Krivoshapov, one of the few old-timers left on the new council, proposed the candidacy of Nina Aleksandrovna Bazilevskaia instead of Volkov, offering that there should be at least one biologist on the Presidium. However, these were new times, and objections from the new claque of careerists forced a hasty withdrawal of the botanist's candidacy. Only Nikolai Borisovich Golovenkov, the scholarly secretary of the Society, was reelected.

Elections to the remaining five slots on the Presidium were similarly conducted under conditions of guided democracy. Avetisian was left on, presumably as a courtesy, and Krivoshapov, Tsitsin, and Dement'ev were named as well; their appointment gave the Presidium a veneer of legitimacy. The remaining choice was the hack Vasilii Ivanovich Egorov, deputy inspector of the RSFSR Ministry of Agriculture's Division of Gardens, Viticulture, Subtropical Crops, and Teas. However, this compromise did not please the extreme anti-academic utilitarian wing, which demanded expansion of the Presidium to include at least one representative of the urban greening group. Another compromise was struck; the Presidium was expanded by two, and the "greener" Aleksandr Filippovich Lukash was elected together with Professor Nikolai Ivanovich Kozhin, a representative of the fishing industry. Two Executive Council members abstained from the vote on Lukash, but they, too, were clearly out of step.[11]

As we seek to understand episodes like these in the absence of full archival documentation, we must always keep in mind the temper of the times. When the republics were faced with repeated assaults on their authority and raids on their portfolios of responsibilities, they tried to defend as much as they could. To a great extent this stance explains the patronage and solicitude of the Russian Republic's government toward the *Russian* conservation movement and the *Russian zapovedniki* when they fell under attack. Although far from liberal, the leadership of the RSFSR played a crucial role in protecting Russia's version of civil society from obliteration. That the RSFSR leadership was willing to defend VOOP in the first place no doubt had something to do with its perception of nature protection as a low-risk issue. It could take a stand, implicitly defending its sense of its own importance in the bargain, without the likelihood of being purged.

With Stalin's death, though, the pressure on the republics from the center eased. It was time to frame new compromises and to blunt the edges of conflict. Patronage of even a remotely dissident conservation movement became counterproductive under the new conditions of rapprochement with Khrushchëv's team. Although the Russian Republic never gave up the goal of restoring its *zapovedniki* and even maintained a certain respect for the old-line elite biologists who had led the conservation movement, it could not allow them to remain in control of a growing organization such as VOOP. Elite biologists could work in subsidiary roles in the RSFSR's Main Administration for Hunting and *Zapovedniki* under politically reliable bureaucrats, but they would never again be allowed to occupy highly visible positions, which only attracted the near-fatal attention of the center to them and to their patrons in the republic's leadership.

The Lakoshchënkov Affair

History occasionally is the story of surprising reversals. Romanetskii, the police bureaucrat who participated centrally in the persecution of the conservation movement, emerged four years later as a naive idealist whose outrage at the Party's abuse of power in the environmental area led him to confront Khrushchëv himself. Another example of how one man's behavior evolved from craven denunciation under Stalin to outspoken resistance under Khrushchëv is the case of Vsevolod Georgievich Lakoshchënkov.

In 1950 Lakoshchënkov was one of several members of the Moscow *oblast'* branch of VOOP who signed a letter denouncing the Society's old guard for promoting corruption and stagnation. The charges were wildly exaggerated and distorted—part of a campaign to remove the independent-minded leadership of the Society and to replace it with a more pliant and loyal Stalinist cadre. Nonetheless, these tactics were partially successful, resulting in the forced resignation of Makarov in 1952 and the eventual take-over of VOOP by Party hacks between 1953 and 1956. Although they failed to eliminate the autonomous, oppositional conservation movement, which migrated to the protection of the Moscow Society of Naturalists, the Party loyalists inherited the expanding machinery of the conservation society.

Lakoshchënkov was a local activist whose star initially rose with the ouster of the Makarov group. Beginning in 1948 he had served on the Presidium and as secretary of the VOOP branch of the town of Perovo, a Moscow suburb. From January 1954 through December 1956 he was a member of the Auditing Commission of the Moscow Regional branch of VOOP, along with V. S. Iukhno, director of the Prioksko-Terrasnyi *zapovednik* and president of the Serpukhov branch, and I. P. Kosinets, secretary of the Leninskii regional branch.

The commission met in October 1954, but the extreme disorganization of the financial records moved the commission to declare that it could not conduct a coherent audit. Although Moscow VOOP branch president A. N. Volkov and the branch's bookkeeper proposed that the commission return in April 1955, by which time the documents were to be put in order, the commission resolved instead to conduct an immediate investigation into possible malfeasance.[12]

The scholarly secretary, S. V. Butygin, had been wearing not one hat, but five, dispersing credits, serving as cashier, and acting as bookkeeper and safekeeper besides. The cash transactions that crossed his desk bypassed the Society's bank account and were therefore never officially recorded. Chaos also reigned in other matters. No membership lists were kept by the regional branches. A close associate of Volkov's, one Korshunova, had been hired as bookkeeper; she simply sat in the office and took the work home to her husband, who was a bookkeeper.

All of this impropriety, Lakoshchënkov alleged in his letter to V. M. Molotov (now USSR minister of state control), was intentional. Preying on Butygin's weakness for alcohol and his illness, as well as his dedication to the Society, Volkov had put the scholarly secretary in an untenable position. Deprived of honest, skilled bookkeeping support staff, Butygin soon was over his head as he struggled to take over those functions in addition to his normal organizational ones. In order to balance the available cash with receipts Butygin at one point had pitched in 2,800 rubles of his own money.

Complications multiplied at a meeting of the VOOP Executive Council on November 12, 1954. Despite the absence of a report from the Auditing Commission, Volkov blamed the messy books on Butygin, whose removal he now demanded. He also demanded the exclusion from the council of P. P. Smolin for his "bungling" of "Bird Day," of M. G. Groshikov for "inactivity," and of P. A. Manteifel' on account of his overcommitted work calendar. Volkov's agenda was not simply to rout the old-line professors and field naturalists. He wanted the field cleared for an even more radical conversion of the Society. Volkov sought to remake VOOP into a profitable business.[13] True, the Society would promote a little greening here and there, but that was all beside the point. The point was profit, and that is why Volkov needed to retire even such personally honest philosophical supporters of the "transformation of nature" as Manteifel'.

As early as mid 1954 Volkov began to assemble his confederacy of wheeler-dealers. P. A. Petriaev was brought on board as scholarly secretary, with two contracts for 5,250 rubles total for "research" and an additional payment of 2,850 rubles for undocumented "lectures" on behalf of VOOP to sweeten the deal. Other Volkov allies, such as P. V. Tsibin, V. S. Iukhno, I. P. Kosinets, and Zemering, head of the Mytishchi regional branch, were also brought into the Presidium.

Commercial activity immediately assumed two lines of action. One was the purchase and resale of DDT for profit by the Moscow regional branch.

Apparently, no financial documents were kept of the transactions within the Society; information and documents bearing on the purchase did turn up in a search of other agencies' files. Nevertheless, testimony was received that the VOOP branch in Mytishchi sold the DDT at more than 200 percent of the average price. Perhaps more shocking was the second commercial operation, which got under way in October 1954. More than any other scandal, it exemplified the ethical rot that accompanied the ouster of the old-timers by the new group. Tsibin and Iukhno were the ringleaders in a scam to uproot 12,000 eight- to ten-year-old linden trees from the Priokskо-Terrasnyi *zapovednik* and resell them for huge sums to interested parties, including the Moscow Telephone Construction Trust. Not only were the trees being illegally pillaged, on the sly, from a nature reserve, but the operation was being masterminded by the reserve's own director, Viacheslav Stepanovich Iukhno.[14] Volkov and his people managed to combine Stalin and Lysenko's development philosophy with the moral vision of the Mafia.

Other unsavory characters were brought in to round out the commercial operation, which by 1955 involved the "sale" of 8,223 trees fetching hundreds of thousands of rubles.[15] Whereas the tree removals commenced in April 1955, official permission for the operation was retroactively provided in May and October by the Main Administration for *Zapovedniki*. Malinovskii himself signed on to the scam. On August 13, the Presidium of the Moscow branch of VOOP awarded V. S. Iukhno 1,000 rubles as a bonus for his successful commercial transaction.

Two more audits were held in 1955, the first conducted by Lakoshchënkov and the auditing bookkeeper, F. K. Alëkhin. Its results, published December 31, 1955, were described as "slanderous" by Volkov, deputy president V. K. Alekseev, and the other regional Presidium members. Then the Moscow *oblast'* Party Committee's Agricultural Sector ordered a second audit. Gagarin, deputy head of the sector, even went so far as to recommend that branch president Volkov not remain involved in the linden tree business, speaking at the Second Moscow *oblast'* Conference of VOOP in 1956.[16] However, the composition of the auditing commission gave one pause; Iukhno, Alekseev, Kosinets, and Butorin (of the All-Union VOOP)—precisely those under the cloud of suspicion—formed its majority.[17] Seeking to explain Lakoshchënkov's absence from the commission, its members asserted that he "declined to serve, giving the excuse that he would be away on business . . . until April 12, 1956." According to Lakoshchënkov's own letter to Molotov, he refused to serve because of his strong objections to the participation of the officials responsible for the alleged abuses.

At the Second Moscow *oblast'* Conference of VOOP where the Party official warned regional VOOP leaders to abandon the tree sales, Lakoshchënkov and Alëkhin were expelled from the Moscow Regional branch of VOOP "for slanderous activities within the Society."[18] The vote was a disheartening 146 to 2.[19]

Repeating essentially the same charges in a letter to Soviet premier Nikolai A. Bulganin written in early July 1957, Lakoshchënkov added an arresting note of emotionality to his appeal for vindication.[20] "You know perfectly well," it opened, "that there is a limit to the amount of pressure that a person can tolerate, and a limit to the social and personal sufferings that the heart is able to bear, especially the heart of a seventy-five-year-old man." Referring to his letter to Molotov, which he enclosed, Lakoshchënkov pointedly accused "the Communists A. N. Volkov . . . , V. K. Alekseev, and G. P. Motovilov, president of VOOP," of a massive cover-up, "denying everything" and "declaring war on all who criticized their improper actions." For Lakoshchënkov, the issue had now expanded from financial and resource-related abuses to the highly political question of Communists' abuses of power:

> They are using their experience and their bureaucratic positions in their struggle against me. . . . Most troubling is that no one has stood up to their attempt to quash criticism. In their actions they, as Communists, have ceased to relate to [us] in a personal, individual, and human way; decency is a basic law of human culture. Are we not Soviet people, even if that fact is unpleasant for Volkov, Alekseev, and Motovilov? As such, we too have the right to a certain amount of respect and the right to defend our dignity. They just do not seem to understand that elementary rule, and, despite my appeals, no one else has yet pointed out their errors to them, either.[21]

The fate of Lakoshchënkov's appeals closely parallels those of other naïve missives of the Khrushchëv and Brezhnev periods. Molotov himself probably never saw the first letter, which was forwarded by his Bureau of Complaints to G. P. Motovilov, president of VOOP. Maintaining the stonewalling, the May 15, 1957 reply drafted by VOOP secretary V. V. Strokov was scathingly dismissive and reaffirmed the decision of the Second Moscow *oblast'* Conference of VOOP of 1956 expelling Lakoshchënkov and three others for "defaming Communist citizen activists."[22]

Lakoshchënkov appealed to Molotov and Bulganin for reinstatement and declared his readiness to submit to a trial over whether his accusations constituted slander. Instead, a hearing was held by the Presidium of the national VOOP, now incensed that Lakoshchënkov would turn whistle-blower. In a confrontation with the N. V. Eliseev, vice president of VOOP and head of the RSFSR's Main Administration for Hunting and *Zapovedniki,* Lakoshchënkov reaffirmed his readiness for a slander trial. Eliseev rebuked Lakoshchënkov for turning to outsiders "with misinformation" instead of to the national Society's Executive Council; apparently, at Lakoshchënkov's prompting, reporters from *Literaturnaia gazeta* even called VOOP and asked why an old and dedicated member of the Society was unjustly expelled and why his complaints were being shunted aside. It was all exceedingly embarrassing and nasty.

VOOP's vigilant trustees in the Russian Republic felt obliged to respond. On November 19, 1957, Motovilov and Eliseev were called in to the office of Deputy Premier Bespalov to discuss the fate of the Society. Although tightening trusteeship over the Society represented an additional burden for the Republic's leadership, no alternative was seen. Bespalov would take overall responsibility for VOOP himself, while an aide, Semikoz, of the Agricultural Section of the RSFSR Council of Ministers, would handle day-to-day affairs.[23]

With time, however, there was a broad "normalization" of the internal workings of the Society; nothing remotely resembling an internal critique against the new line was to be heard within VOOP, and the state trusteeship was lifted after a few months. For its part, the RSFSR was glad to get this responsibility off its hands. Nikita Khrushchëv was implementing his notorious plan to create putatively self-contained economic regions to replace the system of branch ministries, and the republics had few bureaucratic resources to spare for such low-priority items as voluntary societies. Accordingly, VOOP was now free to pursue its new agenda unhindered, or so it seemed. The lush bureaucratization and commercialization of VOOP swung into full gear. The Lakoshchënkov flap highlighted the degree to which VOOP had become unrecognizably different from what it had been only five years earlier.

1955–1960

Of all the indicators that a new ethos had taken hold in VOOP none was more vivid than the proliferation of a network of profit-oriented commercial outlets—the *Priroda* (Nature) stores. This chain of stores required a large amount of start-up capital from the parent society, as provincial conservation-entrepreneurs all tried to get in on the act. In July 1956, the Leningrad City branch of VOOP asked the central leadership for a loan of 100,000 rubles to be repaid by January 1, 1957. It was approved.[24]

Another emblem of the new approach was an indiscriminate campaign to recruit new members. This increasingly involved the induction of so-called "juridical members," entire factories or schools, for example, that joined as institutions. During the discussions of the budget for VOOP for the coming year at a Presidium meeting of January 19, 1956, Vice President Volkov proposed a cut in the publishing expenditures of the Society and a revved-up membership drive instead. Egorov, seconding this, proposed no less than a 50 percent increase in membership, to 300,000.[25] President Motovilov concurred, adding only that the Society should further recruit 200,000 additional Young Naturalists, for a grand total of 500,000.

The new line also demanded leadership even more in tune with its bureaucratic-entrepreneurial goals. Only a year and a half after the imposition of a new leadership, Motovilov was complaining that, of all the Presidium members, only Volkov, Krivoshapov, and V. V. Strokov, the new secretary who replaced Golovenkov, were satisfactory.[26] Kozhin and Avetisian were denounced as ineffectual deadbeats, and they soon left the leadership.[27] The interregnum was over.

The Annual Report on VOOP Activities for 1957

In his presentation of the annual report, A. N. Volkov, deputy president, made the customary complaints about insufficient funds and organizational shortcomings. But internal factors were not the only impediments to the Society's meeting its goals. Despite the relatively small number of individuals involved, the defection of the old-timers to the Moscow Society of Naturalists (MOIP) posed a perceptible threat to VOOP's claim to represent nature protection.

Susanna Fridman, in a letter to Vera Varsonof'eva written late in 1958, again throws light on the deep wound this loss of a social "home" caused her and the old guard. Commenting on the departure of the "exiles," as she termed the old-timers, Fridman ventured that they should not have left so quietly: "It was absolutely necessary to have written an 'acerbic' letter to the new Presidium concerning our departure. . . . Our whole group should have signed such a letter; let the document remain as testimony in the Society's archives. It is too easy simply to beat a retreat. I would have typed up a letter and sent it to the newspapers."[28] Fridman also informed Varsonof'eva that she had saved an old postcard from Grigorii Aleksandrovich Kozhevnikov recommending her for membership in MOIP. In asking Varsonof'eva to admit her to membership, Fridman confessed that she "could not bring anything useful to the Society." Nevertheless, in her last months she only "wanted to be alongside you [Varsonof'eva] and Aleksandr Petrovich [Protopopov]."[29] Better evidence for the poignant place of their societies in the hearts and souls of nature protection activists would be hard to come by.

Although the defection of the Makarov-era activists was almost inevitable, given the changes in VOOP from 1952 on, Volkov had underestimated their mettle; it was difficult for hacks to grasp the intensity of the old-timers' commitment to their values and their capacity for autonomous organization.

"It is entirely incomprehensible to me," admitted Volkov, "how conservation work has been going recently. I don't like the intrusions of MOIP [into our area]," he continued.

> The Moscow Society of Naturalists is a respected organization, but MOIP is convening a conference on *zapovednik* problems, has called a conference on conservation problems generally, that is, MOIP has gotten involved in those issues which are the province of our Society. And we are not concerning ourselves with those issues that we should concern ourselves with. [Conservation] is not the prerogative of MOIP, but a group of activists has appeared there and they are not performing badly.

At that point, a voice from the hall dared to state the obvious: "Those are our former activists!" "Right you are!" concurred Volkov, who added wistfully that "they are moving ahead while we are standing on the sidelines, . . . not only not initiating [these conferences] but not even taking part." That left the field open to the elite biologists, who, "at these conferences, dump on us, as a Society, without compunction."[30]

Cleansed of *nauchnaia obshchestvennost'*, the Society now sought to rejoin the international conservation movement. This time, domestic obstacles were significantly reduced. Khrushchëv's foreign policy emphasized reintegrating the Soviet Union—in a managed way—into the world's economic and diplomatic systems. And the VOOP leadership was now composed exclusively of dependable Communists or those close to the Party. Accordingly, on August 11, 1958, the Presidium sent a memorandum to the Central Committee of the Communist Party of the Soviet Union (CPSU) asking permission to join the International Union for the Conservation of Nature and Natural Resources, which was affiliated with UNESCO, and to attend its conference in Athens and Delphi. Professor N. A. Gladkov of MGU, a Stalin Prize laureate, would represent the Society.[31] A delegation of three, headed by Gladkov, had attended the Twelfth International Ornithological Congress in Helsinki in March 1958.[32]

The new leadership had put particular emphasis on building membership. By 1958 membership was up to 242,624, an increase over the previous year of 100,000; still, it included only 80,261 adults.[33] In response a number of strategies were advanced. One emphasis was to attract more juridical members, which now numbered 1,106: another was to lure individual members with contests and prizes.[34]

By 1959, when the Society's Second Congress convened in Moscow, one year late, membership had swelled to 916,000.[35] The staffs (including both the center and the affiliates) had grown commensurately from 24 paid staffers in 1956 to 306 in 1959, and consisted of bookkeepers, scholarly secretaries, clerks, typists, and instructors/lecturers.[36] If disbursements, including staff salaries, climbed in this four-year period, income also rose, from

1,559,500 rubles in 1956 to 3,584,900 in the first nine months of 1959.[37] Of this, membership dues accounted for only 245,294 rubles, or 7 percent of all income. Despite this impressive growth, it was calculated that to break even the Society would need 9 million members (3 million adults and 6 million youths); even the lucrative *Priroda* stores and the postcard, album, and literature sales could not generate enough profit to keep the operation growing.[38]

A breakdown of the delegates by age, length of membership in the Society, education, and Party membership told the story of the restructuring of VOOP. Of the 316 voting delegates, plus 37 with consultatory status who attended, three-quarters were Party members (243), Komsomols (8), or Pioneers, with only 101 non-Party delegates. The loss of the old guard was even more dramatically highlighted in the tiny number (14) of scientists with degrees of *kandidat nauki* or higher or with the title of professor. Finally, those who had been in VOOP prior to 1954, when it merged with the Green Plantings Society, constituted less than 25 percent of the delegation (81 in all).[39]

Much of the discussion at the Congress, therefore, was tame or even trite. There was much talk of gardening techniques, which pesticide to use on orchards, new hybrid flower varieties, and other horticultural issues. Some of the livelier moments concerned how to make VOOP a financially viable operation. Nature protection was almost an afterthought. Nevertheless, a few voices still reminded the Society of its ostensible mission. One of them was that of Vera Aleksandrovna Varsonof'eva (see figure 11), one of the Society's oldest members and vice president of the "competition," the Moscow Society of Naturalists (MOIP).

As one who "began working in the Society . . . in the first years of Soviet power," Varsonof'eva sought to claim Lenin's endorsement for a stouthearted stance for nature protection. "V. I. Lenin understood well," she asserted, "that with the development of the young socialist state a colossal exploitation of natural resources would be required," but he also knew that "for proper exploitation it was essential to understand all the complicated interrelationships that exist among elements of the landscape . . . .On this realization was based that grand scientific program . . . that was pursued in the *zapovedniki*. VOOP, in its original form, participated broadly in the scientific work." she continued, in an implicit rebuke to the new direction of the Society.[40]

However, now was hardly the time to slacken one's vigilance. She noted that in ten years almost 21 percent of the Carpathian forests of Ukraine had been cut, and the woodlands would last only another ten to fifteen years at that pace. In Siberia, the Siberian stone pine (*Pinus sibirica—kedr,* or "cedar," in the Russian vernacular) was disappearing, while pollution was engulfing more and more formerly pristine rivers and lakes, such as the Chusovaia

Figure 11. Vera Aleksandrovna Varsonof'eva (1889–1976). Vladimir Nikolaevich Sukachëv (1880–1967) is seated at right.

River in the Urals. Part of the problem was that planners and bureaucrats failed to consult with scientists, and the results were not only pollution but disastrous agronomic-engineering schemes such as that which was leading to the desiccation of Lake Sevan. "One would think," she remonstrated, "that the Conservation Society would put precisely this kind of problem at the top of its list of priorities. For this reason it is wrong to view the two questions—of nature protection and of urban greening—as equally pressing. The question of urban greening is linked with that of human health and it is doubtless important." However "it is ill-considered to view it as equal in importance to the urgent and great problem of nature protection."[41]

Varsonof'eva explained that preserving nature's "untouched baseline territories" was not for the sake of an abstract Nature but for living people, and not simply for material well-being but for a more transcendent aspect of human existence: the "restoration of the moral forces of the human being." "We must preserve standards [*etalony*] of the beautiful age-old nature of our Motherland," she continued, "and there, where life forces us to alter its visage, we must not leave a defaced, deformed wasteland. We must pass on to our descendants monuments of nature in their original beauty. . . . The most urgent task of our society is—the protection of nature."[42]

A Leningrad delegate, Georgii Ivanovich Rodionenko, was more direct:

> I would like to pose the following question to the members of the Central Council. Have they raised even *one* problem of national scope, such as the fate of Lakes Sevan or Baikal or of a large *zapovednik?* Nothing was uttered about these problems either in the [official] report or in the other announcements. It seems to me that we must elect to the new Central Council, in addition to those who are adept at organizational work, specialists with a broad field of vision. Without their help it will be difficult to raise questions having national import.[43]

One speaker, V. V. Tarchevskii, a delegate from Sverdlovsk *oblast'*, raised the relatively new problem of air pollution. Cheliabinsk made the problem not only visible but inescapable. "Over all the cities in Sverdlovsk *oblast'*," said the delegate, "and there are 101 of them, lie permanent clouds of smoke. The atmosphere is polluted with toxic wastes dangerous to human beings. For this reason the question of the protection of individual elements of nature, especially the atmosphere, is extremely urgent."[44] Painting a ghastly picture of cities in the Urals surrounded by "deserts of life . . . for dozens of kilometers out from the city perimeters, where there is no vegetation," Tarchevskii complained that already in the *oblast'*s third largest city, Kamensk-Ural'skii, "it is impossible to breathe" owing to the waste belched forth from the monster Urals Aluminum Smelting Plant. He described clouds of asbestos and enormous, exposed waste dumps in the city of Asbest. The *oblast'* branch of VOOP sought to plant them over, but what was really required was a massive national campaign to rehabilitate mined-out and degraded land and, especially, to clean the air.[45]

Another delegate, from Astrakhan', informed the Congress about bacterially contaminated rivers of her *oblast'*, and the writers Oleg Pisarzhevskii and E. N. Permitin cautioned that socialism ipso facto did not guarantee "safe" industrial working conditions.[46]

Despite these few brave words, the activities of the Congress displayed a monumental complacency, reflected in the election of the new president and Presidium. The Russian Republic minister of forestry, Mikhail Mikhailovich Bochkarëv, was selected to lead the Society for the next three years, while the politically reliable Andrei Grigor'evich Bannikov, a mediocre zoologist but regime loyalist, was elected first vice president. Nikolai Vasil'evich Eliseev was named, more or less ex officio, as was past president Motovilov. The only pre-1955 faces were those of Avetisian and Gladkov, who were unlikely to oppose the further commercialization of the Society.

The archive contains Vera Varsonof'eva's secret written ballot for the Central Council of the Society. Fifty-five names were listed as candidates for the Council and fifty-five individuals were ultimately elected to that body. But

Varsonof'eva only placed approving check marks next to twelve, not counting herself—the only real old-timers. Indeed, the only Presidium members she considered voting for were Gladkov and Avetisian.[47]

A few of the Society's publications did address some of the major issues of environmental ruin. An article in the Society's journal *Okhrana prirody i ozelenenie* (*Nature Protection and Greening*) was remarkably candid about the extent and location of water pollution in the USSR and even identified some point sources with descriptions and amounts of their effluents.[48] A much more extensive brochure, authored by the botanist G. G. Bosse and the population geneticist and ecologist Aleksei Vladimirovich Iablokov and designed to coach the Society's lecturers, underscored the problems of biotic conservation that were largely ignored at the conference while also defending the aesthetic side of nature protection as an expression of patriotism.[49] However, these were the rare exceptions to the flood of pamphlets about gladiolus varieties, ornamental trees, and new pesticides for apple orchards.

Although politically, morally, and intellectually stagnant, the Society grew like topsy. By 1962 its membership had ballooned to nine million. VOOP—in 1959 it had regained its old name—had not only become the largest nature protection society in the world, but also one of the largest non-state businesses in the Soviet Union.

CHAPTER TEN

Resurrection

A highly unusual conference on the nature reserves was convened in the spring of 1954 by three voluntary societies, MOIP, VOOP, and the Moscow branch of the Geographical Society of the USSR (MGO). The *zapovednik* conference is a watershed in the history of the Russian and Soviet conservation movements for a number of reasons. First, with almost geological force it thrust up the seething, formerly self-censored passions of the scientific intelligentsia to the surface of public life: its anger, its sense of wounded dignity, its unrelenting claim to a decisive role in public policy, its bitterness at the expropriation of "its" archipelago of freedom—the *zapovedniki,* its disdain of the values and utilities of the Stalinist bureaucracy, and its unrequited patriotism.

Second, the conference ushered in a period of ascendancy in the movement's history of the Moscow Society of Naturalists and the Moscow branch of the Geographical Society, and with it, a new, highly visible place for geographers and geologists. This occurred against a backdrop of chaos and a leadership interregnum in VOOP, so recently rocked by financial difficulties, dissension, regime persecution, and Makarov's retirement and death.

Third, even while the majority of the old activists were still determined to restore the status quo ante and hence equated nature protection with the *zapovedniki,* new voices were heard at the conference and new concerns were tentatively expressed that prefigured a broader agenda for the movement: the issues of pollution and of resource management outside the reserves system.

* * *

Malinovskii's new Main Administration quickly began to reorient the scientific work in the rump *zapovednik* system. The main orientation became developing means of "increasing" nature's productivity. For example, at the Voronezh reserve, work intensified on replacing the "unproductive" aspen forest with other tree species.[1]

Of the total area of 1,328,700 hectares that remained in the twenty-eight reserves of Malinovskii's system, twenty-three reserves with an aggregate area of 915,600 hectares contained forests, of which actual forest cover accounted for 671,500 hectares. Of these, 58 percent were characterized by Malinovskii as very mature or old-growth and another 13 percent as mature.[2] Although forestry measures in twelve of the twenty-three forested reserves were limited to fire control and anti-poaching measures, in the other eleven, some of which were still quite large, "forestry measures . . . were being conducted in full measure."[3] "Full measure" included such "biotechnical means" as clearing trees from the enclosed bison range in the Belovezhskaia *pushcha* in order to plant new forest browse, or clearing black elm in the floodplain of the Usman' River, which had been impeding the growth of willows, the tree of choice for the local beaver.[4] Malinovskii's vision of the role and function of *zapovedniki* mirrored that of the Stalinist theorists of the 1930s—Arkhipov, Boitsov, Veitsman—who saw the reserves as experimental areas to create the lush, superproductive "Communist" nature of the future.[5]

Biotechnics

"Biotechnics," the technical means to achieve a "reconstruction" of "first nature," embraced an array of disparate measures: predator control, pesticide application, the introduction/acclimatization of exotic species of plants and animals, supplementary feeding and the provision of salt licks, and the removal of existing vegetation in favor of another species mix. Motivated by a single-mindedly economic yardstick of benefit, measured in currently identified resources, this reconstruction of nature was also wedded to a voluntaristic perception of existing nature as backward, unplanned, and not having reached its productive potential. Soviet biotechnics would correct all that.

Ever since the first epic battles over acclimatization, especially those fought at the 1929 and 1933 nature protection congresses, biotechnics had acquired intense symbolic meaning for the two opposing sides. This was particularly the case regarding proposals to carry out acclimatization and other biotechnical measures in the *zapovedniki.* For the Stalinist nature-transformation enthusiasts, these measures were weapons in their war against the prerevolutionary, indeed counterrevolutionary, inertia of old Russia. No community—neither human nor ecological—would be permitted to stand aloof from the complete refashioning of one sixth of the earth's surface into a gleaming, rationally planned socialist commune. Nothing would be allowed to "go its own way."

For the nature protection activists acclimatization meant much the same thing, but the proposed "great transformation" elicited not enthusiasm but horror, disdain, and ultimately resistance. Acclimatization, especially in *zapovedniki,* threatened the last little islands of "inviolability," beauty, and purity in the swirling and profane sea of Stalinist changes. True, the threat was ecological—portending the spread of parasites and the transformation of acclimatized species into pests and public menaces. But it was also symbolic, marking the intrusion of the Party-state and its machinery into the last holdout of the scientific intelligentsia. It was a struggle over whether there was to be any kind of "geography of hope" in the Soviet Union.

Makarov and Smidovich in the 1930s decided rhetorically to capitulate to the nature-transformers, renouncing the principle of the "inviolability of the *zapovedniki*" and admitting the permissibility in those reserves of "biotechnical measures." Yet their concession was an exercise in protective coloration, and Makarov tried to limit the actual implementation of many of these nature-transformation schemes for the *zapovedniki* to the best of his ability.

Nevertheless, it was politically impossible to stay aloof from some high-profile campaigns. Regarding the acclimatization of the muskrat, the raccoon-dog, the sika deer, and some other large game animals (especially ungulates), there was almost no choice. Although the elimination of wolves and other large predators in the reserves had greater internal support, particularly if the reserves also harbored endangered herbivores, there too the authorities exerted uncontestable pressure. All the while, campaigns against the wolf were raging outside the reserves; the last family of wolves was exterminated in the Okskii *zapovednik* in 1954 and in the Voronezhskii in 1955. And, depending on the reserve, foxes, bobcats, wolverines, bears, cormorants, seagulls, marsh hawks and other hawks, and owls also found themselves at the wrong end of a gun.[6]

Acclimatization intensified under Malinovskii, although not every attempt resulted in a thriving population (and consequently a biotic disruption). Nine musk deer were released in Denezhkin kamen' in the Urals, for example, but by 1959 all the animals were dead. "The results of the acclimatization of the sika in *zapovedniki* were various. However, in all cases where the deer survived, regular winter feeding and other biotechnical measures were maintained," wrote Filonov. Ironically, when some of the reserves to which the sika had been acclimatized, such as the Buzulukskii bor and Kuibyshevskii *zapovedniki,* were liquidated in 1951, the unforgiving hand of natural selection also carried off the deer.[7] The only successful sika introduction was in the Khopërskii *zapovednik,* where twenty-seven animals released in 1972 grew to 1,800 by 1977.[8] More "successful" attempts involved the raccoon-dog (*Nyctereutes procyonides*), the muskrat, and the American mink, but much of these efforts were now largely conducted outside of *zapovedniki,* in forest plantations or areas designated for legal hunting.[9]

Other measures also continued or were expanded. Hay mowing for supplemental feeding or for feeding the Mordvinian *zapovednik*'s own stock from 1954 through 1967 reached 9.4 tons per year, and in Il'menskii (from 1937 to 1960) averaged about 24 tons annually. Twig bunches (*veniki*) were collected on a massive scale in some reserves, such as the Mordvinian, Il'menskii, and Okskii.[10]

But Malinovskii's beloved preoccupation in his new reserve system was forest management. From this perspective, ungulates were as much a pest to be eliminated as an economic amenity to be promoted. In quite a few *zapovedniki* such as the Crimean and Voronezhskii, deer were shot or, after 1952, captured and relocated.[11]

In reserves that had been "liquidated" and turned over to the USSR Ministry of Forestry, commercial logging soon began. There were important exceptions, such as the reserves of Lithuania, now classified as "watershed" forests protected from lumbering, and areas where the commercial potential was particularly low. The former Troitskii forest-steppe *zapovednik* constituted such an area, and it had the relative good fortune to be handed over to Perm' State University, which rechristened the territory an "Instructional-Experimental Forest Plantation." Here, the supportive local *oblast'* Executive Committee declared the area a *zakaznik* (a protected territory established usually for a period of five or ten years) until 1961, with all economic activities or alterations of the natural conditions prohibited. Thus, with the connivance of the local political authorities, Troitskii de facto remained a *zapovednik,* but now of Perm' University. Perhaps the most visible change was in the kinds of research pursued. More emphasis was placed on developing strategies for pest control, reclamation of salt pans and salt meadows through targeted afforestation with appropriate tree species, and studying the relationship between tree species and soil chemistry. Basic research continued to be pursued vigorously as well.[12]

The contrast between *zapovednik* management in the pre- and post-Malinovskii eras, although significant, has perhaps become exaggerated in the memories and perceptions of partisans of nature protection. True, acclimatization and predator control were conducted as protective coloration under duress during the Makarov years, whereas Malinovskii promoted those policies with enthusiasm. Yet the ecological consequences of acclimatization and predator control were not discernibly different before and after 1951.[13] In the memories of scientist activists, understandably, there has been a tendency to picture the *zapovedniki* before 1951 as idyllic and during the Malinovskii period as degraded. Certainly, from the perspective of scientists' input and autonomy, not to mention the more mundane question of employment, that portrait of the reserves system reflects indisputable realities. Regarding acclimatization and the extermination of predators, however, the truth is not nearly as clear-cut.

The Academy of Sciences Commission on *Zapovedniki*

Bright spots such as Troitskoe or Lithuania were only local responses. The first *coordinated* response of the scientific community following the August 1951 calamity was not long in coming. With the quiet blessing of the new president of the Academy, the chemist Nesmeianov, who had just succeeded the late Sergei Ivanovich Vavilov, and of the Academy's scholarly secretary, A. V. Topchiev, a major new commission was created on March 28, 1952, attached to the Academy's Presidium: the Commission on *Zapovedniki.*[14] Like Vavilov before him, Nesmeianov had to walk a fine line between official obeisance to regime policy and his own vision of the welfare of science. This is well illustrated by a visit paid to him in early summer 1952 by Aleksandr Leonidovich Ianshin (see figure 12) and Vera Aleksandrovna Varsonof'eva in their capacities as co–vice presidents of MOIP. Their goal was to try to convince the Academy president personally to join the fight to restore at least some of the *zapovedniki.*[15]

Varsonof'eva started to speak about the importance of the Kondo-Sos'vinskii reserve on the eastern slopes of the Urals and the Barguzinskii *zapovednik* on the eastern shores of Lake Baikal in restoring the population of sable. Perhaps exploiting his status as a chemist, Nesmeianov replied to the geologist: "Vera Aleksandrovna, why do we need to worry about breeding all those fur-bearing animals these days? With the help of chemistry we can produce fur of any quality, any color, and any degree of beauty. We are now living in the century of synthetics and not natural products," he concluded, refusing help.[16]

Looking back, Ianshin was convinced that Nesmeianov was using a little protective coloration of his own to avoid the opprobrium of scientific public opinion. A Party man, indeed, a member of the *nomenklatura,* Nesmeianov was obliged to obey and fulfill the instructions and decrees of the Party once they were adopted. For that reason he could not be openly associated with the struggle against the 1951 Party decision. Yet, his honor as a member of scientific public opinion was called into question by his inability to join this crusade. Hence the need to present his position in terms of personal aesthetics, colored by his background as a chemist, so as to avoid an embarrassing admission that to protect his position he had no choice but to refuse assistance.[17]

Figure 12. Aleksandr Leonidovich Ianshin (1911–).

Nevertheless, the Academy president allowed Vladimir Nikolaevich Sukachëv, dean of Soviet botanists and director of the Academy's Institute of Forests, a surprising degree of freedom to use both his Multidisciplinary Scientific Expedition on Problems of Shelter Belts as well as the new commission as havens for out-of-work *zapovednik* staff and activists in the area of nature protection.

With Sukachëv as chair of the Commission on *Zapovedniki,* Makarov and Dement'ev were named two of his four deputies, the others being the academician Andrei Aleksandrovich Grigor'ev, a geographer and conservation stalwart, and Nikolai Evgen'evich Kabanov, a biologist working in the Institute of Forests. The remaining membership was no less distinguished.[18]

Barely two weeks later, the commission had already roared into action, convening the first meeting of its executive Bureau. Preoccupied with the continuing political troubles of his interdisciplinary Shelter Belt Expedition as well as an unexpected initiative, probably with its source in the Central Committee, to move his Institute of Forests to eastern Siberia (Krasnoiarsk), Sukachëv was unable to attend. Indeed, according to his close friend and deputy director of the Expedition, Sergei Vladimirovich Zonn, Sukachëv's blood pressure was so consistently high during those days that his doctor did not know whether the academician would live to see the next morning.[19] Happily, Sukachëv had a coterie of brilliant and dependable associates whom he had either attracted to his Institute or rescued from persecution by Lysenko and others, and to them he could confidently delegate some of his important scientific-political responsibilities. One of these was Nikolai Evgen'evich Kabanov, who in the early years of the commission more often than not sat as acting chair and convener.

Under Kabanov's direction the commission developed a work plan for the first half of 1952. Among its central responsibilities was examining the scientific research plans of Malinovskii's new Main *Zapovednik* Administration of the USSR Council of Ministers, particularly because the new decree on *zapovedniki* of August 1951 specifically assigned research-related "methodological leadership" to the USSR Academy of Sciences.[20] The dogged persistence and cunning of scientific public opinion now placed oversight of *zapovednik* research in the hands of Malinovskii's enemies: the old guard nature protection activists and elite field biologists of the nation. Scientific public opinion would not allow its "free territories" to be dispossessed, even if it meant a protracted and grueling guerrilla war. And a guerrilla war is what the central authorities got.

At a meeting of the Bureau on July 9, 1952, with Malinovskii's deputy director for scientific research Aleksei Ivanovich Korol'kov present, the forestry plans of the Main Administration came under fire. One of the most eloquent defenses of the special function of *zapovedniki* as *etalony* was made by Makarov, who insisted that the reserves must find a way of pursuing forestry under conditions of *zapovednost'* (inviolability): "Here [in Malinovskii's plans] a mistake has crept in. [Research] needs to be conducted not [only] within *zapovedniki,* but under conditions of *zapovednost'*."[21] Forestry needed to promote the "natural" regeneration of "natural" forests.

Malinovskii's plans now came under accelerated attack in the commission. At a December 10, 1952 meeting, A. P. Protopopov, who was asked to testify, demonstrated that he had lost none of his acuity or his mettle as he subjected Korol'kov, who then held the rank equivalent to a deputy minister, to inconvenient questioning:

> I want to receive an answer from the representative of the Main *Zapovednik* Administration how we should critique [his plans] in the future. A question has emerged: "What kinds of institutions are we looking at here? What, in fact, *are* the Main Administration's *zapovedniki?*" The Main Administration uses the term *zapovednoe khoziaistvo* [management of a *zapovednik* oriented toward the exploitation of its resources, even if experimentally]. What *are* the *zapovedniki,* scientific-research institutions or *zapovednye khoziaistva?* This term elicits incomprehension. As I see it, *zapovednoe khoziaistvo* is an impossibility. There can only be *khoziaistvo zapovednika* [administrative management of a *zapovednik*].[22]

Korol'kov was equally outspoken:

> I am shocked by the question "What is a *zapovednik?*" The USSR Council of Ministers has already settled this question. Comrade Protopopov will find a exhaustive response [to it] in the statute [on *zapovedniki*]. . . . It must be kept in mind that there is a whole group of objects of economic interest in the *zapovednik.* Experience has shown that it is impossible to practice forestry without cutting and treatment [of trees]. Economic measures must be carried out.[23]

When the commission finally drafted its official assessment of the scientific work plan of Malinovskii's reserves system for 1953 it noted that the Main Administration had taken some of the criticism received at the December 10 meeting into account, which improved the plan. However, the commission continued, "a second look at the plans sent to us shows that the Main *Zapovednik* Administration has still not adequately taken to heart the observations and recommendations of the Commission on *Zapovedniki.*"[24]

The target of the commission's displeasure was the entire section "Scientific and Scientific-Technical Measures for Implementing *Zapovednoe Khoziaistvo.*" "This whole section deserves the most comprehensive and critical discussion at the Scientific-Technical Council of the Main . . . Administration," the report noted. While projected studies of the ecological effects of the flooding of the shores of the Rybinsk reservoir were praised, the attempt to call a whole slew of managerial and technical measures "fundamental research" was roundly opposed.[25]

Malinovskii sought to keep his Scientific-Technical Council completely isolated from any contacts with the old guard on the commission, a state of affairs bitterly condemned by Geptner.[26] By March 1953 the Academy of Sciences and its commission were so frustrated by Malinovskii's lack of cooperation that they tried to get relieved of their responsibilities for oversight of the scientific work done by the Main Administration. Of course, they also no longer wished to be held legally responsible for that research, a responsibility overseen by the USSR Ministry of State Control.[27]

When the nature reserve system was "reorganized" there had been *zapovedniki* that were already subsumed under either the USSR Academy of Sciences or one of the Academy's republican affiliates. With the "liquidation" of most of the reserves, the Academy inherited an additional contingent that Malinovskii rejected for his own system, largely because the reserves lacked significant forest cover. Thus, by February 1953 the Academy system controlled fourteen reserves.[28]

By May 1952, Sukachëv, together with corresponding member I. V. Tiurin, director of the Academy's Institute of Soil Science, tried to roll back the decree of the previous year, beginning with the case of only one *zapovednik,* the Poperechenskaia steppe. Writing directly to Malenkov at the Central Committee Secretariat, the two scientists argued against the transfer of the *zapovednik* from the Penza Pedagogical Institute to a nearby collective farm, since the total area of the reserve, 200 hectares, would hardly represent an appreciable gain for the "Proletarian" collective farm (6,000 hectares). Meanwhile, those 200 hectares were among the last parcels of undeveloped northern forest-steppe. At a meeting organized by the Penza *oblispolkom* of March 5, 1952, they noted, a great many local workers and specialists spoke out in defense of continued protection for the area, as did the Penza branch of VOOP and the Biology Division of the Academy.[29]

Malenkov examined the letter ten days later, and marked in the margins that A. I. Kozlov, head of the Agricultural Department of the Central Committee, should look into it. However, Malenkov significantly made the further notation, "We must act in accordance with the decision of the Government concerning *zapovedniki.* Report back."[30] On June 13, Kozlov's deputy V. Iakushev wrote back to Malenkov:

> The head of the Main Administration, . . . Malinovskii, considers it ill advised to reexamine the decision of the USSR Council of Ministers of October 29, 1951 . . . because pristine, unplowed lands continue to be preserved in the Tsentral'no-Chernozemskii *zapovednik* . . . where practical scientific work on problems of the generation of strong black earth soils is being conducted. I also spoke with the secretary of the Penza *obkom,* Comrade Lebedev, who informed me that the *obkom* . . . did not support the recommendations of Comrades Sukachëv and Tiurin.

Neither did the Agricultural Department.[31] This time, however, Sukachëv and Tiurin did not fold, taking their case to Academy president Nesmeianov. Another attempt was made the following year.

For the short term, Stalin's death on March 6, 1953 worsened the situation of *zapovedniki* in the Main Administration. In the immediate aftermath of the dictator's death there was a significant rearrangement of ministerial responsibilities at the USSR level. Stalin was succeeded by Nikita S. Khrushchëv as first secretary of the Central Committee of the Communist

Party of the Soviet Union. Georgii M. Malenkov replaced Stalin as chairman of the USSR Council of Ministers (premier). One change entailed the abolition of the Ministry for Supplies (temporarily, it turned out), which was merged with Agriculture, with A. I. Kozlov (also temporarily) replacing Benediktov as minister of the enlarged superministry of Agriculture and Supplies. Although Malinovskii was not removed as head, his Main Administration was demoted from its status as an all-Union ministry and subsumed as a somewhat minor department under the Ministry of Agriculture and Supplies. As had happened in the 1920s, *zapovedniki* were trapped in the unfriendly embrace of the "economic commissariats." Little help could be expected from the minister, Kozlov, whose prior position made him Malenkov's right-hand man on the Secretariat for agricultural and land-use matters, and who was partly responsible for the reserves' current plight.

That seemed to leave only one route: to have the Academy system somehow become the nucleus of a new expansion. In his last major speech before his death, Makarov expressed the hope that the Academy would indeed prove to be the savior of his life's work. After all, the *zapovedniki,* he told a convocation of directors of nature reserves of the Academy system in the spring of 1953, "serve the general goals of the development of science."[32] Despite the importance of each reserve preserving its own personality, Makarov also stressed the need for common goals, common scientific perspectives, and common methods, so that the results of research at the various reserves could be compared. Here, too, the Commission on *Zapovedniki* had already begun work, drafting an overall statute for the Academy system "independent" of the statute on *zapovedniki* drafted by Malinovskii one year earlier.[33] It was crucial always to remember that *zapovedniki* were "a huge natural laboratory, a laboratory *of* nature" rather than a laboratory *in* nature, as Malinovskii would have it.[34] Making the obligatory rhetorical bows to "Michurinist biology" and "Pavlovian physiology," Makarov concluded by emphasizing the crucial role of the reserves also in the preparation of graduate students in their *aspirantura* (graduate training) and as a research base for those mature scholars seeking the degree of doctor of science (*doktorantura*) both inside and outside the Academy systems.[35] A little over two months after giving this speech, Makarov died, and the commission was spurred to even more energetic activity to honor his legacy.

The new draft of the statute on the Academy's reserves was completed on September 18, 1953. The reserves were declared to be "independent scientific research institutions" of the Academy systems, with their own staffs of scientists and technical and support workers, and with goals that highlighted the twin missions of protection and fundamental research.[36]

Despite the political confusion following Stalin's death, the country also had a new atmosphere of guarded hope and of greater freedom. True, no one knew who was on top—Khrushchëv or Malenkov—but for the scientific intelligentsia that was not a major preoccupation. As 1953 glided into 1954, the scientific intelligentsia mobilized to reclaim scientific autonomy and restore the "geography of hope," in many ways vastly outpacing their colleagues in literature who were creating the first "thaw."

The Rise of MOIP as a Center of Resistance

When Varsonof'eva and Ianshin went to see President Nesmeianov about enlisting him in the fight to restore the *zapovedniki,* they came as representatives not of VOOP but of MOIP, the Moscow Society of Naturalists, Russia's oldest scientific society.[37] To understand why they presented themselves in this fashion and to understand how MOIP came to represent a center, and later *the* center, of scientific public opinion, we must survey the history of that society from 1948 on.

As late as the early 1940s, MOIP still had a deserved reputation as a sleepy academic society.[38] Its library, adjacent to the Gor'kii Library of MGU's old campus opposite the Manezh, was frequented largely by older men and women poring over biological arcana under the stern gaze of a huge stuffed owl and equally lifeless early nineteenth-century portraits of MOIP's founders. The society's president was the ancient and revered Nikolai Dmitrievich Zelinskii, a chemist and prerevolutionary relic who still favored the round, brimless black academician's cap, which vaguely resembled Central Asian Muslim headgear, the *tiubeteika.* In a word, MOIP was quaint.

Two features distinguished it from all other Soviet societies. MOIP had maintained an almost unbroken tradition of non-Communist leadership, from Menzbir to Zelinskii (although Sukachëv joined the Party in 1937, he was clearly heterodox) and now to Ianshin. Even such venerable and progressive societies as the Geographical Society of the USSR or the Mineralogical Society, two others that survived the early 1930s and that were almost as old as MOIP (f. 1805), were obliged to select Party members as their presidents. The difference was that whereas they were chartered within the system of the USSR Academy of Sciences, MOIP was tucked away under the aegis of Moscow State University, almost out of bureaucratic view.[39]

Like VOOP, MOIP united the scientific, preeminently biological and geographical-geological intelligentsia across Russia and even the Soviet Union, despite its local name. Though the society had no organizers, branches emerged on local initiative in Kalinin (Tver'), Riazan', Sverdlovsk (Ekaterinburg), Tomsk, L'vov, Uzhgorod, Alma-Ata (Almaty), Aral'sk, and Sukhumi, among other places, constituting an informational network across the country. As Nikolai Nikolaevich Vorontsov noted recently, an entire history could be written about science "on the periphery" in Russia. This periphery was a product of many factors, including the flight of many field biologists to distant *zapovedniki* and antiplague stations, to pest-control stations, and to remote research and teaching institutions, partly in the hope of avoiding the repressions that were continually sweeping the "center."[40]

In the words of Oleg Nikolaevich Ianitskii, "The Moscow Society [of] Naturalists was arguably one of the few long-established social organizations that was not taken over entirely by the state. In any case, its official structure differed from the state-determined model. . . . The organizational principles set out in its constitution were democratic, and its members did not have to be professional scientists, but were merely required to be involved in the scientific life of society."[41]

Like VOOP under Makarov, MOIP institutionally, as a community of like-minded members, embodied the ideal of scientific public opinion. Plenary meetings of the society were usually held in the Old Zoological Auditorium in the Zoological Museum on Herzen Street. It could accommodate more than 200 participants, and its acoustics were among the best in Moscow. Just before it was closed as a fire hazard, Zelinskii, then ninety-two, gave his last lecture in the auditorium. Fearing for the president's health, one of the society's vice presidents asked Zelinskii if he wanted to sit down while giving his talk, a fully justifiable break with the society's traditions. To that, Zelinskii firmly replied, "If I gave my talk sitting down, that would be a mark of disrespect to my audience," and then lectured for nearly two hours on the developmental physiology of the Mexican alpine salamander. For the old guard, breaking with its cultural rituals—rituals that embodied and symbolized the dignity of the scientific intelligentsia—would be to renounce one's social and personal identity. Zelinskii remained on his feet to the end of the meeting.[42]

In 1950, when, in connection with the construction of the new Lenin Hills campus of Moscow State University, MOIP was offered the opportunity to move to newer quarters, Zelinskii sent word from his sick bed of his intense opposition. Zelinskii reasoned that MOIP was a scientific organization whose membership was drawn from a number of different institutions and workplaces. Aside from tradition, therefore, fairness, convenience, and its social role dictated that the society remain in the center of the city.[43] The society's Presidium "categorically" insisted on remaining.[44]

Inescapably, because of what MOIP embodied, political risks and dangers were thrust upon that society just as they had burst through the thin defensive perimeter of VOOP's countercultural community. As it was for VOOP, 1948 was the decisive turning point, disrupting the serene and dignified routine of the society. And if VOOP was launched on a trajectory that would ultimately lead to its reincarnation as a huge, Party-dominated business enterprise, MOIP's fate was happier, for it was transformed within a few years into the theoretical center for biology for one-sixth of the globe and into the new headquarters for nature protection activism. Rarely have organizations had to shoulder so much responsibility as MOIP, and rarely have they risen so successfully to the challenges that faced them.

After the August 1948 session of the Lenin Agricultural Academy, Lysenko and his allies, with Stalin's blessing, established a reign of terror in Soviet biology. Their influence permeated every school, every university, every public meeting. Even *zapovednik* directors were required to convene meetings to weed out "Weismannist-Morganist-Mendelian" perspectives and to replace them with "Michurinist" biology. No corner of the vast country was exempted from the new rituals of obeisance to Lysenko.

The most tragic consequences, described at great length in other works, included a wholesale purge of instructors, teachers, professors, and researchers from educational and research institutions large and small.[45] N. N. Vorontsov estimates the number of expelled university professors at 3,000. Some, such as the eminent physiologist D. A. Sabinin, committed suicide.[46] Even the Academy of Sciences was powerless before the personal endorsement of Stalin.

In this atmosphere of tragedy and calamity, the president of the Academy, Sergei Ivanovich Vavilov, called on Zelinskii. For Vavilov, the tragedy was multiple. In 1941 he had lost his twin, Nikolai Ivanovich, one of the USSR's most eminent geneticists, who had been arrested the year before as a result of his attempt to defend classical genetics against Lysenko's increasingly aggressive and ignorant claims. After Nikolai died in Saratov in near-isolation, Sergei, a prominent physicist but not yet Academy president, was forced to hold his tongue even as his twin's body was dumped into an unmarked grave. Even with the partial recovery of genetics' fortunes in 1946–1947, Sergei was compelled to continue to keep silent. Now, he was coerced into a purge of his own institution, the Academy. It was as if Stalin and Lysenko were forcing Vavilov to put a gun to his own brother's head. Vavilov had been covertly trying to help some of the marked individuals and recognized that he was faced with a terrible dilemma: he could preserve his honor and resign, thereby sealing his own fate as well, or he could remain and try to use his position to do as much covert good within the system as the situation might allow. Happily for the scientific intelligentsia, Vavilov opted for the second course, although his reputation is only now beginning to reflect the wisdom and humanity of his difficult decision.

Distraught, Vavilov turned to the superannuated Zelinskii for help, telling him that he, Vavilov, was forced to "liquidate" genetics in the Academy. He was not even permitted to allow the word *genetics* to remain in any institute. "I know this is antiscience," but owing to the Academy's position, Vavilov confessed, there was little he could do at the moment. However, he had a plan. Could not MOIP, with its freer atmosphere, organize a section for genetics to provide at least an intellectual haven for his brother's colleagues?[47]

Zelinskii summoned the members of the Bureau of MOIP's Presidium, recounted the conversation with Vavilov to them, and asked for support. The response was unanimously enthusiastic. Here at last was a constructive way in which the scientific intelligentsia could respond to Lysenko. As Ianshin, who was at the meeting at Zelinskii's apartment on Gor'kii (Tverskaia) Street near the Main Telegraph, recalls it, the members of the Bureau immediately called N. P. Dubinin, B. L. Astaurov, I. A. Rappoport, and V. P. Efroimson and asked them to organize the new genetics section. "We could not provide them with a lab," Ianshin later explained, "but we gave them a roof over their heads and an opportunity to read foreign journals, to hold symposia, and to give talks." Thus S. I. Vavilov helped to keep genetics alive during the darkest of years for Soviet science.[48]

It was inevitable that Lysenko would hear of MOIP's Section on Genetics. In mid-1950, when Zelinskii was still alive, news of Lysenko's awareness of the section's existence filtered back to MOIP. Hastily but without panic, Zelinskii convened a meeting of the Presidium in his apartment. Rumors have spread, he told the Presidium, that we were giving "refuge" to "Weismannist-Morganists" and that they were meeting at MOIP nearly every week. At this Vera Varsonof'eva excitedly objected: "What are you suggesting, to close down the section on genetics?" Zelinskii reassured her: "No, no, calm down. We will under no circumstances close it down. We must invite Trofim Denisovich to talk to our society. Then he will see that there is nothing subversive going on." In order to pull this off, Zelinskii had to make sure that everyone knew his or her part, for one mistake could ruin the stratagem. Over a period of weeks, groups of Moscow's leading biologists—all members of MOIP—filed into Zelinskii's musty apartment to hear the instructions: "Don't hoot, don't whistle, don't ask trick questions! We are doing this to save the section."[49]

In December 1950 the "people's academician" paid his visit to the citadel of scientific public opinion (see figure 13). Chairing the vast meeting—171 MOIP members and 600 nonmembers attended—was venerable zoologist Sergei Ivanovich Ognëv, one of the society's vice presidents. Lysenko chose his own theme, the transformation of one species into another, but despite the extreme intellectual provocation his remarks presented, there was not a hoot, not a whistle, and only one recorded challenging question. Ognëv, in fact, graciously expressed "the deep gratitude on the part of the members of the society for a thoroughly interesting and extraordinarily substantive talk on the problem of Michurinist theory."[50] Zelinskii's strategy worked seamlessly, and, for a time, MOIP continued to be viewed by the Stalinist camp as a collection of *chudaki*—harmless and marginal oddballs.

With Zelinskii's death, the baton of leadership in MOIP was passed to Vladimir Nikolaevich Sukachëv, who was concurrently director of the Academy's Institute of Forests as well as head of the Interdisciplinary Expedi-

Figure 13. Nikolai Dmitrievich Zelinskii (seated) and Trofim Denisovich Lysenko.

tion for Shelter Belts, chair of the Academy Presidium's Commission on *Zapovedniki,* and editor of the *Botanical Journal* (as president of the Botanical Society). In the secret balloting conducted by the society on February 10, 1951, with 126 members present and 115 voting, fifty-seven candidates vied for forty seats on MOIP's Executive Council. In a bizarre coda to Lysenko's appearance at the society, Lysenko was placed on the ballot for a Council seat. Apparently, MOIP's leaders had decided to go with a winning strategy; why not embrace Lysenko with open arms? Perhaps then he might forget about them. Sensibly, the voting members gave Lysenko 100 affirmative votes, less than the totals for Varsonof'eva, Obruchev, Ianshin, Ognëv, Deineka, and Sukachëv, but enough to win him a seat on the Council. His allies, though, did not fare as well. Koshtoiants and Davitashvili ended up with forty votes and thirty, respectively.[51]

With the presidency of MOIP, Sukachëv also inherited de facto control over the society's *Bulletin,* concentrating in his hands leadership of virtually all of the surviving institutions of scientific civil society. Delegating responsibilities for the Expedition to S. V. Zonn, for the Academy Commission to Makarov, Dement'ev, and Shaposhnikov, for MOIP to Varsonof'eva and Ianshin, and for the Botanical Society to D. I. Lebedev and others, and resigning as director of the Institute of Forests when it was transferred to

Krasnoiarsk, Sukachëv used his own relatively unimpaired political reputation as a shield to protect his vulnerable colleagues.

If there was one issue on which Sukachëv gambled his political capital, it was his monumental battle with Lysenko beginning in 1951. Recent scholars have sensibly argued that Sukachëv would not have even considered such a risk had not Stalin's quizzical article on linguistics plausibly signaled, albeit indirectly, that Stalin had felt that *all* would-be arbiters of Soviet science—Lysenko included—had overreached themselves and were attempting to set themselves up as authorities independent of the Party, that is, of Stalin.[52]

Nevertheless, it was a big risk. But Sukachëv's own articles and those published in the *Botanical Journal* and the *Bulletin of MOIP* in 1952–1954 so resonated throughout Soviet scientific society that by 1956 Lysenko seemed to be on the ropes.[53] The example of Sukachëv's civic courage had an inestimable effect on other scientists, especially on impressionable young biologists and students just coming of age in the early and mid-1950s. A 1955 letter to the botanist from a young zoology student at Moscow University and future USSR minister for the protection of nature, Nikolai Vorontsov, illustrates this intergenerational link as well as the affection in which Sukachëv was held:

> Deeply Esteemed Vladimir Nikolaevich!
>
> I warmly congratulate you on your coming birthday. . . . Permit me, a young biologist, to express my deepest gratitude to you for that struggle against Lysenkoism in biology that you have led during the difficult conditions of 1952 and which you continue to lead to the present.
>
> Despite administrative pressure, despite the fact that from their university chairs A. N. Studitskii [here follows a list of Lysenkoist instructors] and others try to enlist us under the flag of "new" medieval views, the majority of conscious university youth, both undergraduates and graduate students, is with you in your struggle for Darwinism, for genuine biology, and against obscurantists in our science. . . . Please know, dear Vladimir Nikolaevich, that in this struggle the ardent hearts of youth are on your side and we will remember those efforts which you have expended.[54]

Under Varsonof'eva's day-to-day leadership, MOIP in 1954 launched a second front in the scientific community's struggle against Stalinist science policy: a public campaign to restore the *zapovedniki.* With VOOP still struggling with political and fiscal problems and, especially after its merger with the Green Plantings Society, internal division, MOIP was the more logical place to establish the campaign's headquarters, especially as it commanded greater prestige among scientists.

The *Zapovednik* Conference of 1954

The 1954 *Zapovednik* conference was completely ignored in the Western media. Most likely it was never even logged in the daily political summaries sent to Washington by embassy staff in Moscow. In the Soviet press there was barely a mention of it. Why, indeed, should anyone have paid serious attention to zoologists and botanists gathering to discuss the current situation and future prospects of *zapovedniki* in 1954? As the cream of Soviet field biology assembled on the morning of May 12 at the Academy's Moscow House of Scholars on Kropotkin Street, however, Moscow was witnessing the first public protest of the scientific intelligentsia, a meeting that affirmed a social identity of scientists-as-citizens sharply at odds with the Kremlin's definition of "citizens," and opposed the dictatorship in science imposed by Lysenko and the Party bosses. To identify the meeting for what it was, an observer would have had to know the behavioral and rhetorical codes and markers of that community, which even its insiders knew only on an experiential, not a conscious, level. Nonetheless, those who participated in the conference remembered it clearly to the end of their lives.

"Comrades!" shouted Vera Aleksandrovna Varsonof'eva triumphantly over the hum, "permit me to declare the joint session of MOIP, the All-Russian Society for the Promotion of the Protection of Nature and of the Greening of Population Centers [VOOP], and the Moscow branch of the All-Union Geographical Society [MGO] open for business!" Following the rituals of Soviet academic arcana, those present elected a Presidium and a chair for the conference.[55]

Leading off, Varsonof'eva cunningly mentioned the February–March Plenum of the Central Committee, the meeting at which Khrushchëv had announced his signature Virgin Lands program, and noted that its successful implementation depended first of all on deepening our knowledge of the natural world. And, she asserted, *zapovedniki,* offering unique opportunities for field study of ecological complexes under natural conditions, were an essential component of the plan. "The failure to study these complexes sufficiently will lead, in some cases, to devastating consequences for the economy," she warned prophetically, alluding to the understudied grasslands of southern Siberia and Kazakhstan. "The unthought-out, monolithic application of the *travopol'e* system in our southern regions may serve as an example," she noted, referring to the ill-fated, dictatorially imposed system of cropping advocated by Stalinist soil science icon G. R.Vil'iams, a figure similar to Lysenko. "I must say," she added, "that these natural geographical conditions are far from adequately understood by us not only [in the Virgin Lands] but in other regions as well."[56] For that reason alone it was necessary to restore the former reserves and even to expand that network to include all biological and physical-geographical zones of the USSR that were not represented in 1951.[57]

As many had argued before her, wild-growing vegetation was the raw material for many commercial crops and applications. The potential value of nature, even reckoned in this way, she observed, was unknown. One example she mentioned was Professor Avrorin's use of wild plants from a number of regions for the greening of the city of Kirovsk on the Kola Peninsula in the Arctic zone. In some cases, the plants exhibited dramatically different physiological responses and potentialities than those observable in their area of natural distribution, which held great interest for biological theory. In light of this, the role of *zapovedniki* as reserves of such wild-growing plants was no less important than their role as pristine natural ecological communities.[58]

Varsonof'eva explained that the conference was the result of the initiative of "three of the most prominent scientific societies linked with the study of natural resources"—namely, MOIP, VOOP, and MGO—which had approached the USSR Academy of Sciences and the USSR Ministry of Agriculture (to which Malinovskii's agency was now subordinated) to send speakers "to inform scientific public opinion about the work being conducted" in the *zapovedniki*.[59] Although it was already highly unusual that voluntary societies in the USSR should request any kind of accounting to them by state agencies, Varsonof'eva took scientific public opinion's claims one step further, asserting the societies' methodological leadership on the question of nature protection. Diplomatically but assertively she concluded by expressing "the conviction that this conference will assist the Main . . . Administration and the Academy . . . to find the way toward future development of this great cause."[60]

Despite the dignified and restrained tone set by Varsonof'eva, it was not easy for Aleksei Ivanovich Korol'kov, Malinovskii's deputy, to face the largely hostile audience, many of whom had lost their positions in 1951. Commencing on a defensive note, Korol'kov announced that he would not discuss the history of the *zapovedniki* before 1952, although he reminded the audience that 200 inspectors were dispatched in 1950 by the Ministry of State Control and that the 1951 decree flowed from the Kremlin's belief that the system of reserves had become "unjustifiably overblown."[61] He also mentioned that the *zapovedniki* had been accused by the government of conducting work that was useless to the larger society, backing up his contention by citing the 1951 decree, which stated plainly that "the scientific research in the majority of *zapovedniki* is pursued in a way disjunct from the practical interests of the economy."[62]

After going into some detail regarding the various research and biotechnical projects conducted in his system, Korol'kov noted a number of quantifiable indicators of improvement. State funding of the system rose from 14 million rubles in 1952 to 17 million for 1954, including an increase from 3.7 million to 4.8 million for research. From 1952 to 1953 the number of candidates of science in the system had increased from nineteen to thirty-six and, equally important in Korol'kov's eyes, their affiliation with the Komsomol or the Party rose from 31 percent to 41 percent.[63] At the end of his speech Korol'kov came to the question that had brought the big crowd together in the first place. "We believe," he pronounced, "that the question of increasing the number of state *zapovedniki* must be decided not as part of a general reassessment of the whole system . . . but on an individual basis," renouncing in advance any support for a strategic reconstitution of the system.[64] Finally, Korol'kov tried to reassure the scientists that the most recent transfer of the reserve system to the USSR Ministry of Agriculture would not be fatal to serious scientific research. Studies would begin in 1955 on intra- and interspecies relationships, as well as comparative studies on reproduction, such as the effect of light. Moreover, there would be experiments on the hybridization of closely related species and an extensive campaign of acclimatization was to commence, with *kulan* (onager or Asiatic wild ass) and pheasant brought to Barsakel'mes Island, musk deer to Denezhkin kamen', roe deer to Prioksko-Terrasnyi, and European bison to the Khopërskii *zapovednik*.[65] But Korol'kov's examples were precisely the kind of science that most disquieted this particular audience: hybridization and acclimatization without regard for either the genetic prerequisites or the possible undesired ecological consequences of these "biotechnical" measures. These programs in field biology vividly symbolized the dictatorship of Lysenko and the Baconian-Timiriazevan-Stalinist vision of total control, which the scientists rejected.

The second major address was delivered by Nikolai Evgen'evich Kabanov, acting chairman of the Academy's Commission on *Zapovedniki*. Kabanov's speech was a strong defense of the scientific intelligentsia's traditional program of nature protection. Invoking the names of some of the founders of that program—Kozhevnikov, Borodin, Zhitkov, Buturlin—Kabanov signaled that the Academy was now ready to support the commission's plans for a radical expansion of the Academy's network of reserves, a stance that implicitly marked the commission's aim to wrest the status of effective center of nature protection from the Ministry of Agriculture Main Administration. Although the preparation of a volume, *Zapovedniki of the USSR Academy of Sciences and of the Academies of the Union Republics*, was a first step,

Kabanov now revealed movement on the more important front of creating or restoring protected territories:

> I wanted to inform you that the Presidiums of the Academies of Sciences remain interested in the status of their *zapovedniki.* Individual Presidiums . . . such as the Estonian now are actively raising the question of creating *zapovedniki* in their republics. Similar information regarding the Latvian Academy of Sciences has reached the commission as well, and there are also reports from the Academy of Sciences of Georgia. . . . All of this clearly shows that the existing network of *zapovedniki* is not the final word on the issue, not bottled up in its present scale, but it will and doubtless must change in connection with the needs of the economy and culture of the various *oblasts* and republics of the Union.[66]

Kabanov asked the audience for the help of scientific public opinion in editing and publishing the rich manuscript materials in the hands of the commission, materials that would not see the light of day if they had to depend on Malinovskii's Main Administration. Further, he asked the meeting to support a group of recommendations, which, taken together, constituted a declaration of war against the Main Administration of the Ministry of Agriculture.[67] He declared, "We hold that any *zapovednik,* be it state or part of the system of the Academy of Sciences, must be by its nature a specific kind of scientific research institution which must develop its specific themes for study . . . linked, principally, with the . . . scientific principles of nature protection in the USSR under conditions of inviolability."[68]

Kabanov explicitly noted that the practical tasks outlined by Korol'kov, particularly under the influence of the Ministry of Agriculture, did not represent "full-blooded" science and that *zapovedniki* ought to be devoted to other tasks. "For that reason," he exhorted, to the applause of the delegates stirred by this bold public rebuke to their enemies, "the time has come when we need to raise the question, 'Doesn't the general regulation of the whole cause of nature protection and *zapovedniki* demand raising the question of organizing under the USSR Council of Ministers an authoritative agency for nature protection and *zapovedniki?*'"[69]

Korol'kov was brought back to the podium to answer questions, which Varsonof'eva had asked to be submitted in written form, evidently to minimize the chance of the meeting getting out of hand.[70] However, many of the emboldened scientists made a point of signing their names to the written questions (which in Soviet practice were usually anonymous), demonstrating that they were submitting written questions in deference to the wishes of their respected colleague, Varsonof'eva, not out of fear.

Korol'kov mentioned that he had received numerous questions all asking the same thing. The most strongly worded were those of Formozov and Nasimovich. Those scientists demanded to know what was happening to the resources—the plants, animals, and minerals—of the former *zapovedniki* liquidated in 1951. Korol'kov admitted that oil was being drilled in the Kronotskii reserve on Kamchatka and that lumbering and hunting operations were going on in other formerly protected territories.[71]

Questioners asked about the scale of logging in the Belovezhskaia *pushcha* and the Khopërskii *zapovednik,* about the amount of income actually cleared by the exploitation of resources in the former reserves and the net amount saved by the government as a result of the liquidation, and whether Korol'kov's talk was preapproved by the Main Administration. Korol'kov informed the audience that in all existing reserves of his Main Administration during 1953 234,000 cubic meters of wood were logged, including 55,000 that went into production (*delovoi les*).[72]

Some questions were acidly or testily phrased, such as the one that asked why Korol'kov mentioned that "the forests were still not inventoried in the Kyzyl-Agach reserve when in fact there was not a single tree there." Another wanted to know why the eider duck was not considered an economically important enough object to support its study in the Main Administration's research plan. But the question that rocked the session was one that challenged the whole political basis for the 1951 liquidation: "What attitude ought we to hold today to the investigations of the activities of the *zapovedniki* carried out under the leadership and on the instructions of the enemy of the Motherland Merkulov?" The implication was that because Vsevolod Nikolaevich Merkulov had been discredited, tried, and shot, the fruits of his political activity should likewise be reevaluated, if not completely reversed.

Korol'kov, however, remained one step ahead of the questioner. "The activities of the *zapovedniki* were investigated by two hundred Soviet people," lectured Korol'kov, "who worked and even now work in the Ministry of State Control. In the government commission worked comrades admired by all, including Comrades Khrushchëv, Kozlov, Bovin, Benediktov, and Chernousov. For that reason the [basis of the] question is simply in error." Korol'kov was able to trump his clever adversary because the system had, as it were, prepared for the contingency by spreading the responsibility for the liquidation of the reserves across a broad spectrum of prominent political figures. Nonetheless, Korol'kov continued to be hammered by questions. One of his most dogged adversaries was A. N. Formozov.

Aleksandr Nikolaevich Formozov (see figure 14) was one of the acknowledged leaders of Russian biology. His monograph on the role of snow cover in animal ecology, published by MOIP in 1946, was nominated for a state prize. Another volume, *An Essay on the Ecology of Mouselike Rodent Vectors of Tularemia,* was also published by MOIP a year later.[73]

At the end of the war the director of the Institute of Geography, Andrei Aleksandrovich Grigor'ev, offered Formozov the opportunity to create a

Figure 14. Aleksandr Nikolaevich Formozov (1899–1973).

department of biogeography in the institute. Formozov accepted, and from March 16, 1945 on he served as the head of the new department, even though his main employment was at Moscow University.

Formozov's postwar research focused on the ecology of the steppe and desert regions of Eurasia. New research problems imposed themselves on him with the announcement of the Stalin Plan. Those who drew up the blueprints for the shelter belts failed to take into account the response of wildlife. It turned out that the massive seeding of oak trees attracted huge numbers of rodents, which were appreciatively eating the acorns.

Formozov's outspoken defense of the importance of natural selection on the basis of intraspecific competition got him in trouble. On November 4, 1947, he was one of three (D. A. Sabinin and I. I. Shmal'gauzen were the others) who appeared before a sympathetic crowd in the huge "Communist" auditorium at Moscow University—the university's biggest—as part of an unprecedented public series of talks and debates pitting Lysenko against his critics. Immediately afterwards, the university published the three talks as *Intraspecific Struggle in Animals and Plants,* a significant show of support for the partisans of Mendelian or classical genetics and the "Great Synthesis" in evolutionary theory.[74] After the August session of 1948, however, the situation radically deteriorated, and on November 4, 1948, Formozov decided to go on half-time status at the university. Before the start of the next academic year he requested permission from the Biological Faculties dean's office to leave altogether, but for his own tactical reasons the new dean, I. I. Prezent, Lysenko's right-hand man, refused, keeping Formozov under his jurisdiction while he awaited an opportunity to discredit Formozov. Nevertheless, the focus of Formozov's energies now was the Institute of Geography, where there were many whom he considered close colleagues and even friends. Formozov was given the opportunity to build up the institute's department of biogeography, and he now hired many experienced biologists, including former college classmates, colleagues, and students. Many who joined the department had lost their positions as a result of the post–August session developments in biology or were unable to get work elsewhere.[75]

Aside from serving as a *zapovednik* for rare and endangered Mendelian-oriented field biologists, the Institute of Geography became a crucible where the two disciplinary intelligentsias formed personal, intellectual, programmatic, and ultimately political alliances, with Formozov and his circle at the nexus. From an institutional point of view, the presence of Formozov, Nasimovich, and a band of other zoologists and botanists in the Institute of Geography as full-fledged members allowed them to enter and authoritatively participate in the Moscow branch of the Geographical Society of the USSR as well. Formozov and Nasimovich particularly enjoyed influence with I. D. Papanin, which continued into the 1960s.

At the 1954 conference on *zapovedniki,* Formozov responded to Korol'kov's address:

> Aleksei Ivanovich [Korol'kov] gave us a detailed talk about the situation in currently existing *zapovedniki,* but the presentation he offered did not deal with the issues we would have liked to hear about. . . . We would have been interested to hear [from Korol'kov] about the future prospects of *zapovedniki* as a whole, what directions they must go, what they are currently lacking, what shape the network should attain in the future, but we did not hear a single word about [the big picture]. . . . We are first concerned about whether any positive results were in fact achieved by the . . . reform of the past few years. We are concerned with the question of what is going on now in the territories of

> the liquidated *zapovedniki.* I posed this question to Comrade Korol'kov and he failed to respond. Evidently, the Administration is completely indifferent to the fate of those territories, [despite the fact that] they hold great interest for science and culture.[76]

Although he had not made a special study of the question, Formozov had received disquieting information from a number of sources about the conditions of the Ministry of Agriculture reserves and the ex-reserves. A full third of the forests of the liquidated Lapland *zapovednik* had burned, and the fires had also spread to the *iagel'* (lichen meadows that served as the principal food base for the reindeer of the region). The *iagel'* requires decades to regenerate in Arctic conditions, and so the reindeer herds that the *zapovednik* had so successfully enabled to recover were once again under threat.[77]

Formozov had also heard that on the territory of the former Altaiskii *zapovednik* a party of geological prospectors slaughtered more than thirty maral deer, taking only a small amount of meat and leaving the bulk of the carcasses to rot. "Is this not an example of barbarous squander of natural resources?" he asked angrily. From the Far East, K. G. Abramov, the dean of zoologists and conservation activists there, had informed Formozov that the goral and sika deer of the former Sudzukhinskii *zapovednik* were on the brink of extinction. "Comrade Korol'kov informed us that the Administration was contemplating establishing a series of game preserves for especially valuable animals," Formozov archly noted, pausing to deliver the punch line. "However, while they are contemplating this there will already be nothing left to protect. For those animals that find themselves in the most threatened situation it will all be over, and this will be an irreplaceable loss not only for the economy but also for science and culture."[78]

Formozov pressed Korol'kov on the issue of whether or not there was any demonstrable benefit from the liquidation of 1951. Practically refuting the idea that the Party-state possessed some kind of superior wisdom, Formozov challenged Korol'kov's repeated assertions that changes in the *zapovednik* network could and should only come about through government decrees and that consequently there was no sense talking about reversing the effects of 1951. "We know that [argument] perfectly," Formozov continued,

> but we are interested in something else: what, in reality, did that decree achieve? Is it not so that recently, for example, a decree on the merger of a group of ministries was issued? However, practice proved that this was ill advised and the decree was replaced with another. And what has practice shown concerning the liquidation of a group of our finest *zapovedniki,* which we had regarded as a significant cultural achievement in the recent past of our country? . . . Are you sure that without those forests [of the *zapovedniki*] our forest industry would be sunk?[79]

Formozov taunted Korol'kov and Malinovskii, alluding to the shortcomings of lumbering and forest culture in the USSR's forests outside the reserves, which would furnish far more usable timber, if lumbering practices were even partly improved, than the available cut in the former *zapovedniki.* "We should have pursued precisely *that* route," Formozov acidly quipped, "and not run after essentially insignificant forested areas of the *zapovedniki.*"[80]

Formozov agreed with Korol'kov on one point: that protected territories alone could not save endangered species. "If we will protect valuable animals on 'postage-stamp tracts' while everywhere else we have picked the land clean, then the result will be pitiable indeed," he warned.

Recounting his experience at the meeting with then USSR minister of state control Merkulov, which, he said, could be confirmed by Professor G. V. Nikol'skii, also in the audience, Formozov revealed that

> all of the determinations of scientific public opinion were rejected without basis owing to his political assignment; the decision to liquidate . . . was taken before any materials were received from the 200 investigators. Even pleas from local authorities were disregarded. Given the situation, it is entirely possible that the government was led to blunder by the deliberately biased materials prepared for its perusal. What the role of the Main Administration was in this is still not clear to me, but history will sort this out and each will receive according to his deserts.[81]

Repeating his accusations that Korol'kov had insulted the audience of scientists by presenting a Pollyannaish "bureaucratic presentation," Formozov called on his listeners to compile a data bank regarding the fate of the former protected territories subsequent to their liquidation. But he closed with a call for struggle: "We must expand the network, increase the areas, and demand that the government reexamine the course that it has adopted." The audience roared its approval; scientific public opinion was standing tall.[82]

For today's readers, the statements made at this meeting may not seem lurid or shocking. For the participants, though, hearing this public censure of officials (some of whom were present) by representatives of scientific society must have been thrilling political theater. And for more than a few, these revelations of Kremlin goings-on told by one of their own—demystified, unawed, uncowed—was a kind of public speech that they had not heard since Stalin's Great Break of the early 1930s.

We have seen how *zapovedniki* resonated for the scientific intelligentsia as symbolic and tangible free territories. Additionally, they embodied and symbolized other values of the intelligentsia, including "responsibility." No speech better underscored that aspect of Russian nature protection than the remarks of I. E. Lukashevich. For him, leaving some areas untouched was a mark of a sense of responsibility to the future, and the fate of what people protected—or failed to protect—today would affect how they were viewed by people in the future. Maintaining this sense of community *in time* was an important value of the movement, and, at least for Lukashevich, our degree of responsibility was the greater because the future was powerless to affect decisions in the present. "However," he noted, "the impression has been created that these questions are not being viewed from that perspective and this has upset many of us." He proclaimed that "an exact and clear depiction of what we have heard today in the first [Korol'kov's] talk" was furnished by Formozov's off-the-cuff descriptor "cold indifference." "This indifference," he cautioned, "cannot lead to anything except the worst possible consequences not only for the present day, but for the far future as well."[83]

Lukashevich argued that scientists were far from knowing the potential of each species both for practical economic benefit but also for increasing scientific knowledge per se. "We cannot even suspect what [secrets] one or another life form carries within itself. Only in the future can the full importance of [these] forms for practice and theory be appreciated; obviously, this cannot be gauged by their current market values."[84]

No less than Formozov, Lukashevich enunciated the *political* claims and political dignity of scientific public opinion. Even while stating that it was the scientists' "sacred duty" to provide the government and the Party with essential information, he also had a different, more radical message: "The obsessive references to the decrees of the Party and the government in no way free the broad mass of scientific public opinion and official authorities [concerned with nature protection] . . . of responsibility not only toward the present generation but to future ones as well. At our cultural level it is already impermissible to hide behind excuses of ignorance, as our predecessors still could." Here Lukashevich injected a remarkable addendum: "We must recall that we are responsible for incorrect decisions regarding *zapovedniki* not only to our own people but to humanity as a whole for all time to come and that to isolate nature behind state boundaries, to be sure, is an incorrect framework of understanding."[85]

Restating the claims of scientific public opinion to possession of the key expertise on the question of nature protection, he insisted that "problems of *zapovednost'* must not be decided by bureaucratic means or by chance people with no connection to the cause. I am greatly agitated because, in contrast to the speaker [Korol'kov], I am unable to treat these questions with indifference," concluded Lukashevich, again to the applause of the crowd. Scientific public opinion had not only had found its voice; it was shouting itself hoarse.[86]

The passionate Lukashevich had warmed the hall. Now, the dryly acerbic and unmovably dignified Vladimir Georgievich Geptner took the podium. "I must state at the very outset," he confessed, "that I was not completely satisfied by the report . . . of the Main Administration." Daring the officials, Geptner sarcastically ventured that he had "counted on the probability that the three years that had elapsed since the reorganization of the system of *zapovedniki* would be enough to be able to demonstrate the superiority of the new system . . . as compared with the old."[87]

As a result of a uniquely developed theoretical framework as well as through trial and error, a number of features came to characterize Soviet *zapovedniki,* explained Geptner. One was that they should represent each and every major natural historical province of the great country. Another was that they should be as large as possible, individually. A third was that the reserves were, first and foremost, to be considered scientific research institutions that would also serve as field research stations for researchers of institutions outside the reserves system. "In that way," emphasized Geptner, "specific features were developed in *our* Soviet *zapovedniki,* in *our* attitude toward *zapovedniki,* which strongly set them apart from attitudes to [protected territories] in foreign countries [emphasis in original]."[88]

Geptner noted that the emergence of this new type of institution, the *zapovednik,* helped to give birth to a new kind of research, whose field observations were station-based, not helter-skelter and performed "on the run." "A whole new branch of zoology was created that hadn't existed before," he added, referring to advances in community ecology and more. "Now, we have grown used to the contributions that the *zapovedniki* have made"; all this, he stressed, "was done by *our zapovedniki,* specific to *our own* conditions, and not by national parks, dedicated to leisure . . . as has been done in other countries."[89] Geptner seconded Lukashevich's assertion that *zapovedniki* were "cultural institutions called upon to preserve models of pristine nature for future generations." Calling the 1951 reorganization "a step backward," Geptner noted that the state-run protected territories accounted for just 0.06 percent of the overall land mass of the Soviet Union. "That is totally unacceptable!" he exclaimed. Indeed, "the very idea of *zapovednost'* has been demeaned. Externally," he continued, "this at the very least has been reflected in the fact that [before] we had an Administration for *Zapovedniki* attached to the [RSFSR] Council of Ministers with good scientific staff, with a scientific-technical Council, and which elicited tremendous interest on the part of public opinion. . . . Now there is merely a department of a ministry and, naturally, the authority of this ministry cannot be compared with that which had been before."[90]

Next, he lashed out at Korol'kov's information about logging in the current Ministry of Agriculture reserves. The assistant director's figures of 234,000 cubic meters, or 0.1 cubic meter per hectare, did not appear too egregious, but statistical averaging might be deceiving: "Sergei Sergeevich and I want to eat, but I eat two sandwiches while he has none. [If you average it out] it turns out that we're both satiated!"[91] "What I'm interested in knowing," he continued, "is how much forest was chopped down in the Belovezhskaia *pushcha.* Averages don't tell us anything. There is a noticeable utilitarianism in your work," he went on, apparently looking at Korol'kov. "I repeat: *zapovedniki* must work in the interests of the economy, this is beyond doubt. However, they may not be objects of economic exploitation, and this must be stated unambiguously."[92]

An unmistakable note of patriotism rang through Geptner's remarks, but it had a rueful tinge. He and fellow biologists wanted to be proud Soviet patriots and to be recognized as contributors to their country's greatness. However, the authorities continued to push them away, accepting patriotism only on the terms of the regime, not those of scientific public opinion.

The zoologist Georgii Vasil'evich Nikol'skii was next to take the floor. Departing somewhat from previous remarks, he offered that "perhaps it is not entirely right to limit our discussion only to questions concerning the system of *zapovedniki*." Nikol'skii saw a larger issue at hand, that of the protection of nature generally. "Unfortunately," he observed, " it must be said without mincing words that we do not have a system for the protection of nature. . . . Water quality protection is inadequate; the situation with forest protection is downright disgusting; the protection of our fisheries is a total fiction; and the situation with the protection of animals and birds is an outrage."[93]

We pour half a million tons of oil and petroleum products into our waters annually, he charged. We lose the equivalent of one quarter of the wood in the forests of the *zapovedniki* in the form of logs that sink to the bottom of our rivers after cutting. "Naturally, such a situation cannot be tolerated," he declared. Nikol'skii told the convention that during the past year he had worked on some questions concerning forestry. "I came across outrageous facts regarding the protected forests of watershed areas." These facts "need to be presented to the public at large." The only way that the country could begin to forge a necessary integrated policy for environmental protection was by creating a new Administration for Problems of the Protection of Nature and Natural Resources under the USSR Council of Ministers.[94]

Sergei Evgen'evich Kleinenberg pointedly began where Georgii Vasil'evich left off, expressing his "enormous sense of satisfaction at the circumstance that, finally, questions about the work of the *zapovedniki* are receiving some kind of airing before scientific public opinion, because in recent years, since the liquidation of the [Scientific] Council of the Main Administration, no information about the activities of the *zapovedniki* has reached *nauchnaia obshchestvennost'* and no one knows what is going on there. This [meeting] is an extraordinarily gratifying fact and for that we must thank our society," he declared.[95]

Kleinenberg, another respected zoologist, tackled the charged question of acclimatization. Twenty-five years after the violent battles over that question of the late 1920s and early 1930s, acclimatization still possessed the symbolic resonance that made it an explosive issue for the scientific intelligentsia. Acclimatization symbolized the wanton, arbitrary, and, for them, wrongheaded way in which the Bolsheviks thought they could disregard all science and social science in order to rearrange nature and human society. Acclimatization had become a symbol of what had happened to the country: its threat to endemic life forms symbolized Stalinism's threat to the "endemic" intelligentsia; its basis in doctrines hostile to genetics symbolized the Party-backed ignorant dictatorship in biology. Acclimatization was an affront not only to the scientists' biological expertise, but to their ethical vision of social relations as well.

Kleinenberg now declared, "I believe that acclimatization is no game. We must approach this technique with extreme seriousness and acclimatization for the sake of acclimatization must not be permitted."[96] Here, Kleinenberg played off the rhetoric of the old accusations against the conservation movement: that it engaged in the "protection of nature for its own sake." More than that, the charge leveled by Kleinenberg of "acclimatization for its own sake" showed a clear recognition of the symbolic and ideological uses that "biotechnics" had acquired in the hands of the Stalinist nature-transformers. Repeating warnings that acclimatization was "pregnant with very serious consequences," some of which were dangerous, Kleinenberg ended with a description of the Main Administration's plans as "hodgepodge, unserious, and ad hoc, . . . the results of which could be quite injurious not only to us but to the economy of the whole country."[97]

At this meeting the terror that had paralyzed a whole country for thirty years had taken a holiday. All were burning to have their say. Georgii Aleksandrovich Novikov spoke on behalf of "hundreds, even thousands of Soviet scientists and patriots of our Motherland" who were "alarmed" at the course of developments in the area of nature protection. "I must say," he declared, "that as a former, old researcher in the *zapovednik* system and having, to a great extent, developed as a zoologist through the . . . system, I have the impression that the current Main Administration is pushing us—*nauchnaia obshchestvennost'*—to the margins and is trying to operate alone, from its own [secret] cells. Perhaps this is more convenient, but I do not think that it is correct." Even here, "face to face with the representatives of Soviet scientific public opinion," the bureaucrats of the Main Administration refuse to speak openly and honestly about their shortcomings, he charged.[98] Calling for the Main Administration genuinely to submit to the methodological leadership of the Academy, Novikov twitted the Main Administration: "Really, now, is it that hard to contact the Leningrad Academy institutes? It's not hard, and it wouldn't hurt you to come over."[99] Then, correcting any false impression about why the scientists had converged on Moscow, he added: "We have enthusiastically responded not to your invitation, but to the call of the Moscow Society of Naturalists . . . and the leaders [of my institute] enthusiastically sent me off to take part in this meeting."

Novikov now took on Korol'kov's rhetorical support for interdisciplinary research and coordination: "You have mentioned Nesmeianov. Well, he says that neighboring disciplines in science must develop together. You, on the other hand, make contracts only with economic organizations. And under those conditions there will never be any sense [to your research program], the more so since the leaders of the Main Administration—I believe for many of those sitting here—the leaders are rather poorly known as scientists." With this withering aside, the hall erupted in laughter, causing Novikov to pause until things were brought back under control.[100] Novikov returned to the attack with a lacerating indictment of the Main Administration's undistinguished publishing record since 1951, noting that even the publications the officials claimed credit for were actually either done before or published by other organizations. "One cannot engage in a living cause with a cold heart and cold hands," he lectured Korol'kov and Malinovskii. "You must radically alter your entire style of operations. Only then, and not by yourselves alone, but together with all of us, may we solve the essential and venerable task of the protection of nature in our Soviet Union," he concluded, again to the enthusiastic hurrahs of the scientists.[101] Not in anyone's memory had high officials been subjected to such a relentless barrage of withering criticism. Scientific public opinion had waited thirty years for this moment, and now it was loath to let the opportunity pass it by.

After V. V. Krinitskii, the director of the Voronezh *zapovednik,* finished his remarks, Varsonof'eva interrupted for an announcement. Fourteen had asked to speak and only seven had done so already, and time had run out. She also had to provide time for the speakers to respond to more questions, for concluding remarks, and for voting on resolutions. The meeting's schedule was getting out of hand. What was to have been a one-day meeting had become a historic event. Someone from the audience suggested extending the conference another day, and Varsonof'eva put it to a vote. Regardless of whether the vote was indeed unanimous as the record states, it seems likely that there was a massive majority in favor of extending the conference; accordingly, Varsonof'eva set the beginning of the next session for 7:30 A.M.

Day Two

The next day brought a headliner to the speaker's platform—Aleksandr Vasil'evich Malinovskii. Having sat through much harsh criticism the previous day, Malinovskii decided to answer in kind. "It must be noted," he said, warming to his topic, "that the debates that unfolded [yesterday] . . . followed an error-ridden line that was adopted beforehand. In fact, those who spoke, as a result of their hotheadedness, tore themselves away from life, based their comments on rumors and on unproven facts, and reach conclusions and recommendations with which we can never agree."[102] Malinovskii challenged his critics to defend their comments in light of their not having set foot in his reserves over the past two years. (If some had visited, Malinovskii was supposed to have known about it, and therefore such furtive visits were "contraband.")[103]

Thus far, the *zapovednik* conference had been distinguished by the extraordinary tone of some of the speakers, bordering on lèse-majesté. As the senior Soviet official present as well as chief intended target of these remarks, Malinovskii had an obligation before the Party to respond.

> People have stated here that the operations of the *zapovedniki* are conducted on an isolated, secretive basis by people with hearts of ice. But what about the comments of Comrade Novikov, who spoke with such a hot heart and a hot head that it was terrifying to listen to him! Allow me to ask you, Comrade Novikov, over the past three years have you been to the Main Administration or to a *zapovednik?* If you have noticed something [some abuse or shortcoming], come to the Administration and tell us. But did you do anything of the kind? . . . Over two years there have been only outcries from this auditorium. . . . I decisively announce that work in the *zapovedniki* has in no way been conducted "in secret" as was charged here, and reject that in the most decisive fashion.[104]

Malinovskii next turned to the question of the boundaries of the reserves. Some had pointed to the Caucasus *zapovednik,* reduced from almost 300,000 hectares to 100,000, as an example of particularly wrongheaded boundary-setting, because seasonal pastures of some of the most important ungulates were excluded. Malinovskii hastened to correct the impression that he was the one at fault:

> One hundred thousand hectares. This was set by a decree of the Government and beyond that, I have neither the right as director nor the duty to audit that decision. People here referred to the decree signed by . . . LENIN. I am in agreement with them. But look, the last decree was signed by Comrade LENIN's pupil and disciple I. V. STALIN, and for you it is no secret that it was he who named the overall figure of 1,300,000 hectares [the total for the surviving *zapovedniki*]. When I increased the area of the Pechoro-Ilychskii *zapovednik,* the [USSR] Ministry of State Control wrote that I had violated the decree of the [USSR] Council of Ministers and that I might not be aware how that might all end.[105]

As he turned to the question of reviving the Main Administration's publishing activities, he noted that the old RSFSR Main Administration's publications were shut down in 1949, and he asked Formozov to tell the conference what had precipitated that. However, Malinovskii was interrupted by a shout from the crowd: "It was closed under you!" Malinovskii rebutted that he had just arrived from his service in the Soviet Administration in

Germany (SVAG) and started only in January 1950. "[That remark] is on your conscience," he retorted to the heckler. The only related material Malinovskii inherited, he claimed, was 120 tons of recyclable paper scrap, which he finally wrote off as a loss. Nevertheless, Malinovskii argued, the publication record of the Main Administration under him exceeded the publications of *zapovednik* research of the Academy system. He understood the scientists' feelings about the liquidation, "but it is impermissible to accuse indiscriminately and to state that the entire system is in ruins."[106] "Why don't you want to look to reality, Aleksandr Nikolaevich?" said Malinovskii, addressing Formozov. "Are we really going to hold up the economic development of the country because there is some *zapovednik* someplace?"[107]

Someone called out, "Then you must [develop] next door!"

Despite the catcalls, Malinovskii wanted to reach out to these scientists. He had already told them that Stalin, not he, was primarily (and personally) responsible for the general contours of the 1951 decree. However, what kept getting in his way was his ideology of transformism and economic development, which was irreconcilable with the vision of development and society embraced by scientific public opinion. Nature and *zapovedniki* were central symbols in this clash of social and economic visions.

Despite the gulf that separated them, Malinovskii did make an important overture: he offered to join in a petition to restore the Lapland *zapovednik.* He also proposed establishing three new reserves: one in the Briansk forest to commemorate the partisans who lived there, one at Shushenskoe, where Lenin lived in exile, and one at the Tul'skie zaseki, which would gain support because it included shelter belts.[108]

Coming to the end of his time, though, Malinovskii could not let Lukashevich's offensive address go unchallenged. "It seems to me that it is impermissible to speak out that way and that it is not allowed to state that a decree of the Government . . . must be reexamined at its roots," he warned the activist.[109] From the floor, however, a listener interrupted: "That is not how it was phrased." Malinovskii responded, "It is nonetheless necessary to have respect when dealing with this situation." He then directly answered Geptner's question of how much timber was cut per hectare in the Belovezhskaia *pushcha:* it turned out to be on average one cubic meter, with an annual increase of biomass of four cubic meters.[110]

Again, in a surprisingly conciliatory mode, Malinovskii promised to consider the "critical comments made here" and to analyze them thoroughly. "I have detained you a long time," he confessed, "but I think that we shall resolve this important question conjointly. . . . I hope that we will be able to come to some agreement." The only deflective note was his expression of the hope that, in the future, their discussions would have less of a "philosophical" hue and would be "closer to real life." Malinovskii proposed that the conference delegate its Presidium to continue a working relationship with him and his agency after the conference, with an aim to generate concrete proposals for improvement and reform of the reserve system.[111]

For historians of the Soviet Union, especially those who study a period still so close to the tyrannical rule of Joseph Stalin, Malinovskii's address is perhaps even more remarkable than the extraordinary speeches made by angry scientists before him. First, the very fact of Malinovskii's appearance to give an accounting of himself and his agency—in Russian, *otchityvat'sia*—to representatives of the mobilized scientific intelligentsia was a striking indication both of the scientists' power and of how much had indeed changed in the space of a year and two months. I would argue that this was the only constituency that was fearless enough and mobilized enough to compel such an appearance at such an early date. Second, Malinovskii's attempts to legitimize his authority, to exculpate himself (here at Stalin's expense), and to make concessions to public opinion are nothing short of astonishing. If Stalin's and Khrushchëv's Soviet Union were nothing but a successful totalitarian regime, such acts of legitimation would be utterly incomprehensible. Yet we know that schools cannot run without teachers, universities without professors, sanitation departments without garbage collectors, and so on. And a *zapovednik* system, even one as imbued with short-term pragmatic economic goals as Malinovskii's, could not exist for long without sympathetic, capable zoologists and botanists. The diehard guerrilla war that scientific public opinion began to wage against his agency was viewed by Malinovskii as a serious enough threat to warrant all reasonable efforts to end it, even to the point of making concessions.

As the rest of the meeting would demonstrate, however, the road to compromise would be long and hard. Lev Konstantinovich Shaposhnikov, academic secretary of the Academy's Commission on *Zapovedniki,* immediately challenged Malinovskii's view of the proper regime and research concerns of those institutions, as set out in his 1953 article "*Zapovedniki* of the Soviet Union."[112] There it was in black and white, in Malinovskii's words: "The basic task of the *zapovedniki* is [resource] management geared to solving problems in agriculture, forestry, hunting, and fishing." Shaposhnikov contrasted this to the classic view of the tasks of *zapovedniki.*

By the middle of the second day, Varsonof'eva was getting weary. When S. D. Pereleshin came to a discussion of 1951 and said: "There is also one very painful, unpleasant, and ticklish question. . . . At this conference someone mentioned the name of Merkulov," he was immediately cut off by Varsonof'eva with a forceful request "not to broach that subject. This is beyond our competence," she continued, "is not our task, and we will not discuss it here."[113] However, the assembled scientists wanted to have their full say. Mikhail Aleksandrovich Zablotskii, a widely respected figure who brought the European bison back from the brink of extinction, restarted the political theme. Countering Malinovskii's appeal to respect government and Party decrees that were already in place, Zablotskii cited recent editorials in the journal *Kommunist* and articles in *Pravda* and *Izvestiia* that called on Soviet people to criticize incorrect decisions. "And this, sad to say," concluded the zoologist, "is still applicable in our Soviet science."[114]

With an unmistakable allusion to Lysenko, Zablotskii began a bold and extended political commentary:

> In particular, we should note that up to now the phenomena of monopolies and of heavy-handed bureaucratic misrule in science have still not been rooted out and still confront us, not to mention a whole slew of other deficiencies. . . . [T]he most recent decisions of the Party and government have subjected to sharp criticism the hackneyed approach to the introduction of the grass-field system of agriculture, whose introduction a while back took place not without the awareness by that same Party and government. You and I have borne witness to the fact that a number of workers who spoke out against the indiscriminate application of [that] system were persecuted and tossed out of a number of agricultural institutions. . . . [T]he shutdown of a number of *zapovedniki* . . . was also carried out according to a decree of the [same] Council of Ministers . . . and that has inflicted a great deal of harm to the cause of protected territories. It seems to me that it is permissible to state here that in that decree too we may find the selfsame hackneyed approach.[115]

Doubtless Varsonof'eva breathed a sigh of relief when Zablotskii ended his overtly political commentary to discuss in depth the implications of the new, smaller post-1952 boundaries for the Caucasus *zapovednik,* which he knew well in connection with his work with the restoration of bison in that part of their former range. The subject of bison gave Zablotskii an opportunity to challenge Malinovskii's attempt to refute the charge that his Main Administration had sealed itself off from *nauchnaia obshchestvennost'*. The director of the Belovezhskaia *pushcha* had spoken to the meeting on the previous day, and he was peppered with all sorts of questions about the status of the bison herd in his reserve. Such dialogue was a good thing, but the flood of questions, explained Zablotskii, could only be explained by the fact that "for the past few years we haven't been able to locate a single informational dispatch concerning the work with bison" there. Zablotskii was able to relate firsthand how, in 1952, the Mammalogical Section of VOOP had scheduled three invited talks on the European bison, including a talk by Zablotskii. However, the Main Administration forbade two of the speakers to present talks, allowing only one presentation from among their staff. "Would it have been so horrible," Zablotskii asked Malinovskii rhetorically, "if the comrades could have heard about bison from the mouths of speakers directly?"[116]

Zablotskii now harpooned the whole work environment of the Main Administration: "The operative principle that underlies the department is that scientists of the *zapovedniki* and the staff of the Main Administration must . . . fulfill decisions [from above] more and discuss and evaluate them less. And those comrades . . . who permit themselves a modicum of independence of judgment, as G. A. Novikov so aptly observed, if their judgments do not coincide totally with the opinion of the Main Administration, risk having themselves cut off."[117] Zablotskii was interrupted by approving cries of "It's an Arakcheev regime" from the audience, the reference being to the rigid, dictatorial style of administration of military colonies under Alexander I's general, Count A. A. Arakcheev, which in Russian parlance had come to stand for the worst kind of bureaucratic authoritarianism.

Resuming his remarks, Zablotskii explained: "Sometimes it's done surgically, sometimes by a whole series of indirect ploys that create for researchers such moral conditions that force older scientists, unable to abide the 'new' demands, to leave by their own decision." Despite his recent face-to-face talk with Korol'kov and Malinovskii's seemingly conciliatory remarks, Zablotskii remained profoundly skeptical about any improvements under this administration.[118]

In his second turn at the rostrum, Formozov advanced ecological arguments against what he believed was Malinovskii's simpleminded approach to the presence of "surplus" resources in the reserves. "We have heard talk here," he began,

> that it is stupid not to clean up fallen trees in *zapovedniki* after a windfall, etc. But this question is not so simple. Under these fallen tree trunks capercaillie and quail make their nests, and marten and sable mate there in winter. I know of instances when ducks and capercaillie have made their nests under one and the same downed tree trunk for a number of years in a row. That means that this trunk is a valuable resource for them. Malinovskii has said here that around Moscow there is not much wildlife. That is true. But it is true because the woods have been cleaned excessively. . . . In forestry plantations often they conduct cleanups of the forests without taking into account the interests of other branches of the economy, which is wrong as well. But in *zapovedniki* such measures are simply impermissible.[119]

For Formozov, Malinovskii would never be able to emancipate himself from his narrow training in commercially oriented forestry. The zoologist drove home his main point. "Your speech, Comrade Malinovskii, proved that those who said that the Main Administration did not have a clear, correct understanding of its tasks were right," Formozov concluded to applause.[120] There seemed to be no letup to this barrage, but then, could the pent-up rage of thirty years be drained in one morning, or even two?

Andrei Aleksandrovich Nasimovich pointed to international practices that, he argued, put Soviet efforts to shame, and indicted the Main Administration for anti-intellectualism as he recounted how portions of manuscripts soggy from improper storage were brought to the Prioksko-Terrasnyi reserve and consigned to the fireplace. Nasimovich and a few others were able to rescue from the flames some valuable materials from the Tul'skie zaseki reserve. "I'm sure you agree," Nasimovich said to the audience, "that this adequately reflects the style of operations of the Main Administration and its attitude toward scientific research." Someone shouted "What a shame!" Malinovskii rose to his own defense, shouting back, "That's not true, they were duplicates." Nasimovich retorted, "That's not difficult to determine," and noted that the deterioration of records had become a universal problem in the system. Geptner jumped in, asking where the other scientific materials from the liquidated *zapovedniki* were being stored.[121]

Among the most passionate retorts was from Lukashevich, whose initial speech had also been among the most barbed:

> There has been an accusation made against me, which I simply am unable to ignore in light of its stupidity. . . . It is difficult for a person at the end of his life, who has worked as a propagandist for decades, to agree with your claim that in my remarks there was some sort of disrespect for the decrees of the Party and the government. . . . The fact is that the all too frequent references to the decrees of the Party and the government seemed to me simply to be an intentional design on the part of . . . Korol'kov . . . to shunt responsibility both from himself and from the official leadership [of the Main Administration] for what they have done.

"That's how we understood you," voices from the audience called out, providing Lukashevich with a needed opportunity. "It turns out that the entire auditorium understood me," he went on, "and only two persons interpreted my remarks tendentiously. Why is that so? It is my opinion that this is a reflection simply of inadequate respect. Who in the world am I? Do I occupy an important position? [Evidently] my name didn't ring any bells for you, and you [permitted yourselves] to fling stupid imputations my way, accusations that people usually do not throw around."[122]

Even the normally more reserved Avetisian, who followed, expressed his "amaze[ment] at the bureaucratic attitudes on the part of the leadership of the Main Administration toward the initiative taken by the scientists. It seems to me," he continued, "that the Administration should be gladdened that such Soviet scientists as Professors Formozov, Turov, Nikol'skii, Dement'ev, Geptner, and many others . . . wish to provide assistance to the . . . Administration. . . . The Administration should heed the voice of scientific public opinion and not hole itself up in its bureaucratic shell."[123] Avetisian, in a long and at times historically referenced address, joined his scientist colleagues in calling for a complete turnaround in reserves policy: not simply restoration of what was lost, but an energetic expansion, while there were still large undeveloped territories:

> There was a period of time when problems of nature protection were not thought to be sufficiently important. Some leading figures in the ministries and a segment of biologists believed that since the goal of the transformation of nature was posed that meant that there was no reason to protect nature. . . . There were even those extremists [*peregibshchiki*] who thought that agitation for nature protection was not necessary either. And it is no accident that when the question of truncating the network of *zapovedniki* was decided, at the very same time some leading [political] figures, not without support from the leadership of the [Main] *Zapovednik* Administration, posed the question of liquidating the All-Russian Society for the Protection of Nature as well.[124]

Avetisian recounted how he was the only dissenting opinion on the Gosplan RSFSR commission that was created to decide the fate of VOOP. "[T]he RSFSR Council of Ministers agreed with that [dissenting] opinion," he triumphantly noted, however, "and I can gladden you with the news that the Society for the Protection of Nature will continue to survive," as a wave of applause rolled through the auditorium again, causing yet another noisy interruption.[125]

Endorsing all of the big demands voiced by the scientist activists, Avetisian concluded that "the time has come" for a decree by the Supreme Soviet on nature protection generally, and that this complex of issues should be institutionally represented by a Main Administration for the Protection and Rational Use of Natural Resources and *Zapovedniki* attached directly to the USSR Council of Ministers. To keep a narrowly focused Main *Zapovednik* Administration under the aegis of the USSR Ministry of Agriculture was like "letting the cat guard the lard and the goat the hay."[126]

The time had now come to try to bring closure to an exceptionally tense two days. Understanding her role as a voice of reason, and afraid, perhaps, that things were now really getting out of hand, Vera Aleksandrovna Varsonof'eva, the conference chair, decided to intervene. She tried in a creative way to make the agenda of the scientists, seemingly so distant from the concerns and methods of Malinovskii and his team, not only intelligible but even attractive to the latter. According to her approach, the bitterly divided camps at the conference were actually mired in an *apparent* contradiction, because what the scientists were proposing was in their eyes just as much for the well-being of the society, the economy, and the state as the program of the Main Administration. The differences lay in methods of activity and in broader or narrower definitions of utility and benefit.

Varsonof'eva sought especially to avoid fatally alienating Malinovskii. Diplomatically, she recognized that "one did not get the feeling that [in] the comments of Aleksandr Vasil'evich there was an effort to oppose so sharply [these differences] or to so sharply condemn the opinion of the scientists." Addressing him directly, she assured him that "we consider it valuable and necessary to attain mutual understanding and to work together in close contact; criticism and self-criticism is one of the methods by which we work in our state. For that reason, no one should be personally insulted by criticism." "Perhaps," she admitted, "there was too much emotionalism in some of the talks and comments. Some questions were even touched on here that do not enter into our competence. It is not necessary to bring up the question of the role Merkulov played in that commission. The government will figure that out on its own. That is not our concern." She took pains rather to emphasize that it was "the sincere desire to help correctly orient this great cause that guided us, and it would be unwanted if our conference and our goals were understood in some other way."[127]

Finally, she said that it was her understanding that Malinovskii was now in agreement on the question of increasing the number of *zapovedniki*. She saw a possibility now to build bridges and a working relationship between the Main Administration and scientific public opinion.[128] She argued against the conference hastily voting on resolutions, especially with the atmosphere so emotionally charged. It would be wiser to elect a Commission and delegate it to propose resolutions, which could then be submitted to the votes of the three societies that convened the conference.[129] Yielding to Varsonof'eva's authority and her image as a kindly but strong old aunt, the conference dutifully elected a Presidium of eleven.

Not all the rancor had subsided, however. Korol'kov had unwisely described the attempts at compromise as a "kind of blackmail," prompting a last-minute speaker from the audience, Tsapkin, to renew calls for a Council of Ministers investigation of the reserves and the Main Administration.[130] This, in turn, elicited the response of Malinovskii, who now retreated to the position that "if criticism is baseless, then we will not accept it. . . . The government has provided the basic principles and the practical direction." Two voices from the audience sought to put Tsapkin's ideas to a vote: that the Council of Ministers should turn to representatives of scientific public opinion for assistance in reexamining the whole question of the *zapovedniki,* which would include a new investigation of the Main Administration and its reserves and also include the *zapovedniki* of the Academy of Sciences.[131] The vote for Tsapkin's proposal carried by a huge margin. In a huff, Malinovskii asked to be removed from the Commission.[132] Arguments continued inconclusively for quite a few minutes before Varsonof'eva ultimately brought the meeting to a close with an expression of thanks to all for their "active participation."

The "Three Societies" Conference of 1954 was a grand stage on which the scientists could brandish publicly their sense of professional dignity, for so long hidden in the recesses of their marginal socium. Central to the drama were the honor, dignity, and integrity of the scientists' and activists' social identity as embodying "certified public opinion." Only viewing the conference in this light may we appreciate the importance of what would otherwise seem petty, the parade of speakers rising on the second day to refute the characterizations of themselves made that morning by Malinovskii, himself rebutting the scientists' prior accusations made against him. Indeed, the battle had only just been joined.

CHAPTER ELEVEN

A Time to Build

Barely a year had passed since the Main *Zapovednik* Administration had been demoted unceremoniously in 1953 from an all-Union ministerial-level authority to a *glavk* (department) within the USSR Ministry of Agriculture and Supplies, with the added responsibility for hunting matters as meager compensation. Yet in the Russian Federation matters seemed to be moving in exactly the opposite direction.

In April 1954 I. K. Lebedev, a deputy premier of the RSFSR, wrote to Premier A. M. Puzanov, arguing that the inclusion of the Main Administration for Hunting of the RSFSR into the system of the RSFSR Ministry of Agriculture in early 1953, paralleling the reorganization on the all-Union level, "had an extremely deleterious effect on game management in the republic, as the experience of 1953 and 1954 has shown." (Almost 80 percent of the game wardens were let go.) With the support of another deputy premier, P. P. Lobanov, Lebedev suggested reattaching the Main Administration directly to the RSFSR Council of Ministers.[1] Nature protection and hunting were domains where the RSFSR was able to stake out a significant sphere of independence in policy, a sphere in which scientific public opinion had more than perfunctory input. This chapter will trace the emergence and early years of Glavokhota RSFSR, that republic's new agency for hunting and *zapovedniki,* as it rapidly tried to reconstitute the RSFSR's decimated network of nature reserves. Additionally, it will examine the activities of the Academy of Sciences Commission on *Zapovedniki,* reorganized in 1955 with a broader mandate as the Commission on Nature Protection. Together the two institutions contributed to an institutional renaissance for nature protection during the mid and late 1950s.

Glavokhota RSFSR

On May 26, 1954, the Bureau of the RSFSR Council of Ministers met to consider the question raised by Lebedev, with Premier Puzanov presiding. I. Kartsev, who had headed up the RSFSR Main Administration both before and after its incorporation into the republic's Ministry of Agriculture, was one of the most ardent supporters of reinstating the agency's (and his) former status, promising in response to one questioner that no new staff were needed and there would be no cost to the republic's fisc. With a green light from the Bureau, Kartsev was given a week to produce a workable draft proposal.[2]

Between May and December 18, when Puzanov signed a new decree de facto creating the new agency, one interesting emendation had occurred; the new agency was designated the Main Administration for Hunting Affairs and *Zapovedniki* of the RSFSR Council of Ministers, usually referred to by its acronym, Glavokhota RSFSR.[3]

Apart from serving as a symbolic rebuke to Malinovskii's Main *Zapovednik* and Hunting Affairs Administration of the USSR Ministry of Agriculture, Puzanov's action seemed to embody an implicit threat as well as a promise: to restore, recover, or create a network of *zapovedniki* in the Russian Republic so that the damage wrought in 1951—for which Malinovskii bore a share of responsibility—might be undone. This meaning was all the more striking in light of the fact that the Russian Republic was still bereft of its own *zapovedniki* at the time of the creation of the new agency.

As we know from the archives, however, Puzanov considered it expedient nevertheless to gain approval for the move from the Kremlin authorities, and on April 28, 1955, he wrote an extended justification of the decision to Soviet premier Nikolai Bulganin: "The RSFSR Council of Ministers considers it beneficial to restore the hunting affairs agencies to direct subordination to the RSFSR Council of Ministers and to the [analogous] executive committees at the *oblast', krai,* and ASSR levels, simultaneously transferring control over the network of state *zapovedniki* to them, as the existing system of authority [i.e., Malinovskii] over these *zapovedniki* is leading to a severance of their work from practical concerns and from the needs of hunting."[4] This had an ironic twist, for Puzanov was using Malinovskii's rhetoric of "practical concerns." It was a clever gambit, for the forest-obsessed leadership of Stalin and Malenkov had yielded to the rule of hunting enthusiasts Bulganin and Khrushchëv, and Malinovskii's almost exclusive emphasis on forestry now appeared to be a narrow-minded deviation.

With Bulganin's blessing given on May 20, 1955, the RSFSR Council of Ministers on July 29 agreed to start up the new administration with a staff of twenty-seven: short of I. Kartsev's request of forty-four and RSFSR

Gosplan's recommendation of thirty-nine, but still enough for a credible kickoff.[5]

And so on August 9, 1955, not even a week before the founding Congress of the new VOOP, Glavokhota RSFSR officially opened for business. Its new head, Nikolai Vasil'evich Eliseev, was instructed to take over sixteen *zapovedniki* from the USSR Ministry of Agriculture, which would be directly supervised by an Administration for *Zapovedniki* and for Renewable Commercial Wildlife. That administration's director would be a deputy head of the Main Administration and would be supported by four staff members.[6] Although we do not yet have information on how the Russian Federation was able to pry loose the sixteen reserves of the USSR Ministry of Agriculture's network located on Russian territory, perhaps the transfer of Ivan A. Benediktov from minister of agriculture to minister of *sovkhozy* on March 2, 1955, the removal of A. I. Kozlov on that same date from all ministerial posts, and the removal of Kozlov's patron Malenkov as USSR premier six days later—all in the aftermath of the February Plenum of the Central Committee—dislodged Malinovskii's chief political protectors, who could have blocked the republic's plans.

To the extraordinary good fortune of the conservation movement, the person designated to head the *zapovednik* section of Glavokhota was Georgii Evgrafovich Burdin, who, for the veteran scientists, seemed a reincarnation of Makarov. Burdin's positive energy and Eliseev's seriousness were reflected in the agency's first months of operation.

In August 1955 Glavokhota had also inherited a game management system that had been gutted by the RSFSR Ministry of Agriculture during the three years in which that agency ran it. Of 362 local wardens and inspectors in 1953, only seventy remained on August 9, 1955.[7] On October 1, the USSR Ministry of Agriculture officially turned over the promised sixteen *zapovedniki*. With all their new units in place, Eliseev and his team immediately activated a surprisingly energetic campaign against poaching, efforts to control which were described as having been "invisible" in a number of *oblasts* prior to the transfer.[8] Aiding the republic's three-score plus game wardens and inspectors was a growing army of 38,800 deputized "citizens' inspectors," up from 34,700 the previous year.[9] Change was in the air.

On January 27, 1956, still on the eve of the Twentieth Party Congress, an all-Union Conference of Directors of *Zapovedniki* and Game Preserves and Ranches opened in Moscow. Significantly, it was organized not by Malinovskii's Main Administration for *Zapovedniki* and Hunting of the USSR Ministry of Agriculture but by the RSFSR's Glavokhota, reasserting the leadership role in these matters once exercised by the old RSFSR Main Administration under Makarov.

The keynote speaker of the conference was the ardent new director of Glavokhota's *zapovednik* system, Burdin. In what must have been an electric moment for the old-timers, Burdin asked the indulgence of the audience so that he could discuss the proper role of *zapovedniki* in light of their history. "This is even more necessary," he explained, "in connection with the transfer of the *zapovedniki* of the Russian Federation to the system of the Main Administration for Hunting and *Zapovedniki* of the RSFSR Council of Ministers, *created anew* [emphasis added]."[10]

Citing at great length the resolutions of the conservation congresses of 1929 and 1933 as well as resolutions of the USSR Academy of Sciences of 1944, Burdin presented an unprecedentedly rousing endorsement for *zapovedniki* as bases for fundamental, not applied, research into ecological and evolutionary dynamics in undisturbed nature. Linked with that was the maintenance of a regime of inviolability. Of the highest priority for *zapovedniki* was "the preservation in their natural state of protected parcels of living nature with all of their component plants and animals, barring the use of the latter for economic goals and creating the conditions that would guarantee the essential processes of natural development of natural complexes in order better to preserve them for scientific and cultural purposes for all time, for future generations of humanity."[11]

Burdin rhetorically asked how the USSR Ministry of Agriculture's former stewardship of the reserves met the goals for them that he just outlined, and his response was hardly a ringing endorsement. In the Caucasus *zapovednik,* before its transfer to his agency, Burdin noted that two European bison had died owing to the spread of hemorrhagic septicemia from domestic cattle illegally grazing inside the reserve, while the poor protection provided by the reduced warden staff accounted for the death of the purebred bison "Beliaka" at the hands of poachers. Similar examples of laxity and indifference characterized a host of other *zapovedniki* of Malinovskii's system. In the Sikhote-Alinskii *zapovednik* in the Far East the attention given to the ginseng plantations "eclipsed the basic tasks of protecting the territory of the reserve and promoting its scientific study." Because they were spending such a great amount of time on the ginseng gardens, the warden-observers were not attending to their patrols, with deplorably predictable consequences. "As strange as it may seem," noted Burdin, "the workers of the *zapovednik* were not the initiators of the recommendation to protect the not very numerous individuals of the Amur [Siberian] tiger and [snow] leopard species that inhabit the region."[12]

Burdin upbraided the previous administration for permitting the "reconstruction" of riparian woodland in the Voronezhskii *zapovednik,* eliminating strips of black aspen. "It is more than in the realm of possibility that this policy has had more than a little to do with the unsatisfactory hydrological regime in the *zapovednik,*" he offered. He also deplored the artificial maximizing of the numbers of protected animals—in the case of the Voronezh reserve, beaver—beyond the natural carrying capacity of the habitat itself through "upgrading" the biological productivity of the reserve's vegetation. "This is not a task for the *zapovedniki,*" Burdin emphasized; "this must not characterize their methods of work." *Zapovednik* workers were "obliged to proceed first of all from the exigency of strict fulfillment of their duty to protect nature so that no harm is inflicted of the protected natural complex." This cardinal rule had been forgotten or deliberately denigrated by the previous administration. "These are all sad facts," contended Burdin, "but they exist in our *zapovedniki* and we are duty-bound to speak about them so that they may be eliminated in the nearest future."[13]

Citing the research done by zoologists, Burdin praised the fundamental research performed in the reserves and promised the full support of Glavokhota RSFSR. He also singled out the nature logs kept by the reserves—the *letopisi prirody*—as a valuable database whose continuity had been threatened or disrupted by unappreciative administrators and staff. Researchers also had obligations, noted Burdin, to keep current in their fields, which included staying abreast of foreign literature, and that presupposed a working knowledge of important foreign languages.[14]

Aware of the need to develop a larger constituency that would at least not be hostile to the reserves, Burdin encouraged developing limited, nondisruptive tourism (in closely controlled corridors) and building interpretive museums attached to the reserves. But, he repeated, that should be accommodated only if it did not disrupt the fundamental natural conditions of the reserves.[15]

Concluding with news that no one had dared to imagine they would hear so soon, Burdin revealed to the conference that, with the support of the USSR Academy of Sciences, Glavokhota RSFSR had already submitted plans for the consideration of the RSFSR Council of Ministers for the restoration of lost territory of the Caucasus, Sikhote-Alinskii, and Barguzinskii *zapovedniki* and for the restoration of the former Gorno-Altaiskii and Sudzukhinskii reserves. An astonished audience gave Burdin a stirring round of applause as the import of his presentation sank in. The day of the resurrection was dawning.[16]

When the conference resumed later that day, it was clear that Burdin's talk was on its way to becoming a watershed event in the history of nature protection in the Khrushchëv era. The first to identify the import of Burdin's presentation was Vsevolod Borisovich Dubinin, deputy chair of the Academy's commission, in the name of which he spoke. "This convocation," he began, "is a scientific form of linkage between those working in *zapovedniki* and scientific public opinion, and for that reason we greet this initiative of the new Main Administration. We hold out the hope that the Main Administration . . . will carry on the way things were before, that is, before Malinovskii, during the good old days when the Main Administration was linked with scientific forces and built its work on that basis."[17]

Going further than the diplomatic Burdin, Dubinin now sought to make explicit that the deficient state of reserve management was

> the result of the harmful anti-state practices of the old leadership of the Main Administration . . . such as Malinovskii and his assistants. . . . It is no accident that in the hallways of the conference reverberated joyous voices with the message that "a fair breeze was blowing." I believe that I am expressing the general feeling if I say that we very much hope that this breeze will fan itself into a mighty gale to whisk away all of the rot that has accumulated during the past few years and that our future will be freshened by the wholesome breeze of positive initiatives, which we already sense in the activity of the new personnel of the [RSFSR] Main Administration.[18]

Dubinin voiced an additional hope for the restoration of a journal dedicated to the scientific publications of the *zapovedniki,* the *Nauchno-metodicheskie zapiski Glavnogo Upravleniia po zapovednikam,* which had been published under Makarov from 1938 to 1950, enjoying wide recognition in scientific circles.[19]

Dubinin closed with rare praise for those who worked or researched in the *zapovedniki:*

> I want to tell the Main Administration that a golden treasure has fallen into your hands: they are the golden cadres . . . under your direct leadership. It is my hope that you will protect them like the apple of your eye. These cadres work in diverse little corners of our nature and a nurturing attitude toward these people, love for them and worry about them, popularization of their works, all these things are needed not only for the *zapovedniki,* but for science itself. For in all of our major surveys and in all of our generalizations we rely on these protected people [*zapovednye liudi*], who have done enormously much.[20]

For the long-suffering, much abused, and indescribably dedicated field biologists, this praise was balm for their wounds. To hear it from the influential Dubinin, the right hand of the great baron of Soviet zoology, General Evgenii Nikanorovich Pavlovskii, was especially sweet.

Echoing the general sentiments, a whole row of speakers acclaimed Burdin's talk. Tit Titovich Trofimov, who had worked with Stanchinskii in the now liquidated Tsentral'no-Lesnoi (Central Forest) *zapovednik,* admitted that during Burdin's remarks "I personally experienced some kind of unburdening, a feeling of breathing easier, like the feeling when you have been long walking through the forest and then come upon a bright open field," and urged the duplication and dissemination of Burdin's speech to all workers in the reserves and in the field of protection of nature. Spurred by Glavokhota's intention to restore the Altai and Sudzukhinskii reserves, Trofimov also made a strong case for the restoration of the Tsentral'no-Lesnoi. Although the Geographical Society had made a strong case for it as the only reserve whose nature represented that of the typical mixed forest of central Russia, Trofimov observed that because it boasted no geological or other natural oddities or monuments "that could knock your socks off," it fell easy prey in 1951. However, he argued, "as an *etalon* of nature, this is one of the finest tracts of land, the more so since the *zapovednik* was located at the headwaters of our Western Dvina, Volga, and Oka rivers."[21]

One of the most poignant reactions to Burdin's speech was from Ivan Osipovich Chernenko, former director of the Laplandskii *zapovednik* and an old activist who served briefly as scholarly secretary of VOOP in the wake of Kuznetsov's removal before resigning in a quarrel with Avetisian.

> We, the old workers of the system of *zapovedniki* . . . took the reorganization of 1951 extremely hard. We similarly viewed the organization of the new Main Administration [Glavokhota RSFSR] guardedly, particularly because hunting was placed ahead of *zapovedniki* in the name of the agency. We feared that the new administration, grounded in hunting principles, was trying to steer the *zapovedniki* in a direction we did not wish to see them go. But after the talk of Comrade Burdin our wariness has dissipated. There is a feeling that the period of the laceration of the system has passed and a new period of restoration, recovery to health, and resurrection of all that was destroyed has now commenced.

Chernenko then touched on the human cost of the destruction of the reserves system, pointedly seeking to honor the man they had felt forced to renounce: "It is just terribly sad that one of the founders of this cause . . . , the most honest and most dedicated of them all—V. N. Makarov—did not live to witness this glorious day."[22]

And so with a mixture of joy and sadness, but most of all hope, the USSR's embattled conservation activists—scientific public opinion—noted an important milestone in their fight for social affirmation and for their beloved cause. When Khrushchëv's epochal denunciation of Stalin stunned the USSR and the world two weeks later, it also put some force behind that balmy breeze that blew through the conference on Burdin's welcome words.

Aleksandr Nikolaevich Formozov, who also spoke at the conference, remembered with pain how "in 1950 and 1951 we were often told: 'your work is no good, your work is of little value.' This we heard from people who neither understood anything about this cause or about [our] scientific work." With the accession of Malinovskii in 1950 people began to "flee" the *zapovedniki*, recounted Formozov. "You've got to wonder how representatives of such a tribe as V. P. Teplov, Zharkovskii, [Oleg Izmailovich] Semënov-tian-shanskii, Mertts, Kozlov, and others were able to survive in such a system. These were literally protected bison [*zapovednye zubry*] who, somehow or other, managed to survive here."[23]

"How did public opinion react to the situation in which the *zapovedniki* found themselves?" asked Formozov. The answer was that it drew on its own reserves of power. Scientists united, forging a "system of MOIP, VOOP, and the Geographical Society" that resulted in the 1954 *zapovednik* conference. "The stenogram of that conference was then replicated," which produced "a shocking impression." "Nevertheless," continued Formozov, "the leadership of the Main Administration at that time not only took no measures to correct its work but erected even higher walls between scientific public opinion and the Main Administration, especially its leadership: Malinovskii, Romanetskii, and [Korol'kov]."[24]

Formozov mentioned the feeling of shame evoked by A. A. Nasimovich's talk at a joint meeting of MOIP and the Geographical Society, where he outlined the status of nature protection and protected territories abroad. (Nasimovich would present his talk to the conference the next day.)[25] The Soviet Union among all major countries was in last place in terms of percentage of national territory under protection, against a backdrop of significant measures undertaken by "capitalist" countries, even after the war. Such shame was experienced by Formozov himself, for his remarks provide us with a rare anthropological glimpse into the thoughts and feelings of a sensitive Soviet participant at an international scientific congress abroad. He himself had recently participated in the eighteenth International Geographical Congress in Rio de Janeiro together with representatives of seventeen other countries, mainly capitalist. In talks "in the hallways," Formozov and his foreign colleagues discussed many of these issues.

> I must confess with complete honesty that I was pained and extraordinarily ashamed for our country because it was embarrassing for me to talk about the status of *zapovedniki* in our country. I tried to keep from talking about it, and all the while [my colleagues] tried to question me about the state of our *zapovedniki* and what the situation is with nature protection here. It was like being roasted on a spit trying to avoid answering that question. . . . For that reason you'll understand me when I tell you it was extremely unpleasant for me to speak about that, since I could not tell people the truth and tried to put the ball back in their court by saying that there are wonderful data available about that matter.[26]

Formozov concluded by affirming that "now the leadership of *zapovedniki* is in secure hands, this leadership is now in hands we can trust. This brings us great joy today because today we need a good and strong line in this cause. . . . We must think about all of the *zapovedniki* of the Soviet Union as we are all patriots of the Soviet Union."[27]

Formozov's remarks provide an example of the often frustrated and unrequited sense of patriotic loyalty among the scientific intelligentsia. Prevented from making a contribution to their country's healthy development, as they saw it, and rebuffed in their efforts to salvage the country's international honor, patriotic scientists such as Formozov and Nasimovich nevertheless did not wish to pass into active dissent. They expressed one of the double binds of scientific public opinion. If their remarks were a cri de coeur to the kindred souls in the immediate audience, there are grounds to posit that there were other, more powerful audiences to which they were simultaneously addressing themselves: the leadership of the RSFSR and the Khrushchëv regime itself. Seeing his only recourse in a public washing of "dirty linen," Formozov's story was a warning to the central regime that authentic Soviet scientists could represent the USSR effectively, productively, comfortably, and in good faith at gatherings with foreign colleagues only if their homeland's policies genuinely inspired pride, not shame. But it was also an appeal to the government of the Russian Republic to step into the breach with unilateral action on nature protection that could compensate for the inaction at the all-Union level. At the beginning of the thaw, this was one way scientists sought to negotiate with the not completely monolithic Party-state regime.

In its discussion of the scientific research of the *zapovedniki,* Glavokhota's annual report—a variation of the speech presented by Burdin—immediately departed from the tone set by Malinovskii. "A series of investigations carried out in 1955," the report advertised, "were of great importance for the development of theory, particularly in the area of wildlife ecology." The report hailed the *letopisi prirody* (nature logs) as valuable innovations.

On the other hand, pragmatic practices that conflicted with scientists' strict understanding of inviolability of *zapovedniki* continued in force: predator control, habitat improvement, acclimatization, fire prevention, and other forestry measures. In addition, many research themes, although scientifically sound, still reflected outside pressures for economic relevance: the study of the natural fertility of the Barguzin sable, "the basis of the correct exploitation of the forest marten in the Krasnodar *krai,*" "results of the acclimatization of the Altai squirrel in the Caucasus *zapovednik,*" and a study on the natural renewal and stimulated growth of the Siberian Stone pine or "cedar" (*Pinus sibirica*), to name just a few.[28] Yet none of these pragmatic concerns disrupted the new tone of respect for scientific researchers, for fundamental research, and for the inviolability of *zapovedniki.*

Early in 1956 Glavokhota was asked to draft its operations plans for the next five-year period, through 1960. Although N. Krutorogov, Eliseev's deputy for hunting affairs, bore responsibility for compiling the draft as a whole, in the section on *zapovedniki* the hand of Burdin and the spirit of *nauchnaia obshchestvennost'* were now fully in evidence. In that section the "liquidation" of 1951 was unflinchingly confronted. "Until 1951 there were . . . forty-six *zapovedniki* on the territory of the RSFSR with an overall area of 9,955,300 hectares, of which thirty-seven with an area of 9,917,300 were under the aegis of the RSFSR Council of Ministers," it stated. And "[a]s the experience of the past few years has shown," it went on, "the elimination of a number of *zapovedniki* and the significant reduction of their overall area . . . inflicted great harm to the cause of the protection of the richest, most typical natural complexes in the various landscape zones [of the country] and to the ability of science to conduct scientific investigations directly under natural conditions, which is a matter of great practical and theoretical importance for science."[29]

Invoking feelings of shame, the report reminded the leadership of the RSFSR that the Soviet Union currently had a negligible 0.06 percent of its national territory under protection as *zapovedniki,* whereas analogous institutions represented far greater proportions of the national territories of many foreign countries, including the USA (1.0 percent), Japan (4.3 percent), Canada (0.74 percent), New Zealand (4.5 percent), and even Switzerland (0.39 percent).[30] Evidently speeches and counsel such as those of Formozov and Nasimovich were having an effect.

The report did not rely solely on shaming, however. In a real break with the past, it listed the civic groups, scientific institutions, and political bodies that had raised their voices for the restoration and expansion of the *zapovedniki* as justification for a fast-track mobilization of Glavokhota's efforts in this area.[31]

Burdin and company proposed the creation of eighteen additional *zapovedniki* in five years to take a major step toward representing the totality of "geographical zones" of the RSFSR, plus the expansion of those reserves whose areas had been slashed in 1951, in some cases restoring them fully to the status quo ante.[32] As Glavokhota director Eliseev told the Presidium of VOOP on May 11, 1956, the RSFSR "was not going to stand in the way," although the USSR Ministries of Agriculture and Fishing were being obstructionist. However, Eliseev added, the USSR Academy of Sciences had supported Glavokhota "in every way" and he was now seeking to mobilize the support of other bases of scientific public opinion. Even VOOP vice president, horticulturist, and Party stalwart A. N. Volkov, who admitted that "in my work I am far from any involvement with *zapovedniki,*" endorsed VOOP's resolution of support for Glavokhota, although the Society would never again be on the front lines of the fight for protected territories.[33]

From 850,100 hectares in sixteen *zapovedniki* inherited in 1955, by January 1, 1960 the network had expanded to twenty-two reserves (still short of the thirty-four proposed) with an aggregate area of 4,256,350 hectares.[34] The centerpiece of the efforts was the reopening of Kronotskii, with 964,000 hectares, and of Zhigulëvskii, with 16,700 hectares, long the object of petitioning by Sukachëv and Tiurin. In one year alone the system grew by over one million hectares.[35] No less impressive was the revival of scientific research and publications in the Glavokhota system, which, as in the Makarov days, again became a Mecca for leading scientists.[36] On thorny hunting problems Glavokhota moved equally resolutely, announcing on February 10, 1956, a prohibition against the hunting of all game birds in the

RSFSR during the spring season of that year with the exception of eight far northern and Siberian *oblasts*.[37]

Perhaps the strongest indication of how much things had changed—or rather had returned to historic patterns—was the response of N. Masterov, a deputy premier of the RSFSR, to V. A. Karlov, deputy head of the Agricultural Section of the Central Committee (in the division of RSFSR affairs) concerning the request of the Voronezh *obkom* and *oblispolkom* to increase the amount of timber cut in the Voronezh *zapovednik*. With the backing of his own Gosplan RSFSR, who instructed that "in *zapovedniki* the entire natural complex must be preserved," and the unflinching insistence of Eliseev's Glavokhota and a number of Academy institutes, Masterov sent Karlov a terse rejection.[38]

Although Glavokhota was unable to restore the defunct *Nauchno-metodicheskie zapiski (Scientific and Methodological Papers of the Main Zapovednik Administration)*, from 1955 on it began to publish jointly with Malinovskii's Main Administration an attractive monthly, *Okhota i okhotnich'e khoziaistvo (Hunting and Game Management)*, which became another powerful voice for nature. Despite the fact that Malinovskii occupied the editor in chief's chair at the journal, which gave the journal a dual personality, many articles hewed to the Russian Republic's more preservation-oriented and ecological line. One article of 1957, for example, condemned the use of DDT, arsenic-based compounds, fluorine-based organics, and other toxic pesticides for inflicting "great harm on wildlife and bird populations." It approvingly noted the passage of a special resolution of the RSFSR Council of Ministers of September 15, 1956, requiring the republic's Ministry of Agriculture within a four-month period to develop a strategy to apply these agents without loss to wildlife.[39] This was implicitly breaking ranks with Malinovskii's USSR Ministry of Agriculture, which was 100 percent behind the use of pesticides. Even more of a polemical challenge to Malinovskii was a provocative article by longtime VOOP activist P. Bel'skii entitled "*Zapovedniki* and Game Management," which also appeared in 1957.[40]

An unbroken united front against the Kremlin's depredations once again stretched from scientific public opinion to the RSFSR Main Administration (Glavokhota) to Gosplan RSFSR to their patrons and protectors on the RSFSR Council of Ministers. Scientific public opinion now used the Academy of Sciences' commission to broaden that united front across republican frontiers.

The Academy of Sciences' Commission Takes Off

With the blessing of Academy president Nesmeianov and, some say, on his initiative, on March 11, 1955 the Academy's Commission on *Zapovedniki* was suddenly transformed into a major player in the USSR's debate over resource use and environmental quality.[41] With the commission's transformation into the Commission on the Protection of Nature, the Presidium of the Academy restated its claim as "scientific advisor" to the regime, noting that the "solution to problems of protection of nature requires the active participation of scientists in developing the scientific bases for that goal, in preparing recommendations for government agencies regarding the protection of animals, plants, forests, water bodies, and soils, and for assuring the protection of nature in regions undergoing extensive development (reservoirs, hydroelectric stations, factories, etc.)."[42] Although certainly not slighting the importance of *zapovedniki* and other protected territories, the reorganization was a recognition that "the growing . . . use of natural resources, however, demands even greater participation by the USSR Academy of Sciences and the Academies of the Union republics in activities to promote the protection of nature in the USSR."[43]

Accordingly, the commission was charged with developing the scientific bases for nature protection and the renewal of natural resources, preparing recommendations for the government, coordinating activities of bodies of both the Biology and Geology-Geography Divisions of the Academy as well as those of the Union republics on issues relevant to nature protection, developing recommendations for new *zapovedniki* and other protected territories, and providing scientific guidance and oversight for research in the reserves. It was given the authority to publish up to three issues a year of its new journal, *Okhrana prirody i zapovednoe delo v SSSR*, with Dement'ev, who became acting chair of the commission, also serving as editor in chief.[44]

The fifteen members of the Presidium of the commission and the forty-six additional members read like a who's who in Soviet biology, geology, and geography. The parasitologist Vsevolod Borisovich Dubinin of the Leningrad-based Zoological Institute of the Academy served as deputy chair. Two geographers who were full members of the academy, A. A. Grigor'ev and Innokentii Petrovich Gerasimov, the new director of the Geographical Institute of the Academy, together with the historical biogeographer S. V. Kirikov, who was given refuge in that institute, gave that branch strong representation in the Presidium. Botany was also well represented, with Sukachëv staying on as the senior representative of the Academy's Biology Division. In addition to the zoologists Dement'ev and Dubinin, the Presidium included ichthyologist G. V. Nikol'skii and Dubinin's mentor, Evgenii Nikanorovich Pavlovskii, director of the Academy's Zoological Institute. A fascinating and complex individual, Pavlovskii, in addition to being an outstanding parasitologist, probably had the most political clout of anyone on the Presidium, owing to his directorship of the Military Medical Academy, also in Leningrad, where he investigated the control of epidemics among troops. Lev Konstantinovich Shaposhnikov, a lackluster zoologist but an honest and energetic individual, found a niche for himself as the commission's scholarly secretary.

Among the distinguished regular members of the commission were geographer David L'vovich Armand, zoologists Geptner, Formozov, and N. E. Kabanov, botanist N. V. Dylis, and soil scientist A. A. Rode. Seeking to be inclusive, the commission counted representatives of almost all of the Academies of the Union republics as well as of the Academies of Pedagogical and Medical Sciences.[45]

Perhaps seasoned by his difficult experience as vice president of VOOP, Dement'ev came into his own as acting chair of the Academy commission. At a plenary session held on April 5 and 6, 1956, Dement'ev presented a one-year retrospective as well as a guide to the future.[46] Nobody hearing it could doubt that once again Russia's scientific community was on the move.

A complicated figure, Georgii Petrovich Dement'ev was born in the St. Petersburg suburb of Petergof on July 5, 1898. His father was a middle-class physician. Dement'ev spoke German and French and acquired the vast erudition typical of the better-educated youth of his generation. From the age of ten he loved birds. After a move to Moscow in 1920 to study at Moscow University he became an authority on the ornithological collection of the Zoological Museum there. A bent toward systematics led him to collaborate with his teacher, S. A. Buturlin, on *The Complete Identification Book for the Birds of the USSR,* which appeared in 1928. Hired on to the staff of the Zoological Museum, he earned advanced degrees in 1936 and 1940 and was brought into the vertebrate zoology department, where from 1956 he headed the Ornithological Laboratory. After completing a monograph on falcons (1951) and one on the birds of Turkmenistan (1952) and directing the six-volume *Birds of the Soviet Union* (1951–1954), his interests shifted increasingly toward the protection of nature.

Although his professional life had been shaped by the valiant "zoological intelligentsia" centered on the Moscow University Zoological Museum, Dement'ev did not have the stomach for a fight. Unlike the quietly courageous Makarov or the fiery Protopopov, he was prone to compromise or even acquiescence (although he was not a tool of the authorities, as were Andrei Grigor'evich Bannikov and Nikolai Gladkov). But in that wonderful springtime of hope that followed Stalin's death and Khrushchëv's secret speech, Dement'ev was able to lead, and, in leading, to help write one of the brightest chapters of postwar Russian conservation history.

Dement'ev parted company with the other old-timers in one other respect. Because he was more loyal to the Soviet regime than the old guard, he was not as fixated on the issue of inviolable *zapovedniki,* which had, for his colleagues, a transcendent significance. Dement'ev was able to approach nature protection more pragmatically and could recognize important emerging new issues and solutions. In this he found crucial support from his deputy on the commission, Party member Vsevolod Borisovich Dubinin, despite Dubinin's long and intimate association with the Astrakhanskii *zapovednik*.

Striving to reorient the wasteful Stalinist economy to sustainability, Dement'ev highlighted the rhetoric of economic self-interest, cautioning that "it is erroneous to suppose that under conditions of the Soviet social system and the presence of a planned economy . . . the necessity for organizing a special system for the protection of nature is excluded." On the contrary, he insisted, "protection of nature is an inseparable and essential part of the planned economy," whose proper pursuit would guarantee expanding quantities of pelts, game, fish, lumber, and agricultural crop production.[47]

Public health issues were also intimately connected with nature protection, asserted Dement'ev. Did not M. D. Kovrigina, USSR Minister of Public Health, state as much in her speech to the Twentieth Party Congress just two months prior, pointing to the health threats of water and air pollution? Because the regime had not yet placed a ban on discussing the subject, Dement'ev also daringly raised the "new and fundamental problem . . . developing ways of protecting living organisms from the harmful consequences of exposure to radiation."[48] "In that way," he continued,

> the contemporary tasks of nature protection extend far beyond the limits of the passive preservation of existing natural resources—animals, plants, and individual parcels of nature—as it was understood in the historically first phase of the movement for nature protection, particularly in our own country; the more so since the goal of preserving the status quo ante in this case is not only practically inexpedient but also technically impossible. For that reason the view that the basic task of nature protection is to promote the creation of *zapovedniki*—that this [emphasis] is the most complete and developed, "highest" . . . form of nature protection—is also an incorrect one. The organization of *zapovedniki* and the improvement of their work is an important and necessary cause, but it is only a small part of the whole problem. What deserves fundamental attention is nature protection in its broader sense—in sites where resources are being exploited and on developing the scientific and economic basis for regulating the use of natural resources.[49]

Dement'ev had no intention of marginalizing the struggle for protected territories, and he returned to the question toward the end of his speech, if only to reassure the partisans of the *zapovednik* cause that he was still a friend. Even so, he reiterated that "it is high time for a decisive repudiation of that view that holds the *zapovedniki* as the 'highest' form of nature protection." Although *zapovedniki* were valuable as *etalony* (baselines of nature), they represented areas that were spared only the "*decisive* intervention by humans." He called the reserves "conditionally natural" (*uslovno-estestvennye*) parcels, a state that could still be compared meaningfully with that of overtly exploited areas. However, he implied that they should not be idealized as either "pristine" or "harmoniously perfect." Nevertheless, he supported efforts to expand the network of the reserves:

> Changes and growth in the economy constantly push to the fore the question of establishing new *etalony* of unexploited, unused nature and of the study of its changes as compared with those . . . in exploited areas. There is neither geographical correspondence nor one of scale in the existing network of *zapovedniki* to the natural geographical regions or zones [of the country]. For that reason the commission holds that the question of developing a rational, scientifically grounded network of *zapovedniki,* taking into account a broad understanding of nature protection, is a major and important task.[50]

On the organizational front, a working group headed by Evgenii Mikhailovich Lavrenko, veteran botanist and former researcher at Askania-Nova in the 1920s, was created to develop just such a network. Dement'ev had raised some issues that had been swept under the carpet as the scientific community attempted to resurrect exactly that status quo ante to which Dement'ev had alluded in his critical remarks. It remained for the political "ecology" of the Soviet Union as well as ecological theory generally to change enough to give these important questions a proper hearing. That would not happen soon.

Decrying the balkanized way in which resource issues were still addressed in the Soviet Union, with each ministry responsible for the slice of nature that provided its raw materials, Dement'ev called for a more integrated, ecological approach, but here too biocenology and population ecology had fallen short, he conceded. As Sukachëv had reminded ecological thinkers, in order to sustain living systems one also had to sustain their habitats, which included nonliving natural features and qualities.[51] That truism was all well and good, but ecological science still had only the sketchiest of road maps.

Dement'ev observed with satisfaction that the Presidium of the Academy had already moved ahead with initiatives designed to support the commission's work. On May 4, 1955, it issued a directive requiring all institutions affiliated with the Academy to identify specific natural sites or living species that required special protection, limitations on use, or measures to restore and increase their stock. Union republic Academies were requested to do the same. Already valuable data and reports had streamed in to the commission.[52]

Another great triumph of 1955 was the organization of analogous commissions in each of the Union republic Academies; the Uzbek commission even managed to organize a conference as early as October 3.[53] Other conferences were held in Tbilisi, Stalinabad, and Ashkhabad, while representatives of the commissions participated in the Congress of VOOP and the

All-Union Ornithological Congress in Leningrad and addressed meetings of the Geographical Society (in both Moscow and Leningrad) and MOIP.[54]

Khrushchëv's overtures to the international community had decisive and heartening consequences for nature protection as well. Conservation activists were among the most vociferous Soviet scientists in demanding access to their foreign colleagues and to the broader international scientific community. Dement'ev put it well when he stated, "The solution of the problem of nature protection in a number of ways requires going beyond the state frontiers of the USSR and requires forging international contacts. As an illustration we may point to problems of an epidemiological nature, questions concerning the pollution of extraterritorial waters, fishing issues and the hunting of marine mammals outside territorial waters, and the question of the protection of birds and fish whose migratory paths take them beyond the USSR, among others."[55] Respecting the protection of the Arctic environment, the 1954 meeting of the International Union for the Protection of Nature's General Assembly in Copenhagen made it clear that "it was not possible for the USSR not to respond to this initiative."[56]

There were other reasons, too, for reaching out to foreign colleagues. With Stalin not only dead but also the subject of unprecedented criticism by the new leader, the old sin of "kowtowing to the West" was quietly dropped as a heinous intellectual crime. It was now possible to acknowledge that useful things could be borrowed from "over there." The commission had lost no time in establishing ties with many foreign counterparts.[57] Literature exchange was begun with conservation organizations in Switzerland and Finland, and the commission even sent an inquiry to the Ornithological Society of the Netherlands about the danger to birds that migrate to Soviet sites from Dutch landfill efforts. Of particular importance was gaining approval, finally, for Soviet representation on the International Union for the Preservation of Nature; the personal invitation by that organization's president to the commission during his visit to Moscow was an altogether promising start.[58]

The commission was also moving on other fronts. In the Far East it had initiated efforts to protect the highly endangered sea otter and sea lions. To protect Arctic fauna, the commission had recommended a total ban on the killing of polar bears and a ban on harvesting walrus for all but indigenous peoples of the North. A *zapovednik* was urged for Novaia Zemlia, reviving the plans of 1949. In light of the deteriorating water quality in the USSR, the commission joined with Glavrybvod (the Main Inspectorate for the Protection of Fisheries of the RSFSR Council of Ministers) and the Ministry of Public Health to monitor untreated waste-water discharges and to recommend ameliorative measures. Dement'ev noted that the Volga basin was particularly polluted, as were those of the Don, Tom', and Northern Dvina

Rivers, the Caspian and Azov seas, and a number of reservoirs. Of particular concern was the crisis situation of Lake Imandra, which was the recipient of naphthalene wastes from the huge Apatit plant.[59]

A few victories had already been won. The Khabarovsk and Primorskii *kraiispolkomy* enacted bans on the hunting of the Siberian tiger at the commission's request. Similarly, Glavokhota had banned the culling of reindeer on the Kola Peninsula. An ornithological station on the Baltic coast, earlier eliminated, was restored after a petition by the commission and the Zoological Institute. Even the RSFSR Ministry of *Sovkhozy* relented and halted the sowing of the Poperechenskaia steppe, a tiny parcel of 280 hectares in Penza *oblast'* that was of great botanical interest. Dement'ev could justifiably conclude that "the initiative of the USSR Academy of Sciences in creating our only general organ for nature protection thus far has facilitated a cardinal mobilization of work in this area."[60]

If 1955 was a good beginning, 1956 and 1957 were years of triumph for the Academy's commission. Its new, broader focus led it to train its sights on air and water pollution, especially in cities. And again, it took the lead in sounding the alarm about radioactivity as a new environmental danger to both people and other living organisms as well as an agricultural threat. Probably owing to Dubinin's forward-looking scientific interests, the crucial importance of radioecology in tracing the pathways of radioactive substances through living nature was given high billing.[61]

Ties with the international nature protection community were finally established in the summer of 1956 when Dement'ev, Shaposhnikov, and A. V. Malinovskii went to Edinburgh for the Fifth General Assembly and the Sixth Technical Meeting of the International Union for the Conservation of Nature and Natural Resources, where the delegates unanimously voted to include the Academy's Commission on Nature Protection as a full member. The novelty of Soviet participation was tremendous. The special exhibit set up by the Soviets (photographs of *zapovedniki* plus a copy of the first number of the Commission's new *Bulletin*) attracted a good deal of interest and a special plenary session was organized to hear the Soviet delegates on June 26. Whereas Dement'ev spoke on protection of fauna and Shaposhnikov discussed the role of *zapovedniki* in nature protection overall, Malinovskii's talk was fittingly titled "The Use of the Protective (Pest-Control) Properties of Forests of the USSR."[62] Doubtless Malinovskii's inclusion in the delegation was the price the Academy commission had to pay to gain Party approval to attend at all, but the awkwardness of the situation must have been all but unbearable to Dement'ev and Shaposhnikov.

An brisk tempo of work continued through the fall, with a major report by Dubinin on the drastic hydrographic and ecological changes of the Volga delta, which again called into question the Astrakhanskii *zapovednik*'s status as a representative, baseline tract (*etalon*). A month later, on October 16, a conference on nature protection in Central Asia was held in Tashkent with Dement'ev and Shaposhnikov again in attendance. Interestingly, the report on the conference notes that "particular attention was paid . . . to the question of the pollution of the water bodies of Uzbekistan and the problems of fishing associated with hydroelectric construction" in the region. Had the Khrushchëv regime allowed the Academy commissions to have a real say in development, perhaps the catastrophes of the Aral Sea, the Amu-Dar'ia and Syr-Dar'ia Rivers, and the Ili River basin could have been avoided.[63] Meanwhile, on the legislative front a major victory was achieved with the enactment of the commission's draft law "On Measures to Protect the Animals of the Arctic" by the RSFSR Council of Ministers in November.[64]

With the new year the commission intensified its work on the threat of radioactivity. "Atomic industry is growing at a furious pace in our country and abroad," wrote Shaposhnikov,

> and . . . the ever-widening scale of prospecting for and processing uranium ore, construction of atomic reactors, and the experimental explosions of atomic and hydrogen bombs are leading to the rapid increase of the radioactive background levels over the whole planet and to the pollution by radioactive substances of particular areas of the globe. The question of protecting animals and plants from the harmful effects of ionizing radiation are acquiring ever greater importance. The Commission is collecting evidence on the influence of ionizing radiation on the animal and plant kingdoms, is studying the literature published here and abroad, and is actively trying to attract specialists to work on solving these problems.[65]

In April 1957 a large Transcaucasian conference in Baku was organized by G. A. Aliev, chair of the Azerbaijan Academy's Commission on the Protection of Nature. To the Azerbaijani capital came large Armenian, Georgian, and Dagestani delegations as well as the Russians Dubinin and Shaposhnikov, who represented the USSR Academy. Owing to the somewhat deceptive yet impressively effective pax Sovietica, these often feuding groups were able to sit down together amicably and reach common positions on a whole range of problems. Some photographs of the meeting saved by Vsevolod Dubinin and his family provide a sense of that optimistic season that now seems to us like a distant warp of time.[66]

In Georgia, a major victory was won when that republic's Council of Ministers on April 10, 1957 adopted the local commission's proposal to restore eight *zapovedniki* liquidated in 1951. Not to fall behind, the Latvian government followed exactly two weeks later with its own decree creating four new *zapovedniki*. However, the fruits of the Estonian commission's efforts were particularly impressive. On June 7, the Supreme Soviet of that republic discussed and approved the first republican law on nature protection generally, drafted by zoologist Erik Kumari and the Estonian Academy's commission. With its provisions for a regular procedure for the establishment of *zapovedniki,* for protection of rare species of plants and animals, for the creation of parks, and for the establishment of an Administration for the Protection of Nature to function under the immediate aegis of the Estonian Council of Ministers, the Estonian law led the way for the entire USSR. In November the RSFSR Council of Ministers came through with the first installment on the restoration of the lost *zapovedniki* of that republic, with legislation that reestablished the Altaiskii, Laplandskii, Bashkirskii, and Sudzukhinskii reserves—over a million hectares altogether.[67] This was against the immediately preceding backdrop of the approval of the Lavrenko plan for "a rational network of *zapovedniki* for the USSR" by the Presidium of the USSR Academy of Sciences with strong endorsements by biology division chief V. A. Engel'gardt and Nesmeianov.[68]

In addition to writing a highly influential article in *Pravda* of July 13, 1957, "The Protection of Nature Is a Matter of State Importance," in which he propagandized the work of the Academy commission but also trained attention on the problems of industrial air and water pollution, soil erosion, and the *zapovedniki,*[69] Nesmeianov took the lead in calling for an authoritative all-Union State Committee for Nature Protection with an Academy of Sciences resolution of March 15, 1957. That having failed, Nesmeianov wrote personally to RSFSR premier Frol R. Kozlov on January 28, 1958 to request the creation of such a committee on the RSFSR level, pointing to Lithuania and Estonia as precedents.[70]

Despite the failure on the all-Union level of 1957, Nesmeianov continued to back the commission's efforts to enact all-Union nature protection legislation and create an all-Union conservation service. A proposal for draft legislation to that effect was submitted to the USSR Council of Ministers again on October 16, 1958. Unfortunately, the bill was handed over to a commission of the Presidium of the Soviet cabinet apparently headed by V. V. Matskevich, USSR Minister of Agriculture, who let it die quietly.[71] Not one to give up, Nesmeianov again tried the route of the RSFSR with yet another letter of January 8, 1960.[72]

After Gosplan of the USSR organized a special commission on December 31, 1957 to prepare recommendations for the long-term development of individual branches of the economy in connection with the prospective seven-year plan, the Academy's Commission on Nature Protection petitioned Gosplan to create another special commission to deal with the challenge of protecting and replenishing natural resources. Historically friendly, Gosplan on February 1, 1958, created such a commission with Academy president Nesmeianov at its head.[73] Parallel to the work of Nesmeianov's commission, the USSR Academy commission together with those of the republics drafted a twenty-year plan for the protection and restoration of natural resources, which they submitted to Gosplan USSR in 1960.[74]

Finally, the Commission for the Protection of Nature organized three large all-Union Conferences on Nature Protection held in 1958 in Tbilisi, in 1959 in Vilnius, and in 1960 in Stalinabad/Dushanbe, in addition to a host of republican and regional conferences. From a staff of four in 1955 the central commission had grown to thirty-seven by 1961, and its budget had grown correspondingly from 79,500 rubles to 660,500 rubles. Dement'ev in 1960 had been elected to the Executive Committee of the International Union for the Conservation of Nature, and Shaposhnikov was chair of that organization's committee on education.[75]

Thus, although Nesmeianov's conversation with Aleksandr Ianshin and Vera Varsonof'eva in 1952 had not seemed to augur well for his support of the cause of nature protection and especially for that of the *zapovedniki,* he was perhaps as good a friend to the movement as anyone who had ever sat in the Academy president's chair.

With Glavokhota and the Academy of Sciences system as invaluable active allies and patrons, the scientific community now accelerated its organizing efforts during this golden period of political relaxation. However, the stalwarts of the movement were not growing any younger; many did not live to see Stalin's body removed from Lenin's tomb or the publication of *One Day in the Life of Ivan Denisovich.* It was only natural that the activists would begin to pay greater attention to the question of who would carry on their struggle. Would the lost tribe of scientific activists die with them, or would they succeed in instilling their values and their social identity in a new generation?

CHAPTER TWELVE

A Time to Meet

Cemented by propinquity, the alliance between MOIP and MGO grew into a powerful force from the 1954 "Three Societies" Conference through the end of the decade. The personalities of the Moscow leaders of geography and geology, such as Andrei Aleksandrovich Grigor'ev, Grigor'ev's successor as head of the Institute of Geography (IGAN), Innokentii Petrovich Gerasimov, the good-natured Ivan Dmitrievich Papanin, and the secretary of the Geographical Society's Moscow branch, Iurii Konstantinovich Efremov, played an immense role in developing these links. Nevertheless, structural and theoretical developments in geography and the earth sciences, together with the influential presence of Nasimovich, Formozov, Leonid Nikolaevich Sobolev, and other zoologists and botanists in the Biogeography Department of IGAN, as well as the rising prominence of the biogeographer Anatolii Georgievich Voronov, an MGU professor, within the Geographical Society, all worked to push geography as a discipline into an intimate embrace of nature protection issues.

Enter the Geographers

By the mid-1950s geography was finding a second wind in the Soviet Union. Gosplan began to seek out geographers for consultations, especially when planning hydroelectric installations, particularly after the unsatisfactory experience with the Rybinsk reservoir, which demonstrated the adverse consequences of doing without solid scientific advice.[1] The "Stalin Plan for the Great Transformation of Nature" and Khrushchëv's Virgin Lands program also created a demand for detailed descriptions and maps of large areas of the USSR, a boon for geographers.

With geology and meteorology now completely separate disciplines, geography—a loosely knit federation of such fields as economic geography, physical geography, cartography, and biogeography—had lost its most "scientific-looking" subdisciplines, intensifying the discipline's crisis of identity. Physical geographers made a bid for control over the discipline and its image, hoping to stabilize geography's prestige by emphasizing the scientistic landform categories of their field. Nature protection afforded those in biogeography and economic geography the opportunity to try to claim for geography the title of the "environmental science." For the discipline as a whole and for the various subdisciplines, this was a complicated interplay that would affect the distribution of resources and the future of research directions.

Theoretically, geography was experiencing a period of ferment as the search for a coherent object of study for the discipline, long elusive, was drawn now to an attempt to identify "natural" units in the overall environment. This search had much in common with the evolution of the discipline of ecology. One major Soviet school in geography was linked to the ideas of Nikolai Adol'fovich Solntsev, who asserted the possibility of identifying coherent units—"landscapes." Basing his ideas on those of V. V. Dokuchaev, who incorporated soil layers into an understanding of larger biotic-abiotic systems, and Lev Semënovich Berg, who was the first to use the term *landshaft* (landscape, from the German *Landschaft*), Solntsev saw the "landscape" as a self-contained molecule, much as ecologists from Kozhevnikov to Sukachëv imagined the biocenosis. Solntsev and his school proceeded from a geomorphological starting point, viewing each landscape as formed by a discrete genetic process, involving a certain originary mother rock and a shared history of uplifting and weathering, in turn determining the vegetation.

A competing perspective, partly influenced by A. A. Grigor'ev's ideas, was held by V. B. Sochava, E. M. Murzaev, S. V. Zonn, and I. P. Gerasimov, who saw the unity of "landscapes" as the product of a process of coevolution: here, the later history of the landscape mattered more than its earliest history, with biota assumed to have greater capacity to influence soils, relief, and ultimately the vegetational cover than in competing theories. Nevertheless, both views agreed on the presence of "landscapes" as holistic units in nature, however defined.

A few geographers, such as David L'vovich Armand, Iurii Efremov, Fëdor Nikolaevich Mil'kov, and others argued for a broader understanding of landscape that was less essentialistic. For them, geography was the broadest study of all—that of the earth's entire envelope of life, akin to Vernadskii's biosphere—and they viewed the scientistic search for a smaller "empirical" unit of study as a misguided narrowing of the discipline's perspective.

Finally, botanical and animal ecologists now lodged in the Institute of

Geography of the Academy of Sciences introduced to geography the perspectives of those who sought to define natural communities—biocenoses—on the basis of spatially bounded processes of mineral cycling and shared rates of productivity. These descended from the approaches of I. V. Larin, L. E. Rodin, and Nina Bazilevskaia, and found expression in the productivity studies and cycling analyses done by the Institute of Geography on the Tsentral'no-Chernozëmnyi *zapovednik.*

Despite their interpretive differences all these groups could unite around a single assertion: the world of life was under threat and needed to be protected.[2]

Geographical propinquity as a factor in human social affairs is not to be dismissed lightly. We have already seen how the occupation of the same building by MOIP and MGO helped forge the powerful alliance between the two organizations during the 1950s. Propinquity also had a hand in the enlistment of geographers into VOOP in the early 1950s, a development that had interesting consequences later on. On the twenty-fourth through the thirty-first floors of the main tower of Moscow State University is the Museum of Earth Sciences (*muzei zemlevedeniia*). Thematically, it suggests approaches to the study of three levels of our environment: local (*kraevedenie*), regional (*stranovedenie*), and planetary (*zemlevedenie*). In 1949, upon his demobilization from the Soviet Army and his return to Moscow (having pioneered the geographical exploration of the Kurile Islands in 1946 while still in uniform), Iurii Konstantinovich Efremov (see figure 15) sought to resume his teaching duties at Moscow University, which the war had interrupted in 1941. Born on May 1, 1913, Efremov apprenticed as a metalworker in 1930–1931 and then attended the Omsk Agricultural Institute, transferring to the Moscow Timiriazev Agricultural Academy until 1934. A stint working with tourists in the Western Caucasus preceded his enrollment in Moscow University, from which he graduated in 1939. Now, Efremov wanted to create a museum of earth sciences. Soviet bureaucracy being as formidable as it was, it took six years to prepare for the opening, which was just in time for the 200th anniversary of the university in 1955.

Gurgen Artashesovich Avetisian had also joined the staff of the museum, and he and Efremov struck up an acquaintance. Shortly after VOOP's merger with the Green Plantings Society, Avetisian, as president of the Organizing Committee of the new merged organization, invited Efremov to attend one of their meetings. Drawing in other geographers, Efremov and his colleagues began to constitute a "fifth column of geographers in a milieu that was rather geographically ignorant." With the departure of much of VOOP's old guard to MOIP, VOOP's *aktiv* now consisted of a group of "gladiolus and strawberry breeders from the outskirts of Moscow" who "had nothing at all in common with ecology, biocenology, and nature protection" as the founding generation understood it.[3]

Figure 15. Iurii Konstantinovich Efremov (1913–).

About that time, the Academy's newly reorganized Commission on Nature Protection had, among its other initiatives, prepared a draft law on nature protection for the Russian Federation. Promoted and largely written by the commission's secretary, Lev Konstantinovich Shaposhnikov, a zoologist of high ethics but juridically and economically naive and narrowly trained, the draft law focused exclusively on protection of biota and protected territories, omitting issues of human health and what we today understand as "environmental quality." Owing to its defects, the draft failed to gain the approval either of legal experts or of Gosplan of the RSFSR, and there the matter rested for about a year.

After the Estonians enacted a pioneering nature protection law in 1957, however, someone—Iurii Zhdanov, head of the Central Committee's Science Department, or Mikhail Suslov, perhaps—decided that in light of the international atmosphere and considerations of the image of the USSR abroad a Soviet law on nature protection should be enacted. It was thought prudent to begin with the remaining republics before enacting an all-Union law; such preliminary efforts would help to "get the lumps out" of the all-Union law.[4]

That such a consideration began to emerge from within the Soviet Party elite about that time is generally confirmed by the recent testimony of a Kremlin insider. Iurii Zhdanov has written that

> still working in the Science Department, I accidentally became aware that few in our country were writing about its nature. Specifically, there were no natural history albums. On the initiative of our department the Geography Faculty of Moscow University under the leadership of I. Gerasimov created the first album, *The Nature of Our Motherland.* Pictures of natural scenes in *zapovedniki* were included in it as well. At the time, the album was still a thin, monotone affair, but it soon was expanded and improved. . . . [U]nder Khrushchëv contacts with foreign figures surged. At such meetings it was the custom to give coffee-table albums on the nature and culture of your country. And we suddenly remembered: aside from our album there was nothing at all. That is when the decision was made to reissue the volume as a deluxe gift album, *Nature in the USSR.*[5]

The same logic applied to the legislation and to the decision to permit the Academy Commission on Nature Protection to join the IUPN about the same time.

Once it was decided that the republics should follow in Estonia's footsteps, the RSFSR Council of Ministers, finding the Academy commission's efforts unacceptable, moved on, forwarding the assignment to VOOP. However, by 1957 that organization was, to understate matters greatly, even less equipped than the commission to draft such legislation. As Efremov described it, "You can imagine at what a loss [the strawberry and gladiolus breeders] found themselves after they had been summoned by the government to prepare a nature protection law for Russia. It was then that they turned for help to the Geographical Society. And we responded."[6]

Together with Leonid Nikolaevich Sobolev and David L'vovich Armand, Efremov took the initiative and decided to craft a law that would shift the emphasis from *zapovedniki* to the realization that "nature was dying" as a result of improper use, greed, and waste. In the absence of anyone else in VOOP capable of completing the assignment, the new president of VOOP, Mikhail Mikhailovich Bochkarëv, who also was chair of the RSFSR State Forestry Committee, had to turn to Efremov's group; appealing to Bochkarëv's "practical" side, they were able to convince him to support them in their "geographical" approach. Efremov's approach emphasized preserving environmental quality even as nature is used and transformed rather than trying to preserve "pristine" nature. To solicit as much feedback as possible, Efremov and his team held "seminars" for planners, economic managers, and legal experts at which they fielded objections. Armand's rhetorical skill made him a standout in these discussions. "We even put the legal experts to shame," gloated Efremov thirty years later, because they had not even considered these questions until they began attending the seminars. The shamed legal experts moved quickly to establish university departments or chairs of conservation law such as the one created at MGU by V. V. Petrov.[7]

Another outcome of Efremov's activity was the penetration of conservation ideas to the Party's elite. Efremov got an unexpected invitation from P. A. Satiukov, editor in chief of *Pravda,* who asked the geographer if he could present some of his ideas to the paper's editorial board. On June 5, 1957, Efremov arrived at the *Pravda* complex near Leningradskii Prospekt. About twelve persons were present at the meeting, including a good percentage of the editorial board.

What Efremov told the assembled Party journalists must have pricked some ears: "The protection of nature is very much impeded by the underestimation of its importance, by a lack of comprehension of its significance, and by lack of respect for this citizens' movement both by the broad masses as well as by leading [Party] figures." Perhaps, he offered, that lack of respect was spurred by the label "protection of nature." "'Protection of nature' for some sounds like something conservative, like a call to some kind of museumlike mummification and to an absurd inviolability of nature generally; it seems like some kind of private cause and concern of a few weirdo do-gooders [*chudaki-blagotvoriteli*], like a haven for those unfit for gainful work." Unlike his field biologist friends, geographer Efremov had his own objections to the term *okhrana prirody* (protection of nature); it sounded too passive. "Rather," he said, "defenders of nature must be on the offensive, they defend not for the sake of preservation and mummification but in order to enrich natural resources. For that reason their banners should read 'Protection and Enrichment of Nature.'"[8]

A wide gulf and feelings of "strain" characterized the two polar positions regarding the use of nature in recent Soviet history, he explained. Some of scientific public opinion supported the extreme position of using bans and prohibitions to preserve nature as it is. On the other hand, some partisans of use "justified their rampages and their rapacious ravishing of nature by such demagogic 'principled' appeals as: 'From whom are we defending nature? From the people? No! Nature must serve the people! It must be subjugated and transformed, not protected!'" This kind of slogan-mongering, argued Efremov, did a lot of damage, especially when combined with a cavalier attitude toward the complexities of the natural world. "It disoriented people, deflected them from the necessity of continuously caring for nature—our mother—and the basis of our productive forces and our economy," he continued. Confronting us, he declared, "is the necessity to overcome first of all the mass disrespect to the tasks of protection and enrichment of nature" and to oppose the equally harmful view of nature protection as "simply this season's big campaign."[9]

Efremov explained that the reason nature protection needed to be a constant concern was that it "was the cornerstone of the present and future well-being of the inhabitants of our planet." Nature protection and enrichment was the permanent rudder that balanced the "inevitable [dialectical] contradiction between nature and its use by humanity, between the inevitability of using up natural resources and caring for their replenishment and further expansion." For that same reason, he argued, now addressing partisans of preservationism,

> it is not necessary to show anxiety when we are witnesses to ever newer intrusions on nature as it is: this is an expression of that same permanently operating contradiction that leads us now and will no doubt lead us many times into conflict. Enterprising neighbors will never stop greedily eyeing protected meadows, or lumbermen the forests, or hunters the game animal. Under the flag of "temporary measures" they will always find justifications . . . for exceeding tempos of exploitation set by science.

For that reason, Efremov repeated, nature protectors had to be eternally vigilant. The dialectic furnished no respite.[10] Efremov held out some hope of social peace between the economic managers and the nature defenders, however. If the principles of true planning gained ascendancy over the principles of resource-grabbing among the economic managers themselves, Efremov assured the editors, "then there would be less necessity for the defenders of nature to engage in sharp struggle from their side. The economic managers must transform themselves into defenders of nature, its friends and enrichers," instructed Efremov.[11] Then all social interests could truly be harmonized.

Although his emphasis was on what we would call "wise use," Efremov was also a supporter of inviolable *zapovedniki,* although for him they had a lower priority than for the hard-core field biologists. Significantly, though, despite the fact that it might sound like the agenda of some "weirdo do-gooder" before this tough audience, he defended *zapovedniki* as they had been originally constituted, that is, as inviolable and established for all eternity. Although later, he noted, "this formulation was thrown out as allegedly idealistic for asserting the existence of 'eternal values,'" Efremov cleverly noted that those critics "forgot that Marxism-Leninism had discovered *eternal* principles; let us recall, for starters, the transfer of land eternally for the use of the collective farms. . . . Setting aside of protected territory only justifies itself if it is for all time. . . . The principle of *eternal preservation* [*zapovedanie navechno*] must be *eternally* restored to our *zapovednik* cause and in legislation."[12]

As a representative of scientific public opinion in the specific form that such a social group assumed under Soviet conditions, Efremov found himself caught in his own dialectic. Despite his disposition toward economic development, he was still at bottom a non-Party scientist who was fighting to defend the dignity of his social identity in a system that was at best disrespectful of it. And in his milieu the struggle for the purity of *zapovedniki* was *the* central means by which this social identity was affirmed and announced. In the bowels of the system, at the editorial offices of *Pravda,* as he struggled to find a common language with the Party *apparatchiki,* Efremov found himself unable to betray his own values or identity. Although the rich and productive tension generated by these efforts permeated his entire remarkable presentation, his proud restatement of his dignified claims to civic empowerment and respect are most vividly profiled in his discussion of nature protection as an ethical endeavor. "The protection of nature is a battlefront," he said, "where the struggle demands valor and decisiveness and a deep conviction of the rightness of the principles being defended. Valor is needed at all levels of this struggle. For the warden guarding the *zapovednik* the poacher's bullet always threatens. Incidents of the heroic deaths of scientists at the hands of vengeful violators of *zapovednik* conditions are familiar to us; we need only recall the fates of Isaev or Kaplanov." In an obvious reference to A. V. Malinovskii, Efremov continued:

> Unfortunately, such courage has not always been shown by the highest leaders of nature protection and *zapovednik* management. . . . It wasn't even the bullet of the poacher that frightened them, but merely the ire of their immediate superiors, little black marks on their high reputations. But their stamps of approval, signifying assent to the ravaging of major natural treasures, to the destruction of *zapovednik* conditions, were more than little black marks. They left an inky streak on the cause of the protection and enrichment of nature, and led here and there already to irreversible devastation and to irrevocable losses.

For the *real* representative of scientific public opinion, the defense of honor and ethical duty, not public office or fear of political retribution, had the greater claim on one's actions. At least that is how scientists wanted to think about themselves.

Efremov tried to shame the *Pravda* editors by noting that the USSR was in last place among major nations in the percentage of national territory under protection:

> The fact that we, with all of the wealth and bounty of our natural resources, have fourteen times less land under protection in percentage terms than even the United States, echoes for us like a reproach among international scientific and cultural public opinion. Still more shameful is the fact of the liquidation and clear-cutting of forests of those *zapovedniki,* for example in the Carpathians and the Transcarpathians, that arose under conditions of capitalism and were assiduously cared for by the Ukrainians and Poles even under Austro-Hungarian rule.[13]

Nor did Efremov neglect the "impermissibly large scale of soil erosion in our country." Wasn't it ironic, he asked, that the erosion-control ideas of physical geographer D. L. Armand, "who has just brilliantly defended his doctoral dissertation," were successfully implemented by the Chinese but ignored by the USSR Ministry of Agriculture, in whose system he worked?[14]

Efremov called for the editors' support for the creation of an all-Union society for the protection and enrichment of nature and an authoritative state committee with the power to levy "severe sanctions" on violators of the law.[15]

Efremov's final point concerned his long-standing passion for *kraevedenie.* Almost until the latter movement's final forcible disbanding in 1937, *kraevedenie* and the nature protection movement had been inseparable civic twins. They had shared much of the same leadership, and a remarkable number of their leading activists were active in both movements. If we look upon both *kraevedenie* and nature protection activism not simply as esthetically motivated but as activities laden with social meaning, the linkage between the two immediately becomes clear: both represented the idea that fragile human social relations, particularly those that affirmed the political, moral, and intellectual dignity and empowerment of the educated citizen within the community, were built up with great sacrifice and over immense political obstacles in Russia. The ecological community and the local cultural region that was the object of the *kraeved's* study both were formed from a long process of coevolution. Both symbolized the ideals of diversity within harmony—a harmony cemented not by hierarchical authoritarian power but by the almost organic ties of mutual dependence, assistance, and duty, old themes in Russian intellectual social thought. Stalin's attempt to make Soviet society uniform through his politics of "leveling" [*uravnilovka*] of all genuine, autonomous diversity (tightly controlled state-sponsored ethnic dance troupes and the like were Stalin's replacement for that diversity) via a "great transformation" of society and nature both was viewed as a mortal threat to the prerevolutionary intelligentsia's ideal.

Efremov's love was old Moscow. After Stalin's death he fought to restore the historic names of Moscow's streets, efforts that have only recently been rewarded. Just as he and other nature protection activists sought to roll back the "vandalism" of 1951, they sought to restore other symbolically resonant landmarks of the old world that Stalin had tried to efface.

Most important, though, was "to guarantee a decisive about-face in the public opinion of the whole country concerning the protection and enrichment of nature." And for that, Efremov concluded, an "authoritative article" had to appear in *Pravda* providing a "corrective in their worldview, the most fundamental attitudes of people toward nature."[16] Satiukov and some of his colleagues were so impressed that he asked Efremov for materials for an editorial. Although Efremov prepared an extensive draft, no editorial resulted.[17] But was it pure coincidence that *Pravda* published Academy president Nesmeianov's powerful essay on nature protection some five months later?

Despite the *Pravda* disappointment, Efremov was later asked to write the speech introducing the Russian Republic's new law on nature protection for Nikolai Nikolaevich Organov, chairman of the RSFSR Supreme Soviet. Invited to witness the enactment of his law by the RSFSR Supreme Soviet, Efremov for the first time in his life entered the mysterious government compound within the Kremlin.[18]

From the perspective of enforcement, the law passed on October 27, 1960 was no more distinguished than other Soviet legislation; liability and enforcement authority were both unclear, and the law's gaps seemed larger than its substance. Much was sacrificed between Efremov's daring presentation before the editors of *Pravda* and the ultimate redaction of his draft law by a team of bureaucrats. Nevertheless, the law contained phrasing that reflected a sophisticated understanding of society and nature as a system. It called for "taking into account the interrelationships among the resources listed [separately] under Article 1, so that the exploitation of one resource will not inflict damage on others"; it called for continuous qualitative as well as quantitative monitoring of what we identify as resources, to be centralized in the Central Statistical Administration of the RSFSR; it prohibited reductions in the size of useful natural areas such as forests, meadows, and bodies of water unless they were specifically approved for alternative uses; and called on economic agents to avoid damage to natural resources during construction. Efremov's law called for a prominent place for science in the planning of nature protection strategies and specifically granted VOOP the right to create citizen inspectorates to monitor compliance alongside the official agencies.[19]

The decade (1955–1965) during which Efremov served as the scholarly secretary of MGO under Papanin's presidency was the golden age of that organization's civic activism. Founded in 1945, the branch grew quickly and only one year later began to publish an influential series of anthologies, *Voprosy geografii* (Problems of Geography), which grew to well over one hundred numbers. By the end of Efremov's term the Moscow branch alone had 2,555 members, about what the entire Geographical Society had had in 1946.[20]

In close alliance with MOIP, where his counterpart Konstantin Mikhailovich Efron (see figure 16) labored equally tirelessly under Sukachëv and Varsonof'eva, and with the Academy of Sciences' Commission on Nature Protection and the Academy's Moscow House of Scholars under Professor V. I. Sobolevskii, Efremov was at the hub of the frenetic organizing activity of the period. After the effective departure of almost all of the old-timer biologists from VOOP, Efremov and Armand were among the handful who served as that society's sole remaining bridge to the older scientists' movement.

Within the Moscow branch of the Geographical Society an important role was played by the Biogeography Commission, founded in 1956. Its very first session, dedicated to problems of nature protection with talks by Dement'ev and Nasimovich, set the tone for the future activity of this unit, which sponsored twenty-seven talks in 1956–1957 alone, attended at times by more than 200 people. Acknowledging the prominence of the new commission, the Moscow branch in 1957 voted to entrust to it the preparation of an entire volume of its anthology. This appeared as number 48 of *Voprosy geografii* in 1960, devoted to problems of biogeography and the protection of nature.[21]

The Geographical Society was, along with MOIP, one of the oldest surviving scholarly organizations in the Soviet Union, having been founded in 1845. Like MOIP and VOOP, it survived in the postrevolutionary period doubtless owing to its aura of venerable quaintness: here was another clan of *chudaki*. Like MOIP and VOOP, through the late 1940s it had a modest membership, although the number of full members had expanded from 896 in 1941 to 3,560 in 1947.[22] Its Second Congress (the Second All-Union Geographical Congress), which took place in January 1947, was attended by 1,600 delegates and guests.[23]

On December 24, 1950, longtime society president Lev Semënovich Berg, an eminent academician, limnologist, and biogeographer, died. Merkulov's minions were turning their attention to the *zapovedniki* and to VOOP, and could just as easily turn on the geographers. Like VOOP, the Geographical Society endured a tense hiatus until the Party finally gave its permission to elect a new leader. Unlike VOOP, the society chose brilliantly. Assuming the presidency on July 23, 1952, Evgenii Nikanorovich Pavlovskii, the eminent parasitologist, director of the Academy's Zoological Institute and the Military Medical Academy, and academic politician supreme, led the Geographical Society for more than a decade, until May 29, 1964. Serving under him as vice presidents were Gerasimov and Stanislav Vikent'evich Kalesnik, a glaciologist who had served for a decade under Berg as scholarly secretary.

Figure 16. Konstantin Mikhailovich Efron (1921–).

Again paralleling VOOP, the Geographical Society was allowed to hold its next Congress (called the Second Congress of the Geographical Society of the USSR) in 1955. The 209 delegates were joined by more than 2,000 guests, now including foreigners from eleven countries. Among the most notable of the 106 talks were those by Sukachëv and by Zonn on the shelter belts, by Armand on soil erosion, and by N. E. Kabanov on the scientific role of *zapovedniki.* With the Moscow branch at the oars and with the sympathetic captaincy of Pavlovskii and his crew, the larger Geographical Society now also began to set a strong course for nature protection.[24]

Within the Geographical Society the following years saw an accelerating series of activities and conferences dedicated to problems of nature protection, but what must rate as the landmark event of the era was a conference so powerful that, at the last moment, publication of its proceedings was halted by the censors.

The 1957 Conference on Rare and Endangered Species

When the old-timers were driven out of VOOP and into the arms of MOIP, they represented an enormous fund of organizing experience as well as passion for nature protection. Although Aleksandr Petrovich Protopopov had already made an unforgettable impression at the 1954 *zapovednik* conference, the feisty activist had an idea for one last campaign that he hoped would dramatize the larger issue of the accelerating impoverishment of the biotic world around us. As longtime former chair of VOOP's Crimean Commission, Protopopov was no stranger to management and organization. Having convinced his friends G. G. Adelung, I. O. Chernenko, and A. A. Nasimovich (see figure 17) to serve with him as a "war council," Protopopov spared no efforts to ensure the success of his proposed all-Union conference on rare and endangered species of plants and animals, set for March 1957. The organizational committee was based at MOIP in the old Zoological Museum on Gertsen Street, where Protopopov now worked in MOIP's newly created Commission on Nature Protection.[25] With the Moscow House of Scholars and the Geographical Society's Moscow branch strongly on board, the four committee members sent letters to scores of colleagues around the country to get out the word.[26]

Real spring was still over a month away when the Conference on Problems of the Protection of Valuable, Rare, and Endangered Species of Plants and Animals and of Unique Geological Objects and Their Rational Use (its cumbersome official title) began in the early evening of March 25, 1957. Into the low white building—a former aristocratic mansion—on Kropotkin Street near Chistyi Lane, in the older part of the Arbat, streamed a huge crowd of *chudaki*. As they pressed to get out of the freezing air, they jammed the lobby and the parklike courtyard all the way back to the wrought-iron gates by the street.

When P. A. Polozhentsev's gavel brought the conference to order, nearly every seat in the stately auditorium was filled. With well over 600 in attendance, this was not only already a resounding triumph for Protopopov, but an astounding demonstration of scientific public opinion. Of this number, 400 had come from out of town, from as far as eastern Siberia and Central Asia. The press, which had decided to cover the event, could not fail to notice this impressive constituency: the cream of Soviet field biology, geol-

Figure 17. Andrei Aleksandrovich Nasimovich (1909–1983).

ogy, and geography. Party and government officials did not miss the significance of such a show of strength, either, which may help to explain the timing of the invitation to Efremov to come to speak to the editors of *Pravda.* "In a certain sense," wrote Efron three years later, "this may be considered a pivotal moment in the struggle for the protection of nature, giving its participants the push to organize a series of local conferences that permitted a new beginning for activism at the periphery and in the center."[27]

Over five days the conference heard 108 talks, beginning with G. E. Burdin's broad survey from his vantage as head of Glavokhota RSFSR's *zapovedniki.* He set a cosmopolitan tone for the mass meeting by characterizing nature protection as "reflecting not only vital national interests of one particular people or state. The solution of these problems concerns all the peoples of the world."[28] Such a public acknowledgment of one's international citizenship had become safer as a result of Khrushchëv's active diplomacy, with its implicit criticism of Stalin's isolationism—isolationism that had been painfully ironic, because Marxism was the "cosmopolitan" ideology *par excellence.*

Much as he had delighted the conference of *zapovednik* directors and workers the year before, Burdin now thrilled the huge auditorium with the announcement that 108,000 hectares had already been restored to the Caucasus *zapovednik* and that his agency had slated fifteen new reserves to be created by 1960, along with the expansion of existing ones.[29]

In a blatant swipe at Malinovskii and the USSR Ministry of Agriculture's approach, Burdin again declared that "the protection of nature, including the forests, must be achieved on a strictly scientific basis, holistically [*kompleksno*], and may not be subordinated to only a single narrow economic goal," a position that completely supported the claim of scientific public opinion to be the arbiter of scientific standards governing resource use.[30] However, he went on, "in the actions of the leaders of those economic agencies which simultaneously bear responsibility for nature protection as well, the economic tasks always are given priority." Part of the problem was that the press and the Party were "generally not terribly interested" in those agencies and the result was that "the protection of natural resources is sacrificed by [the agencies] for narrow economic interests, which are guided by the attitude toward resource availability: 'a hundred years is plenty for us.'"[31]

Of all the remaining formal talks, though, with the possible exception of A. A. Nasimovich's review of foreign literature on the protection of nature in which he brought up the issue of the harmful effects of pesticides, the presentation that pushed the political limits furthest was that of A. A. Peredel'skii, who worked in radiation ecology. "In the majority of cases," stated Peredel'skii, "human activity is the causal factor of the extinction of species." Now, however, there was an "unusually serious" and growing threat, not simply to individual species "but to all life itself, including humanity," the threat of radiation.[32]

With the proliferation of radiation a new branch of biology had emerged—radiation ecology—Peredel'skii informed his listeners:

> A child of the atomic age, radiation ecology at the present time is in its first stages of infancy. However, the basic outlines of its future profile are already sufficiently defined. . . . In first order facing radiation ecology is the task of assisting with epidemiological controls in the atomic age, to identify the biological pathways by which radioactive isotopes on those abundantly poisoned areas of dry land, water, or air are disseminated further. These are the result of military and experimental explosions of atomic and hydrogen bombs and of the activity of nuclear reactors and other nuclear-related industries, including extraction and enrichment, and also the broad use of natural and artificial radioisotopes in technology and in scientific research, medicine, and agriculture.[33]

Although the diffusion of isotopes through soil erosion, wind action, and the circulation of water, and the roles of temperature and precipitation, were all well studied, the migration of these isotopes through the tissues of living and dead organisms remained mysterious. First-level concentrations of radioisotopes in the organs of living creatures, Peredel'skii suggested, were thousands of times greater than in the general environment. When the organisms' carcasses were eaten, these concentrations substantially increased, rising in tandem with the place of the consuming organism on the food chain. "What emerges are dauntingly complex ecological 'radiation food chains.'"[34]

Although bodies of water could be cleansed by some bacteria or by insect larvae, which concentrate the isotopes and then disperse them when the insects reach their mature, airborne stage, the problem was one of scale. "We might well ask," he said, "what is the nature of the threat to cetaceans [whales and dolphins] . . . in the oceans?" The Pacific was now polluted by radioactive dust and ash from the explosions by the United States in the Marshall Islands, he warned, and isotopes could collect in plankton and cause a catastrophic extermination of whales.[35]

And while radiation was a threat to all species, he averred, it was a particular threat to those rare and endangered species already on the brink of disappearing. In a highly unusual final appeal Peredel'skii concluded: "Nature protection activists must . . . address the governments of all countries, insisting on the prompt attainment of strict bilateral and international legal measures to protect nature from pollution by radioactive isotopes."[36] It was plain that Peredel'skii was making no distinction between "good" Soviet socialist radiation and "bad" capitalist radiation.

After the first evening's plenary session the conference broke up into more specialized sessions and workshops, reconvening for a final discussion on the penultimate evening of the great meeting; the final day's plenary would be devoted to resolutions. After four days of horror stories, the mood was militant. One listener submitted a written question to L. K. Shaposhnikov of the Academy commission, that began: "Comrade Shaposhnikov, isn't it time to move from words to action? Nature is disappearing catastrophically."[37] Another charged, "While you are creating your very own bureaucratic system of commissions within the Academy of Sciences nothing will remain of nature. There won't be anything left for you to protect. Have you truly heard the 'cries from the soul' that welled up from the audience?"[38] The well-meaning Shaposhnikov was hurt: "I was astonished by the first lines [of the question]. I don't believe, after all, that the Academy of Sciences deserves that kind of characterization."[39]

What the questioner did not appreciate, though, is that Academy president Nesmeianov was an expert at the "game," and knew that the appointment of a firebrand as scholarly secretary of the nature protection commission could have knocked the entire game board over and his job into the bargain. At each level—from activist to Academy president to chairman of the RSFSR Council of Ministers—the trick was to play the game as close to the threshold of political permissibility as possible without going over the edge. For each level of player there was a different set of rules and a different threshold that defined political life or death, promotion or demotion, freedom or imprisonment. Worse yet, these thresholds were always in motion; they could narrow abruptly at the whim of the Party bosses. For players, the object was not only to play as close to the edge as possible but also to engineer the retreat of that edge: that is, to gain playing ground and political "space" for their side.

However, unlike the politicos in the RSFSR cabinet or even the Academy president, who was still a loyal member of the Communist *nomenklatura,* authentic scientific public opinion constituted a special class of players who had their own rules and their own goal line at the rear of their playing field, retreat beyond which would fatally compromise their social identities and their self-respect. Scientific public opinion continually had to balance between compromising its ethical injunctions in order to keep itself in business—a genuinely valuable social goal in an authoritarian political regime such as that of the USSR—or acting on its sense of entitlement to full civic and political rights and risking curtailment of privileges or even obliteration. Neither choice was easy or satisfactory.

Burdin was next to answer the questions of the audience. One question dealt with the touchy subject of acclimatization and culling of animals within the Glavokhota *zapovedniki,* as these measures were viewed by activists as the very symbols of the hated transformation of nature and as a deep profanation of the purity of the reserves. Here, too, the RSFSR had come through for the activists; Burdin told an ecstatic audience that both measures had been ended in *his* reserves system.[40]

The next question, though, introduced a note of disquiet into the otherwise triumphal gathering. Someone asked whether there was any truth to the announcement yesterday at the conference that the Crimean and Belovezhskaia *pushcha zapovedniki* were to be reorganized into game management preserves. Burdin responded that Glavokhota did not have any official information on this, although he did have a copy of the letter Pavlovskii had sent to the political leadership protesting these changes.[41]

After the question and answer period, there was a final round of statements before people dispersed for the night. Ecologist G. A. Novikov, though a Party member, was sharper than most in his criticism of the political leadership:

> For us here today, as well as for many others, . . . it is completely clear that the question of the protection of nature in the Soviet Union is in an extremely grave state. . . . The broad Soviet public, whom we represent, has for a long time already been demonstrating its deep concern over this matter. . . . If those comrades who were placed by the Party and government to direct this cause [Malinovskii and his aides] had acted as befitted Bolshevik leaders and

> worked together with the masses instead of avoiding them like the plague, then we would not be in the shameful situation in which we now find ourselves. I will speak candidly about this, naming names, as was done a number of years ago. There was a time—1950—when the cause of *zapovedniki* was prospering. There then followed a period of sharp deterioration. Who is guilty of this? Malinovskii. I, as a Soviet scientist and as a Communist, cannot, I must confess, understand for the life of me why our leading political institutions have not listened to what broad public opinion has been saying about this matter and about this person [Malinovskii]. I cannot understand how this person, who has compromised himself from the bottom to the top before the Party and the state is still occupying his position![42]

Novikov's harangue set off a demonstration in the hall. When the cheering and clapping died down, Novikov resumed his unexpectedly explicit remarks: "How is it that he still heads the [USSR Ministry of Agriculture's] Main Administration of Hunting and Nature Protection?! To let Malinovskii defend this cause is the same things as letting the wolf guard the sheep or letting the elephant into the china shop!" Again the hall erupted in applause and laughter. Novikov's talk had detonated four days of growing feelings of tension, anger, concern, frustration, and militancy.

Malinovskii, observed Novikov, came to the opening of the session with a bored and weary look on his face, sat down and then left, "vividly testif[ying] to Malinovskii's credentials as a 'zealot' of this cause." Again to applause he expressed his hope that "our cause will soon be rid of such grief, of such a leader." Even the unkind words directed at Malinovskii in 1954 paled before the abuse hurled by Novikov that evening. By contrast, Novikov did extend to Burdin his best wishes for success. Before this audience, the juxtaposition could not have been more effective.[43]

Not to be outdone, Nasimovich insisted that in the resolutions the conference should go on record as holding "a sharply negative view of the reorganization of the system of *zapovedniki* and . . . to our own home-grown Herostratus of this cause, A. V. Malinovskii," also provoking laughter and applause.[44] Nasimovich also could not resist recording his "amazement at why this man who more than anyone was responsible for the destruction of the system [of reserves], more than anyone represents the Soviet Union at international conferences on problems of the protection of nature! This is a shame!" Again, applause thundered through the large hall. Nasimovich urged the conference to write a letter to USSR minister of agriculture Matskevich calling for Malinovskii's ouster.[45] No one could remember any other unsponsored, unchoreographed appeal for the removal of a highly placed member of the Soviet government.

Decidedly calmer heads prevailed when it came time to draft the resolutions to the conference.[46] Although the reorganization of the *zapovedniki* was condemned in resolution 13 as having "the most harmful consequences,"

Malinovskii personally was not mentioned. Also glaringly omitted from the resolutions was mention of the danger of radiation, although all of the other major issues—botanical, zoological, or geological—were accommodated.

Still, there was a heady feeling when the conference closed on March 30. Because of MOIP's formal ties with Moscow University, it was able to gain the assent of the university's press to publish the proceedings. Leaving aside content, the bulk of the tome—477 printed pages—entailed a major commitment. With Protopopov assuming hands-on editorial control of the volume and Fëdor Nikolaevich Petrov, "the oldest living Bolshevik," acting as editor in chief, the volume was ready for publication by early 1959, with an interesting, heavily historical introduction by Protopopov. Galley proofs were prepared, and the volume was only days away from production with a print run of 2,000 when the entire project was shut down by the Moscow censor, the movement's old "friend" Tsyriul'nikov.[47]

An order, similar to Lysenko and Prezent's order to smash the frames of Stanchinskii's book at the compositor's, had gone out to destroy the prepared materials of the conference. Was it the article on radiation that upset the censor? Or was it the composite impression of the volume? Or was it an order from a higher authority yet—Matskevich or even Khrushchëv? We still cannot say for sure. Fortunately, the head of MOIP's publishing operations, Grigorii Naumovich Endel'man, had his wits about him. Rescuing one copy of the galleys, he cut them, rebound them, and hand numbered them, depositing the unpublished volume, including the title page, in the MOIP archives.[48]

The excitement and feelings of solidarity generated by the conference spilled out into a range of new public relations initiatives. A week after the close of the conference, activists were able to criticize the 1951 "reorganization" in the central press for the first time. "In Defense of *Zapovedniki*" was published in the April 6, 1957, issue of *Izvestiia,* which concluded with the now customary recourse to shaming the regime into action: "The protection of nature and the organization and support for a network of *zapovedniki* is a matter of the honor of the entire Soviet people!"[49]

The All-Union Conference on *Zapovedniki*

Flushed with their success, the MOIP and MGO activists immediately began planning the second big nature protection conference, set for one year later. Again held in the Moscow House of Scholars under the chairmanship of V. I. Sobolevskii and V. G. Geptner, the All-Union Conference on *Zapovedniki* began its work on the evening of March 17, 1958, with 473 in attendance representing 159 institutions and organizations.[50]

Geptner welcomed the large gathering, reminding the audience that "our public opinion always was passionately concerned with the cause of the protection of nature and always devoted special attention to the *zapovedniki.*" Those who knew the history of this cause in Russia were aware, he continued, that a very large percentage of *zapovedniki* were organized at the initiative of citizens' organizations, including local *kraeved* societies. The state had always relied on civic initiative.[51] Although the changes introduced in 1951 rejected and denigrated this tradition, "in the recent past we have been living through a new period," he said. "The expansion of the rights of the Union republics, the organization of economic regions, . . . etc. have created an entirely new situation for the protection of nature and for our *zapovedniki,*" he added, noting the restoration of previously eliminated *zapovedniki* and the creation of new ones by Russia and the other republics. It was a desire to exploit this new, optimistic climate of opinion and these new opportunities for influencing policy, Geptner explained, that moved MOIP, MGO, and the Moscow House of Scholars to call the present conference.[52]

One of the more dramatic moments at the conference, which was noticeably more sedate than those in 1954 and 1957, was the announcement by Geptner that there would be a special talk dedicated to the work of V. N. Makarov in nature protection by S. S. Turov, Makarov's successor as director of the Moscow University Zoological Museum. When Geptner asked that the audience stand in Makarov's memory, the great hall heaved as 473 naturalists came to their feet to remember the cause's great leader with a standing ovation.[53]

Stretching the bounds of political criticism, the botanical ecologist S. Ia. Sokolov for the first time raised the issue of "highly placed poachers," pointing the finger at the militia and prosecutors of *raion*-level governments. "They style themselves Louis XVI," governed by the slogan "Après moi, le déluge." Sokolov added that "they kill off wildlife in the most merciless fashion, sometimes shooting senselessly." If we cannot fight against this, he noted, particularly when the violators are figures on the *oblast'* level or up, prospective members of our cause will shrug and say: "This is a fool's errand; we will be protecting nature, and Comrade Zver'ev will come along and destroy everything."[54] Sokolov graphically described how Zver'ev, a colonel and director of a Noril'sk *kombinat* (multiprocess factory), went out with his comrades to the Nairna River and shot several hundred wild reindeer that were trying to ford the river. Then they abandoned the carcasses and drove away. When *kolkhoz* workers came by and collected the deer, they were able to use them only for bait for trapping foxes.[55]

Winding up on March 21 with a very long list of resolutions, the conference repeated the calls made by the 1957 Rare and Endangered Species and Western Ukrainian Nature Protection Conferences and by the Congress of the All-Union Botanical Society of the same year for an all-Union agency to ensure the protection of nature, for a network of *zapovedniki* representing every landscape zone of the USSR, and for an all-Union law on nature protection and the rational use of resources. In addition, it called for a range of new measures:

1. that all republic- and *oblast'*-level government and Party organs should discuss the current state of nature protection and *zapovedniki* in each economic region (Khrushchëv had just reorganized the territorial units of the USSR into *sovnarkhozy*, or economic planning units);
2. that the press should become more involved with propagandizing nature protection;
3. that the journal *Okhrana prirody* and the *Nauchno-metodicheskie zapiski* should be revived;
4. that full courses on nature protection should be introduced at the university level, including at teacher-training colleges, and that for lower grades relevant materials should be integrated into the curriculum plans; and
5. that any plan for acclimatization of exotic flora and fauna should be submitted to the Academy's Commission on the Protection of Nature for approval.[56]

A letter sent to the RSFSR Council of Ministers and to N. N. Organov, chairman of the RSFSR Supreme Soviet, from Gosplan RSFSR member V. Domrachev on July 16, 1958 illustrates the effect of this massing of public opinion. Reacting to a letter of June 17 from MOIP to the RSFSR Council of Ministers urging the adoption of the conference's resolutions, Domrachev's memo, speaking for Gosplan RSFSR, found it "exigent" that Glavokhota RSFSR before January 1, 1959, after consultations with all interested parties, submit concrete proposals for the organization of new *zapovedniki* and the expansion of existing ones.[57]

Once again, an emboldened movement took hope. A full list of Soviet congresses and conferences at which the protection of nature was the exclusive or a prominent theme for the period 1957 through 1960 would be surprisingly long.[58]

Zoologists and biogeographers constituted a proselytizing force for nature protection and *zapovedniki* with the Academy's Institute of Geography and the Geographical Society. MOIP was already multidisciplinary, although by the end of the 1950s, chemistry and physics were completely marginal areas and there was even discussion of disbanding those sections.[59] As a result of these processes and of the intimate collaboration of the two societies in nature protection, by the end of the 1950s a remarkable interdisciplinary culture extended from academic geology through geography and botany to zoology—a broad front of organized scientific public opinion.

Winning the Young: MOIP, KIuBZ, VOOP, and the Young Naturalists

Perhaps because field naturalism was viewed as both a craft and a worldview, the Moscow Society of Naturalists from the beginning had a place for young apprentices, called *pitomtsy* (fledglings).[60] The "fledglings" were trained not only in technical methods in science and in frameworks of scientific analysis, but also in the proper way to conduct scientific discussions and disputes, and in a whole world of other values besides.

One of the most important values to be inculcated was the "autonomy of science" from political authority. It was not merely coincidence that MOIP vice president Vera Aleksandrovna Varsonof'eva, who in 1955 published an extensive history of MOIP, noted that the society maintained as active members Decembrists and those fallen from official favor: "These facts graphically testify to the fact that the leaders of the Moscow Society of Naturalists in the oppressive years of the reign of Nicholas I did not fear to attract politically 'untrustworthy' people to their milieu."[61] Aleksandr Ivanovich Gertsen (Herzen) was elected to the "youth" section in 1830, later becoming a full member. The Society did not flinch from listing him as a full member in 1842, when he was already in exile in Vologda. Varsonof'eva's commentary conveys how deeply entrenched these proud traditions were:

> The authentic face of the Society was revealed, of course, not in official meetings and pompous receptions but in its scientific activity and its attitude toward the representatives of the progressive intelligentsia. In science as well there was a struggle between the new and the old. . . . In this struggle the Moscow Society of Naturalists took an identifiable position. Doubtless, within the society were individuals with reactionary inclinations but the majority of the members were on the side of the progressive materialist teachings, of evolutionary ideas, and later defended Darwinism from the attacks of reactionary scientists. This attracted the Decembrists and A. I. Herzen to the Society as well. . . . In turn, the Society continued to value its "fledgling" Herzen even when he became a political exile.[62]

With the eclipse of MOIP from the 1860s through the 1920s, the role of training "fledglings" had been preempted by a number of organizations. In this regard the first decade of Bolshevik power was a particularly fecund time, with the creation of a youth section within VOOP as well as the emergence of KIuBZ (the Circle of Young Biologists of the Moscow Zoo) and the more "loyal" Young Naturalist movement.

Pëtr Petrovich Smolin, affectionately called "PPS" by generations of children and students, was smitten with a love for field biology as a child, reading Brehm's *Life of Animals* at age five. After the revolution, Smolin found employment at the Moscow Zoo, where he was central to protecting the animals during the Civil War. Along with zoologist V. G. Dormidontov, in 1923 Smolin organized KIuBZ. However, Smolin's objections to keeping animals behind bars moved him to transfer the headquarters of the Circle to the K. A. Timiriazev Central Biological Station in Sokol'niki. There, KIuBZ and the Young Naturalists functioned as one, with Smolin representing the Biological Station of the Young Naturalists at the First All-Russian Congress for the Protection of Nature in September 1929. Smolin, however, was called away to the far North to organize a commercial game procurement station in Arkhangel'sk in 1930. There, along the coast of the Arctic Ocean, with the help of Biological Station "graduates," Smolin worked to identify the richest regions for fur-bearing mammals and other game. From one extreme of the country he traveled in 1935 to the other, to the Crimean *zapovednik,* where he worked until 1939. He then returned to Moscow to work in the Darwin Museum as an interpretive guide. With the coming of war he cut short his stay in Moscow to enlist in the army, becoming the commander of a platoon as well as an instructor on the military use of dogs.[63]

In Smolin's absence, leadership of KIuBZ fell to the zoologist Pëtr Aleksandrovich Manteifel', who had been a co-organizer of the group; according to one account, Manteifel''s son was in the Boy Scouts before the revolution, and Manteifel' wanted his son's outdoor education to continue. KIuBZ became one of the Soviet-era equivalents to scouting.[64]

Manteifel', whom Varvara Ivanovna Osmolovskaia (a zoologist who had been a member of KIuBZ in the 1930s) once termed a "natural Lysenkoist" owing to his passion for the transformation of nature, also deeply loved nature and was a formidable naturalist and a hunter. Like Smolin, he was charismatic and, despite his belligerently anti-preservationist ideology (which led him to ally himself with Lysenko against the vast mass of field biologists), he inspired his young charges with the excitement of conducting serious research and observation in undisturbed nature. Despite her subsequent strong scientific and ethical opposition to Manteifel''s views, Osmolovskaia recalled with nostalgia traveling to the Altai in 1934 with a group from KIuBZ to catch marmots, which the students then introduced to alpine habitats in Dagestan (at that younger age she was caught up with Manteifel''s vision of rearranging nature by means of acclimatization).[65] During the 1930s, perhaps the golden age for KIuBZ, a host of future zoologists received inspiration and training at his hands, as reflected in a photograph of a reunion of *kruzhok* graduates (see figure 18).[66]

Of course, enthusiasm alone does not make knowledgeable scientists,

Figure 18. KIuBZ fiftieth anniversary.

and here Manteifel"s leanings toward dubious doctrines of nature transformation sometimes did his protégés a disservice. As Konstantin Mikhailovich Efron remarked, although Manteifel' was a great leader for those in the sixth grade or younger, he had a great capacity to confuse the developing minds of older youths.[67] In the opinion of Elena Alekseevna Liapunova, a cytogeneticist who was a member of Smolin's VOOP *kruzhok* (circle) in the early 1950s, "PPS was incomparably more interesting than Manteifel'." On the other hand, after the August 1948 calamity, Manteifel', despite his own hostility to classical genetics, remained personally supportive of "his" fledglings such as E. D. Il'ina, who was fired for embracing a "formal genetics worldview."[68]

With peacetime Smolin first taught at the Institute of Furs and Pelts at Balashikha, just east of Moscow, and then in 1948 returned to the Darwin Museum. Simultaneously, from 1946 Smolin headed the youth section of VOOP as well as returning to the directorship of KIuBZ. However, a falling-out with the leadership at the Moscow Zoo led Smolin to resign from KIuBZ in 1949, taking a number of youngsters personally loyal to him over to the VOOP *kruzhok,* where he now invested all of his efforts. Among the members was the future population ecologist Aleksei Vladimirovich Iablokov, who was elected the first president of the VOOP circle. Another member of that impressive cohort was the priest Father Aleksandr Men', who was later murdered. Based first in the premises of the Moscow *oblast'* Pedagogical Institute and from 1966 in the Darwin Museum, the VOOP *kruzhok* was weak until Smolin took over the reins.[69]

Although the atmosphere of the groups was more alike than dissimilar, the VOOP group was more traditionally scholarly than KIuBZ. If we discount his personal animus against Manteifel', Smolin's observations from 1951 illuminate this difference: "The existing Young Naturalist institutions and the children's collectives grouped around them exhibit a one-sided agrobiological ['Lysenkoist'] tendency. . . . Knowledge of nature stands at a very low level not only among the so-called Young Naturalists but among their leaders as well [a reference to Manteifel']."[70]

All of these groups, which were formally united under the umbrella of the Young Naturalists, had a profoundly democratic spirit. As Liapunova put it, they exuded a spirit of *grazhdanstvennost'* (citizenship). Loyalty to the group, which embraced those in grades five through ten, was important, but so was the exercise of individual responsibility.[71]

In the VOOP *kruzhok* in particular there was a "strong feeling of community." Aleksei Andreevich Liapunov, Elena's father and an eminent mathematician with a strong interest in biology, often invited the group over for discussions with light refreshments in their apartment. They even had a definite schedule. On Tuesdays there were lectures held at the Lenin (Potëmkin) Pedagogical Institute in the central Frunze district of Moscow. PPS often invited famous scientists to these; in those days they came willingly. On another weekday the *kruzhkovtsy* met by themselves and read their own lectures; these meetings were sometimes held at the Darwin Museum, where PPS worked. On weekends there were excursions to natural areas in the vicinity such as the Prioksko-Terrasnyi *zapovednik*, Zvenigorod, Lake Kiëvo, and other interesting places. Everyone got up at dawn, and PPS led the members along trails and identified the various birds they encountered. Every once in a while he hinted at opposition to Lysenko's ideas, but never explicitly, not wishing to place the children in a situation where they would have to lie.[72]

When general meetings were held, the members experienced the feel of a "real" scientific society on the model of MOIP. There were membership inductions. To be accepted, the young woman or man had to make a scientific presentation based on her or his own research with photos and a dossier, already a sign of seriousness; until you did that you remained a *soiskatel'*, or candidate member. Elena Liapunova, for example, traveled to the former Verkhne-Moskvoretskii *zapovednik* and worked on a method of censusing beavers (including a demographic analysis) by examining the size of gnawed sections of trees. With induction immediately came membership in VOOP. Whenever the group held elections of officers, members had a "real democratic feeling."[73]

We cannot overestimate the importance of these groups, particularly the VOOP circle, as crucibles for the formation of the leading field biologists of the USSR. Those of the VOOP circle had the highest success rate of all students in the MGU Biology Faculty, and the graduates of both circles—VOOP and KIuBZ—demonstrated a solidarity within their ranks (especially among those of the same cohort) that endured for decades. That solidarity enabled them to organize outside of any institutional framework. One example was the organization of Nikolai Vladimirovich Timofeev-Resovskii's first lecture in Moscow (1955) since his deportation from Germany at the end of the war, which was held at the Liapunovs' home. The audience was composed largely of a group of university students who had all been members of the VOOP circle. This lecture was an important milestone in the revival of formal and population genetics in the Soviet Union.[74]

Following the 1958 MOIP-organized All-Moscow Conference on the Role of Youth in the Protection of Nature, MOIP joined the quest for the hearts and minds of the naturalistically inclined youth. MOIP was aiming at university students, too old for the other groups. Under the chairmanship of Fëdor Nikolaevich Petrov, who from 1954 was chair of MOIP's new Section on the Protection of Nature, and through the efforts of Nikolai Sergeevich Dorovatovskii, one of the section's vice presidents, and of Konstantin Mikhailovich Efron, the society's scholarly secretary, a Student Subsection was organized, of which the most active members later became zoologists and conservation activists: Maria Cherkasova, Boris Vilenkin, and V. Baranov.[75] Even MOIP, however, proved too tame for the university students. Within two years the students, centered at the Biology Faculty of Moscow University, had amicably gone their own way, creating an entirely new kind of organization, the *druzhina po okhrane prirody* (Nature Protection Brigade). Nonetheless, it was no accident that the *druzhiny*, with their roots in MOIP, became the standard-bearers of future Soviet field biologists and of future scientific public opinion.

Finally, any discussion of how the founding generation of scientific public opinion sought to perpetuate its worldview, values, and vision of science must mention the last of the young naturalist organizations, the Circle of Young Naturalists of the Section on Nature Protection of MOIP. To a certain extent, this group was MOIP's compensation for having lost the university students. Founded in the early 1960s by zoologist Anna Petrovna Razorënova, this circle arguably became the most successful of the youth groups of the next two decades. Single, chronically ill, and also caring for a sick father, Razorënova could not work, and instead for twenty-five years she selflessly devoted all her energies to the circle, organizing weekly seminars for children in grades six through ten. The earliest seminars were attended by Boris Fëdorovich Goncharov, Nikolai Aleksandrovich Formozov, Vadim Mokievskii, Arkadii Tishkov, and other now prominent natural scientists; more than 250 of Razorënova's "graduates" went on to become natural scientists. Some years' cohorts had up to twenty-five members.

On holidays and Sundays the MOIP *kruzhok* would convene on a farm outside Moscow. Later, Razorënova rented and then bought a house near Myshkin on the Volga, using it as a base for excursions. Like the other groups, the MOIP circle organized trips to *zapovedniki,* and the older students participated in animal censuses. A special treat was the trip to the Black Sea coast during winter vacation.[76]

The generation of founders of the Russian nature protection movement was schooled at a particular time and place. The imposing level of erudition that generation attained was as much a product of the prerevolutionary familial environment as of tsarist-era educational possibilities. These were strongly colored by the class structure of the era and by the values of both the traditions of the landed gentry and the emerging commercial and professional culture of Russia's cities. It is also important to recognize that science was still "small science" and that it was possible for an individual to gain prominence by investigating some of the great uncharted areas of the natural world. It was also a time when it was fashionable to advance grand theories and when there was greater faith in science. Finally, it was a time before scientific paradigms, ideas, and "facts" were called into question as perceptually driven artifacts resting on ultimately arbitrary or unprovable premises.

Almost all of those conditions had radically changed or were changing by the 1960s. The magnificently rich prerevolutionary education that wellborn and even middle-class children received in the home and at elite schools gave way, at least in school, to a routinized, rote, and dulling education, particularly after the early 1930s. It was far more difficult to affirm in public an identity based on the ideals of civic dignity and the autonomy of science, and the idea that scientific public opinion possessed special knowledge that conferred on it the right to intervene decisively in some public policy areas. Finally, in the era of emerging "big science" it became difficult for anyone to achieve the status of titan. Scientific authority itself would soon be questioned—later in the Soviet Union than in the West, but in time to cast a shadow on those coming to maturity in the 1980s and 1990s.

Taken together, all of these changes foreclosed the possibility of replicating the generation of the founders. And considering that the political and moral authority of those figures in part flowed from their sense of scientific accomplishment, reproducing that basis for political and moral authority among the next generation of scientists—who were less erudite and who practiced more what Thomas Kuhn has called "normal science" than paradigm-shaping—was problematic. There were limits to the effectiveness of the youth circles in socializing future naturalists to become just like their forebears. Yet much of that older spirit was passed along. Considering the political, cultural, and socioeconomic environment in which the founders were fated to work, that itself must be considered a monumental achievement.

CHAPTER THIRTEEN

More Trouble in Paradise

Crises of the Zapovedniki *in the Khrushchëv Era*

Few leaders have embodied as many contradictions as Nikita Sergeevich Khrushchëv. These are reflected in the leader's gravestone, composed of nearly equal quantities of black granite and white marble. Russians and the world remember him with gratitude as the man who courageously informed us of the crimes of the Stalin era, freed perhaps eight million prisoners from the labor camps, allowed the publication of Solzhenitsyn's *One Day in the Life of Ivan Denisovich,* and opened the USSR to the world community, even if only haltingly. On the other side of the ledger, it is impossible to recall Khrushchëv without a shuddering remembrance of the Cuban missile crisis, the Berlin Wall, the suppression of the Hungarian Revolution, his personal support of T. D. Lysenko, and his mismanagement of Soviet agriculture. Ironically, the man who liberated a terrorized nation from the legacy of Stalin's tyrannical "cult of personality" ended up imposing a less terroristic but palpably injurious cult of his own.

A product of the Stalin era and of Stalin's political machine, Khrushchëv carried over more than the cult of supreme leader from Stalinist political culture into his own period of rule. Among these continuities were a suspicion and, at times, active disdain for basic research, a hostility to protected territories "sequestered" from active economic use, and an inconsistent toleration (not unlike Stalin's) for the increasing reward of the *nomenklatura* (the Party elite) with privileges and perquisites. These trends came together in Khrushchëv's first years of power, when he turned a blind eye to the organization of illegal hunting in protected territories controlled by the USSR central government. Later, when his active diplomacy created a need for relaxed venues to host visiting dignitaries, Khrushchëv's own penchant for hunting suggested a solution: transform game preserves or even *zapovedniki* into well-appointed hunting lodges, available only to the highest ranks of the Party and state leadership. Lower-level minions would have to content themselves with freer opportunities for poaching; only the top elite would go deluxe.

Like Stalin a mixture of the Communist romantic and cultural anticosmopolite with a profoundly bucolic view of the world, Khrushchëv held a ploddingly materialist vision of the Communist utopia. Much in the spirit of the American labor leader Samuel Gompers, Khrushchëv's vision of progress was driven chiefly by one idea: more. Communism would prove its superiority over capitalism by outproducing it. Communism would produce more milk, more meat, more wheat, more electricity. Its hydropower stations would be the biggest, and its rockets the heaviest. The idea of demonstrating Communism's *qualitative* superiority over capitalism had escaped the unimaginative Soviet leader. The closest he came was his belief that the Soviet people were morally superior to those of the West because they were prepared to endure privation in the present to guarantee plenty in the future.

Even Khrushchëv, however, recognized that there were limits to the Soviet people's capacity for sacrifice. His policies, therefore, combined continued heavy investments in industry and the military (though not as onerous as those under Stalin) with simultaneous attempts to establish decent living standards for the masses through frenetic housing construction and gargantuan agricultural campaigns. These attempts, however, were marked by what Brezhnev and Kosygin termed "adventurism" and "voluntarism," and by what Gorbachëv described as "extensive," as opposed to intensive, development. What Khrushchëv's critics meant was that he made decisions without sufficient political and especially scientific consultation, that they involved a considerable (and, retrospectively, an unacceptable) element of risk, and that his policies were attempts to increase output on the cheap, by expanding existing patterns of production or sown areas instead of changing industrial or agricultural processes to make them less wasteful and more productive.

Additionally, the Soviet leader felt the need to appear on an equal footing with foreign leaders. The old Stalin-era tunics gave way to tailored suits and fedoras; Nina Khrushchëv, though usually well in the background, began to accompany her husband on his international forays. As we know from Iurii Zhdanov, colorful coffee-table books featuring the landscapes of the USSR became essential as presentation gifts to foreign guests or hosts, and, with the melting of Soviet isolation, attention began to be paid to suitable places to take important foreign visitors for a few days of relaxed conversation and recreation.

In addition to reopening the Soviet Union to the outside world, Khrushchëv's liberalization unleashed creative energies in high culture, popular culture, science, and everyday life. Khrushchëv returned to the traditions of

1917 and began to transfer some responsibilities for maintenance of public order and justice from the government bureaucracy to voluntary (but supervised) social organizations. In this spirit the offices of citizens' inspector for fishing, hunting, and nature protection were created in the 1950s, open to members of VOOP, and detachments of *druzhinniki* (ultimately numbering some seven million) were formed in factories and neighborhoods to maintain public order and safety.

Finally, Khrushchëv's tenure marked an expansion of a culture of Party corruption, although the Soviet leader himself displayed ambivalence about that development, at times even going so far as to prohibit the personal use of official vehicles and the like. Nevertheless, with the rare exceptions of a few highly visible chastisements (and a handful of executions) for crimes against state property, the Khrushchëv era was governed by the philosophy that in the forward march to Communism, the hard-working ranks of the Party-state *nomenklatura* should be well rewarded.[1] In the Khrushchëv era the enjoyment of perquisites was expected to be discreet and on a relatively modest scale, a standard that weakened under Brezhnev and later leaders. The boundaries that separated the condoned from the punishable were still being drawn, however, and were extremely fluid under Khrushchëv, a situation the almost total laxity of the Brezhnev era did much to clarify.

Stalin himself, as early as the late 1920s, had condoned the organization of a system of recreational perks for the Party hierarchy, as Vladimir Boreiko has uncovered. Pioneering the future system of *spetsokhotkhoziaistva* (restricted hunting grounds) for republic-level Party moguls was the one organized northwest of Moscow at Zavidovskoe by Klim Voroshilov for the Red Army's high command in 1929. In that same year *Vechernii Kiev* reported that illegal hunting and fishing outings were being organized for "the select few" at the Koncha-Zaspa and Askania-Nova *zapovedniki*. By 1934 the pretense, at least regarding Koncha-Zaspa, near Kiev, was dropped; the minuscule *zapovednik* was transformed into a special *sovkhoz* to accommodate the recreational desires of members of the elite.[2]

In the RSFSR a legal basis was laid for what Boreiko has labeled "special safaris" with the 1930 law on game management, in which article 5 allowed for the creation of "special hunting areas . . . set aside for the pursuit of model game management with the application of special measures for the protection and breeding of animals and birds and with restrictions on those permitted to carry out hunting."[3] Boreiko notes that in the first decade of this system, some discretion was employed; "palaces were not erected, lakes were not lighted with lamps, . . . concrete was not poured to create helicopter landing pads."[4] By 1940, though, in Ukraine there were already six hunting preserves, disguised by the designation of "republic-level *zakazniki*." Working in the Ukrainian state archives, Boreiko unearthed "a document unique in its immorality," as he describes it: a decree of the Ukrainian

Council of People's Commissars of February 24, 1945, allocating funds for the restoration of staff and ancillary buildings in the state game preserves of that republic at a time when the war was still raging and the human suffering there was beyond expression.[5] A neat 250,000 rubles was earmarked for the immediate preparation of a few of these preserves for the hunting season that was due to start on August 1. Nikita Sergeevich Khrushchëv was Party secretary of Ukraine at the time.

Although as early as 1951 a speaker at the Congress of the Ukrainian Society of Hunters and Fishers spoke out against "shameful doings in our system, when state *zakazniki* are organized and turned into places for hunting for certain individuals," demanding that this be ended and that his remarks be included in the stenogram for the edification of Party leaders, his words had absolutely no effect. Others complained of local Party bigwigs hunting with the aid of automobile headlights, but the Party *nomenklatura* did not break ranks. By 1956 the number of such special "*zakazniki*" in Ukraine had doubled to twelve.[6]

In the summer of 1955 Khrushchëv had made one of his most important early foreign ventures, to Yugoslavia to repair ties with Josip Broz Tito. Khrushchëv's gesture of reconciliation was marred by his attempt to cast Beria, and not Stalin, as the author of the rupture between the two countries. In his Secret Speech, though, Khrushchëv redeemed himself, graphically recounting Stalin's attempt to destroy Tito:

> I recall the first days when the conflict between the Soviet Union and Yugoslavia began. . . . Once, . . . I was invited to visit Stalin who, pointing to the copy of a letter lately sent to Tito, asked me: 'Have you read this?' Not waiting for my reply he answered, 'I will shake my little finger—and there will be no more Tito.' . . . We have dearly paid for this 'shaking of the little finger.' . . . Tito had behind him a state and a people who have gone through a severe school of fighting for liberty and independence, a people that gave support to its leaders.[7]

In the summer of 1956 a second trip by Khrushchëv and Mikoian followed. Tito, who was now far more cordial than a year earlier, took the Soviet leaders hunting on the island of Brioni, where he had a residence. The Soviet leaders reciprocated in the fall, playing host to a delegation of Yugoslav leaders headed by Tito.

On the eve of Tito's visit, Khrushchëv's thoughts turned to finding a suitably impressive place where the two leaders could enjoy their common passion, hunting, while continuing the delicate work of political bonding. As Khrushchëv tells it in his memoirs: "Once we invited him to the Crimea for a few days of rest and for a hunting trip. The hunt, of course, had been used for centuries as an opportunity for leaders of two or three different countries to get together and discuss issues of mutual interest and importance.

The atmosphere of my discussions with Tito during our hunt together was warm and friendly."[8] Khrushchëv failed to disclose in his memoirs that the solution he hit on for hosting Tito was to organize a hunting vacation in the Crimean *zapovednik,* "a crude violation of nature protection legislation," in the words of Vladimir Boreiko. Almost coterminously, on September 27, 1956, "on Mikoian's initiative or, more likely than not, that of Khrushchëv himself, the Central Committee of the Party directed the USSR Ministry of Agriculture to prepare a plan for the organization of first-class 'aristocratic-style hunting opportunities' [*barskie okhoty*]." Boreiko further informs us that while the Moscow authorities now began to weigh various *zapovedniki* for that purpose (those originally proposed included the Crimean, the Caucasus, Kyzyl-Agach, and the Belovezhskaia *pushcha*), the Ukrainian premier, Kal'chenko, losing no time, issued a technically illegal directive on January 10, 1957, "On the Organization of Game Management Facilities in the Crimean State *Zapovednik.*"[9] The directive ordered the construction within six months of an electrical generating station, hotel, restaurant, and roads in the heart of the reserve.[10]

There is another prehistory to the conversion of the Crimean *zapovednik* into the Crimean international hunting lodge. Immediately after the 1951 events, local political leaders in the Crimea, eager to use the *zapovednik*'s land for logging, urged a severe culling of the local Crimean red deer, which were allegedly impeding forest growth and regrowth. Although permission was granted, the issue of exploiting the reserve's territory lingered.

On April 8, 1955, F. Krest'ianinov, deputy head of the Agricultural Section of the Central Committee with responsibility for the Union republics except Russia, issued a long memo on the "Improvement of Forest Management in the Crimean State *Zapovednik.*" The memo documented that an investigative commission on this question had been organized, consisting of an *instruktor* of the Central Committee's Agricultural Department, Alisov, an inspector of the USSR Ministry of Agriculture's Inspections Bureau, and representatives of both the Crimean *obkom* and the Ukrainian Party's Central Committee.[11] The memo supported the conclusion that the drying up of some of Crimea's mountain rivers was linked to deforestation of the upland watershed (110,000 hectares or 33 percent of all woodlands) over seventy-five years, an assertion that heartened conservation activists who hoped to put a damper on logging in the reserve. But the solution proposed was a drastic cutback—through shooting and capture—in the population of Crimean red deer and roe deer, "preserving them only as representatives of a species." After clearing overmature growth and eliminating the ungulate threat, replanting of 1,500 hectares of the 30,200-hectare *zapovednik* could begin in the next fiscal year. "Fortuitously," this plan provided the "scientific" justification for converting the Crimean *zapovednik* into a hunter's paradise.[12]

Thanks to its rather remarkable network of informants—that thousand-eyed army of *kraevedy* and local activists, in Efremov's words—those at the center of the nature protection movement almost immediately learned of the ominous plans being cooked up at the USSR Ministry of Agriculture. Turning to one of its biggest political guns, Lieutenant General Evgenii Nikanorovich Pavlovskii, the Academy's Commission on the Protection of Nature had the zoologist send a letter, cosigned by commission deputy chair Vsevolod Dubinin, directly to First Secretary Khrushchëv: "The Commission . . . has received evidence of a proposal to organize special hunting management facilities of special designation . . . some of which will be sited on the territories of four existing *zapovedniki*. . . . [T]he Commission for the Protection of Nature . . . considers [their] organization unthinkable [*nevozmozhnym*] on the territory of these or other *zapovedniki,* with the organization of regular hunting expeditions on these protected territories all the more unthinkable."[13]

An urgent state telegram from the USSR Academy's and the Belorussian Academy's Commissions on the Protection of Nature to Khrushchëv followed on October 27, 1956, and was equally direct: "The collective of scientific researchers of the Belovezhskaia *pushcha zapovednik* consider it impermissible to organize a game management facility on the base of the oldest internationally known *zapovednik*. . . . The decision about the organization was taken without discussion and without the approval of the collective of scientific researchers and the scientific forces of the country. We ask you to intervene and to stop this."[14] Ironically, appeals to history in this case were on the side of those seeking to convert the reserve to a giant hunting ground; for hundreds of years before it was declared inviolable, the Belovezhskaia *pushcha* had been a *tsarskaia okhota* (royal hunting preserve) for the Lithuanian grand dukes and their successors, the Russian tsars.

The authors of the letter were notified by the Agricultural Department of the Central Committee, its director P. Doroshenko informed the Central Committee Secretariat in a memo of December 6, that the draft legislation for organizing fifteen special state hunting management facilities, including five *zapovedno-okhotnich'ia khoziaistva* (on the basis of existing *zapovedniki*), had already been submitted for consideration to the USSR Council of Ministers with the approval of the Agricultural Department. Doroshenko assured the worried letter writers that "the regime of inviolability would be preserved with the exception of an entirely limited amount of hunting."[15]

Finally, on the heels of the historic Conference on Valuable, Rare, and Endangered Species of Plants and Animals of March 1957, a long letter to Khrushchëv was sent on April 24, 1957 in the name of the conference and of MOIP by Fëdor Nikolaevich Petrov (see figure 19), the conference's chairman. Much of the letter was a political reminder from Petrov, who served under Lenin, that Khrushchëv, who claimed to be restoring the

Figure 19. Fëdor Nikolaevich Petrov (1876–1973).

authentic Leninist legacy, needed to study that *entire* legacy, including the Lenin-era decrees on the protection of nature and on *zapovedniki.*

Noting that the conference, "having heard and discussed a whole series of reports from all corners of the Soviet Union, with the very greatest feeling of alarm went on record as declaring that the situation with the protection of nature . . . is extremely grave," and that "existing laws, rules, and directives are not being carried out," Petrov called on the political leadership of the country to pay attention to the voices of scientific public opinion. The ancient revolutionary closed by invoking the interests not just of "today's needs, but those of future generations." Copies of the letter were sent by Petrov to Voroshilov, Molotov, Bulganin, and D. T. Shepilov.[16]

Despite these high-level and vigorous protests, the needs of the new diplomacy and of the *nomenklatura* for appropriate recreational facilities took precedence, and on August 9, 1957, just in time for the fall hunting season, the USSR Council of Ministers under Premier Nikolai Bulganin's signature approved the establishment of twelve elite hunting facilities, of which three (Belovezhskaia *pushcha,* the Crimean, and the Azovo-Sivashskii) were *zapovedniki* converted to the new status. Perhaps the protests reduced the number of affected *zapovedniki* from five to three, but that is still a matter of conjecture.[17] Understandably, provincial leaders of the Ukraine and Belorussia were more than willing to disregard the petitions of their local scientists with the prospect of attracting to their republics important visitors (including the first secretary and his entourage) at venues where they, too, could play host. That explains in part the haste with which Kal'chenko ordered the conversion of the Crimean *zapovednik* months before the all-Union decree was published. Four new facilities were created on the territory of the RSFSR, but none of the Glavokhota RSFSR *zapovedniki* were touched, perhaps because of expected resistance. The four new hunting reserves had a combined staff of 243, a combined area of 227,000 hectares, and five hotels among them.[18]

The reorganizations of 1957 were still a relatively minor setback against the backdrop of the energetic efforts of Glavokhota RSFSR and of other republics to restore the reserves eliminated in 1951 and to create new ones. However, the 1957 reorganizations were an augury of a much larger crisis for the reserves.

The Squirrel that Destroyed Thirty Nature Reserves

The morning of Tuesday, January 17, 1961, was an ordinary workday for Moscow. Aside from an eight-day Central Committee plenary meeting and a cultural agreement between the "USSR-Japan Society" and its Japanese counterpart, not much of particular interest was going on that day. Two days earlier, there had been big news from scientists in Siberia—linguists and anthropologists—who convened in Novosibirsk to explain how they had broken the mystery of the Mayan hieroglyphs with the aid of computers. But on that Tuesday, Nikita Sergeevich Khrushchëv, speaking to the Plenum, had other Siberian scientists on his mind.

The first secretary evidently was in a jocular mood. First, he turned his wit on the partisans of historical preservation, who were engaged in a battle to save some of the most remarkable architectural monuments from his antireligion campaign.[19] "Now let's say you are a cultured individual," Khrushchëv began, warming up the delegates with an immediate figure of fun and a swipe at the Russian non-Communist intelligentsia's continuing pretensions to privileged knowledge: "You know and understand the importance of this monument of antiquity. You, and only you, can evaluate this historical monument. You are the one who understands and can assess the importance of the fact that such-and-such a famous person strolled in this spot; you can tell us that this is where he sat and thought up his projects, and this is the place where, in a fit of anger, he spat on the ground." Khrushchëv's humorous anti-intellectualism brought the otherwise dull-faced Party bosses to life, and a ripple of laughter rolled through the Kremlin hall. Khrushchëv continued: "I am not exaggerating, comrades. These kinds of outrages indeed exist."[20] The first secretary milked the rich theme of historical preservation for several minutes more before turning to another tribe of *chudaki,* the nature protection activists (many of whom were also historical preservationists). "And now about one more thing," the Soviet chief began. "There are a great many *zapovedniki* that are being organized all around. I, and no doubt you, saw the documentary film on the *zapovednik* in the Altai Mountains. The film was made very well. The film showed how this person exuding good health, most probably a scientist,—if it's a *zapovednik,* then they all must be scientists there [the jocund clucking and chuckling in the hall temporarily interrupted Khrushchëv's anecdote]—lying on a rock and observing through his binoculars how a squirrel is gnawing an acorn. Then he shifts his gaze to watch a bear moving along." Khrushchëv now came to the punch line:

> What is this thing called a *zapovednik?* It is a *zapovednik* for those who live there. They also graze there, graze and browse better than the bears and the squirrels. Isn't it a fact that even if those people weren't there, the squirrel would still be gnawing on that acorn? It's all the same to the squirrel whether there's a scientist around or not. But the difference is in the fact that now it is gnawing acorns under the observation of a scientific researcher, and that researcher is receiving money for that, and good money to boot!

The ridiculous picture of field biological research painted by the first secretary provoked another round of chuckling in the hall, as Khrushchëv finally zeroed in on his more serious conclusions:

> What is this thing called a *zapovednik?* It is the nation's wealth, which we must preserve. But in our country it frequently happens that *zapovedniki* are organized in places that do not represent anything of serious value. We must impose order on this business. *Zapovedniki* should be located where it is essential to preserve valuable corners of nature and to conduct authentically scientific observations. Certainly our country has these kinds of *zapovedniki* already. But a significant proportion of the *zapovedniki* currently in existence represents—a contrived operation.

"What will happen in the forests if *zapovedniki* won't be established in them?" Khrushchëv asked rhetorically. "Nothing. It is necessary, of course, to protect nature and care for it," he concluded, "but not by organizing *zapovedniki* everywhere with large staffs."[21]

Vladimir Boreiko comments:

> Knowing the explosive, irrepressible character of Khrushchëv and his not terribly great intellect, it is entirely possible to hypothesize that the film *Zolotoe ozero* [or *Altyn kol'*—"golden lake"—in Altaic, apparently the actual name of the film] about the Altai *zapovednik* and Lake Teletskoe, accompanied by appropriate commentary by [Khrushchëv's] cronies was in fact the thing that provoked the new disaster. Adding to that, Nikita Sergeevich already had acquired some good experience, having participated in the pogrom of the *zapovedniki* in 1951.[22]

Boreiko could also have mentioned Khrushchëv's personal involvement in the conversion of the Crimean *zapovednik,* which helped to keep the issue alive in his mind. Whatever the mix of precipitating causes, the first secretary's seemingly spontaneous remarks at the January 1961 Plenum were already planned one month before. Exploring the archives of the Presidium of the USSR Council of Ministers, Boreiko found the following decision taken on December 31, 1960: "Assign to Gosplan USSR (Comrade Zotov) together with the USSR Ministry of Agriculture, the USSR Ministry of Finances, the USSR Academy of Sciences, and the councils of ministers of the Union republics to investigate the network of existing *zapovedniki* and state game management facilities [*okhotnich'ia khoziaistva*] and within one month submit a report and suggestions to the USSR Council of Ministers, keeping in mind the need to eliminate the excess in this area of activity."[23] The order was signed by Khrushchëv himself, who had recently also assumed the post of USSR premier. Could he have forgotten that only two months earlier he had authorized the final passage of the RSFSR law on nature protection?[24] On the other hand, the fact that the item "On Setting Right the Situation in the *Zapovedniki*" ranked as that day's twenty-sixth order of business reveals just how marginal the Kremlin leadership considered the *zapovedniki* and nature protection generally.

Thirteen days after the Plenum, an expanded session of the Commission for the Protection of Nature of the Academy of Sciences met to deal with the unexpected blow. The commission was not going to give in without a fight. Cleverly, on the model of the response of the Baltic republics to the 1951 liquidation, the commission proposed converting the Altaiskii (named in Khrushchëv's speech), Teberdinskii, "Stolby," and Mariiskii *zapovedniki* into national (*narodnye*) or natural (*prirodnye*) parks, to be run by the trade union central organization. Under the new management, care would be taken not to destroy the natural amenities and research could still be conducted by visiting researchers; there would be no permanent scientific staff.[25]

The Commission tried to save other reserves such as Denezhkin Kamen' by supporting their transfer to universities as research and teaching bases for their students. They needed the help of Gosplan USSR and the USSR Ministry of Higher and Specialized Secondary Education, to whom they appealed. Further, they tried to save yet other *zapovedniki* by merging them as "branches" with others less under threat. Accordingly, the Commission recommended that the Lapland become a *filial* (branch) of the Kandalakshskii *zapovednik* and the Khopërskii a branch of the Voronezhskii *zapovednik*. Conceding the possibility of reducing the area of the Kronotskii reserve, the commission nonetheless drew the line on twenty-six *zapovedniki* listed in the document, whose preservation was "essential." It also drew the line on making the reserves self-financing (*khozraschët*) and even criticized the overly accommodating staff cutbacks proposed by Glavokhota RSFSR. Finally, the commission noted, against the tide, that the existing network of reserves was still inadequate and that it was necessary to create new *zapovedniki* in the tundra, southern taiga, steppes, and semidesert regions.[26]

Word of these changes ignited a firestorm of protest from scientists. Hundreds of impassioned letters arrived at the offices of V. P. Zotov, deputy chairman of Gosplan USSR and Stalin's minister of the food industry for a decade (1939–1949). Letters were also sent to the councils of ministers of the individual republics and to Nesmeianov at the Academy. In the RSFSR, Deputy Premier Aleksandr Semënovich Bukharov was given responsibility over the *zapovednik* question and he, too, was the recipient of numerous letters. A letter to him and to Zotov by the acting dean of the Biology and Soil Science Faculty of Moscow University, V. F. Riabov, set out in great detail the importance of the *zapovedniki* for research and especially for the training of specialists of a broad range of disciplines. Many letters repeated the argument that the Soviet *zapovedniki* were conceptually unique.[27]

While conceding that the reserves might indeed harbor a few malingerers, a letter from Moscow University professors emphasized that they were only a few bad apples. The rest were "enthusiasts, without exaggeration selflessly working in extremely hard conditions such as the taiga and the desert." What made the first secretary's attacks even more unfair were the "completely insubstantial funds" that the network absorbed, less even than the game management facilities of the USSR Ministry of Agriculture.[28]

Some local activists and scientists tried to save individual reserves, such as a group from Kuibyshev (Samara) that included representatives of the local branches of VOOP, the Botanical Society, the Geographical Society, the Union of Hunters, and the Pedagogical Institute.[29] Other letters came from the *zapovedniki* staffs, including one from V. V. Krinitskii, the director of the Altaiskii, which had been specifically mentioned by Khrushchëv in his sarcastic remarks.[30] Most of these tried to make the case that although other *zapovedniki* perhaps might corroborate the accusations made by the first secretary, their own clearly was beyond reproach. As in 1951, local authorities occasionally intervened to protect (as well as to destroy) the reserves; the most vivid of these came from the deputy chairman of the Stavropol' *kraiispolkom*, V. Chumakov, who, in a letter to Academy president Nesmeianov called the liquidation of Teberdinskii *zapovednik* "unacceptable in the eyes of our province [*krai*]."[31]

Primed by the preceding five years of activism, scientific public opinion quickly mobilized to combat Khrushchëv's New Year's surprise. I. D. Papanin of MGO asked the RSFSR government to include representatives of his society in any discussion about the reserves system.[32] When the Presidium of MOIP met on February 14, the issue was right there on the agenda. Konstantin Mikhailovich Efron, as rapporteur, read a draft of the letter to Khrushchëv to be sent in the name of the society, though he suggested a meeting with Academy president Nesmeianov prior to sending it to insure that MOIP and the Academy had coordinated positions and strategy. Accordingly, the Presidium decided to constitute a delegation of Varsonof'eva, Ianshin, and botanist B. A. Tikhomirov to meet with Nesmeianov, while approving pro tem Efron's draft, which would undergo a final editing by Varsonof'eva and F. N. Petrov.[33]

When the Executive Council of MOIP met three days later with Varsonof'eva chairing, the MOIP Presidium thought it wise to marshal the support of that larger body. Varsonof'eva herself asked A. L. Ianshin, another vice president of the society, to read the draft letter aloud. Following a discussion in which some politic editing changes were suggested, the Council voted its approval of the Presidium's decision to send the letter to the first secretary.[34] It went out eight days later over the signatures of Varsonof'eva and Fëdor Nikolaevich Petrov.[35]

This letter informed Khrushchëv that only thanks to the *zapovedniki* was the sable rescued as a commercially exploitable species. Now the annual take was about 100,000 skins, each fetching 70 rubles (for a total of approximately $7,770,000 per year). The same situation held for the beaver. Additionally, *zapovedniki* were responsible for introducing more than twenty practical recommendations for the improvement of forestry and agriculture, including pioneering the introduction of pasture crops in high alpine regions and the successful acclimatization of ginseng. The scientific research conducted in the reserves was so highly regarded that *zapovednik* studies were widely used in dozens of the best regarded handbooks and textbooks.[36] Varsonof'eva and Petrov reminded Khrushchëv that "the great successes of the USSR in the area of nature protection and *zapovedniki* helped to elevate the international authority of [the country]." These benefits loomed large, they argued, against the paltry sums of money spent on their upkeep (two million [new] rubles annually for the twenty-nine RSFSR *zapovedniki* in 1960) and the minuscule percentage of the territory of the country they occupied (0.26 percent).[37]

The MOIP letter closed with a set of recommendations: First, any elimination or creation of *zapovedniki* should be implemented only if it fit the master plan for *zapovednik* development proposed by the Lavrenko commission and adopted by the Presidium of the USSR Academy of Sciences on September 13, 1957 (and reaffirmed by the March 1958 conference on *zapovedniki*). Second, better qualified scientific staffs needed to be hired and the budgets for science in the reserves needed to be increased. Third, the authors called for the creation of a single, authoritative Committee for Nature Protection attached to the USSR Council of Ministers as well as for "the participation of broad scientific public opinion in the discussions concerning critical issues of *zapovednik* activity." Finally, the authors pointed to the need "to fortify international links in this area, gaining a leading role for the USSR in international nature protection organizations" despite the good start made by entry into the International Union for the Protection of Nature in 1956 and subsequent participation in a number of conferences.[38]

Through internal memoranda prepared by aides, the Presidium of the RSFSR Council of Ministers was kept abreast of the scope of the public outcry.[39] Even before the main mass of letters of concern began to pour in, the Presidium of the RSFSR Council of Ministers met on February 11. At that meeting Gosplan RSFSR and Glavokhota RSFSR presented their plans for "rectifying" the reserve network. Terrified by Khrushchëv's barbed attack and desiring to propitiate the mercurial supreme leader, Eliseev, the head of Glavokhota RSFSR, proposed the outright elimination of eleven of the twenty-five *zapovedniki* in his system and the truncation of three additional ones. That would reduce the overall area of *zapovedniki* in the Glavokhota system by 72 percent, to 1,329,400 hectares. Overall staff would be reduced from 1,749 to 1,062, with scientific researchers reduced from 202 to 163. Taken together, aggregate savings would amount to 867,000 rubles a year, almost half the current budget of 1,898,000 rubles.[40]

The Presidium of the RSFSR cabinet approved Eliseev's proposal and sent if off to Gosplan USSR's Zotov. V. P. Zotov, an old Stalin-era politico, was a savvy political survivor who had taken kindly to the nature protection cause. In early March he called a meeting and invited everyone who had any serious connection with the nature reserves. By contrast with Merkulov's meeting with the nature protection activists, this was a huge affair that drew representatives of the Academy of Sciences, higher educational institutions, institutes connected with various ministries, the various *zapovednik* systems, MOIP, the Geographical Society, and other interested civic organizations.[41]

Zotov, who chaired the meeting, allowed all who wished to speak to do so, and the atmosphere he created was surprisingly respectful toward scientific public opinion. By this time the proposals by N. V. Eliseev, who had lost his nerve, were already reasonably known by this audience, and this provided Zotov with a springboard for his remarks. "Of course," said the politically wise Zotov, "the remarks of Nikita Sergeevich oblige us to close down the Altaiskii *zapovednik,* but which other ten or so should we include as well?" he asked. Eliseev in his panic had proposed eliminating eleven in the RSFSR for starters and twenty-four more in the other republics. As Iurii Konstantinovich Efremov, a participant, vividly recalled, the wily old Stalin-era veteran Zotov "chided [the Glavokhota chief] as one would a small child" for buckling too fast under political pressure. He really "shamed him," added Efremov.[42]

Zotov began to "instruct" the packed auditorium on how to salvage as many *zapovedniki* as possible, making as few concessions as feasible to Khrushchëv's fit of pique. He brought up the case of the Teberdinskii *zapovednik,* located in the North Caucasus. How could Gosplan USSR keep the reserve on Eliseev's list of those slated for elimination, he asked, in light of the 400-odd letters of protest received, including those of the *kraiispolkom* (which implied support from the *kraikom* of the Party)? The case for saving the gargantuan Kronotskii *zapovednik* was more problematic, however, despite the very large number of signatures collected on petitions to save that great reserve of active volcanoes, geysers, and boiling mud springs; many letters were collective expressions of protest, and therefore Gosplan had received only ninety separate envelopes. The Central Committee looked at the pile of envelopes, not the number of signatures, so individual letters were more effective than petitions.[43]

Efremov and his colleagues understood that what counted in the Gosplan bureaucracy was the number of envelopes and whether or not the local Party and state authorities sent protests as well. As scholarly secretary of MGO, Efremov was well positioned to begin a new emergency campaign. Wasting no time, he sent out instructions to each of the 2,000-odd members of the branch to send a postcard to Zotov at Gosplan. On the basis of this strategy Zotov was able to save many of the reserves. This episode also showed that there were some intelligent people in the bowels of the apparatus who knew how to assess and circumvent disruptive orders from above.[44]

Zotov's ultimate proposal in the Russian Republic eliminated the giant Altaiskii and Kronotskii *zapovedniki* as well as the smaller Denezhkin Kamen', Mariiskii, the ill-fated Zhigulëvskii/Middle Volga/Kuibyshevskii, and the tiny Khostinskii. Spared outright were the Volzhsko-Kamskii and Bashkirskii reserves, which the frightened Eliseev had included on his list, as well as the Sudzukhinskii, Khopërskii, and Laplandskii *zapovedniki,* which were each "eliminated" as separate units but then merged with relatively nearby reserves left untouched. Zotov's intermediate proposals for reductions in area of three additional reserves—the Darvinskii, the Pechoro-Ilychskii, and the Sikhote-Alinskii—came to 727,500 hectares as opposed to Eliseev's proposed 868,000-hectare reduction; Zotov later reduced the cuts to 313,200 hectares. In the other Union republics, Zotov selected an impressive number of relatively small reserves to eliminate but remerged many of those into reserves still standing, particularly in Ukraine. On paper, the number eliminated—thirty-two—looked significant, but of these ten continued under assumed identities"—a cunning bureaucratic sleight-of-hand. Taken together, the reserves systems declined from 6,360,000 hectares on the eve of this "reorganization" to 4,046,700 hectares in late 1961, largely a result of the elimination of the two huge Siberian reserves.[45]

In his letter to the USSR Council of Ministers, Zotov justified his limited cuts with a deft display of "toughness," noting that Gosplan firmly rejected the requests of the USSR Academy of Sciences and professors Formozov, Bannikov, and G. V. Nikol'skii to preserve the Altaiskii and Kronotskii reserves. On the other hand, Zotov averred that "the materials at hand and the information we have heard at the conference in Gosplan USSR from representatives of the USSR Academy of Sciences and the academies of the Union republics testifies to the fact that the *zapovedniki* have accomplished a great deal in preserving sites, in studying them, and in pursuing specific problems that have significance for practice and science both."[46] Even more exceptional is his mention of the "fundamental works" (*kapital'nye raboty*) written on the basis of research conducted in the *zapovedniki*, "which have made a great contribution to the development of Soviet ecology, a science that now occupies a prominent place in the system of the biological sciences not only in our own country but abroad."[47]

Gosplan USSR had always been a haven for bright, civic-minded staffers, and for reasons still largely unstudied remained an oasis of relative liberalism, despite episodic purges such as that of Groman and Bazarov and their associates in the notorious Menshevik Trial of 1930–1931. Soviet environmental history reveals this clearly; during both *zapovedniki* "liquidation" crises, the leaders of Gosplan, Saburov and Zotov, tried to mitigate the blow.

Vladimir Boreiko was able to discover equally interesting information about the reception of Khrushchëv's speech in Ukraine. When the Ukrainian Council of Ministers received the December 31, 1960 instructions from Moscow to examine the network of reserves and to identity where cuts could be made, the Ukrainian government turned to the Ukrainian Academy of Sciences for advice. The Academy's vice president, N. Semënenko, on January 21, 1961 responded that there was no fat in the system: the Ukrainian reserves were all small and injured no economic interests. Despite this, Ukrainian deputy premier Grechukha decided to convene a conference to decide—just in case—which reserves to place on the chopping block. On February 1 the conference was held in Kiev; noted Ukrainian nature defenders Ivan Grigor'evich Pidoplichko, Pëtr Stepanovich Pogrebniak, M. A. Voinstvenskii, and G. N. Bilyk were in attendance.[48]

Looking for somewhere to start, Deputy Premier Grechukha proposed the Chernomorskii *zapovednik* for elimination. "I am not against keeping the wardens," he said, "but there are four scientists there and the director makes five. What are they all doing?" Professor Voinstvenskii responded that the four scientists should be transferred to his institute, but that the wardens and some sort of administrator should remain in place. Grechukha seemed content with that. "Let the wardens stay, and make the senior one among them responsible. From January through March, to insure that five or ten are students present, let them organize practicums there; from April, instead of five, let there be fifty students. All these questions are practical ones and may be solved, while in principle we can raise the question of liquidating the Chernomorskii *zapovednik* [at least in name]." To that someone at the meeting shouted out, "Not under any circumstances!"

As Boreiko explains, Grechukha, sensing resistance, now turned the discussion to a smaller reserve, the Strel'tsovskaia steppe, where he proposed a more acceptable course of action: to liquidate the reserve as an independent unit but to turn it over to the Ukrainian Academy to be preserved as a *zapovednik* under the Academy's auspices. The scientists put up an unyielding resistance to any de facto elimination of any of the Ukrainian reserves. Evidently, their solidarity had an effect on the deputy premier, who wrote back to Zotov:

> In response to your telegram . . . of January 10, 1961, the Ukrainian Council of Ministers informs me that we have five *zapovedniki* [*sic;* the correct number was four] with an aggregate area of 13,000 hectares and a staff of forty-four, whose funding runs to 65,500 rubles a year. . . . The *zapovedniki* . . . listed are very important to the protection and restoration of natural resources of the republic and owing to their size and the expenses they incur do not represent superfluous items in Ukraine's economy.[49]

The ultimate resolution followed the lines suggested by Grechukha: the small steppe reserves were combined into one unified Ukrainian steppe *zapovednik,* while in the Chërnomorskii reserve staff and budget were reduced by 10 percent. This gave the appearance of cutbacks without vitiating the system.[50] The all-Union decree, signed by Kosygin and issued on June 10, 1961, despite having been softened by the protests of scientific public opinion and some local authorities and by the helpful maneuvers of Zotov, nevertheless did impose one new and serious limitation: in the future, all new *zapovedniki* of whatever system had to be approved by Gosplan of the USSR. Unfortunately, there was no guarantee that enlightened and vigorous individuals would always be found in that institution's leadership. As Boreiko notes, Ukraine had to wait seventeen years to be able to establish any new reserves.[51]

Sensing political benefit, some quickly tried to make hay by targeting the *zapovedniki* in the press. A particularly objectionable satire, "Zapovednye pni" (Protected Tree Stumps), was featured in *Komsomol'skaia pravda* on February 2, 1961 and was directed against the Teberdinskii *zapovednik.* That was followed on March 10 in the humor weekly *Krokodil* by a piece of similar sarcastic slant, "Redkii ekzempliar" (Rare Specimen), and finally by a satirical essay "Chik-chirik" in the newspaper *Sel'skaia zhizn',* directed against the Tsentral'no-Chernozëmnyi *zapovednik.*

What the enraged representatives of scientific public opinion could not write to Khrushchëv himself they permitted themselves in letters to the editors in chief of the offending publications. Because the articles subjected their very social identity to ridicule, the response of the scientist activists verged on fury. For example, Geptner wrote to the editor in chief of *Komsomol'skaia pravda:*

> I read the satirical essay "*Zapovednye pni*" . . . with a feeling not only of revulsion but also of deep sorrow. Regrettably, as we see, in our midst are still some "journalists" who are prepared, for a modest reward, to besmirch [*oplevat'*] anything at all, especially if they think they can "get into the good graces" of higher-ups. . . . In the last analysis, to write or not to write such . . . vulgar and stupid caricatures is the personal decision of such petty and talentless little people as Voinov and Oganov [the authors]. But how could *you,* the leader of one of the largest and most popular Soviet newspapers, allow this not only stupid and ignorant but, worse yet, harmful drivel into print?[52]

Assuming that the editor's lapse was ultimately explicable by his lack of familiarity with nature protection issues, Geptner gave him a crash course in the rationale for and status of protected territories in the USSR. After noting that, shamefully, the Soviet Union occupied nearly last place among all major nations—including Burma, Chile, and Ceylon—in the percentage of its territory under protection, Geptner explained that now it had become "a very delicate question." "To come crashing into this [debate over *zapovedniki*] with farcical ridicule," Geptner remonstrated, "showed extremely bad timing."[53]

Articles of this kind, Geptner cautioned, could wreak additional harm on a cause already under siege: "Among short-sighted administrators and economic bureaucrats . . . are many of our enemies. As one who has been familiar with our . . . cause for forty years already, I can assure you that with the publication of the satire . . . these people will spring to life and that attacks on our *zapovedniki*—which as it is do not have it easy—will intensify everywhere."[54]

Geptner was particularly pained by the satirical targeting of "good people, honestly carrying out their considerably arduous work," referring to the scientists of the Teberdinskii *zapovednik.* Despite *Komsomol'skaia pravda*'s past years of support for nature protection, which merited praise, the publication of the satire, said Geptner, had wiped the slate clean. "I would even go further," said he. "*Komsomol'skaia pravda* has positively compromised itself and, believe me, not just in the eyes of a few but in the eyes of very, very many. It will be a long time until you live down that stream of filth which poured down from the pages of your newspaper on this pure cause."[55]

Geptner concluded by stating that his purpose was not to vent his emotions but to prevent a similar mistake from being committed in the future. "I hope that the publication of '*Zapovednye pni*' was an accidental and thoughtless misstep," he offered, holding out an olive branch.

> But if this signifies a change in the line of the newspaper regarding the cause of nature protection, wouldn't it have been better initially to invite knowledgeable people to the editorial offices and quietly discuss [this issue] with them? I am sure that none of those who hold our nature and its future dear would have refused to take part in such a discussion. For the staff of your newspaper this would doubtless be beneficial and perhaps could prevent future such "disruptions" as the publication of "*Zapovednye pni.*"[56]

No less infuriated by the articles, Professor Aleksandr Nikolaevich Formozov wrote to the MOIP Presidium and to N. V. Eliseev, demanding "a decisive and serious rebuttal," and going straight to the top—to the press and science departments of the Central Committee. Formozov chose to delve deeply into the issues of biology raised by the satirical piece of E. Andreev, "Chik-chirik" (Cheep-Cheep) in *Sel'skaia zhizn'*. Noting that the author relied for his "ignorant attacks" on a textbook (Brehm) that was decades out of date, "thereby revealing the primitive middlebrow level of his biological knowledge," Formozov defended the study of the competition for prime nesting sites between sparrows and other insectivorous birds being pursued in the Tsentral'no-Chernozëmnyi *zapovednik.* In no other country, complained the noted zoologist, were researchers of ornithological stations subjected to such mockery.[57]

The question of how to attract and keep insectivorous bird populations from attenuating, asserted Formozov, was becoming particularly poignant in light of recent disturbing trends linked with the massive application of pesticides. That means of controlling insect pests "had crucial drawbacks," the scientist explained. Pesticides also killed off many beneficial predators, songbirds, and smaller insects that preyed on crop pests themselves, wiping out a potent array of natural pest controls. More than that, the poisonous agricultural chemicals ran off into the subsoil water regime, flowing into rivers and accumulating ultimately in the organs of many fish and animals. All over the globe, he went on, heavy application of pesticides was more and more frequently accompanied by the irruption of pest populations immune to the agent's toxins and exhibiting even higher fertility than the initial population. The problem had become so serious that it was the subject of discussion already at a number of international congresses such as those of the entomological society, IUPN, and other bodies.

Further, there was the question of whether field sparrows were in fact inferior to other songbirds in their ability to control insect pests. Years of observation in an apple orchard, where birdhouses that were put up were primarily settled by permanent families of field sparrows, showed that they were highly effective in controlling the insects that infest the fruit trees, said Formozov. "It is curious that *Sel'skaia zhizn'* [*Agricultural Life*], a newspaper duty-bound to be aware of and assess the experience of the most progressive pioneers in production, allowed itself to publish the illiterate article of Andreev," he concluded.[58]

Following Formozov's recommendation, the leadership of MOIP drafted a letter that it sent to the Central Committee's departments of science and the press on May 16, 1961. It was signed by the F. N. Petrov, still chairing MOIP's Section on Nature Protection, and by Vice President Varsonof'eva. In surprisingly strong language, they reproved the editors of the newspapers that published the offending satires for misleading the trusting Soviet public. Constructive criticism was always welcome, they emphasized. "But if under the pretense of criticism there are really attempts at totally baseless blanket smears, then this is already not help but rather the commission of enormous harm verging on criminality," they wrote. Additionally, Petrov and Varsonof'eva in the most powerful way they could tried to explain that the *zapovedniki* were not "warm little spots" where researchers lived the good life but were places that mostly suffered great material privation; working there was a sacrifice endured for the sake of service to nature and to the Motherland by those who dedicated their lives to research in *zapovedniki*. Sometimes they even lost their lives, noted Petrov and Varsonof'eva, and they added:

> Dear comrades! You only have to think about this to realize that before you in all its nakedness is the tragedy of their situation. What must it mean for a human being who is honestly giving his or her whole self to a chosen, constructive cause, when he or she is globally painted a malingerer, when the label "parasite" or "sponger" is pinned on him or her? . . . Is it any wonder that as a result there has been a colossal flood of letters to our leading authorities and to the editors of newspapers and journals protesting these unheard of accusations against workers in the cause of nature protection and against the state *zapovedniki* themselves?[59]

Petrov and Varsonof'eva informed the Central Committee departments that the editor in chief of *Komsomol'skaia pravda* responded to Geptner's and other letters only after a month, and then merely announced that the editorial board would "conduct a supplementary review of the facts" and would convey to the letter writers its "final opinion" in short order. "How is it possible to talk of any kind of review" in such a clear-cut case where "from the beginning to the end there was not one word of truth in the whole article?" asked the MOIP leaders. "Evidently the editors of the newspapers and of the journal *Krokodil* have forgotten Lenin's injunction about the need for a caring, sympathetic, and attentive attitude to the human person if they allow filth to be poured on totally innocent people in the pages of their publications." They concluded: "Deeply upset and insulted by the behavior of the editorial boards . . . the Bureau of the Section on Nature Protection of MOIP decisively protests against this blanket slandering in the press of the work of the state *zapovedniki* . . . and requests that you give appropriate instructions to the editors and oblige them to respond to the letters of workers."[60]

The aftershocks of Khrushchëv's speech and the second "liquidation" continued to be felt throughout the year. Movement leaders were kept busy fighting holding actions, such as the letter of MOIP president Sukachëv, vice president Ianshin, and secretary Efron of November 27 protesting the Kazakh republic's intention to transfer the northern half of the Aksu-Dzhabagly reserve to a collective farm, or that of M. A. Lavrent'ev, president of the Siberian branch of the Academy of Sciences, which forcefully urged canceling the elimination of Kamchatka's Kronotskii *zapovednik*.[61] The events were a paradoxical reminder of both how much and how little things had changed since Stalin's time.

The Fate of the Academy of Sciences Commission on Nature Protection

One aftershock that altered the administrative lineup of nature protection was a decree of the Central Committee and of the USSR Council of Ministers no. 299 of April 3, 1961, transferring the Commission on Nature Protection from the USSR Academy of Sciences to Gosplan USSR. The new president of the USSR Academy, M. V. Keldysh, agreed to the move.[62] Although we do not yet have archival evidence to explain why this happened, one plausible hypothesis is that V. P. Zotov initiated the transfer in the hope of providing his personal support and patronage to the cause of nature protection in the wake of Khrushchëv's January outburst.

By February 1962 a new charter had been drafted as well as a new roster of members. While Dement'ev was still left as chairman and L. K. Shaposhnikov as scholarly secretary, a Gosplan functionary, A. D. Ponomarëv, deputy head of that agency's forestry division, was made deputy chairman (initially, botanist E. M. Lavrenko was asked). Compared with the previous membership, specialists in energy, public health, and air and water pollution were better represented on the Gosplan commission. On the other hand, only a handful of the old elite of the movement—Sukachëv, Efron, Lavrenko, and Voronov—were left, isolated in a sea of sixty-four other "new" people.[63] Nevertheless, the commission was impressive: eleven full or corresponding members of the Academy, sixteen doctors of science, and sixteen candidates of science.[64]

Testament to the supportive circumstances that the commission encountered within the all-Union Gosplan is a letter from Shaposhnikov to botanist Evgenii Mikhailovich Lavrenko written on December 24, 1961. Shaposhnikov first alluded to the months and months of delay waiting for the official issuance of a new charter for the reorganized commission. This was no trifle, he explained,

> for, without a charter it is impossible to bring to life the work of the Plenum and the Bureau of the Commission. Georgii Petrovich [Dement'ev] and I have invested a huge amount of sweat and time to speed up the time when this charter sees the "light of day." Just recently we have had some important successes. Besides that, the current business of the Commission is going ahead full steam. As far as the functionaries at Gosplan are concerned, we are exclusively encountering attitudes of good will and great—I would even say generous—assistance as far as material and technical support of our work is concerned. The possibilities here are incomparably greater than within the Academy. Come visit us. We will all be happy to see you and to consult with you.[65]

Lavrenko, though, was apparently less interested in continuing his central involvement, and in a note to Shaposhnikov of April 17, 1962 asked to be removed as deputy chairman of the commission and made an ordinary member "because I am otherwise occupied and owing to the condition of my health."[66]

The Gosplan commission was able to hold on through 1962, but by the late spring 1963 it, too, had attracted the suspicious eye of the increasingly arbitrary Khrushchëv. Rumors of a new "reorganization" began to flow. Although exhausted from the seemingly continuous defensive campaigns to save its modest scientific and civic world, scientific public opinion once again rallied to the cause. On May 25, 1963, three heavyweights, F. N. Petrov, Sukachëv, and the eminent chemist and defender of genetics N. N. Semënov, along with Dement'ev, wrote to USSR deputy premier K. N. Rudnëv asking for a final transformation of the beleaguered commission into the long sought-after State Committee for Nature Protection attached to the USSR Council of Ministers. They cited foreign examples. They cited the examples of Estonia, Lithuania, and Belorussia. They referred to the resolutions of the all-Union conferences of 1958 (Tbilisi), 1959 (Vil'nius), 1961 (Novosibirsk), and 1962 (Kishinëv). And they got nowhere.[67]

Two weeks before the publication of the decree eliminating the Gosplan commission, which was signed by Khrushchëv on October 2, 1963, a new wave of desperate letters began to flow to Kremlin addressees. One letter interesting for its emphasis on the image of the USSR abroad was from V. S. Pokrovskii, deputy secretary of the Commission and the head of its Laboratory for Nature Protection, to the Presidium of the USSR Council of Ministers with a copy to Foreign Minister Andrei Gromyko.[68] "From the moment of its creation in 1955 the Commission . . . has been making great efforts to enhance the influence of Soviet scientists in international [conservation] organizations," he began. Two Soviet initiatives on economic development and nature protection were adopted unanimously by the XVIII session of the General Assembly of the UN and by UNESCO. Further, the Commission, with the support of the USSR Ministry of Foreign Affairs, had been devising strategies of cooperation with the COMECON (Council of Mutual Economic Assistance) countries in the area of environmental protection.[69] With the aid of the ministry, the Commission had been able to obtain valuable information from Soviet embassies abroad on nature protection activities around the globe; already, the card file of the Commission contained the addresses of 350 organizations and scientists who regularly exchanged literature. The Commission, noted Pokrovskii, had recently submitted the findings of the National Academy of Sciences' report to President Kennedy to the Academy's Siberian division head, M. A. Lavrent'ev, to review what might be relevant in the USSR. Not only was the Commission a source of goodwill and a positive image of the USSR abroad, argued Pokrovskii, it was also a source of information about the outside world. For instance, its analysis of international legal norms in the area of resource conservation enabled the Soviet delegation to be more effective in the talks surrounding the study and use of Antarctica.

> It is clear that the proposal, advanced recently, to eliminate the Commission may negatively affect the position attained through such hard work of the USSR among the progressive international movement for the rational use of natural resources of the earth, and could lead to the weakening of ties between Soviet scientific specialists in nature protection and their foreign colleagues. There is the danger that such a step would be greeted with incomprehension in the IUPN and among scientific public opinion of foreign countries.[70]

Another letter, one of F. N. Petrov's last (he retired as head of MOIP's Section on Nature Protection in 1964), was addressed to the Presidium of the Central Committee of the Party. Despite his age, Petrov showed that he was still following public affairs. Raising the argument that both socialist and capitalist countries alike had seen the need for authoritative agencies for the protection of nature, Petrov chose a highly unusual example to make his case: "Particularly great attention to developing the scientific bases for the protection of nature is being paid in the United States of America where, for example, on the request of President J. Kennedy to the Congress, the National Academy of Sciences began special studies by the most eminent scientists in America on the condition of natural resources in the U.S."[71] With the contemplated dismantling of the Gosplan commission, warned Petrov, the contrast with the Soviet Union's archrival would be striking and not to the USSR's advantage. At the end of his life, Petrov decided that he could afford to dispense with niceties:

> In the Soviet Union as a result of the liquidation of the Commission for the Protection of Nature of Gosplan USSR and the transfer of its Laboratory, the development of scientific bases for the rational exploitation of natural resources will be brought to an end. The Soviet Union will lose official representation and will be deprived of any links in the international arena in the area of nature protection. Active state oversight over the rational use and reproduction of the entire complex of resources of the USSR will be liquidated. On the basis of the above I ask you to reexamine the draft decision of the USSR Council of Ministers prepared by the State Committee for the Coordination of Scientific Research of the USSR.[72]

The letter ends with no attempt at cordiality or propitiation. It was Petrov's last big campaign, although he lived for nearly another decade. As for the liquidation, it went ahead on schedule, and the Commission's Laboratory of Nature Protection Research was transferred to the USSR Ministry of Agriculture's Glavpriroda, Malinovskii's old outfit.[73] The Party chose not to heed one of its last links to Lenin.

Beaten down just when they had allowed themselves to regain hope, the older activists now perceived the extents of their social weakness and isolation. No longer in their prime of life, and unable to influence social events to their satisfaction, they clung to one of their few tangible achievements: the preservation of MOIP as the independent institutional locus of their social group.

A poignant series of letters from Boris Evgen'evich Raikov, longtime member of MOIP, to Vera Aleksandrovna Varsonof'eva, the society's vice president, a conservation activist, and Raikov's close friend, suggest that at least some members of the pre-*perestroika* nature protection movement were aware of its importance as a hidden site of opposition to the dominant official social and economic vision. Written on July 17 and November 18, 1963, the letters from the eighty-one-year-old historian of science reveal Raikov's fear that Varsonof'eva's care for her ailing sister could fatally interrupt her scientific and, especially, her civic work. "Your last letter devastated me," wrote Raikov.

> It is positively tragic. That your relative has died is, of course, sad; but we all die sooner or later. But the situation with your sister is worse than anyone could have imagined. Worse for you, because she scarcely is aware of her own con-

> dition, and remembers still less. But for you to spend time with her several times a week . . . is to doom yourself. I *beg* you straight out to stop this. . . . You *do not have the right* to sacrifice yourself in the name of a relative. . . . I am in complete sympathy with your views and feelings about the desecration of the Volga. But I do not have merely a feeling of sadness about these "refashionings" [*peredelki*] of nature, but a sharp feeling of anger [*negodovanie*]. Anyway, there is no sense writing about that! [emphasis in the original][74]

In his next letter, Raikov renewed his admonitions:

> You have surrounded yourself with several sick charges . . . but surely there are others who could and even, perhaps, must take on part of your load. I have not even come to the question of MOIP, which is doing work of enormous importance, because this is the only scientific institution that has maintained its *civic dignity* not only in Moscow, but in the entire [Soviet] Union, and which by some kind of miracle has so far retained its integrity amid all the other state-dominated ones. And you are so needed there, even indispensable, precisely as a guarantor of scientific public opinion.[75]

Ask any of the veteran members of MOIP about its golden age in the 1950s and 1960s, and they will tell you the same thing: politically insulated by its loose subordination to Moscow State University and by its physical location within the Moscow University Zoological Museum—the citadel of old guard nature protection activism—MOIP remained virtually the only "voluntary society" in the land that could claim scientific, intellectual, and even political autonomy. This was because MOIP, like the elite conservation movement as a whole, existed at the distant margins of Soviet life. Perhaps the high hopes engendered by Khrushchëv himself set the stage for the feelings of disillusionment and extreme social isolation experienced by this lonely outpost of the scientific intelligentsia. However, the double bind of serving the ideal of "science" as the activists defined it and trying to be loyal and patriotic state servitors remained; the scientists were not ready to join the still invisible, sparse ranks of dissenters from the system.

CHAPTER FOURTEEN

Student Movements

Catalysts for a New Activism

Alongside the cresting activism of scientific public opinion, the Khrushchëv years saw the emergence of student environmental activism. For some, such as the Estonian students at Tartu State University, this activism was tinged with ethnonationalist feelings from the start. For others, particularly in biology programs at the elite universities, it revived prerevolutionary traditions of the *studenchestvo* (students' special social identity). Exploiting the moral authority traditionally enjoyed by students, activists took direct action against such social "ills" as poaching and also participated in the planning and staffing of protected natural territories. Finally, for a third group concentrated in the somewhat less prestigious engineering and higher technical schools, environmental activism allowed the students to put their newly gained technical knowledge in forestry and other areas to patriotic use, in trying to circumvent the cumbersome, wasteful, and conservative Soviet bureaucratic system. Only later were some members of this group, disillusioned by the bureaucracy's opposition to their efforts, drawn to an ideology of Russian nationalism.

As the professional scientists did, the students used nature protection as the nucleus for their sense of mission and social identity, but they were motivated more by youthful impetuousness and by love of nature than by a sacred ideal of Science. This was true even of the students in the elite biology programs. Despite the fact that the Moscow University Student Brigade for the Protection of Nature was originally sponsored by MOIP, the ideals, social identity, and practice of the new organization diverged from those of the older activists. This divergence demonstrated the impossibility of reproducing the social identity of the scientific intelligentsia under Soviet conditions.

The students' efforts to curb the abuses of the system and to implement a more conservation-oriented approach to resource management provided an object lesson. Although suffocated by bureaucrats, the students' naive efforts revealed unrecognized contradictions between the Soviet system and widely shared social values. The experience of the students catalyzed the link between nature protection and an awakening antimodernist, xenophobic Russian nativism, helping to bring that larger movement into being.

The Moscow State University Biology–Soil Sciences Student Brigade for Nature Protection

The first university students' nature protection circle in the USSR was founded in Tartu on March 13, 1958, uniting students from Tartu State University and the Estonian Agricultural Academy. "Taking up the initiative of Tartu University," as a recent history put it, the students of the Biology–Soil Sciences Faculty (Biofak) of Moscow State University in 1960 founded the first *druzhina* (nature protection brigade).[1] The curiously anachronistic designation *druzhina* was not idly chosen. The chronicles tell us that Prince Vladimir, "who loved his *druzhina,*" consulted with it about affairs of the land and of war. The *druzhina* was the circle of closest warriors and counselors of the princes of Kievan Rus', the first line of defense of the Russian lands. "All this is known, of course, only by historians," writes modern-day *druzhinnik* Ksenia Avilova.

> Nevertheless, when the Komsomol enthusiasts, having decided voluntarily and without any thought of gain to defend nature against doltish assaults, called themselves the *druzhina* as of old, they immediately and precisely defined the sense and thrust of their work: *active* activity [*aktivnaia deiatel'nost'*], struggle, and the protection of Russian nature from evil, from soullessness, and from a lack of care for it. In that way the word *druzhina* acquired its contemporary meaning, distinguishing itself from all other circles, clubs, and societies that frequently exhibited only the external facade of solidarity.[2]

Vadim Tikhomirov, the *druzhina*'s faculty adviser, commented on this lexical matter as well: "The very word '*druzhina*' was a happy choice. . . . In it we see reflected the striving for *active* efforts, for struggle, and not for meditations on nature protection themes. It presumes a certain level of organization and solidarity and, more than anything, a strict sense of civic responsibility [*strogaia obshchestvennaia otvetstvennost'*] even while preserving the conditions of voluntary enrollment."[3]

That account, however, omits a much more recent antecedent, the *druzhinniki,* who had just been brought into being by Khrushchëv as part of his vision of a transition of the Soviet Union from a "dictatorship of the proletariat" to an "all-people's state." The Party Program that followed the Twenty-first Party Congress in February 1959 promised that organs of state power would gradually become organs of public self-administration, in keeping with the withering away of the state that Marx and Engels had promised.[4] One of the first areas selected for this transition by the Soviet chief was the maintenance of public order, and squads of *druzhinniki* with their telltale red armbands became ubiquitous at sports events, parades, and even university entrance checkpoints following a decree of March 2, 1959, although their experimental precursors date to 1957.[5] The Moscow University *druzhina* could point to Khrushchëv's *druzhiny* as a potential source of legitimation while creating a social space for civic activism and public self-administration far more independent than what the state had intended.[6]

The origins of the MGU *druzhina* actually go back to 1959 as well, when a university student subsection was created under F. N. Petrov's Section on Nature Protection, with the active patronage of the president of MOIP's Biology Division, Nikolai Sergeevich Dorovatovskii, and its secretary, Konstantin Mikhailovich Efron. At the time, the most active student members were Boris Vilenkin, Maria Cherkasova, and V. Baranov. "The numerous excursions and trips organized were motivated by the urge to apply the members' personal efforts in the defense of living nature," recalled Vadim Nikolaevich Tikhomirov, one of the subsection's organizers. The students were looking for "*active* efforts" and "struggle," motivated not by anti-Soviet attitudes but by a fierce impatience with the imperfections of the system. Many were members of the Komsomol who had acquired credentials as "citizens' inspectors" for nature protection. The first student inspections throughout Moscow *oblast'* to combat poaching and trips to the forest to prevent the logging of fir trees for the New Year date to this MOIP period of the movement.[7] The students proclaimed: "We've had enough talk about purity. Let's start cleaning up!"[8]

Aside from those dramatic undertakings, the students addressed numerous groups, from teachers to factory workers, and organized seminars to upgrade their own knowledge of conservation issues and biology. Through MOIP, they were able to attract speakers of the stature of soil scientist David L'vovich Armand. This exposure to scientific advocacy for nature protection, with all of its customs and rules of polemics and conversation, socialized the students to the social identity of *nauchnaia obshchestvennost'*.[9] For many students, however, devotion to the older cult of science with its rules and ethical norms had less appeal than the tug of adventure and a life of action.

Spurred by news of independent student organizations in Tartu and Astrakhan (Iu. N. Kurazhkovskii's "For a Leninist Attitude toward Nature," founded earlier in 1960), the students soon went their own way, with the reluctant blessing of the MOIP leaders.[10] Geography played a role in this turn of events. The headquarters of MOIP was on the old campus, in the

Zoological Museum off Manezh Square. However, the students lived and studied at the new campus, miles away on Lenin (Sparrow) Hills. This physical separation facilitated the creation of a distinct student identity.

The development of an autonomous society was quickened by the emergence of a group of exceptionally independent-minded first-year students in Biofak in 1960. As Tikhomirov relates it, "In the majority, they were shaped by having been in the various young naturalist circles: KIuBZ, the VOOP circle led by Petr Petrovich Smolin, and the circle within MOIP led by Anna Petrovna Razorënova. These students led by Evgenii Smantser sought out teachers who were also deeply troubled by problems of nature protection. As a result of the combined efforts of students and faculty came the birth of the *druzhina*."[11]

Much like field biology itself, a region of natural science that seemed to draw in those who experienced a special need or delight in studying life in its unfettered condition, the *druzhina* attracted the most independent, self-reliant, and, it appeared, most sensitive students. They were also, in their own way, those most aware of social wrongs and, on some still dimly conscious level, of the possibility that the Soviet vision had taken a drastically wrong turn. It is likely that no one will improve on the portrait provided to us by *druzhina* veteran Ksenia Avilova: "It is natural that at the origins of this unique coalescing of youth, born as it were as a sign of the times on the eve of the RSFSR law on the protection of nature, stood individuals who were out of the ordinary. . . . [I]n the far gone days of 1960 this kind of activism with respect to nature was a reflection of an improbably daring, even dangerously bold view of life."[12]

At the October 1960 Komsomol conference of the members in the Biology–Soil Sciences Faculty, the activists secured support for the creation of a *druzhina po okhrane prirody* (nature protection brigade). On December 13, 1960, at a meeting of the Komsomol of the faculty and nature protection activists, the "fighting brigade" was formally christened. Its first "commander" (*komandir*) was biology student Evgenii G. Smantser. Biofak dean Nikolai Pavlovich Naumov named *dotsents* Vadim Nikolaevich Tikhomirov and Konstantin Nikolaevich Blagosklonov, who later coauthored the first university textbook on nature protection and conservation biology, as the *kuratory* or faculty advisers of the new group.[13] Both were active in MOIP and VOOP and they were well positioned to convey the old-line tradition to the up-and-coming generation of elite field biologists.

As the "zoological" half of the faculty leadership, Blagosklonov trained the students in the skill of observing bird behavior. The author of an internationally known handbook on the protection and attraction of birds, Blagosklonov had organized "Bird Day" in the 1920s with some Young Naturalist groups. He also supported youth programs such as the Zvenigorod summer schools, KIuBZ, and VOOP's youth section, in which he continued to be active despite the transformation of VOOP's leadership into a sinecure for retired Communist operatives. Blagosklonov's other great strength was in organization. For more than twenty years he organized biology "Olympics" at Moscow State University, summer camps for nature protection, and a host of other events. He was also an eager propagandist for nature protection, writing innumerable articles and appearing on radio and television.

As the "botanical" half, with particular expertise in floristics and taxonomy, Tikhomirov (see figure 20) was an invaluable guide to the world of vegetation. His students were treated to exhaustive but also exhausting training: two months in the field in Mordvinia on the upper Volga after the first year, geobotanical field work and ecology in the Moscow region after the second, a transzonal field journey south to the Caucasus or Crimea together with soil scientists, and then a 10,000-kilometer odyssey around central Russia, including the southern steppes. Various *zapovedniki* served as training bases. Nature protection based on a profound scientific knowledge of vegetation was the emphasis. Those who sought to spend their summers sunning themselves on the beaches in Batumi were encouraged to select another area of specialization. The students had to love botany enough to consider a grueling field trip a vacation. Tikhomirov brought more than technical knowledge to the *druzhina;* he was its moral compass, with his militancy, personal daring, and charisma.[14]

In those first few months, the *druzhina* began with a band of forty-two stalwarts, an impressive number for such a risky and dubious cause. In that first cohort, two members were involved with organizing and providing lectures and two more worked with the youth group at MOIP, a residual link to the parent society. Three served on the editorial commission and three more started a separate section concerned with the fight against water pollution. Almost half were involved in the antipoaching patrols, which were later (1974) christened "Operation 'Shot'" (*Programma 'Vystrel'*). Instruction and training were a central part of preparing the *druzhinniki* for this dangerous work. Because the *druzhina* from the outset was under the aegis of both the Komsomol and VOOP, as members of the latter, *druzhinniki* were able to get credentialed as "citizens' inspectors for nature protection," which allowed them to apprehend violators of Soviet nature protection and hunting laws and to confiscate their equipment and catch.[15]

Regular antipoaching operations began in earnest in the autumn of 1961. Paraphrasing Smantser's memoirs, Sviatoslav Zabelin described the scene in those days (see figure 21): "In the evening upon the *druzhinniki*'s arrival at the base, the atmosphere was jolly around the far from sober dinner table. But from early morning on followed the inspection, which involved a profusion of confrontations with violators who had no suspicion about the existence of hunting and fishing regulations and relied in all questions on the

Figure 20. Vadim Nikolaevich Tikhomirov (1932–1998) and Tat'iana Bek, *druzhina komandir* (leader), mid-1960s.

Figure 21. *Druzhinniki* inspecting hunting documents, mid-1960s.

power of their fists and their lungs."[16] On rare occasions the raids ended in tragedy; at least six *druzhinniki* were shot dead during these hunting checks, creating an aura of danger and responsibility around the *druzhiny*. The group also conducted raids on the Kalitnikovskii open-air market, detaining those who were putting songbirds up for sale.[17]

Targeting ordinary citizens, albeit lawbreakers, reflected the political immaturity of the *druzhina* movement. From the standpoint of theory, the movement failed to develop a thoroughgoing socio-politico-economic analysis of the roots of the destruction of natural amenities in their society. From the standpoint of tactics, these raids alienated the great majority of the workaday public from the students. In many cases those who were fined or whose merchandise was confiscated were those toward the bottom of the Soviet social ladder, trying to earn a few rubles or bring back a partridge for the family pot. Despite the physical danger of flushing out poachers, the students were going after relatively minor offenders while major bosses, who were not only poaching but poisoning the rivers, lakes, streams, and air, were conveniently ignored. No wonder the antipoaching campaign alienated the masses from these elite students, as Tikhomirov obliquely recognized: "The attitude to its work on the part of the population was, as a rule, hostile. The activities of the *druzhinniki* often met with total incomprehension, especially when they affected the personal interests of citizens regarding the use of forests, hunting, or fishing. . . . It was a rare encounter in the forest that concluded without the use of force or some other sort of extreme action, and there were shoot-outs."[18]

The authorities had little use for the student raids, either. "We would frequently hear such outbursts [from them] as 'You *what?* Defend nature? From whom? From our Soviet person?'" Many officials even considered their cause "harmful."[19]

Even in the Biological Faculty itself, the stronghold of scientific public opinion, some questioned the wisdom of letting the students go off into the woods to catch poachers. Rebutting these misgivings, Tikhomirov emphasized the larger social meaning of the students' activities: "But it was precisely [these questions] that demonstrated that [the critics] had completely failed to understand that our foremost task was the socialization of these future specialists, forging their intellectual and political outlooks as well as molding them as citizens."[20]

By the group's official first anniversary in December 1961, a conference had been organized on the emerging problems surrounding Lake Baikal, a rather large and impassioned meeting held on the forestry experiment "Kedrograd" in the Altai, and four wall newspapers published. And Tikhomirov thought that it was time to create *druzhiny* in other higher educational institutions around the USSR.[21]

The next year's activities saw the New Year's tree campaign move to

Moscow railroad stations where, with the knowledge and cooperation of the stationmaster and militia, buyers and sellers alike were apprehended; in all, 800 trees were seized. It also marked an inconclusive attempt by the Komsomol of Biofak to disown the *druzhina* by preventing it from delivering its annual report. This and an investigation in November 1962 by the Party Committee were potentially damaging.[22] But the *druzhina* weathered these squalls, mainly thanks to the support of Naumov and his faculty, again demonstrating the ideals of *obshchestvennost'* in action.[23]

By the following year, the Party Committee of Biofak was showering the *druzhina* with compliments despite a few uncomfortable moments in early 1964 when apparently drunken *druzhinniki* on a trip to the Kyzyl-Agachskii *zapovednik* in Azerbaijan even engaged in some poaching themselves. Routines developed as *komandiri* and other posts were rotated every year. The campaigns netted increasing numbers of violators (see figure 22), and the *druzhina* became a visible and important institution in Moscow University's student scene, even if membership remained modest.

Like its parent, the *druzhina* was a counterculture. Dmitrii Nikolaevich Kavtaradze, a former leader of the group during the early 1970s, compared it to a "military unit." "Some people were influenced by it to such an extent that the whole course of their lives was altered."[24] Group loyalty was especially great in the *druzhina* (see figure 23)—partly because so many of its members had already been socialized in the VOOP and MOIP youth groups and in KIuBZ. The geologist Pavel Vasil'evich Florenskii, who had been in KIuBZ, speculated that this powerful bonding ethic had its origins in the brotherhood of the Tsarskoe Selo Lycée and in later traditions of solidarity within university student culture (*studenchestvo*).

In the words of Oleg Ianitskii, the *druzhina* and its later offshoots were spreading "the 'small is beautiful' virus" in a system based on gigantomania. Inevitably this generated tension between the *druzhiny* and their official sponsors in the Komsomol and VOOP. As Ianitskii notes, "Under our conditions, these contacts did not in any sense amount to a compromise between the two sides; each was playing its own game. The people from the system considered that the independent organizations could be held in check, and that they were a useful valve for letting off the steam of popular dissatisfaction. The club leaders hoped that as they gained more muscle, they would gradually reconstruct the System from within."[25]

The Kedrograd Experiment

In Leningrad another influential youth group arose during the Khrushchëv "thaw" of the mid to late 1950s. Ever since the first Five-Year Plan, the Soviet state had been taking large numbers of workers' and farmworkers' children

Figure 22. V. N. Tikhomirov detains poachers.

Figure 23. *Druzhinniki* at leisure, perhaps singing songs of Bulat Okudzhava.

and propelling them into new lives after their training in technical and engineering schools. As Sheila Fitzpatrick has observed, despite the privations of their school years, these students were often grateful to the regime for the opportunity for significant upward mobility. They were the system's, and Stalin's own, loyal constituency.[26] A discernible sociological gulf emerged between the mass of poorer students at the technical universities and the *jeunesse dorée* and hereditary intelligentsia at Moscow and Leningrad state universities and other elite schools.

At the Leningrad Forestry Technical Academy, one of these students of modest background was called "the dreamer." Described as a serious, "grey-eyed, slightly wild-looking Siberian who, according to his passport, [was] called Fotei," Sergei Shipunov was rechristened by his classmates, who asserted "that such a name did not exist."[27] Although a silent, intense young man habitually "glued motionless to a book in the evenings,"[28] Shipunov became popular because after graduating from technical high school (*tekhnikum*) he had worked a stint as a forester and knew the forests better than any of his peers. He also accrued greater authority owing to his independence and to his severely categorical judgments. Classmates also envied his habit of taking from books and lectures only what he needed. It was machismo with an angry Russian twist.

Shipunov and his friend Vitalii Feodos'evich Parfënov were elected to the faculty bureau of the Komsomol, where Shipunov became secretary. Soon, though, Shipunov was generating sparks. "He was excessively unbending and dealt with people too highhandedly," wrote the journalist and novelist Vladimir Chivilikhin. "He wanted to remake too much his own way and this frequently generated conflicts in the faculty." It was not long before he was called into the dean's office because he had announced that the students were being taught subjects that had little bearing on their future practical work and even had leaflets printed and distributed at nearby forestry plantations. "You're taking a lot on yourself, Shipunov!" snapped the dean. "In short, we have given the order to return your underground leaflets from the *leskhozy* [Soviet forest plantations]."[29] Shipunov did not bend. His response was to publish a biting article in one newspaper on how forests were being improperly cut in Leningrad *oblast'* and on the poor preparation of new foresters.[30] His friends tried to tone down his contentiousness, but it proved resistant to alteration right to the end of his life.

In the fall of 1957 Sergei was elected to the academy's Komsomol Executive Committee. By this time, he had become inextricably associated with his "dream" to establish a model forest plantation in the Altai Mountains in order to harvest the secondary products—squirrels, sable, deer, and some game fowl, as well as mushrooms, berries, and pine nuts—of the Siberian stone pine or "cedar" (*kedr*) forest there. Only sick trees would be logged to preserve the health of the forest complex. Although some derided his dream as "utopian," others such as Parfënov, Lesha Isakov, Kolia Novozhilov, and Vladimir Ivakhnenko shared it.[31]

More than thirty years earlier, there had been similar plans for the sustainable utilization of the Altai's "cedar" forests. Sometimes called "*chudo-derevo*" (wonder tree), "*khlebnoe derevo*" (the bread tree), "*derevo-korova*" (the tree-cow), or "*derevo-kombinat*" (the multiproduct/multiple services tree), the Siberian stone pine has long been known to be a good source of pine tar, *bal'zam* (resin), vitamins, nuts, wood, bark for corks and pigments, roots for wickerwork, and a prime habitat for economically valuable plants and animals.[32]

Cedar nuts constituted 50 percent by weight of all the trade traffic heading toward the *iarmarki* (fairs) of Irbit and Nizhnyi and constituted more than one quarter of the traffic by weight on the Trans-Siberian Railroad before the First World War; average yearly shipments were 189,000 puds (6,840,000 lbs.) during the period 1899–1908.[33] The "wonder tree" had also attracted the attention of curious lay polymaths such as V. Tatishchev in the eighteenth century—he described them in his journals—and later of a number of professional botanists and academic forestry specialists, including Sukachëv. Obviously no ordinary tree, the "cedar" had already become a symbol of Siberia's economic potential and the richness of its natural resources by the time the Bolsheviks assumed power.

The Bolsheviks' early rhetorical commitment to rational resource management emboldened those who sought to restrain the accelerating tempo of resource exploitation.[34] At the end of the 1920s inside the system of consumer co-ops seven multiple-use cedar plantation-complexes were planned but only one, in the Gornyi Altai (Altai Mountains)—the Karakokshinskii cedar plantation—was organized, and that only lasted two years, going down in 1933. One reason for the failure of the venture was that there was not enough work year round; the project's planners had failed to diversify.[35]

When, after a hiatus of more than twenty years, the students of the Leningrad Forestry Technical Academy returned to the problem of the sustainable use of the cedar forests, they did so armed with their professional training as foresters but also with the knowledge of the cause of the previous venture's failure. Their plan for multiple use, which they dubbed "Kedrograd" (cedar city), included tapping spruce sap (for turpentine), grinding pine-needles to produce vitamins, bee-keeping, gardening, some agriculture, and limited logging, predominantly sanitary. This was the base for year-round activity on which the superstructure, based on harvesting nuts, berries, mushrooms, and pelts, would be erected. Logging would be allowed only if it enhanced, rather than undermined, the overall sustainable economic regime, and could not be done when it would disrupt the reproduction of wildlife. "All this was impossible to attain given existing practices, where the numerous forest users . . . worked autonomously, guided only by their entrenched bureaucratic interests," noted Parfënov; he and his young colleagues were going to reform the system, however, and help it regain the true path to Communism.[36]

Professionally, the students' ideas about forest structure were strongly influenced by the holistic ideas of Georgii Fëdorovich Morozov, Sukachëv's teacher and a proponent of the idea that there existed "forest types" that were relatively closed and self-reproducing, not unlike Sukachëv's later elaboration in his concept of the "biogeocenosis."[37] The cedar forest, accordingly, was considered one such bounded "type." Unlike Sukachëv, however, the students absorbed the post-Stalin Leningrad Forestry Technical Academy's commitment to the principles of rational resource use (sustainable use) and rejected the idea of "placing nature under lock and key." This pragmatic orientation distinguished the graduates and students of the Leningrad Technical Forestry Academy from the *druzhinniki* of MGU Biofak.[38]

Shipunov's "dream" was not subversive; he and his fellow students were seeking only to make Communism arrive faster by making production less wasteful and more efficient. Shipunov's vision of nature was a workshop, not a temple. His dream was that of a Soviet patriot. Seeking support,

Shipunov went to Moscow. Not wanting to alienate the ardent young students, officials extended their support; they included A. F. Mukin, head of the Forest Division of the RSFSR Ministry of Agriculture, as well as P. F. Kaplan of Gosplan RSFSR. Sergei Andreevich Khlatin of the Main Forestry Administration of the RSFSR in particular became the student's patron.[39]

Back at the Leningrad Forestry Technical Academy, some professors also gave the students warm support. One respected and popular professor of geodesy and surveying, Gubin, offered one month's salary to help pay for the reconnaissance expedition. In addition, Professor Gubin wrote a letter supporting the expedition to select the specific territory for the forest plantation. Alas, Gubin, who was too solicitous of the students' living conditions, went over budgetary allowances in his division and it was decided to remove him for "gross violations of financial rules."[40]

Sergei Shipunov, as a member of the Komsomol Committee of the Academy, decided either from "pigheadedness" or from "inexperience" to defend Professor Gubin. Despite intense pressure from the head of the Central Committee of the Komsomol, Oleg Maksimovich Poptsov (who later became editor of *Sel'skaia Molodëzh'* and after 1993 served for a time as director of Russian TV), as the member of the Committee with responsibility for student life Shipunov continued publicly to speak out. In fact, he provided his own "interpretation" of the meeting of the Komsomol Committee, making it appear that the committee was much more militantly opposed to Gubin's dismissal than in fact it was. Responding to Shipunov's call and organized from within the faculty where Parfënov was Komsomol representative, within twenty-four hours at an agreed-upon time almost all the students walked out of classes in protest at the professor's dismissal. Some instructors walked out with the students.[41]

However, the case did not end there. The district committee of the Komsomol intervened, "frightened that, in the academy, some sort of student 'circle' [*gruppok*] had organized. Not getting down to the details, the bureau of the *raikom* expelled Sergei as well."[42] A week later, Shipunov's baccalaureate thesis defense was scheduled to take place. When he arrived, he read an order on the door expelling him from the academy, and was told to report to the Komsomol district committee. He was asked to surrender his Komsomol membership card, refused, and disappeared. Even his father did not know his whereabouts.[43]

A "political conspiracy" was alleged. "This is what the thaw has turned into," complained irate administrators as they kicked Shipunov out of the academy. Parfënov was given a warning from the deputy director of the USSR Federal Forest Service, who was on the faculty bureau. Gubin was falsely accused of "incitement."[44]

After graduation Parfënov organized the expedition alone; all the other students received their diplomas and went home. However, all was not lost. They indicated a readiness to go to the Altai when called. So, when Parfënov arrived in the Altai in the summer of 1959, Sergei Shipunov showed up in September along with Ivakhnenko. And when Sergei Khlatin traveled to Choia to supervise, the young graduates knew that things were really underway.[45]

At first, local officials in the Altai as well as the faculty were supportive, either out of conviction or because they sought to humor the students and graduates, thinking that nothing would ultimately come of their scheming and dreaming. Among the genuine supporters was Roman Aleksandrovich Dorokhov, deputy chairman of the Altai *kraiispolkom,* who soon became the provincial Party first secretary of the Gorno-Altai *oblast'*. Sadly, he died in 1963 and was replaced by an enemy of the project.[46] Another important source of support was the Scientific Council of the faculty of the Leningrad Forestry Technical Academy, which met on March 6, 1959 and gave the project its blessing too after hearing a presentation by Shipunov.[47]

Of course, there were some early nay-sayers, such as academic forester A. D. Kovalevskii, who worked in the Central Black-Earth *zapovednik,* who wrote an article in *Nash sovremennik* that Kedrograd was all too theoretical and abstract.[48] This was countered by Feliks Kuznetsov, who defended the students (he later became the head of the Union of Russian Writers), and by Parfënov, who was given space in the journal two issues later.[49] Articles defending Kedrograd appeared in the student newspapers of the Moscow Aviation Institute and the Sverdlovsk Technical Forestry Institute, to name two.

As the saga of "Kedrograd" gained wider notice in Komsomol circles, in part owing to Shipunov's notoriety, the science editor of *Komsomol'skaia pravda,* Vladimir Alekseevich Chivilikhin, a Siberian, decided to cover the story himself. While Chivilikhin was preparing his story for press on December 28, 1959, the RSFSR Council of Ministers issued order no. 8285-R, setting aside 71,400 hectares for the experimental plantation. It was a huge victory for the students.[50]

Meanwhile, romance intervened to commingle the two very different traditions of student activism exemplified by the Moscow and Leningrad groups. At an all-Union conference of biology students, held at Moscow State University, which Parfënov and Shipunov attended, Shipunov made the acquaintance of Maria Valentinovna Cherkasova, a twenty-one-year-old zoology student at Moscow University.[51] Cherkasova, daughter of an engineer and a music teacher, was a child of the Moscow intelligentsia and was socialized to scientific public opinion. "When I was ten years old," she told Oleg Ianitskii in an interview, "I joined KIuBZ, of which Aleksei Iablokov, Nikolai Vorontsov, and other well-known biologists were members. My first teacher was the unforgettable Pëtr Smolin, who had an amazing knowledge of birds and a great love for them—all my life I've remembered the excursions we made with him into the woods in springtime."[52] It was almost inevitable that she should enter the Biology and Soil Sciences Faculty of Moscow University after graduating from high school. She could not have picked a more intense time of intellectual ferment, and Biofak was at its epicenter. Recalling her conversion to nature protection activism, Cherkasova said: "My enlightenment came during the so-called Khrushchëv 'thaw.' I attended the lectures of David Armand, who had returned from imprisonment and had quickly published his first brilliant book on nature conservation, *For Us and Our Grandchildren*. These lectures were a revelation to me. I'm also greatly indebted to Vadim Tikhomirov, who played a huge role in educating the students of biology at the university."[53]

Shipunov's family, on the other hand, was decidedly nonelite; his father was a forester in Siberia and lived simply. Despite their differences in background, they became romantically involved. In February 1960 Cherkasova led six other biology students from Moscow University to Uimen', the plantation's first "capital," where they conducted a census of maral deer, registering 200. In that group of Moscow University students were some of the core organizers of the *druzhina*, including Cherkasova, which was officially inaugurated in December.

Official approval was not enough to secure the project. It took seven visits to Moscow by Shipunov, plus countless trips to Barnaul and Gorno-Altaisk, the capital of the province (*krai*), to get the boundaries set and the local loggers off the territory. But after the publication of Chivilikhin's passionate saga of the birth of Kedrograd, "Roar, Taiga, Roar!" in the February 14, 1960 issue of *Komsomol'skaia pravda*, public opinion began to respond in an unexpectedly big way. Letters came pouring in from scientists, hunting experts, students, and workers, as well as members of the military.[54] Gifts came too, such as the 100 rubles from an anonymous engineer "to sweeten things up a little for the young *kedrogradtsy*."[55]

On April 7, 1960 at Moscow University's Biology and Soil Sciences Faculty a small conference was convened on Kedrograd with talks by S. A. Khlatin, Maria Shipunova-Cherkasova (who had married Shipunov), and G. V. Kuznetsov, attended by the academician A. S. Iablokov of VASKhNIL (the Lenin All-Union Academy of Agricultural Sciences), a big supporter.[56] Soon thereafter, in the fall there was a "tumultuous" Komsomol meeting on Kedrograd at Moscow University, where future *druzhina* members dominated with speeches supporting the experiment.[57] When the *druzhina* of Moscow University was formally organized later in the year, it officially pledged its methodological assistance and leadership (*shefstvo*) to Kedrograd, a move that, although welcomed, may also have been perceived by the Leningraders as annoyingly patronizing.

Particularly striking was the reaction Kedrograd stirred among students of the USSR's other engineering and technical schools. That first year, students of the Moscow Aviation Institute organized an aid convoy to the Altai, while the next year the baton was passed to the Moscow Energetics Institute. Technical schools in Leningrad, Biisk, Voronezh, Krasnoiarsk, and other cities rallied as well. Thousands participated in the "Movement to Help Kedrograd." Summer student brigades formed autonomously, arranging travel to the Altai through the Central Committee of the Komsomol. It was nicknamed the "Virgin Lands of the Taiga" in the spirit of Khrushchëv's much-touted campaign to the west. Competitions were held to select the students best at wielding an ax and a rifle. Hundreds came during the summer. Nikolai Pavlovich Telegin, a talented forester, transferred from Perm' and was named official project director for three years by the government. Iurii Nikolaevich Kurazhkovskii, an Astrakhan' professor who had pioneered a movement there for "rational resource use" (*ratsional'noe prirodopol'zovanie*), worked in Kedrograd as deputy director for science in 1960–1961, and in 1962–1964 as an instructor in the Gorno-Altai Pedagogical Institute, at the invitation and pleading of the *kedrogradtsy*. Kedrograd was a social phenomenon capable of motivating established professionals as well as students to uproot their lives and to live in the most rudimentary conditions.[58]

Paradoxes abounded in the opening year of Kedrograd's operation. After the *komandiri* of the Brigades to Aid Kedrograd of the Moscow Aviation Institute and of the Forestry Technical Institute traveled to Uimen', they, like Shipunov, were thrown out of the Komsomol. Yet funding for their travel came from the Komsomol's Central Committee.[59] On the surface, the forestry bureaucrats supported the plan, but they often failed to adopt practical measures to effectuate it, revealing their true attitudes. "After the appearance of Chivilikhin's 'Roar, Taiga, Roar!' the really serious complications in our lives began in earnest," recalled Parfënov. "The bureaucrats understood that they would either have to accept the blame for mismanaging forest resources or destroy Kedrograd." Consequently, the years from October 1961 to 1975 constituted a protracted "bloody war."[60] There were a number of causes of the "progressive paralysis" (Chivilikhin's expression) of Kedrograd, "but the main one was that the essentially progressive idea of multiple use of the cedar forests could not fit into the organizational-planning structure of the economics of that time, when all around sectorial-bureaucratic monopolism held sway in the area of resource exploitation."[61]

However, the young forestry graduates' education about the system had only begun. Not even a year had gone by when the first "knockout punch" was delivered by the bureaucrats in December 1960. One frosty day a commission drawn from *oblast'* organizations showed up at the "Kedrograd" encampment at Uimen' and announced a lawsuit against Kedrograd for illegal nut-harvesting. First, the commission charged that Kedrograd was paying workers a higher price per ton than the rate set by the *oblast'* government, a disparity that could draw workers away from neighboring plantations and constituted illegal competition. (Strangely, the commission did not take into account that the only harvesters were members of Kedrograd or those who resided in its territory.) Second, the commission pronounced that "foresters" had no legal right to harvest ancillary forest products such as pelts because harvesting was the monopoly privilege of consumer cooperatives (and later, from 1961, of Glavokhota RSFSR and its local agents). With the backing of provincial authorities, the commission sequestered the harvest of cedar nuts in Kedrograd's storage sheds and the project's income flow dried up.[62]

Chivilikhin reported the incident with quiet fury in "The Taiga is Roaring," published in 1961. He let the actions of the authorities speak for themselves:

> In Uimen' there were some violations of harvesting regulations. Rather than correct these mistakes of the young komsomols in timely fashion the *oblispolkom* kept silent until December. And then suddenly there was a decision: to sequester all the nuts and to impose a fine of 400,000 rubles on the enterprise, in the meantime seizing the 257,000 rubles that the *kedrogradtsy* had in their account. The acting director, Anatolii Malakhovskii, and chief engineer, Vitalii Parfënov, went to Moscow—to the State Arbitration Bureau [*Gosarbitrazh*] in the Ministry of Finance. After a lengthy review, the violation was found to be trivial and an order was given to erase the fine and rescind the sequestration. But how to return the 257,000 rubles now that the financial year had ended? "Clever" folks knew when to impose a fine. The youths did not receive their salaries for two months and organized debt lists in the cafeteria. For them it is a time not willingly recalled.[63]

Indeed, one participant recalled the incident thirty-five years later only with great pain:

> The consequences for the *kedrogradtsy* were tragic. The deep wound bled for many years and it is difficult to overestimate the moral blow inflicted on these young people who had come to tame the "virgin lands of the taiga." The legal action initiated by the financial organs of the *oblast'* led to the immediate imposition of a fine of Kedrograd's entire property and money on hand—404,000 rubles. As it later was revealed, this money went to pay employees of the *oblast',* while the *kedrogradtsy* were left without a cent on the eve of the New Year and hadn't received salaries in months. . . . Hunger began to stalk this inaccessible taiga settlement in the mountains. Lack of experience and deep snow cover thwarted their efforts to catch maral deer for food. With no way out, the youths were driven to catch and eat dogs.[64]

With no expectation of this kind of persecution, the *kedrogradtsy* were at first bewildered and terribly hurt: "There was no answer to their questions: Where were the authorities? Where was the Komsomol? Where was the concern about the human being, propagandized at school? Where was common sense in the capricious actions the authorities permitted themselves? Attempts to demonstrate the stupidity of the bureaucratic claims . . . and to convince the leaders of the *oblast'* to rescind the fine and return the money, to not allow the faith in Kedrograd and the [youths'] patriotism to be lost came to nothing."[65]

As the representative of the Kedrograd Komsomol Committee, Parfënov had to go to Moscow, where a conference attended by the press and the public pressured the State Arbitration Bureau to rescind the fine and to return the nuts and the money. "Justice seemed to triumph, but the financial organs [of the *oblast'*] never did return the money, noting that the financial year had already ended," commented Parfënov. The *oblast'* authorities "exploited their monopoly of bureaucratic power to enrich their coffers." Dozens of disillusioned youths quit the project and abandoned the taiga.[66]

Such an eruption of aggressive local bureaucratic opposition—from forest plantations, other land users, and provincial bosses—proved to be the first phase of the real, "hands-on" education in Soviet political economy for these ardent, sincerely devoted Young Communists:

> This history, however, permitted us—and not only *kedrogradtsy*—to reach a number of fundamental conclusions. The awareness that in the depths of the conservative economic system based on bureaucratic foundations exist economic, social, and moral contradictions demanded a focused analysis of decisions taken in the area of resource use. It was obvious that Kedrograd touched on bigger questions than simply a responsible approach to the resources of the cedarwoods taiga, and could not succeed without help from the center.[67]

The *kedrogradtsy*'s faith in the system, in the "center," in the existence of a "good tsar" was still unbroken. That faith was sustained by a number of decisions taken in Moscow. The RSFSR Council of Ministers now ruled that the all-Union consumer cooperative Tsentrosoiuz had to allow Kedrograd to harvest and sell ancillary products of the taiga as an exception to its monopoly. At the same time, Uimen' was legally given over to Kedrograd from the neighboring Karakokshinskii plantation. Kedrograd's territory was increased to 298,000 hectares and later that year to 400,000. Protests from loggers were dismissed by the Russian Republic government.[68]

Infuriated, the local bureaucratic interests wanted more than ever to eliminate Kedrograd. Desire for revenge intensified after Chivilikhin's reportage, which inflamed public opinion and created a wave of sympathy for the students. A flood of letters inundated the Gornyi-Altai *obkom* provincial committee) of the Party protesting the unfair treatment of the youths. Even more impressive, monetary and material contributions were sent to the *kedrogradtsy* from people across the USSR. One woman from Voronezh wrote that her family circle decided to send 2,000 rubles in savings to the youths "because we love our country's nature." From contributions a library of 3,000 books was assembled.[69] At Moscow State University, other universities, and a host of engineering schools, committees sprang up in defense of Kedrograd. Perhaps the most flamboyant gesture of support came from cosmonaut Iurii Gagarin, who selected "*kedr*" as his "handle" in his first flight (a gesture doubtless lost on foreign commentators and intelligence gatherers). "Gornyi Altai unexpectedly became the center of attention of the whole country, and the Altai cedar the symbol of an honorable relationship with nature."[70]

But the center's support for Kedrograd lacked conviction. The year 1962 should have been very profitable. When the *kedrogradtsy* went to total up their first profits, however, they found that the bank account was empty: their earnings had been expropriated by the deputy director of Glavleskhoz RSFSR (the RSFSR Main Forestry Administration), Nikiforov, to support a group of specialists from Moscow working on a "minor problem" involving the use of cedar forests. As a result, the enterprise sustained a small loss for the fiscal year.[71] At the start of the season, the Altai Regional Forestry Administration, in whose jurisdiction Kedrograd had been placed by the RSFSR Main Forestry Administration, cut off all operating funds. By way of "compensation," the head of the regional forestry administration, Vashkevich, dispatched about two hundred people to Kedrograd "from steppe forest plantations who had not laid eyes on a cedar since they were born. The nut harvest was subverted," or at least that was Vashkevich's hope.[72]

In Barnaul, Vashkevich tried to abort an interdisciplinary conference of scientists, foresters, and planners to put the basic elements of the technical plan for Kedrograd into final shape. "Vashkevich flatly announced: 'There will be no such "plan." Everyone go home!' The conference, of course, took place anyway. However, Vashkevich did not give up."[73]

Aside from this act of resistance, there were defectors even among Vashkevich's subordinates. One specialist, N. Zhideev, volunteered to become the director of the new enterprise "so that I will have done something good for the forest before I retire," he told Chivilikhin. Even in the face of Vashkevich's attempt to undermine the nut harvest, Kedrograd registered a profit of 78,000 rubles in 1963 as against losses of 150,000 and 400,000 rubles for the neighboring timbering concerns.[74]

This kind of success was hard to ignore, and chief forest engineer Parfënov was awarded a certificate of merit by the Altai *kraikom* (regional committee) of the Party. (Shipunov, who balked at accepting the post of deputy director of Kedrograd, had already parted company with his brainchild.) Parfënov was even the star speaker at a major conference in Moscow on the cedar forests along with experts such as Prof. Boris P. Kolesnikov, who warned that those forests would soon be wiped out if practices outside of Kedrograd did not change.[75]

Speaking for the collective, Parfënov proclaimed Kedrograd's resolve to resist in the national press. Writing in *Komsomol'skaia pravda,* he described chopping down a living cedar as "the same as doing in a cow for the sake of its bones. For that reason it has lately become the symbol of the struggle for the rational exploitation of the taiga."[76] "One particular complication," he explained, "has been the fact that from the get go we have been directed to fulfill a production program." Even so, he noted, with skillful adjustments the settlers had been able to make the timber cuts, hunt, and harvest nuts, taking 1,000 sable and 10,000 squirrels in just the calendar year 1962.[77] The *kedrogradtsy* still pinned their hopes on the central authorities, who they hoped would rein in the local logging interests and eliminate the production quotas. Ultimately, they saw Kedrograd as a heroic model for forestry, "a laboratory in nature, a base from which innovations can spread all across Siberia." However, they were beginning to suspect that the central authorities might not be as sincere supporters of Kedrograd as they had once assumed:

> If the Main Forestry Administration does not want or cannot immediately organize one or several multiuse plantations in Siberia, it is still within its power to put an end to the attempts of several local authorities . . . to extend logging in the cedar taiga, to force us to increase our timber cuts at the expense of living trees, to cut us down at the knees. . . . And is it not high time for Glavleskhoz not only to take notice but to take steps to defend the cedar taiga from the saw and the ax?[78]

Parfënov's article ended with Komsomol bravado: "Whatever may come, our Kedrograd will live, for we are now firmly on our feet." But not even the most attentive of *kedrogradtsy* suspected how institutionally isolated they really were.

"At the very moment that V. Parfënov addressed the conference," wrote Chivilikhin, "Kedrograd for all practical purposes had already ceased to exist. The head of the RSFSR Main Forest Administration, Comrade [Mikhail Mikhailovich] Bochkarëv, signed a decree directing the transfer of the richest cedar taiga to logging enterprises. The brand new settlement, the technology, the roads, and, most important of all, the marvelous cedar groves were handed over to the Karakokshinskii forestry plantation."[79] These stands were densely stocked with squirrel, sable, and maral deer; their loss devastated Kedrograd's ancillary hunting sector. "Despite protests, and pleas from the *kedrogradtsy,*" wrote an embittered Chivilikhin, "Comrade Bochkarëv was unshakable. The order was signed and discussion was closed." The only compensation offered by Bochkarëv was to permit Kedrograd to relocate to Koldor, a place of inaccessible cliffs and a swampy delta—"no place to even pitch a tent."[80] At that point Parfënov asked to relocate to Iogach, which had run up the 400,000-ruble deficit. "M. Bochkarëv graciously acceded," noted Chivilikhin facetiously.

The bureaucrats tried to force the *kedrogradtsy* to abandon their experiment by tormenting them in every possible way. A week after Bochkarëv agreed to the relocation, his local vicegerent Vashkevich showed up, removed the sympathetic Zhideev as director and demoted Chief Engineer Parfënov to a humiliatingly minor position in Iogach. "Then he set to work on the other specialists," wrote Chivilikhin.

> He called them in one at a time, spreading slander and using threats and flattery, he offered them higher salaries and bigger apartments, but . . . in other *leskhozpromy* [forestry plantations] of the region. Not going along, our fellows stood like a rock. Here in front of me is a declaration signed by thirteen engineers of Kedrograd. They turned down these higher salaries and apartments because they "came to Gornyi Altai in order to create a multiuse enterprise" and ask (ask!) that they be allowed to work together in one place. It is impossible to hold such a document in one's hands without becoming incensed. My goodness, we should be nurturing such people, not breaking their spirit![81]

To add insult to injury, "Vashkevich haughtily served the [Kedrograd forest engineers] an infeasible production plan of cuts, forcing them to shave bare the upper reaches of the Bii River and part of the Teletskoe lakefront."[82] He was able to do this because the Altai *krai* was the only region of Siberia where commercial logging was carried out by forestry organs, which were normally supposed to concern themselves with forest *protection*. In the Altai, where the protective and extractive bureaucracies were merged into one, the logging mentality thoroughly dominated. Politically dependent on the major economic and political bosses of the region, local papers were pressured to label Kedrograd "a kindergarten for adults" (*vzroslyi detskii sad*) and other names.[83]

To justify the expropriations, demotions, and harassment, Bochkarëv charged that Kedrograd was unable to pay its own way, despite its track record of the first three years. He even encouraged a correspondent from the popular newsmagazine *Vokrug sveta* to go to the Altai and to write a story on the experiment, casting doubts on its viability. The reporter asked to be able to spend six days in the field on the lower slopes. He was shown everything and, once back in Moscow, he decided to write what he saw. As a result of his personal revolt of conscience he was fired from *Vokrug sveta* and went unemployed for a number of years.[84]

Despite the privations, the alliance between local and central bureaucrats, the relocations, demotions, and expropriations, the young forest engineers hung on. They were supported by a massive wave of public opinion, led by the journalist Vladimir Chivilikhin. And they increasingly understood their struggle to be one between "good" and "evil."

Whereas initial local opposition to their project had provided an introduction to the political economy of the Soviet system, the collusion of Mikhail Mikhailovich Bochkarëv and the RSFSR Main Forestry Administration, which was supposed to be the forests' defender, raised the *kedrogradtsy's* education to a more advanced level. In a biting piece on the fifth anniversary of Kedrograd, published in *Literaturnaia gazeta* in January 1965, Chivilikhin raised the troubling possibility that the entire system was incapable of organizing the truly rational use of the country's resources:

> How could it happen, for instance, that forests—which are the property of the whole people and the state—have now become parceled out in an almost unmonitored state to republican, inter-*oblast'*, and *oblast'* organizations and into the hands of specialized logging enterprises and co-ops? And why was it several years back that half a dozen forestry *vuzy* and many *tekhnikumy* were closed down? Can it really be that the astronomical figures of annual forest growth, which the logging agencies up to now have officially manipulated so as to justify current rapacious levels of logging, have convinced us all that the Russian forests will never be exhausted?[85]

Reticent to criticize the center, Chivilikhin still had to point the finger at the local extractive interests that were manipulating data. Yet, who gave them the latitude to create these pernicious fiefdoms? Who allowed the institutes and technical schools to close down? How deep did the sickness go? In his 1967 speech to the fifteenth Komsomol Congress, Chivilikhin revealed that the once idealistic youths of Kedrograd had also begun to ponder why their experiment was foundering on the shoals of Soviet realities. Their conclusions, it seemed, now pointed to a pervasive malaise of Soviet official culture:

> The forest engineers of Kedrograd, serious, hardy lads fully devoted to our ideals, write to me: "On the basis of our six-year experience we have come to the definite conclusion that no 'cedar problem' exists in Siberia, but there *is* a problem of institutional narrowness and bureaucratism, a struggle with those who hide from taking responsibility for their actions and with the spinners of red tape. That is, the scientific and economic problem is fused with a social one."[86]

Bureaucratic obstructionism and, at times, outright malice had taken the luster off Khrushchëv's attempt to breathe new life into the Communist ideal. Idealists frustrated or crushed by the system now dared to question the structure of the Soviet social order. For many of the *kedrogradtsy* and their supporters, the "battle for Kedrograd" catalyzed their eventual transformation from Soviet patriots and Communist idealists to Russian nationalists and even embittered chauvinists. The best embodiment of this redirection of loyalties may be found in the subsequent career of Vladimir Chivilikhin, who wedded the protection of the taiga, archetypal "Russian nature," to the preservation of a Russian culture thought to be under mortal threat.

Vladimir Alekseevich Chivilikhin

"Have you turned your attention to the way in which Vladimir Chivilikhin ends his essays on the Siberian woods?" asks Aleksandr Petrovich Kazarkin, a critic and docent at Kemerovo State University. "Double and triple afterwords and epilogues—that is, a chronicle of ever-mounting calamity. Is that not why the two-volume *Pamiat'* exploded in the popular consciousness, because the novel struck a nerve regarding a superproblem—the prehistory of the ecological crisis?"[87]

Like the ethnographer Lev Gumilëv, Chivilikhin believed that the major sources of life and hope and meaning are the people's national memory, especially their shared environmental experience. "The Russian people have never lived without forests and can never do so," insisted Chivilikhin.[88]

Chivilikhin was not always the Russian nationalist–environmental determinist of his later works, particularly *Pamiat'*, which won him a USSR State Prize in 1982. He began as a Soviet patriot, in the very thick of the Komsomol movement—a journalist and then editor of the newspaper *Komsomol'skaia pravda*. Nonetheless, his provincial Siberian background provided the seeds of Russian chauvinism. Born in the coal-rich Kuzbas of southwest Siberia, he studied in Mariinsk and Taiga before completing his education at Moscow State University. One of his earliest literary heroes and models was Leonid Leonov, who began to smuggle in themes of "Russian" nature from the late 1940s.[89] Chivilikhin was already influenced by an incipient body of works in Russian letters voicing the tragic trope of the desecration of the Russian land and of heroic efforts to save that land. But, like Leonov, Chivilikhin had not yet disentangled the two not always compatible ideologies of Soviet patriotism and Russian nationalism. Only as a result of the bruising struggle over Kedrograd did the *Russian* element come to full consciousness.

Later, Chivilikhin would identify Leonov as the fount of his new ideology of literary Russian environmental nationalism: "In the novels of Leonov we may first notice the linkage between national character and the forest. . . . The books of Leonov breathe 'Russia' . . . and in them are the cast and logic of the Russian mind."[90] One only need look at the list of contributors to the various anthologies dedicated to and honoring Leonov's opus to appreciate his position as the godfather of this current.

Like Leonov, Chivilikhin was not actually against exploiting the taiga; the question was *how:* "Logging the taiga is necessary: there are trees rotting in it, and priceless national wealth is going to waste—marvelous construction materials, irreplaceable chemical raw materials and food supplies. But the time has come when we need soberly to weigh the resources of the taiga and to give serious thought to how to operate in that environment so that the taiga will produce the most benefit for the people."[91]

Kazarkin writes that "the works of Chivilikhin from the mid-1950s have sketched a scene of a thoughtless and therefore terrible process of the destruction of forests over a great territory from Arkhangel'sk to Vladivostok." By the late 1960s, Chivilikhin began to see this as nothing less than a struggle for the cultural and physical survival of the Russian people: "This foundation is the reserve of national ecological ideas, the people's perceptions about the land as their fate and about history as a link between the generations."[92] The deciding battle would be fought in Siberia.[93]

Like Gumilëv, Chivilikhin developed a notion of "the ecology of culture." Such an ecology was, in the words of Kazarkin, "that which insures its stability, a reserve of resilience of its way of life, an unsullied consciousness of one's identity, which is oriented toward things vital and permanent. One wants to call his historical conception a 'forest' conception."[94] The forest, for Chivilikhin, was the key to the survival of the Russian people during the years of Mongol-Tatar rule. Only forested Rus' preserved the pure genotype of the Russian people and their cultural heritage. Vladimir Chivilikhin was the "writer-intercessor . . . sent by the Siberian forests to plead the case for living nature." "The natural environment creates what, poetically, we call the soul of the people and in reality determines the salient characteristics of national culture. In preserving our traditional natural environment the people can count on preserving their creative originality. A writer as far back as N[ikolai] Leskov said it—the Russian character is impossible to imagine without [Russia's] expanses of forest."[95] This struggle to preserve the alleged aboriginal arboreal environment of the Russian people also took place in Siberia, according to Chivilikhin; he held that as far back as the first centuries of this millennium proto-Europeans there (Di, or Dinlins) had been in conflict with the Huns. Their descendants today, Chivilikhin claimed, are the Ket.[96] If the Russian people were to survive, they needed to preserve not one but two key elements undergirding Russian culture: the (Siberian) forest and cultural memory. For Chivilikhin they were intertwined, for at the center of the people's memory was the memory of the forest. And when a people forgets its folkways, Chivilikhin believed, echoing Gumilëv, it becomes a "a rapacious mongrel-group" (*khishchnaia khimera*) bringing environmental (and then cultural) collapse upon itself.

By 1967, when he was awarded Komsomol's special medal for his reportage on Kedrograd, Chivilikhin openly paraded his urgent concern for the survival of the Russian people. At the time, it took a bit of daring to cast the ethnic Russians as an oppressed group, particularly in an organization officially dedicated to promoting *Soviet* patriotism, which strove therefore to replace ethnic particularism with a "supraethnic" *Soviet* nationality, even if the cultural forms of that nationality, including language, were derived in good part from the Russian one. "The past of our people, our fathers and mothers," the Russian past—the embodiment of historical memory, that organ of national survival—was under attack from within and without, warned Chivilikhin. With pain and resentment he spoke of "attempts to insult and denigrate, to devalue that which is dearest to us."[97]

Chivilikhin did seek to soften the Russocentric core of his message. "I am introducing this subject," he continued, "because in several works of literature, and, unfortunately, not only in antisocial [*podonochnye*] underground publications, there is a tendency to paint Russians, for example, as a meek people, passively enduring torments, weak, without will and dull-witted, incapable of attaining the heights of culture and at the same time nationally self-centered, and there are attempts as well to belittle other nations inhabiting our Motherland."[98] Above all, Chivilikhin was concerned to refute the "Western," cosmopolitan assessment of Russia as backward and especially as weak:

> At the beginning of this century my people, allegedly willing to put up with any suffering, under the leadership of the Bolshevik party and the great Lenin, together with other peoples . . . made three social revolutions and [then] saved the world from fascism. The Russian people gave the world Pushkin and Lenin, built Rostov and Kizhi, and in our day the Soviet people . . . built Magnitka and Dneproges, Bratsk and Rudnyi, . . . and were first to go to space! Meanwhile, heroes of stories and films pronounce even such words as *ancestor* or *patriot* with a kind of loathing snickering intonation![99]

To these snickers of the cosmopolites, Chivilikhin quoted Voronezh poet Vladimir Gordeichev's response:

> And when over the ashes of patriots
> Foreign wits amuse themselves
> I stand up to meet their barbs
> Baring my boils unflinchingly.[100]

Later, after his epic *Pamiat'* appeared, Chivilikhin provided an emotional credo in response to an interviewer who wanted to know why he had "focused precisely on the history of the *soul* of the people":

> I am a Russian and my heart overflows with love for my homeland, for the path she has trod, and I am grateful to her for the happiness of living on Russian soil. Our people—builders and warriors—has something to be proud of. It is the only people on the face of the earth that has withstood three world-scale invasions. And my duty before the past, before the land that has sustained me and raised me, is to dedicate myself to studying the history of my people.[101]

The highest contribution anyone could make was to preserve and disseminate the nation's history. Sounding like a Stalinist cultural boss of the late 1940s, Chivilikhin said that the historian's task was to remind the people of Russia's greatness, of its innumerable priorities: "Memory is one of the strongest weapons on earth."[102]

Throughout Chivilikhin's writings the tincture of an anti-steppe, anti-steppe peoples, and anti-Asian bias is discernible, as is Chivilikhin's conviction that ethnic differences are deeply engraved: "Yes, humans have only one Earth, but if we try to apply this standard to our theme, then what disorder and confusion we discover in our common human home, the biosphere, what a complex, variegated, and changeable picture of the world emerges, what striking dissimilarities exist among the historical, geographical, social, and other conditions of life for every people!"[103]

Contra Gumilëv, whose accounts softened the destructive impact of the Mongols, Chivilikhin restores the Mongolian invasion to the level of an epic historical trauma.[104] One region, however, escaped the burden of Russia's traumatic history. During an interview the journalist Ol'ga Plakhotnaia once told Chivilikhin: "Vladimir Alekseevich, I know that you have a special feeling for Siberia, your homeland." Chivilikhin's response again reflected his belief that the Siberians were the purest, the most "Russian" of Russians, to the extent that they had evaded the effects and aftereffects of the Mongolian yoke, serfdom, and the taint of Western invaders and immigrants: "Siberians are a punctual, hardworking, and knowledgeable *narod*. . . . Almost every summer I come down with a 'Siberia' attack and travel to my homeland, to Baikal, the Sayans, the Altai." Siberia, a land of "strong characters and uncorrupted language," with its forests, was the new hearth of Russia.[105] As a prototype and standard of Russian nature, Siberia remained at the center of Chivilikhin's concerns even while the "battle for Kedrograd" was still raging.

After Chivilikhin's first articles, a flood of letters came to Kedrograd from all over the USSR complaining of other abuses. Among the topics most frequently brought up by his correspondents was that of the threats to Lake Baikal. In 1962 while still in Kedrograd, Chivilikhin wrote his "Sacred Eye of Siberia" (*Svetloe oko Sibiri*) dedicated to the lake's problems, one of the first wake-up calls on the threat to Baikal from military-related nylon and cellulose mills on the lake's southern shore.[106] Chivilikhin also turned his attention to the problem of land use, seeking to publish a long essay called "Land in Trouble" (*Zemlia v bede*). Here, however, Chivilikhin began to run up against the hand of the censor, who banned half of the manuscript and the title besides. The remainder of the essay was eventually published under the title "The Land—Our Food-Giver" (*Zemlia-kormilitsa*), an alternative suggested by the helpful censor.[107]

With his critique of economic structures such as the Baikal plants and the *sovnarkhozy* it would seem as though Chivilikhin were inching toward a critique of the Soviet system based on an analysis of its political economy. However, a culturally and ethnically based critique proved easier and more attractive to him (and others). In contrast to a seductively slick cultural model based on "Western cunning," Chivilikhin praised the honest young people like the volunteers of Kedrograd, who were continuing the fight for the soul of Russia:

> The young folk don't complain; whining and skepticism is alien to their nature. . . . Our everyday heroes think, they struggle, and they are accumulating experience in the social defense of our natural resources. They don't intend to . . . use cunning or chemical trickery to win their cause. . . . To remain on the moral high ground, to maintain their lifelong youthful ardor for work, to keep their principled political attitudes—that is the task for them and for all of us! And meanwhile I have faith that the economic reforms taken on the initiative of the Party will be extended to other spheres, in particular to that of resource use, or else we shall impoverish our native land and consequently impoverish ourselves, both materially and spiritually. . . . Love of nature, like love for the Motherland, is not only in the sphere of feelings but in the sphere of deeds as well. And here, facing the Komsomol, is an enormous unplowed field, virgin lands in every direction.[108]

Despite his Komsomol background and his opposition to the "fetishizing of nature," Chivilikhin's attitudes toward modern mechanized society remained ambivalent:

> Does the introduction of such good things as electricity and residential neighborhoods obligatorily have to be accompanied by the crushing of the flowers? Must industrial beauty *replace* natural beauty? . . . Why then were the people of Krasnoiarsk able to preserve a large tract of "wild" taiga right in the middle of their city? Why haven't they leveled the taiga, then, in Angarsk and Akademgorodok, but instead integrated their residential areas into it? . . . All of that, however, amounts to a few small islands of good relations with nature in a sea of evil.[109]

In another work Chivilikhin's antiurban feelings were more explicit, as he quoted Le Corbusier's observation that "cities were dangerous and unworthy machines for life in our epoch." Indeed, Chivilikhin himself added, "the specter of urbanization hangs like a black shadow on the horizon."[110]

The question was how to allow those islands of good to triumph over that sea of evil. Was it simply a matter of culture, or was that evil embedded somehow in the structural aspects of the system? "My lifelong and difficult love—the cedar—the symbol of powerful and generous Siberian nature . . . to this day mercilessly is being logged out with impunity all across Siberia despite a special clause banning that in the Law on Nature Protection," he protested.[111] True, writers could make a difference; "by its urgency the problem of nature protection is the theme of the age," he declared in 1978.[112] Yet, as he complained to Leonov, the rapacious bureaucrats and managers did not read, or at least were not affected by what they did read.

To improve the environment, in the last analysis, human societies needed to be harmonized and humanized, argued Chivilikhin. The "fullest development of the human personality," as he understood it, needed to take precedence over industrial production. That meant taking the road back to the *Volk*. "The aggression of 'mass culture,' the total illiteracy of almost a billion people, the standardization of life, violence, the spirit of acquisitiveness [*priobretatel'stvo*], the forgetting of the principles of humanism, cosmopolitan stereotypes in art," as well as the arms race and the greed of the well-off countries were at the root of the global environmental and cultural crisis. Beckoning as a lone, arduous way out was "the tormented processes of national, creative, self-expression . . . the only guarantee of the spiritual development of the world."[113]

Chivilikhin's ideological odyssey was repeated by Fatei Shipunov and resembled the attitudes of Soloukhin, Rasputin, Viktor Astaf'ev, Proskurin, Shukshin, and a host of others. These represented a new set of social actors—journalists, writers, foresters, engineers, and other ordinary people—distinct from the "lost tribe" of ecologists, botanists, zoologists, and geographers who were still fighting on behalf of "pristine" nature and the *zapovednik* ideal. This new group was composed of upwardly mobile beneficiaries of the system who had conformed but who felt disillusioned and betrayed. What pushed these otherwise average Soviet subjects into environmental activism was the sense that their environmental "homeland" was being destroyed and that the system on its own would not stop it. Whereas the Moscow- and Leningrad-based naturalists looked to their Western colleagues for information, solidarity, and new approaches, seeing themselves as part of an international community tackling global problems, the new stratum of activists concerned over the despoliation of Russia regarded that cosmopolitan, "Western" orientation as one of the main sources of the problem. For ethnic Russians, ironically, the nationalist-environmental movement was more democratic and inclusive; ethnicity, not erudition, was the only criterion for membership.[114]

United in their outrage over the bureaucrats' wanton and heedless attitudes toward such rare and disappearing habitats as the Altai "cedar" forests and Lake Baikal and over their treatment of the students, the cosmopolites and the nationalists joined together to give these struggles unusually high visibility in the early 1960s. The MGU *druzhina* and the *kedrogradtsy* worked together, despite the strains generated by their vastly different backgrounds. These "camps" continued to cooperate on such other major issues as the river-diversion project of the late 1970s and early 1980s, but this cooperation tended to occlude an important underlying reality: the existence of not one, but several environmental movements. The divorce of *kedrogradets* Fatei Shipunov and *druzhinnik* Maria Cherkasova is a metaphor for the eventual fate of the temporarily unified strands of the Russian environmental movement.

CHAPTER FIFTEEN

Three Men in a Boat

VOOP in the Early 1960s

The Khrushchëv years were quite a passage for the Soviet conservation movement. While the venerable All-Russian Society for the Protection of Nature was colonized by Communist bureaucrats, the movement veterans migrated to the shelter of the less exposed Moscow Society of Naturalists and from that redoubt managed to nurture the emergent student movement, the *druzhiny*. And while the idealism of the Khrushchëv years for a time clouded perception of the intractable, oppressive features of the system, voluntary activism in the area of conservation proved to be a university for learning about the ways of power in the USSR.

In a humorous episode, the "Bochkarëv affair," the three nature protection cohorts converged around a common enemy, exposing Mikhail Mikhailovich Bochkarëv, the RSFSR minister of forestry and the president of VOOP, as a poacher. For Vladimir Georgievich Geptner and the old guard, the affair brought revenge for the Communists' takeover of VOOP and its conversion into a despicably corrupt business. For the scientists' student protégés in the Moscow Biofak *druzhina*, it was an opportunity to show public bravado and to hold a heady political mass meeting. Finally, for Vladimir Chivilikhin and the *kedrogradtsy*, it was revenge for Bochkarëv's central role in undermining the Altai forestry experiment. Although veterans of Kedrograd, the student movement, and the old professoriate each claim the lion's share of the credit for Bochkarëv's ultimate downfall, Bochkarëv must be given his due for his own weighty contribution to his political endgame.

From Khrushchëv to Brezhnev: Learning to Read the System

The "Bochkarëv affair" of 1964–1965 was a culmination of the natural process of decay set in motion by the colonization of VOOP a decade earlier.

The first scandal, involving complaints to Politburo members Molotov and Bulganin by one Vsevolod Georgievich Lakoshchënkov, an old VOOP activist, about corruption and commercial abuses by the Society's leadership, ended with his ouster from the Society in 1957. In July 1962 a second scandal broke, also involving commercial doings. This time, thirty-two VOOP retail outlets in its chain "Priroda" (Nature) were accused of "gross violations of the rules of trade" by the Russian Republic's State Trade Inspectorate. The stores, which were officially chartered by the RSFSR Council of Ministers on April 10, 1960 to provide gardening and animal-care supplies to VOOP's members, were found to exhibit "a brazenly commercial character." That is to say, they were in business to make healthy profits, which meant setting arbitrary purchase and sales prices for flowers, failing to display prices, and failing to provide receipts to customers. One store manager's wife in Northern Ossetia was doing a thriving business in black market pet fish. In neighboring Krasnodar *oblast'* "Priroda" staff were buying up flowers and plants and reselling them for over 30 percent more to Moscow municipal flower outlets. Nor did the Moscow branches of "Priroda" take a back seat to the provinces.[1]

Although such abuses were commonplace and, indeed, inevitable, even essential for the functioning of the system, other groups of bureaucrats—in the state inspection system—also needed to justify their employment by periodically uncovering and disciplining these abuses. These imperatives drove the endless petty dramas of Soviet justice such as the inspection of VOOP's "commercial abuses." In this case, a conference of high *apparatchiki* in the RSFSR ministries of justice, trade, finance, and state control with a VOOP representative present recommended a disciplinary decree from the RSFSR Council of Ministers to demonstrate the state's concern and resolve, especially as the abuses "had already become widely known in the Society's branches."[2]

For depth of hypocrisy, not to mention appeal as journalistic copy, however, these prior scandals paled before the "Bochkarëv affair." Moreover, the affair floodlit the gap between "regime conservationism" and the environmental creed of the old guard.

Professor Vladimir Georgievich Geptner (see figure 24) of Moscow State University, a field zoologist trained in the grand tradition and still technically a "citizen's inspector for nature protection" of the All-Russian Society for the Protection of Nature, had been taking summer vacation trips to the Oka River near Izhevskaia pristan' in Riazan' *oblast'* for fifteen years. A stern man of German heritage, with a dry sense of humor, Geptner was a veteran of the *zapovednik* wars of the preceding four decades and counted among the closest scientific advisers of VOOP during its pre-Communist period (pre-1953). The zoologist had grown even bolder with age.

On August 22, 1964 at four in the afternoon it was bright and sunny on

Figure 24. Vladimir Georgievich Geptner (1901–1975).

the Oka. Geptner had taken his wife and son, Mikhail, a budding marine biologist, along on the annual trip, and they were cruising upriver in their motor launch. It was to be a relaxed, intimate field trip for parents and son. The Geptners were on the lookout for osprey and other wildlife, which were common then on the Oka. They did not anticipate finding bigger game.

Unexpectedly, they saw two fishing boats up ahead, illegally suspending a homemade drift net between them. As he neared the first boat, the professor saw, with a shock of recognition, that one of the three men in the boat was none other than Mikhail Mikhailovich Bochkarëv, head of the Russian Republic's Main Forestry Administration, of Kedrograd notoriety, and president of the All-Russian Society for the Protection of Nature. If anyone emblemized the degradation of the once valorous and honorable All-Russian Society for the Protection of Nature, it was this government minister and Party *apparatchik* now in control of the hijacked Society.

Lifelong habits of steely nerves and iron discipline while tracking wildlife now stood Geptner in good stead. With his son steering the boat up to the poaching president's vessel, Geptner demanded to know why Bochkarëv was fishing with an illegal net. Bochkarëv, according to Geptner, responded: "I have permission to do so." In the text of the complaint he filed with the Fishing Inspectorate, Geptner added that Bochkarëv claimed he was given oral permission from a fishing inspector "whose name [Bochkarëv] did not know." Sure of his own ground, Geptner asked Bochkarëv to pull in his net, which he did—a virtual admission of culpability. Geptner also managed to capture the whole scene on film. In a few days the photographs were developed; the aperture and shutter speed were perfectly set, and the excited professor's hands were steady. It was a masterpiece of cinema vérité. Back on shore, Geptner drew up a legal complaint, witnessed by his wife and son, and sent it to the Riazan' *oblast'* Fishing Inspectorate.[3]

Geptner's version of the story was not the only one presented. At a meeting of the Party fraction of the Presidium of the Executive Council of VOOP that met to discuss the matter on October 13, 1964, Bochkarëv provided his explanation. He had been in Riazan' *oblast'* on a business trip, he began, and had gone to the river to go bathing. At that time, a group of fishermen with a net approached the shore. Bochkarëv claimed that he had asked them if they had permission to fish and that they answered in the affirmative. "Not a fisher myself," he continued, "out of curiosity I decided to watch and see how they used the net and what they would turn up in it." No sooner had the group set out to fish than a motorboat approached with Professor Geptner sitting in it, "pointing a camera at me," alleged the cornered president. Disputing Geptner's account, Bochkarëv argued that he had answered Geptner's question with a more general "There is permission to fish." After Geptner insisted that fishing with a dragnet was still poaching, Bochkarëv decided to check the locals' permits. When he discovered, to his surprise, that they had none, he himself demanded that they end their fishing and ordered that they immediately inform the Fishing Inspection about it and bring the ringleader to justice. Bochkarëv's self-portrait was that of good-natured innocence abused. Once alerted to the illegality, he had demonstrated his true civic conscience and conservation concerns. Bochkarëv did admit to one oversight: not having checked documents at the outset.[4]

The Party fraction meeting circled the wagons to protect an embattled bureaucratic colleague. L. V. Ross averred that he had known Bochkarëv for fifteen years and that it was "stupid" (*nelepo*) to accuse him of poaching. Indeed, Geptner had acted badly in "raising such a ruckus," which ultimately "did no benefit to the Society." Another member, G. I. Kulinskaia, accused Geptner of intentionally trying to nail Bochkarëv and of irresponsibility in writing to the press. This accusation jolted Bochkarëv's own further recollection of the incident; the VOOP president now remembered that Geptner threatened him on the river: "I'll bake you to a crisp! Just you wait!"

Not all the conservation bureaucrats believed that stonewalling was the best way to deal with an angry and wily *zubr*.[5] Already the rumors were flying all over Moscow, and anonymous letters were arriving at the Society's headquarters. As D. P. Proferansov added, someone might even raise the question at the forthcoming VOOP General Congress. He and V. E. Golovanov recommended doling out evenhanded criticism—of Bochkarëv for his carelessness and of Geptner for failing to turn over his evidence to the Presidium and going through channels. A slap on each wrist could allow the Society to put the matter behind it.

Biologist N. A. Gladkov was one of Geptner's few defenders, expressing his shock and regret at Bochkarëv's accusation that Geptner's actions were the result of a "personal grudge." Further, Gladkov rejected the notion that Geptner purposely delayed examination of his charges, noting that he was legitimately on vacation. More ominously for Bochkarëv, A. P. Kasparson, a politically congenial figure, began to worry aloud about the effect on the Society of exculpating the president. In particular, he was concerned that further facts might come to light in the press overturning an endorsement of Bochkarëv's claim of innocence.[6]

Kasparson's fears were countered by yet another loyalist, I. A. Khomiakov, who reassured his jittery colleague that, after all, they had official letters from the Fishing Inspectorate reaffirming Bochkarëv's innocence. Indeed, one letter went much further than Bochkarëv's own defense in embellishing his conduct:

> Having seen the illegal poaching by [Andrei Fëdorovich] Frolkov, [Bochkarëv] forbade it, an order which Frolkov then obeyed. As he was leaving the district, Comrade M. M. Bochkarëv issued a directive to the director of the forestry collective enterprise, Comrade I. I. Krylov, to inform the Fishing Inspectorate so it could bring those guilty of poaching to justice. [This was done], which initiated [our] investigation. . . . Comrade M. M. Bochkarëv had no involvement in the poaching incident of August 22, 1964. On the contrary, he contributed to putting it to an end.[7]

Unfortunately, Khomiakov's logic had a fatal flaw. The meticulous Geptner had kept copies both of the complaint he filed on August 22 and the response of the Fishing Inspectorate to him two days later, denying that anyone from their office had given Bochkarëv oral permission and acknowledging that Geptner's letter had served as the signal to initiate an investigation into this affair. But Geptner was saving his most powerful ammunition for the critical moment.

It was clear that the Party fraction had to come to some decision. Any course of action would be costly. The Party fraction believed that absolving Bochkarëv represented the lowest cost and called his presence amid unlawful fishing "accidental." To mollify the disgruntled elements who still took the Society's conservation mission seriously, VOOP's Party leaders unanimously adopted a resolution that chided Bochkarëv for being "excessively trusting," which by its wording almost effaced the censure to create a sympathetic portrait of a kindly, but uninformed, regular guy. Bochkarëv was urged to be more aware in the future. The Party regulars had put this latest embarrassment behind them. Or so they thought.

Even before taking this symbolic action, the leadership of VOOP had moved swiftly to limit the damage caused by Bochkarëv's slip up. As early as September 28 a letter went out to *Literaturnaia gazeta* and *Izvestiia* (to which Geptner had already written), denying charges of premeditated poaching by Bochkarëv. The letter was accompanied by the September 1 and September 26 reports of the State Fishing Inspectorate.[8] Those reports insisted that a local man, Andrei Fëdorovich Frolkov, was the chief instigator and that Frolkov himself, in his own confession of guilt, confirmed Bochkarëv's innocent participation. The hapless Frolkov was subjected to civil punishment, and his net was confiscated.[9] The system had delivered for its own, or so it seemed.

Events, however, overtook these bureaucratic efforts. One of the most widely read publications in the USSR during the 1960s and 1970s was the humor biweekly, *Krokodil.* Bitingly funny cartoons and rib-tickling satires made it one of the few outlets for exposure of the foibles and inefficiencies of the system. With the January 10, 1965 issue, Bochkarëv's luck ran out. The full-page exposé of Bochkarëv—"Get a Load of Those Goldfish!" (*Vot kakie karasi!*)—could not have been more embarrassing (see figure 25). Immediately attracting the reader's eye was the juxtaposition of two photographs of the VOOP president. At the upper left hand of the page, just under the magazine's logo of a devilish, smirking crocodile running at full tilt and aiming a pitchfork—this all perched over the feature's rubric "A Pitchfork in the Side"—was a photograph of a sententious, self-satisfied Bochkarëv in suit and tie, speechifying behind a lectern with a microphone. Immediately beneath the title of the article, which divided the page horizontally, was a second photograph of the proverbial three men in a boat. The man on the left, a corpulent middle-aged man in a white undershirt, was wielding an oar. On the right, judging by the one leg and arm visible, was a second, thinner oarsman. And in the center, also in a white tank top and clutching the handle of the homemade dragnet, was M. M. Bochkarëv. The

бы неводом? Это — браконьерство, понимаешь? Хищническое истребление рыбы.

— Избави господи! Зачем истребление? Разве мы злодеи какие? Что вы!

— А отчего, по-твоему, происходит уменьшение запасов рыбы в реках?

Денис усмехается и недоверчиво щурит глаза.

— Ну! Уж сколько лет в волейбол играем, и хранил господь, а тут истребление... уменьшение запасов... Ежели б я бомбу в реку уронил, ну, тогда, пожалуй, истребил бы, а то — тьфу! — невод!

— С прошлого года рыбы в Оке стало меньше,— говорит

ты, Денис, отвечай... Председатель! Надо председательствовать умеючи, не зря... Хоть и высеки, но чтоб за дело, по совести...

* * *

Вот так, почти по Чехову, мог бы пойти разговор между председателем президиума Центрального совета Всероссийского общества охраны природы М. М. Бочкаревым и браконьером, изображенным на втором снимке. Председатель Бочкарев — человек и впрямь строгий, и законы знающий, и решительно выступающий против браконьерства со своей высокой трибуны (еще раз см. первый снимок).

ВОТ КАКИЕ КАРАСИ!

РАЗГОВОР МОГ БЫ ПОЙТИ ПРЯМО ПО ЧЕХОВУ

Перед председателем президиума Центрального совета Всероссийского общества охраны природы М. М. Бочкаревым (первый снимок) стоит немаленький, плотный человек. Он хмур.

— Денис Григорьев! — начинает председатель.— Подойди поближе и отвечай на мои вопросы. Двадцать второго числа прохожий застал тебя на берегу Оки близ Ижевской пристани за вытаскиванием из воды невода. С каковым неводом тебя и сфотографировали (показывает второй снимок). Так ли это было?

— Чаво?

— Так ли все это было?

— Знамо, было.

— Хорошо. Ну а для чего ты тянул невод?

— Коли б не нужно было, не тянул бы,— вздыхает Денис, косясь на потолок.— Мы из их волейбольные сетки делаем.

— Кто это мы?

— Мы, народ... Московские мужики то есть.

— Послушай, братец, говори толком. Нечего тут про волейбол врать!

— Отродясь не врал, а тут вру...— бормочет обвиняемый, мигая глазами.— Да нешто, товарищ председатель, можно без сетки? Ежели ты команда на команду играешь, разве без сетки обойдешься? Вру... Черт ли в нем, в мяче-то, ежели он просто так летать будет! Вон в пинг-понг и то через сетку играют.

— Для чего ты мне про пинг-понг рассказываешь?

— Чаво? Да ведь вы сами спрашиваете. У нас и мастера [illegible]

председатель.— [illegible] почему...

— На [illegible] и обра[illegible] чтобы понимать, [illegible] наши... Люди знали, кого в председатели выдвигать. Вы вот и рассудили, что и как, а прохожий тот — мужик без всякого понятия, хватает фотоаппарат и снимает... Ты рассуди, а потом и фотографируй!..

— Послушай... Закон об охране природы гласит, что за хищническую ловлю рыбы запрещенными орудиями лова... Понимаешь, запрещенными! — виновный несет строгое наказание.

— Конечно, вы лучше знаете... А мы нешто понимаем?

— Все ты понимаешь! Это ты просто прикидываешься!

— Зачем прикидываться? Спросите у людей, коли не верите. Без сетки только в шашки играют. На что хуже — баскетбол, а и там сетки требуются.

— Ты мне еще про пинг-понг расскажи! — улыбается председатель.

— В пинг-понг мы не играем. В подкидного — случается...

А браконьер со второго снимка, уличенный в ловле рыбы плавной сетью, действительно [illegible] разыгрывал потом из себя чеховского злоумышленника Дениса Григорьева.

Но только никакого разговора между председателем Бочкаревым М. М. и браконьером не было. Дело в том, что фамилия браконьера — тоже Бочкарев. И он тоже М. М. А попросту сказать, сам председатель, только не на служебной трибуне, а в часы досуга.

И Рязанская областная инспекция рыбоохраны вняла речам не грозного Бочкарева — председателя, а лепету Бочкарева — Дениса Григорьева. Рыбоохрана решила: да, очень даже может быть, вероятно, председатель президиума не знал, что ловить рыбу сетью запрещено. Человек отдыхал на досуге от своих председательских забот и вполне мог что-нибудь запамятовать.

А потому рыбоохрана составила документ, в котором сказано, что «к браконьерству тов.

Figure 25. "Vot kakie karasi!" ("Get a load of those goldfish!").

publication of Geptner's unflattering photo had stripped away the dignities of office, revealing a pathetic and banal lump of Soviet humanity.

Although this photo unquestionably did the most damage, the text was no less scathing. Through the vehicle of a fictitious reconstruction of Bochkarëv's subsequent "confrontation" with the local "instigator"—all done in the Chekhovian mode—Bochkarëv's claims of innocence were turned into a farcical mush. The piece also implicitly condemned the casting of the local patsy, Frolkov, as the sole villain.

Krokodil's pitchfork pierced the thick hide of the VOOP bureaucracy, and it proved almost impossible to dislodge.[10] Letters expressing worry (from VOOP officials) and betrayal (from members) flooded the Society's mailroom. One letter, signed by a number of members of the Society's Executive Council, including the satirist Natalia Il'ina and longtime member (since 1936) S. Nazarevskaia, noted that "we, just as many other members of our Society, expected that the Presidium of the Executive Council would immediately react to the publication of the satire. We expected, first of all, that the public would be informed about those conclusions or decisions reached by the Presidium as a whole and by Bochkarëv in particular."[11] The writers further noted that a full month had gone by since the feuilleton and a half-year since the incident itself, "while public opinion—we mean here the active membership of the Executive Council—has been given no information." That in itself was described as "abnormal."

Finally, the signatories of the letter did not omit the largest political issue of all: that a poacher remained at the helm of the nation's official nature protection society. "Under these circumstances the only correct course of action acceptable to broad public opinion is to remove Comrade Bochkarëv as president . . . and to publish that decision in the press," they insisted. Not only had Bochkarëv's own lapses become the subject of growing interest in the press and society, but also, as the letter's authors described it, the entire "thoroughly bureaucratized [*kantseliarsko-biurokraticheskii*] style of operations, the gulf between the Society and scientific public opinion, the gulf between it and the broad masses of members, and the feebleness in solving the . . . pressing issues of conservation." To fail to address these problems "would be a violation of the elementary requirements of Soviet democracy," the letter warned. The three-page letter concluded with a call to elect a person of unimpeachable reputation as the new president; it urged full publication of materials about the "Bochkarëv affair," and called for "reinvigoration and *perestroika* of the entire activity" of VOOP aimed at converting the Society "into an authentic defender of natural resources in the interests of both the present and future generations."

Other letters mentioned Bochkarëv's poor conservation track record in forestry, recently spotlighted in the press. Perhaps exploiting political opportunities in light of Khrushchëv's recent fall, the reformist wing of the

Soviet press engaged in the closest analog there to a feeding frenzy. In the space of two weeks in late January and early February, *Literaturnaia gazeta* published two pieces—one by Vladimir Chivilikhin (a piece by him also appeared simultaneously in *Komsomol'skaia pravda,* a mass circulation daily) and one by Oleg Volkov—exposing Bochkarëv's role in dooming the idealistic attempt in 1957–1960 to manage sustainably the cedar forests of Eastern Siberia ("Kedrograd") and in degrading the woodlands around Lake Baikal in the same area. Konstantin Blagosklonov, a longtime member of VOOP and member of its Council, added from personal knowledge that Bochkarëv was the only member of the Council to "categorically reject" the proposal that the RSFSR, on the model of other republics, establish a ministry-level State Committee on Nature Protection.[12] Blagosklonov also independently noted that the "operational style of the Society had changed," now being marked by "a tendency to be cut off from scientific public opinion" and characterized by a "bureaucratic" flavor. "Scientists well-known for their scientific activism in conservation continue this work in complete isolation from the Society," confirmed Blagosklonov, "and it is precisely these folks that created the Society to begin with. The initiative for the break began with the Society," he observed, "and not with these scientists." The only comfort to be found in Blagosklonov's letter was that he did not favor legal proceedings against Bochkarëv for his poaching, since, after all, "that would doubtless . . . inflict [even more] harm . . . to the Society."

On February 24, 1965 the Presidium of VOOP met to deal with the unraveling crisis once again. First vice president N. G. Ovsiannikov led the discussion of how to respond to the *Krokodil* piece and its spreading, swirling aftershock. Once again, the official resolution generated by the meeting represented a decision to uphold Bochkarëv's version of the events. It was a last shot, averring that there were no profit motives involved and that the episode was a case of bad judgment, for which Bochkarëv had been suitably reprimanded.[13]

The noose around Bochkarëv's neck began to tighten, however, when his Communist cronies in the Society's leadership began to feel the pressure themselves. It was time to abandon the stonewall defense and move to "human sacrifice." Two influential members of the leadership, V. Zharikov, president of the Oversight Commission of the Central Executive Council, and his deputy, A. Kasparson, both Presidium members, demanded convocation of "an extraordinary session of the Central Executive Council of the Society" to resolve the issue—with Bochkarëv's resignation as the expected outcome.

Geptner, in the meantime, had been waging war on all fronts. A new enemy had emerged in the guise of the shamelessly complicit Fishing Inspectorate. With lawyerly acuity, Geptner responded to the cover-up with his own carefully crafted letter to the Fishing Inspector:

> In particular, I believe that you have not fully assessed the strange contention of A. Frolkov that Bochkarëv ordered him to stop fishing with the dragnet. It was Bochkarëv himself who was personally holding the net. It was certainly an ambivalent position for the person giving the order to stop, and I assume that you will certainly not accept those kinds of "conclusions" that exculpate Bochkarëv.
>
> I hope that you . . . uncover the identities of the other violators besides Frolkov. There were six of them, and I would be very appreciative if you would inform me about who they are and what steps you are taking to bring them to justice.[14]

Unable to bring the Baikal pulp and cellulose projects to an end or to mitigate any of the other growing environmental crises in the Soviet Union, conservation activists made the most of their moral victory. Outside VOOP, Geptner could count on the support of the student movement represented by the Moscow State University *druzhina po okhrane prirody.* The students convened a two-day conference on March 15 and 16, 1965, to discuss the piece in *Krokodil.* The highlight was Professor Geptner himself showing no fewer than ten photographs of Bochkarëv wielding the net. The conference went on record as rejecting Bochkarëv's explanation as an insult to intelligence. Nor did the dishonorable complicity of the Riazan' Fishing Inspectorate go without comment. Finally, the conference expressed its astonishment that the Presidium of VOOP's Central Executive Council could give any credence to the report of the Fishing Inspectorate. Rejecting the description of Bochkarëv's actions as "careless" and "overly trusting," the students boldly characterized them as "gross abuses of his social role," as "amoral," and as indicative of a "contemptuous [*naplevatel'skoe*] attitude toward those who gave him a position of trust."

The great Russian painter Il'ia Repin had once brilliantly depicted the glee with which defiant Cossacks composed a presumably insulting response to an ultimatum of the Ottoman sultan. It is easy to imagine that same spirit of mirthful defiance as the students now set about crafting their official resolutions in the auditorium of the Biology and Soil Sciences Faculty. In all they proposed five:

1. To affirm that the account in *Krokodil* conformed to the facts of the case,
2. To thank the editorial board of *Krokodil* for subjecting Bochkarëv to the court of public opinion,
3. To censure the conduct of the poacher, M. M. Bochkarëv, and to demand his immediate removal as president of VOOP and his expulsion from the Society, . . .
4. To ask the authorities to identify all of the participants . . . , including those who perjured themselves, and bring them all to strict justice, and
5. To express our opposition to the decisions of the Presidium of VOOP.[15]

Copies of the resolution were circulated to *Krokodil,* the Presidium of the Central Executive Committee of VOOP, the Science Section of the Committee for Party and State Auditing of the Central Committee of the Communist Party and the USSR Council of Ministers, the Riazan' Fishing Inspectorate, the Commission on Party and State Auditing under the *oblast'* Committees of the Riazan' Communist Party and regional government, the Presidium of the Moscow Society of Naturalists, the Moscow branch of the Geographical Society of the USSR, *Literaturnaia gazeta,* and Professor Geptner. The seditious document was signed by the faculty sponsor (*kurator*) of the *druzhina,* then candidate of biological sciences Vadim Nikolaevich Tikhomirov, and by the leader (*komandir*) of the student group, Sergei Nikolaevich Ivanov.

On July 20, 1965, again during the height of the summer field research season (to exclude largely dissident working naturalists), VOOP's Central Executive Committee held a plenary meeting at the clubhouse of the USSR State Committee on Problems of Labor and Wages. Chairing the meeting was acting president N. G. Ovsiannikov; Bochkarëv had recently "resigned" on April 13. The official agenda listed only three topics: preparing for the Fourth Congress of VOOP, awarding honorary membership in the Society, and "an organizational question."[16] The "organizational question" was none other than the Bochkarëv affair. As in 1955, a representative of the Russian Republic's government came to oversee the rectification of the Society's internal affairs. It was the RSFSR representative, P. V. Minin, who announced Bochkarëv's resignation as president to the eighty-six delegates in attendance. The official explanation, promulgated by Minin, was that Bochkarëv was already too overburdened by the press of his primary responsibilities as the manager of the republic's forests. The delegates unanimously ratified Bochkarëv's departure and just as unanimously endorsed the election of Ovsiannikov as the new president.

* * *

A few words are in order about how the civil degradation of Bochkarëv could have occurred. First, this episode occurred during a time of political transition. Khrushchëv had been removed in mid-October 1964 and all expectations (wrongly, it turned out) were that Brezhnev and Kosygin would continue, if not expand, Khrushchëv's liberalization policies while eliminating the capricious aspects of his rule. Indeed, in the late autumn of 1964 the new leaders had initiated the removal of Lysenko and turned to the community of elite biologists to organize the rehabilitation of classical genetics. It was that impression, evidently, that emboldened reformers in the press (*Krokodil, Komsomol'skaia pravda, Literaturnaia gazeta*) to publish the damaging articles about the high official Bochkarëv and permitted the extraordinary spectacle of his public ridicule by the student assembly at Moscow State University. Although these propitious conditions were soon brought to an end by the new rulers, the memory of the episode persisted.

In the short term, very little of substance changed in the way that the All-Russian Society for the Protection of Nature did business or the way that natural resources and environmental amenities were managed in the Soviet Union. But in the long term, the scandal damaged the system's legitimacy.

Three important consequences flowed from the cumulative experience of the conservation movement in the 1950s and 1960s. The first was that this band of elite biologists and their followers in educated society and the student population came to an important realization: that they were almost unique in representing a reasonably autonomous, cohesive, self-actualized movement of a portion of the citizenry in opposition to important economic policies pursued by the regime such as the heedless prosecution of industrial and agricultural development in environmentally fragile areas.

The second important outcome was the conservation activists' improved grasp of the real workings of the system. Although these elite biologists, like other vast sections of the Soviet population, were moved by the spirit of idealism and hope that Nikita Khrushchëv stoked in the mid and late 1950s, their experiences of struggle for conservation values began to erode, and finally canceled out, these hopes. From the state-dictated removal of elected officers of VOOP in 1955 to Khrushchëv's anti-intellectual and ill-considered "second liquidation" of *zapovedniki* in 1961 (justified by an unanticipated attack on "useless field biology"), from the scandals involving the retail stores ("Priroda") of VOOP to the front-page civil disgrace of M. M. Bochkarëv, activists began to derive an understanding of how politics was played. They began to understand the game of Soviet justice, where, to legitimize the embezzlement, corruption, and black marketeering that was indispensable to economic performance and delivery of products, another set of bureaucrats—themselves dialectically dependent on this corruption to justify *their* jobs—provided the theater of investigation and auditing. Highly publicized slaps on the wrist and occasional scapegoating (such as the execution in the early 1960s of a number of Jewish "economic criminals") assuaged public anger at the system's uneven access to goods and power while altering nothing fundamental. Indeed, these staged scandals even served in a Darwinian way to reward the best players and weed out the weaker ones.

With respect to the Bochkarëv affair, although the system was unable to suppress public indignation or save Bochkarëv, its damage control measures were successful, and Bochkarëv was replaced with another politically reliable administrator. But this short-term victory was achieved at the price of exposing the corrupt network of Communist bureaucrats from the government of the Russian Republic to VOOP to the Fishing Inspectorate. It also revealed that within the press were reform-minded elements that, given the proper conditions, would come to the activists' assistance. Indeed, in this episode the press played a critical role. Passionate and more serious attacks on Bochkarëv had appeared several years before the Geptner episode. Chivilikhin's articles on Kedrograd, for instance, had already drawn blood. Public opinion in the reform wing of the central press had already formed a negative perception of Bochkarëv that prepared the way for "Get a Load of Those Goldfish!" That stunning body blow set the stage for devastating articles two weeks later in *Literaturnaia gazeta* and *Komsomol'skaia pravda.* And although the pieces in the two papers "were not knowingly coordinated, neither was it a complete coincidence," in the words of Kedrograd leader Vitalii Fëdorovich Parfënov.[17] Of course, Semënov, editor of *Krokodil,* ran less of a risk; like a medieval court jester, his publication was a safety valve and was allowed a certain license to poke fun at the regime. But the editors of the other journals risked high personal stakes.[18]

Third, the experiences of the 1950s and the early 1960s proved to be a university for conservationists in the praxis of activism. Faced with the expropriation of their society, VOOP, by the regime, they found alternative, safe institutional protection in the Moscow Society of Naturalists and the Botanical Society, immune from direct regime pressure or interest, and then used those bases to expand their influence into the crucial student community in 1958–1960. They learned to use the press and to exploit the moral victory of Bochkarëv's resignation as a piece of activist folklore; memory of that symbolic victory was passed down as late as the 1980s. Nevertheless, they continued to see the solution in replacing uncouth bureaucrats with more cultured ones and in their own inclusion into policymaking. They understood the system better but could not part company with it.

Two years after the first scandal, in 1967, Bochkarëv was removed from his remaining government post for "personal abuses of power." A state dacha he had built (cutting corners) burned down. It is unknown whether he pleaded naïve ignorance of building codes in connection with this scandal.

Kedrograd Coda

Bochkarëv's other legacy was the disruption and evisceration of Kedrograd. In October 1966, with the experiment standing on one leg, Vitalii Parfënov received a telegram from Moscow with instructions to come quickly to the RSFSR Ministry of Forestry, then led by Ivan Emel'ianovich Voronov. With Chivilikhin present, Voronov offered Parfënov a job in the ministry. Still nourishing a belief in the ultimate reformability of the system, Chivilikhin urged Parfënov to accept, arguing that Kedrograd was thirty years ahead of its time and that in Moscow Parfënov could help prepare the ground.

Parfënov, however, fell ill with tuberculosis and, thinking that he would not survive, wrote his book *Kompleks v kedrovom lesu* (Complex in the Stone Pine Forest), which was awarded a Komsomol Prize. Ultimately, he did accept a position in Moscow with the forestry authorities, but he could not prevent the disbanding of the experiment by 1976.

In 1981 longtime Tomsk Party first secretary Egor Kuz'mich Ligachëv, later second in command under Gorbachëv, revived the idea of Kedrograd in his own *oblast'*. A conference was held and a film on the original Kedrograd, shot by a studio in Novosibirsk, was shown. Parfënov warned Ligachëv, however, that everything depended on the quality of workers attracted to the project, explaining that the original Kedrograd was based on well trained and committed volunteers. Seeking a dramatic political success, the obdurate Ligachëv went ahead with his plans regardless, thinking that he could airlift a bunch of foresters into the taiga and the project would take care of itself. As Parfënov predicted, the project did not take off, and Kedrograd was not resurrected.[19]

Yet another conference on restoring Kedrograd was held in July 1987. However, the experiment was not restored, and in 1993 the new Russian forestry agency Rosleskhoz changed the plantation's name, "thus eras[ing] it from memory and [consigning it] to oblivion like the Church of Christ the Savior, like thousands of historical and cultural monuments of our people," in the embittered words of Parfënov. To Parfënov Kedrograd represented a gallant attempt to preserve national memory through the preservation of "native" landscape. "For myself," he wrote, "I believe that the authentic cause of the unexpected elimination of Kedrograd is rooted in something . . . very serious. Our society is increasingly ravaged by the disease of 'Ivanov, who cannot remember his ancestors,' which has arisen as a result of the flowering of selfish, individual interests and a lack of respect for the labors of previous generations, on the indifference to everything, including the future of our own children."[20] Foresters especially are supposed to look decades and hundreds of years ahead, and it is a particularly poignant evil when they betray this responsibility. "The unique cedar woods," Parfënov concluded, "are a national treasure and not an object for the vagaries of the anarchy of the market."[21]

If the Bochkarëv and Kedrograd affairs were learning experiences for those involved in those struggles, what participants learned about the system may have been similar, but the morals they drew varied. While there was a general agreement that self-interested bureaucrats were damaging the interests of the country, the remedies of the various activist camps were hardly identical. For the field naturalists and their *druzhinniki* protégés, what the country needed was more input from the scientific and university communities; for the *kedrogradtsy* and their literary patrons, what was lacking was Russian patriotism, because "historical memory" had been ignored or even defiled. Each camp created its own mythology of these affairs. Where field naturalists remember Geptner's single-mindedness and civic courage, Parfënov sees the *Krokodil* piece as a sideshow, damaging for Bochkarëv, but no more than a farce. For him, it was obvious that Chivilikhin's polemical forays were the force that ultimately brought Bochkarëv low.[22]

Just as this grimly amusing episode shows that the various nature protection groups were drawn to some of the same issues and appeared to be working in tandem, it also shows that the same cause could have an entirely different resonance and meaning for each of the groups. This was also the case with the fight to prevent the pollution of Lake Baikal.

CHAPTER SIXTEEN

Storm over Baikal

More than any other issue, projects that threatened the integrity of Soviet, and especially Russian, waters elicited the passionate opposition of all varieties of environmentalists, from the high intelligentsia to the newer Russian nationalists. From the 1930s on, the megalithic water projects beloved by Stalin had been managed by the GULAG administration under the immediate supervision of the infamous major general S. Ia. Zhuk, head of the Main Hydrological Construction Agency of the People's Commissariat for Internal Affairs (later, the Ministry of the Interior and the Ministry of State Security). Beginning with the Baltic–White Sea Canal, the desolate banks and byways of tens of watersheds yielded to the weary thrusts of hundreds of thousands if not millions of *zeks* (prisoners) whose shovels and bare hands recarved the land into new waterways, inland seas, and hydropower stations. The orgy of hydroelectric and canal construction, which only began to subside during the 1950s, left a legacy that rivaled the Tennessee Valley Authority in scope: the Moscow-Volga Canal, the Volga-Don, the Rybinsk hydrostation and reservoir, and the "reconstruction" of the Dnepr river. It also left another, darker legacy. Perhaps 120,000 died on the White Sea Canal project alone.

At least some nature protection activists made the connection between Stalin's violent transformation of the land and his violent, instrumental treatment of humans. Anton Struchkov provides the example of Andrei Petrovich Semënov-tian-shanskii, whose close friend, the sixty-seven-year-old Andrei Dostoevskii, nephew of the writer, was one of those deported to the White Sea Canal project. "Happily," writes Struchkov, "after a year of labor Andrei Dostoevskii managed to return to Leningrad, but this experience was enough for his friend to realize that violence to nature and violence to people literally went hand in hand."[1] Semënov-tian-shanskii went on to publicly oppose the massive hydroelectric projects at a special conference on wildlife management in February 1932, an act of high civic courage.[2]

For the intelligentsia, therefore, even after Khrushchëv dismantled the forced-labor system in the mid-1950s, hydroelectric stations, canals, and other gargantuan earth-moving projects continued to resonate as icons of Stalinism, recalling both the repression and the megalomania associated with the dictatorial system. After the "thaw," though, it became possible to speak out against them.

For first-generation educated Russians the most objectionable aspect of the big water projects was their threat to Russian villages and historical monuments. Whereas twenty years earlier newly minted engineers or young "proletarian writers" would have looked on the huge dams, canals, and reservoirs with patriotic pride, by the 1960s these same kinds of people had become increasingly distressed that "progress" and "modernity" had been purchased with the destruction of their spiritual home. For people of this background, the taiga represented the last remaining hearth of Russian culture, and it was on the northern and Siberian forests that the dam-builders and planners had focused their attention from the mid-1950s. At the center of it all was Chivilikhin's "luminous eye of Siberia," Lake Baikal.

Unlike any previous environmental struggle, the fight to protect the vast lake from physical alteration and industrial pollution embraced not only the various branches of committed nature protection activists but a broader public as well. Such public participation imbued the struggle around the lake with a larger meaning as an incipient general protest against the rulers' abuses of power.

* * *

For those who have seen Baikal, adjectives fail. One of the best short descriptions of the lake has been provided by geographer Philip R. Pryde:

> Lake Baikal, located . . . just north of the Mongolian border, is perhaps the most remarkable freshwater lake in the world. Geologically, it lies in a huge graben (a structural depression between two parallel fault systems), and is approximately 700 kilometers long. The lake has the distinction of being the most voluminous freshwater body in the world (23,000 cu. km.), and the deepest as well (1,620 m.). Lake Baikal's main significance is not just its size, however, but rather its biology. In its unusually pure waters can be found over 800 species of plants and about 1,550 types of animal life. More importantly, the majority of these are endemic, making Lake Baikal an object of worldwide scientific importance.[3]

S. Ia. Zhuk's hydraulic empire was the first project to threaten large-scale changes to the natural conditions of the lake. "In the bowels of this, one of the country's most politically potent institutions [the S. Ia. Zhuk

All-Union Hydrological Planning and Scientific Research Institute—Gidroproekt], an authentically diabolical plan was hatched and developed," wrote novelist, journalist, politician, and Baikal activist Frants Taurin: "to blow up the mouth of the Angara and to lower the level of Baikal by several meters."[4]

Chief engineer N. A. Grigorovich of the Angara Sector of Gidroproekt intended to detonate an explosion—50 percent larger than that at Hiroshima—to allow greater water flow from Baikal to the hydroelectric stations downstream on the Angara River, into which the lake drained. The plan is described by Paul Josephson in his study of the Siberian branch of the Academy of Sciences.[5] Josephson also authoritatively identifies the important role played by Siberian scientists and writers in opposing this and other schemes. Exploring the story in detail Josephson became perplexed by an apparent paradox: scientists with impeccable political credentials, a strongly patriotic outlook, and plain (peasant) social origins, such as geologist Andrei Alekseevich Trofimuk, assumed key leadership roles in the defense of Baikal. We must disaggregate the different iconic meanings of the lake to appreciate why the reactions to its despoliation were so broad and passionate.

Grigorovich's plan met its first resistance at the August 1958 conference on the development of the productive forces of Eastern Siberia sponsored by Academy of Sciences' Council on Productive Forces (SOPS), held in Irkutsk. The giant conference, with 2,377 scientists and others in attendance, was preceded by a week-long series of regional miniconferences, in Chita, Krasnoiarsk, Ulan-Ude, and other localities, with an aggregate attendance of 5,690.[6]

Mikhail Mikhailovich Kozhov, an academic specialist on the fish and mollusks of Lake Baikal, was one of several to rebut the arguments laid out by Grigorovich, who also spoke at the Irkutsk conference. While arguing on scientific grounds that the Baikal fishery would suffer from lowering the lake's average level five meters, a likely consequence of the Grigorovich plan, Kozhov concluded with an admonition colored by an aesthetic, moral, and perhaps nationalist sensibility: "We don't have the right to destroy the harmony and beauty of this unique gift of nature." Kozhov was backed by the massive assemblage, which voted to support *zapovednik* status for the lake and a ten- to fifteen-kilometer radius of woodlands, and which also noted that "the unique value and significance of Lake Baikal with its unique nature and . . . the need to transform Baikal into a health resort and center of tourism" mandated its "strict protection . . . from pollution by industrial wastes."[7]

The outspoken opposition of Siberian scientists and even politicians did not go unnoticed by the Party watchdogs, who reported to the Central Committee: "In a portion of the presentations, especially in the talks of a number of enterprises and institutions of Eastern Siberia, regionalist tendencies came to the surface, for example in the discussions over questions of energy production linked with Lake Baikal."[8]

In Irkutsk, the largest city near the lake, city leaders raised not a whisper of protest to the Grigorovich plan. "Many of the local notables even approved of the project," recalled Frants Taurin, who himself once served as chair of the municipal soviet (mayor) of Iakutsk in the early 1950s. "However," he went on, "in a complete surprise for those in power, public opinion came alive. For those days, that was, let us say, a completely atypical development."[9]

Taurin attributes the vigorous response of significant segments of Irkutsk society to the city's unique urban traditions. Much like the social milieu of the field biologists, in Iakutsk as late as the 1950s the city's intelligentsia still survived largely intact and in significant concentration. These were the descendants of generations of tsarist political exiles: Decembrists, Poles deported in 1830 and 1863, populists, anarchists, Bolsheviks, Mensheviks, and Socialist Revolutionaries. "These were people who imbibed the idea of personal honor and a feeling of their own dignity, so to speak, with their mother's milk and were able to uphold this dignity through the black years of Stalin's arbitrary rule," judges Taurin, although he notes that many perished in the process.[10]

Covering Eastern Siberia in 1958 for *Literaturnaia gazeta,* Taurin and a colleague from TASS, Aleksandr Gaidai, decided to help the Iakutsk civic activists bring their message to the national press. After consulting with Grigorii Ivanovich Galazii, head of the Academy's Siberian branch's Baikal Limnological Station (later Institute) on the lake at Listvennichnoe (Listvianka), who equipped the journalists with scientific data and arguments, Taurin and Gaidai drafted a letter to the editor of the *Literaturnaia gazeta* and circulated it to collect signatures. The fact that hydroelectric engineers, including the director for construction of the Iakutsk Hydroelectric Station and deputy to the RSFSR Supreme Soviet, the station's chief engineer, and the project's Party committee head all signed gave the letter unusual weight. Galazii, Taurin, Gaidai, and eight others were also signatories. And on October 21, 1958, "In Defense of Baikal" appeared. The letter was evidently based on Galazii's projections of how the Grigorovich plan would affect fisheries, marine fauna and flora, the Angara floodplain, water supplies, and even railroad bridges in the area. Journalistic and literary flourishes were kept to the bare minimum, but the letter powerfully asserted that Baikal "belongs not only to us but to our descendants." The closing appealed for the active backing of "broad public opinion."[11]

The appeal touched a nerve in the newspaper's readers. Within a month more than one thousand letters from all over the country poured in to *Literaturnaia gazeta,* many bearing the signatures of whole groups and collectives.[12] Dazzled by the reaction, the paper's editorial board lauded Taurin's "material" as the best item published that month.[13]

Beginning in July 1956 *Literaturnaia gazeta* had led the way in the press's discovery of environmental issues, and in February 1957 it even sponsored a large conference to discuss the state of the *zapovedniki* and of game management.[14] That same year even the *Pravda* editors expressed interest in the issue, inviting Iu. K. Efremov to make a presentation before their board. For reform-minded editors environmental issues carried relatively low political risk while at the same time bolstering Soviet citizens' tentative efforts to find their civic voices and to exercise political initiative. Before the reform press's skewering of Mikhail Mikhailovich Bochkarëv there had been a good eight years of press involvement on environmental issues: *zapovedniki,* Kedrograd, and, most crucially, Baikal.

Other new dangers to the lake were now being identified publicly. Also in 1958 an influential book was published in Iakutsk: *Okhraniaite prirodu!* (*Defend Nature!*) Written the previous fall by game management specialist Vasilii Nikolaevich Skalon, the book, while concentrating on the depletion of Siberian forests and game, was probably the first to publicize threats to the lake. True, Skalon only pointed out the risks of continuing to allow huge quantities of cut logs to sink while being floated to port, but the book's importance lay in its recognition that Baikal and Siberia were neither too large nor too remote to withstand the advance of modern industrial society.[15]

Finally, in 1958 the public learned of plans for a major military-industrial installation to be built on the shores of the lake. The proposal—to build two factories for making viscose cord for airplane tires on Baikal's southern shore and main tributary, using the lake's ultrapure water—is now notorious. Like the story of the desiccation of the Aral Sea or the Kara-Bogaz Gol, it has become a parable of the inflexibility and myopia of the Soviet production system.

Where the Grigorovich plan threatened the lake's biota through an alteration of water levels, especially near the shoreline, the military factories raised the specter of chemical and thermal pollution, which could wipe out many of the species of fauna in the lake. Of course, the public was not told about the strategic nature of the original proposed factory. Rather, the plant was depicted as one dedicated to producing high-quality paper goods. The nature of the product made the environmental risks and costs seem that much more intolerable. Even those who knew the plant's true original purpose were constrained to refer to it in their public criticism as a "paper and pulp" plant, which, ironically, it eventually became.

The battle was waged, for the first years, on the pages of the *Literaturnaia gazeta.* Indeed, Taurin learned of the "cellulose" project at the same editorial meeting that lavished praise on his drafting of "In Defense of Baikal."

Editor in chief Sergei Sergeevich Smirnov boldly declared that Baikal was "our story," and approved travel expenses for the journalist to pursue the story. While in Moscow Taurin arranged for an interview with Nikolai Nikolaevich Nekrasov, chairman of SOPS, the Academy's Council for the Study of Productive Forces.[16] Nekrasov revealed to the journalist the true military purpose of the factory and explained that although Lakes Onega, Ladoga, and Teletskoe had equally pure water, the pine forests around the first two were largely logged out, and around Teletskoe the predominant conifer was spruce, whose molecular structure was unsuitable for the tire cord.

To date no single author of the plan has been identified. All we know are the names of the chief engineer, Boris Aleksandrovich Smirnov, of Sibgiprobum (the Siberian Planning Institute for the Paper Industry) and of G. M. Orlov, who had been serving as the Soviet minister of the pulp, paper, and woodworking industries (variously renamed) since World War II, under whose auspices the plants were built. Taurin insinuated himself into Smirnov's confidence, feigning ignorance of the scientific issues involved with the viscose plant (although Taurin had once studied organic chemistry at Kazan Polytechnical Institute in the fur and tanning department). Smirnov allowed Taurin (who worked under a pseudonym) to examine all of the technical documentation under the supervision of the engineer's assistant, one Viacheslav Maksimovich. On parting, Smirnov asked Taurin whether the assistant was able to resolve any misgivings Taurin may have had about the project. Taurin replied, "Viacheslav Maksimovich provided highly knowledgeable answers to my questions. He raised my chemical literacy to a completely new level. I believe that, basically, the situation is now clear to me." Still unsuspecting, the engineer effused, "That's just wonderful!" as Taurin left for Moscow.[17]

Taurin's hard-hitting and technically literate piece "Baikal Must Become a *Zapovednik*," published on February 10, 1959, took on all of the lake's enemies at once. "With all due respect to the work of the planners," Taurin wrote, "it is impossible to renounce the conviction that they are pursuing an erroneous and harmful 'line.'" How much difference in water quality, asked Taurin, could there be between the lake itself and some downstream point of the Angara, into which Baikal empties? Weren't there forests enough in Siberia without laying hands on the lake's watershed? Interestingly, noted Taurin, those from Grigorovich's Gidroproekt who responded to the letter "In Defense of Baikal" supported the idea of turning the lake into a vast *zapovednik* and condemned the prospect of industrial pollution of the lake, overfishing, logging the watershed, and killing the freshwater seals. Their only criticism of the letter was the signatories' "wrongheaded" opposition to detonating the mouth of the Angara. Conversely, employees of Giprobum, which held responsibility for planning the viscose plant, were equally adamant in their opposition to Gidroproekt's explosion. Taurin facetiously mused that, had there been time to elicit the position of the logging interests, they would doubtless have opposed both the industrial pollution of the lake and blowing up its outlet. "In the abstract, everyone stands for protecting Baikal," wrote Taurin, "under one critical condition—that such protection would not affect their bureaucratic interests. But that is the crux of the matter. Concern for Baikal, with preserving its nature, is not a matter to be entrusted to bureaucratic agencies to resolve; it must be resolved by the entire people." Noting that at the conference on productive forces of Eastern Siberia "our country's best scientists . . . spoke out for declaring Baikal and adjacent zones a state *zapovednik,*" Taurin concluded his article with the admonition that the lake "not only belongs to us, but to our descendants," making its preservation a moral duty and not simply a practical exercise.[18]

Shortly thereafter, *Literaturnaia gazeta* published a collective letter "To the Defense of Baikal!" from the leaders of the USSR Academy of Sciences' Commission on the Protection of Nature, Dement'ev, Shaposhnikov, A. A. Grigor'ev, and staffers.[19]

Perhaps under pressure from this combination of scientific public opinion and a newly awakened press, in April 1960 the USSR Council of Ministers adopted a law on water pollution requiring that pollution abatement technologies be in place and working before new factories began operations. On May 9, the RSFSR passed specific legislation designed to insure the protection of Lake Baikal and its basin, prohibiting the start-up of the Selenginsk and Baikal'sk factories until waste purification installations were operational.[20]

The issue refused to go away, however. In July 1960, Balzhan Buiantuev, a Buriat geographer with the East Siberian branch of the Academy of Sciences, published a brochure critical of development of the lake's basin, while the Fourth All-Union Conference on Nature Protection opposed any discharges of waste water into Baikal, even after treatment.[21] Moreover, by 1961 the Komsomol daily paper, *Komsomol'skaia pravda,* was in a bidding war with the *Literaturnaia gazeta* to be *the* defender of the environment within the Soviet press. It also needed to make moral "reparations" to the nature protection community for the tactless publication of the crude satire "Protected Tree Stumps" following Khrushchëv's January 1961 speech. The answer was Galazii's "Baikal Is in Danger," published as a letter on December 26, 1961. As Josephson notes, Galazii's letter "spilled the beans" on the nature of the chemical effluents that the two plants would discharge.

Literaturnaia gazeta lost an opportunity to recapture its story by failing to publish Gennadii L'vovich Pospelov's "Musings on the Fate of Baikal," which the senior researcher at the Institute of Geology and Geophysics at Akademgorodok offered to the newspaper in a letter of February 8, 1963. It was snapped up by *Sibirskie ogni,* which ran it in its June issue. Pospelov's accompanying letter to the editor of *Literaturnaia gazeta* was unexpectedly revealing, for it confessed that the emphasis on economic arguments against the cellulose plants in his article was "to make the argument more effective." His letter was also the first in print to alert the general public to the danger of the siting of the plants along the virtual epicenter of the Mongolian–Sea of Okhotsk seismic zone. (These arguments had been deployed in April 1962, when a prestigious group of academicians including Sukachëv, M. A. Lavrent'ev, and others wrote to the USSR Council of Ministers warning of the devastating consequences of an earthquake at the future plant site. They were soon joined by Academy president M. V. Keldysh, who sent his own letter.)[22] The real objections of the author, however, who claimed to speak for *nauchnaia obshchestvennost'*, was that the pollution of Baikal represented a loss to science and a blow to the international prestige of Russian science.[23]

If indeed Pospelov sought to downplay scientific public opinion in the article as it appeared in *Sibirskie ogni,* he was not entirely successful. Readers, the author doubtless hoped, would find themselves as infuriated as he by the wanton disregard for scientific expert opinion. Neither the conclusions of an all-Union scientific conference of late summer 1962 chaired by geographer I. P. Gerasimov nor the repeated objections of both the USSR and Siberian branch of the Academy of Sciences had the slightest effect on the prosecution of the project.[24]

"The whole matter turns on our different understandings of benefits to the state and duties of the state," ventured Pospelov in a rather daring political challenge to the Party's claim to a monopoly on political "vision." "We are a society actively creating the future of humanity, and for that reason we must place on the scales of benefit and harm to the state the interests of our descendants as well," he continued. That included "the interests of all people on Earth." "The scientists who passionately spoke out against the faults of these projects, in opposition to the declared interests of the state, were in fact carrying out their moral duty and duty to the state." "Baikal is our national pride," he concluded, "and we must not permit bureaucratic mentalities to doom one of the greatest one-of-a-kinds on Earth."[25]

By 1963 yet other publications had joined in publicizing the lake's plight, and *Oktiabr'* in its April issue published what became the most famous of all the essays on Baikal, Vladimir Chivilikhin's "Luminous Eye of Siberia."[26] Reflecting the general awareness that Chivilikhin's article had generated a massive outpouring of letters, a letter to the journal's editor in chief, V. Kochetov, from an aide to MOIP president Sukachëv explained that for years now a Baikal Commission had existed within the society and requested that the commission's scholarly secretary, Baikal activist Nikolai Pavlovich Prozorovskii, be permitted to acquaint himself with the correspondence, particularly from scientists.[27]

Oktiabr' had an even better idea. In a special section in October 1964 the journal published a selection of letters it had received on Baikal following publication of Chivilikhin's piece.[28] In a commentary prefacing the excerpts from this correspondence, the journal editors noted that "today the average person, wherever he or she may live, wants to know everything that is happening in his or her country. As a master of his/her fate, the average Soviet person often demands that his/her opinion, too, be taken into account."[29] Aside from quoting excerpts from the letters of such academic experts as Lev Zenkevich, a Moscow University ichthyologist, or A. S. Iablokov, an academic agronomist in VASKhNIL, the piece mentioned the article by Gennadii Pospelov in *Sibirskie ogni* and drew a parallel to the fight against the Lower Ob' Hydrostation, which found expression in a debate in the pages of the Party's theoretical journal *Kommunist*. To dramatize the depth of "public opinion" on the Baikal issue, the article listed some of the many scientific conferences and organizations from 1958 through 1962 that had gone on record to defend the integrity of the lake, as well as letters and public statements from a number of prominent scientists.[30] In a slap in the face to this scientific public opinion, "the organizations under criticism have not even found it necessary to respond," wrote *Oktiabr'*. "We don't know what they are thinking in Giprobum, Gidroproekt, the Irkutsk Sovnarkhoz, or the State Committee on the Pulp, Paper, and Woodworking Industries of Gosplan USSR with respect to the numerous outcries in defense of Baikal, but there is not even *one* lone document that disputes the conclusions of 'The Luminous Eye of Siberia.'"[31] Seeking support for its positions in the removal of Khrushchëv and the criticism of his style of governance, the journal announced that "now the Party has condemned arrogance and voluntarism in the solution of the serious problems of our economic development" and that decisions should now firmly rest on true planning and on science.[32]

The piece closed with a portrait of the economic system's continued irrationality. Informing the readers that even the justification for the plants had become obsolete, because viscose tire cord had been replaced by synthetics, "Once Again on the Subject of Baikal" noted with outrage that the factories were prepared to start operations—within the year—without pollution control equipment. "The poisoning of this Siberian sea, it may be said, has become inevitable," lamented the article. In an implicit challenge to regime claims that there was a rule of law in the USSR, the article noted that laws were on the books prohibiting the polluting of Baikal, but questioned whether they would be enforced: "Will the enterprises really be able to begin operations without pollution abatement equipment? Legally, they do not have that right. But what will happen in actuality? This remains an open question, and public opinion is waiting for an answer."[33]

With the ouster of the cantankerous Khrushchëv, the reform press and the critics of Soviet environmental policies both became bolder. Even *Pravda* published an adventurously philosophical piece by correspondent A. Merkulov, who challenged the economic calculus of the "planners" with vivid descriptions of the social price tags of externalities and with a land-ethic perspective remarkably close to that of Aldo Leopold. "Is contaminating the Angara and turning it into a sewage canal really a measure without a cost?" he asked. "Supposedly nature takes no part in the debate between the planners of chemical and wood-chemistry enterprises and the defenders of the purity of waterways," he continued. "However, every time nature's interests are disregarded, it answers with dead rivers or lakes." Anticipating Christopher Stone's idea that nature had intrinsic interests that needed to be defended during development decisions, Merkulov added:

> When the planning of enterprises on Baikal began, representatives of many agencies assembled. Some defended the interests of the fish, others the interests of the forests, and a third group the interests of the mining industry, but there was not a single representative to defend the interests of Baikal itself, to speak on its behalf. . . . The rivers and lakes await a law [to protect them]. They await a single master who will vigilantly stand watch protecting the interests of the state and the people.[34]

Literaturnaia gazeta got back in the game in February 1965 with a hard-hitting piece by the writer Oleg Volkov titled "A Fog over Baikal."[35] An individual of rare resilience, Volkov (see figure 26) had once worked at the Greek Embassy in Moscow during the 1920s. Approached by the GPU to spy at the embassy, Volkov refused, whereupon he was arrested and served twenty-seven years in prison and the camps. Rehabilitated by Khrushchëv, Volkov began a successful writing career. He soon found his voice in the defense of living nature. A prerevolutionary-vintage *intelligent* himself, Volkov became the literary voice of the field biologists/nature protection activists. Boldly, Volkov cited U.S. trade figures that showed the declining production (and obsolescence) of cellulose-based cord, now replaced by nylon and especially polyethylene. Further, he proposed the heretical idea that no single ministry or agency, even representing industry, should have the right to determine the highest (indeed, exclusive) use of the lake. That was more properly settled by a broad public discussion of the issues.[36]

Baikal came up as well at the Second All-Russian Congress of Writers, which met from March 3 to 10, and in the meeting's wake a group of writers—three Muscovites and five Siberians—published a collective letter in *Literaturnaia gazeta* on March 18. Otherwise unremarkable, the letter contained one paragraph that stood out in its linkage of Baikal, "the perfect model of the unduplicatable beauty of our own Russian nature," with Soviet—and Russian—patriotism.[37]

Figure 26. Oleg Vasil'evich Volkov (1900–1996).

What next occurred was rare in the annals of Soviet politics: G. M. Orlov, Stalin's old minister of the paper and pulp sector and its boss as chair of the USSR State Committee on the Pulp, Paper, and Woodworking Industries, felt constrained to answer his critics from science, letters, and the press publicly, in the pages of the *Literaturnaia gazeta.*[38] Singling out for criticism the articles by Volkov and other Russian writers and *Literaturnaia gazeta* for publishing articles that "incorrectly portray the situation," Orlov made one notable concession: he claimed to support the same environmental values as his critics, promising that the factories "will not impair in the slightest either the purity of this unique volume of fresh water with its flora and fauna, or the cultural and esthetic value of the lake."[39]

Volkov immediately wrote a rebuttal to the minister in *Literaturnaia gazeta.*[40] Listing a series of earthquakes that had struck the Baikal region, Volkov again asked the project's defenders to speak to all the scientific objections, which he recatalogued in impressive detail. But the bottom line was the illogic of the situation; if cellulose cord was obsolete, then the Baikal plant had no strategic importance. Paper could be produced anywhere, so why place the lake in jeopardy? Orlov's attempt at damage control only inflamed and emboldened his opponents.

Two days later A. A. Trofimuk let loose with a sharp rebuttal to Orlov, also in *Literaturnaia gazeta,* entitled "The Cost of Bureaucratic Intransigence: A Reply to USSR Minister Orlov":

> It would seem that Comrade Orlov as chair of the State Committee for Forest, Pulp and Paper, and Woodworking Industries . . . ought to respond attentively to the recommendations of scientists. But, as we see, Comrade Orlov is preoccupied with other things. Trying to save the honor of his uniform, he has tried to show with remarkable energy that all the questions of construction . . . were solved correctly and that the recommendations of the USSR Academy of Sciences do not deserve attention. . . . Comrade Orlov has gotten so carried away in his struggle that he doesn't even pause when spouting bald disinformation.

Expressing shock, Trofimuk confessed that the "methods [by which Orlov has been conducting this debate] are simply incomprehensible to me. To describe them is beyond my scientific capability." One thing, though, was certain: Orlov and his colleagues were "indifferent to the fate of the lake." Indeed, added Trofimuk, "I will even go one step further; in their actions I do not get the impression that they have any genuine concern either for the fate of that branch of industry entrusted to them. The fog of bureaucratic optimism, washing the shores of Baikal, must be swept away," he concluded, "and the sooner, the better."[41]

Taken together with the Bochkarëv affair, this spate of public denunciations of high officials by scientists was virtually unprecedented since the early thirties. Both Khrushchëv's thaw and his ouster go far toward explaining why this eruption took place in the late winter and spring of 1965. At that early date, most neither could see nor wished to admit the true conservative, repressive face of the Brezhnev regime.

Consequently, scientific public opinion had mobilized vigorously to defend Baikal. As early as February 25, 1963, Vera Varsonof'eva and N. P. Prozorovskii wrote to G. V. Krylov, chair of the Siberian branch's Commission on Nature Protection, informing Krylov that the Presidium of MOIP had authorized an on-site visit to the lake by a member of its Section on Nature Protection and sought the Siberian Commission's help in getting the MOIP representative in contact with *nauchnaia obshchestvennost'* in Irkutsk and in the Buriat ASSR. An urgent response, by air mail, was requested.[42]

As the start-up date for the Baikal factories approached, MOIP reasserted its central role as the voice of scientific public opinion. A chiding letter went to USSR minister of agriculture Vladimir Vladimirovich Matskevich in January of 1966 over the signatures of Varsonof'eva and Prozorovskii. Reminding Matskevich that within his ministry was lodged the Main Administration for the Protection of Nature (the direct descendant of Makarov's and then Malinovskii's Main *Zapovednik* Administration), the MOIP officials wrote that "we consider it our duty to bring to your attention the views of scientific public opinion concerning the problem of Baikal, as the situation, which continues to evolve thanks to the 'projects' of Giprobum, threatens with its consequences both the population and the survival of huge productive territories on land and in water that could be lost to the country."[43]

Every step of the project was based on bad science, argued the authors, from a lack of guarantee that the plant could actually produce a pure polymer needed for the viscose cord, to logging plans, and finally to pollution controls. Not wanting to indict the Party or the system of governance, however, Varsonof'eva and Prozorovskii blamed a six-year campaign of deceit by Giprobum, with the collusion of radio and the press, intended to mislead the public and the government. Consequently, they noted, the conclusions about the "project" reflected in the decrees of the Supreme Council of the National Economy of the USSR of July 27 and September 20, 1965 were also in error, reflecting a "contempt for the law" and an exercise of "arbitrary power."[44] Those decrees "did not accord with the opinion and recommendations prepared for the Supreme Council . . . by specialists and scientists who worked on that precise problem," complained the letter's authors. "In other words," they continued, "the leaders of the Supreme Council . . . *proceeded undemocratically* and at the same time demonstrated their lack of competence [emphasis added]."[45] Here again, we see the idea that decisive expert input into policy-making was congruent with "democracy" in the understanding of the high scientific intelligentsia, which continued to hold to the "Progressive era" notion that political problems such as the "appropriate" use of natural resources *could* be resolved by scientific experts.

"The USSR Supreme Council of the Economy by its two decrees considered it possible to offer the natural wealth of Baikal up to adventurers for the conduct of criminal experiments," Varsonof'eva and Prozorovskii continued, reflecting the old-line scientists' visceral revulsion against Soviet "transformism." Like acclimatization, Baikal became a symbol of officialdom's desire to turn the entire Soviet Union—nature and people—into a guinea pig for its ignorant and short-sighted experiments. And like the *zapovedniki,* the lake was regarded as one of those rare places that had somehow miraculously retained its pristine qualities despite the system's thirty-year military-style campaign.

The MOIP letter aspired to discredit those whom natural scientists viewed as compromised "specialists," reserving for natural scientists the right to represent both science and society. Some geographers and especially economists "for some reason are unable to break out of their bureaucratic agency-minded narrowness" to recognize the greater truth spoken by natural scientists, the MOIP officials explained, and demanded that the authorities heed "the opinion of scientific organizations representing more than 20,000 scientists and specialists, as well as that of the All-Russian Society for the Protection of Nature, which embraces broad masses of progressive Soviet public opinion."[46] They attached nineteen pages of resolutions adopted by *nauchnaia obshchestvennost'* that demanded solving the problem of Baikal "in accordance with the scientific and moral bases of our epoch," which the Giprobum proposals decidedly did not do.[47]

With the exception of a few regime collaborators like organic chemist N. M. Zhavoronkov, secretary of the Chemistry Division of the Academy, who chaired a committee that gave its stamp of approval to the factories, much of the scientific and literary establishments, following the lead of their respective activists, strongly opposed the projects. A collective letter to *Komsomol'skaia pravda* of May 11, 1966, "Baikal Waits," signed by Academy vice president B. P. Konstantinov, academicians L. A. Artsimovich (physics), A. I. Berg (cybernetics), B. E. Bykhovskii (zoology), I. P. Gerasimov (geography), Ia. B. Zel'dovich (physics), P. L. Kapitsa (physics), G. V. Nikol'skii (zoology), V. N. Sukachëv (botany), and A. A. Trofimuk (geology) among other full members of the Academy, plus scientists Galazii and Zenkevich, Old Bolshevik F. N. Petrov, writers Chivilikhin and Leonid M. Leonov, MOIP Baikal Commission chair Prozorovskii, actors, and others, embodied this unusual coalition. Terming the decision to build the plants a "mistake" by the Gosplan USSR Committee, the letter described the decision-makers as having "taken a risk of unheard-of scale, turning Lake Baikal into an experimental basin for the trials of a pollution abatement system that has never been tested in actual production conditions and which is not suited to the severe climatic conditions of the Transbaikal region."[48] "Those who created and encouraged the realization of these virulent projects should pay for their vulgar errors and oversights, for their arrogance toward the voice of Soviet scientific public opinion and for the suppression of criticism," the letter boldly demanded; "it is intolerable that the whole country pay for their mistakes."[49]

In its editorial commentary on the letter, *Komsomol'skaia pravda* asked, "Have we really learned nothing from the countless examples when economic bureaucrats in the name of the plan devastated waterways and lakes and poisoned their currents?"[50] "Readers may well ask: What the heck is going on?" the editors continued. "Why have the powerful voices of protest and the well-argued positions of science remained only a matter of conscience for an alarmed public opinion?" Even Gosplan's new commission, headed by chemist N. M. Zhavoronkov, reported the editors, showed no hint of taking the concerns of science and society seriously. "Comrade Zhavoronkov categorically refused to comment now in any form on the work of the commission in advance of a final decision," reported the editors, who explained that they had tried to elicit his views. To the editors the style of Zhavoronkov's treatment of the press and public opinion now became important in its own right. "This calls forth a sense of shock," they wrote. "Why is an agency whose decision is awaited by thousands conducting its activities in such a secret atmosphere?! Perhaps it is the case that the conclusions of the commission will follow a fait accompli, and the problem of construction on Baikal will be decided without prior consultation."[51]

Meanwhile, MOIP set about organizing a more public protest, reminiscent of the "three societies" meetings of the 1950s. A big scientific conference, "The Future of Baikal," was convened on June 13, 1966 at the Moscow House of Scholars. In advance of the meeting MOIP scholarly secretary Nikolai S. Dorovatovskii wrote to N. N. Mesiatsev, chair of the USSR Council of Minister's Committee on Radio and Television, asking that the radio station "Iunost'" be allowed to tape the speeches and musical intermissions for rebroadcast.[52] Expectations of a major event were justified. Once again, the aristocratic mansion was filled with the energy of an aroused scientific intelligentsia. Officially, 618 attended, rivaling the mass convocations of 1957 and 1958.[53] Judging by the resolutions adopted, the atmosphere at the conference was militant.[54] The collective letter to *Komsomol'skaia pravda* probably played a role in encouraging the large turnout and the belligerent stance.

Immediately after the conference Prozorovskii was again sent to Baikal, and on July 5 letters were sent to Brezhnev, Kosygin, and Podgornyi addressing them not as "General Secretary," "Chairman of the USSR Council of Ministers," and "Chairman of the USSR Supreme Soviet," respectively, but simply as "Deputy to the USSR Supreme Soviet" (as they were referred to in the conference resolutions), with a conspicuous lack of deference.[55] The letter accused the new plants of "violating the basic principles of proper resource use" and of "leading . . . to the loss of Baikal and its adjacent alpine region." The MOIP letter was uncompromising in its opposition to the plants. Referring to a recently proposed technological fix, Varsonof'eva and her colleagues wrote that "these mistakes cannot be remedied, even by means of building a pipeline from the outflow point of one of the factories to the Irkut river," which did not flow into the Baikal basin. "Only dismantling both cellulose plants and relocating them outside of the Baikal basin can rectify these dangerous mistakes," concluded the letter, citing a resolution to that effect supported by the mass meeting of scientists.[56]

One of the last letters sent by V. N. Sukachëv in his capacity as president of MOIP, cosigned by vice president A. L. Ianshin and by Prozorovskii, was to Viktor Borisovich Sochava, director of the Institute of Geography of Siberia and the Far East.[57] The aim of the letter was to galvanize geographers to more active opposition to the development of Baikal. Reiterating the point made by the 1964 article in *Oktiabr'*, Sukachëv and his coauthors noted that since the 1958 conference on productive forces of Siberia "about forty meetings, symposia, conferences, gatherings, and congresses of scientific public opinion in our country—embracing about 20,000 scientists—have . . . expressed a negative assessment of the scientific basis of the project of Giprobum." Added to these was the disapproval of the project by VOOP, with its ten million members, and the applause elicited by the author Mikhail A. Sholokhov when he criticized the project at the Twenty-third Party Congress in 1966. Further, materials from the Conference on Soil Erosion in the Buriat ASSR, held in 1963, already testified to the serious problem of deforestation in Eastern Siberia and its associated problem of soil erosion.[58]

Invoking a moral imperative, the letter argued that scientists did not have the right to condone catastrophic mistakes simply because they were only following orders:

> It would be a mistake to think that the state can remove the responsibility of a scientist for mistakes of bureaucratic projects and their consequences, for the potentially tragic fate of Baikal. On the contrary, we hold that at present, when the fate of Baikal is subject to the whim of the wrong people and the wrong institutions, . . . the responsibility of scientists has correspondingly increased. . . . We are inclined to support a course of action where scientific public opinion, independent of bureaucratic strictures, should get the Academy of Sciences system to adopt its set of recommendations for solving the problem of Baikal . . . taking into account the views of scientific public opinion.[59]

Sukachëv and his colleagues specifically asked for Sochava's support in passing a resolution at the Conference of Geographers of Siberia and the Far

East, scheduled for September 23, 1966, and for a discussion within the Geographical Society of the USSR.

This first full-scale coalition of the scientific intelligentsia and Soviet writers (the division between *liriki* and *tekhniki* does not reflect the significantly more complicated social identities of the educated stratum), including the reform press, gave no quarter to the "planners." Nearly simultaneously with the great MOIP conference of June 1966, *Komsomol'skaia pravda* published a follow-up to the collective letter of scientists and writers. "A mound of responses to this letter lies here on the editor's desk," wrote the editorial commentary, and "the voice of the letters to the editor is unanimous—we must head off the catastrophe threatening Baikal as quickly as possible." Letter writers, we are told, criticized the Ministry of Forest, Pulp and Paper, and Woodworking Industries of the USSR, "which even now, after the protest in the newspaper, after the alarm expressed by broad scientific public opinion, continues to insist on going ahead with the construction . . . at Baikal."[60] Even the Economic Commission of the Soviet of Nationalities of the USSR Supreme Soviet, not to mention USSR Gosplan's own Council for the Study of Productive Forces report recommended relocating the cardboard plant being constructed in Selenginsk to another region, with the Economic Commission also endorsing a total ban on wastewater discharge by the Baikal plant into the lake (and its transfer by pipe to the Irkut River, which flowed into another basin).[61] The newspaper editors concluded the piece by insisting that the authorities take into account the letters received, that is, public opinion, as well as the scientific conclusions advanced by the authors of "Baikal Waits."[62]

On June 22, the venerable physicist Pëtr L. Kapitsa spoke at a joint meeting of the Collegium of Gosplan of the USSR, the Collegium of the USSR Council of Ministers' State Committee for Science and Technology, and the Presidium of the USSR Academy of Sciences. To make his point, Kapitsa recalled an earlier American debate of the mid-1950s about atmospheric pollution by radioactive isotopes. Then, as now, he noted, there were two sides to the debate. One side was exemplified by the physicist Edward Teller. That group believed that atmospheric atomic testing would add only a small increment to existing ambient radioactivity. "They tried to prove their beliefs, of course, with long calculations" designed to demonstrate that the demands of public opinion to stop the tests were baseless, "which served a certain group of militarists and businessmen in the USA."[63] On the other side were scientists like Linus Pauling who held that "a formal, quantitative approach such as Teller's was not applicable in this instance . . . because we need to take into account the influence of these pollutants on the mechanism of biological functions." Sounding like Formozov or Geptner, Kapitsa asserted that

> in nature there exist very precisely established biological equilibria whose mechanisms contemporary science does not yet know. But we do know that comparatively small factors can destroy this equilibrium and that then the consequences are capable of assuming a catastrophic scale. For that reason Pauling believed that even a small change in the nature of atmospheric pollution by radioactive elements could . . . elicit substantially large changes in living nature and even fatal consequences for humans. Life has shown that Pauling was right.

Kapitsa found Zhavoronkov's expert commission similar to the virulently anti-Soviet Teller in its approach to risk and the value of life.[64]

The theoretical physicist's remarks also took a swipe at the pride of Soviet educational policy: the engineers. Doubtless, Kapitsa's view was colored by his elite disdain for the *vydvizhentsy* (social upstarts) who represented the great majority of the engineers and technicians as well as the *nomenklatura* bureaucrats of the postwar USSR. "From the debate the extremely low level of engineering and technical culture among workers of our paper industry has become painfully obvious. One example that shocked me was a statement by a worker in the paper industry during a consideration of . . . the question of the essential level of water purity for discharges from the cellulose plant." Kapitsa related that the official had announced that there should not be more than 0.5 milligrams of silicon per liter of water. When the official was asked how he arrived at that figure, he responded that that was the figure offered by the foreign firm supplying the equipment.[65]

This extraordinary and vociferous protest ultimately failed. In 1966 the Baikal'sk plant started up, and the Selenginsk plant was running by the following year. The concerted protests of scientific and literary public opinion had failed to stop the plants, to force them to delay production until all pollution abatement facilities were running optimally, or, as had been proposed by some members of the Academy of Sciences, to divert the wastes by a long pipeline to the basin of the Irkut River, where they would flow into the Arctic Ocean rather than into the lake.

The story of the continuing opposition by Siberian scientists to the operation of the Baikal mills has been admirably chronicled by Josephson. The later unsuccessful attempts by various Academy of Sciences commissions—under I. P. Gerasimov, a geographer, and Vladimir Evgen'evich Sokolov, a zoologist—to compel the regime to revisit the issue have been sketched by a number of scholars.[66] In all likelihood for Sokolov, whose strengths were in administration and not scholarship, and perhaps also for Gerasimov, involvement in these high-profile commissions was a means of enhancing their standing both internally and abroad. Fighting for nature made one a hero. Certainly for Sokolov, who was a member of the Academy Presidium for many years, his strong stand on Baikal, his nature protection advocacy generally, and the use of his personal influence to protect active researchers and activists within his Institute for Ecology and Evolutionary Morphology, were a kind of currency that purchased acceptance by more reputable scholars; they treated him with outward respect and appreciation in exchange for his political support and protection. This bargain may have soothed science bosses who obtained their high positions through nepotism but still suffered from bad consciences.

Did all of this clamor produce any tangible results for Baikal? After decades of decrees, resolutions, and calls for a "general plan," experts remain skeptical of their efficacy. The USSR Academy of Sciences cautioned in 1977 that Baikal was facing irreversible degradation. No one knows precisely the tolerances of the lake's myriad life forms for toxic effluents, thermal changes, and changes in dissolved oxygen and other gases. As Pryde has noted, in addition to the two big cellulose plants, "a hundred smaller enterprises around the lake still discharge untreated effluent. Thus, the lake's fate appears to remain undecided."[67] Pryde is being cautious in his judgment, but it is hard to find any experts who are optimistic about the future of the lake's biota.

For the dynastic intelligentsia, Baikal was another metaphor for the ability of rude, ignorant bureaucrats and their slightly less tainted engineers and technical advisers to usurp power and scientific credibility and to impose their devastating transformist experiments on society and nature. Some men and women of letters identified with the old-fashioned, scientific intelligentsia led by the field biologists, which now had attracted to its cause a significant number of theoretical physicists such as Kapitsa and other powerful intellects.

However, a discernible and increasing number of writers—also involved in the struggle to save Baikal—saw in the lake's ordeal a somewhat different story of Russia's purity threatened. In some cases, notably that of Sergei Pavlovich Zalygin, they melded various narratives to make new combinations. That nature protection issues could have such resonance testified to their evocative power in a modernizing society, on the one hand, and to the special status they enjoyed as a "protected" area of dissent in the neo-Stalinist state, on the other.

CHAPTER SEVENTEEN

Science Doesn't Stand Still

Because of its role as the scientific justification for an expanding network of *zapovedniki,* a holistic understanding of the natural community or biocenosis retained its hold on Russian ecologists and field biologists longer than in the United States or other countries. Through the 1980s advocacy (support for inviolable *zapovedniki* as institutions) continued to be dressed up as science (biocenology, the study of ecological communities). That linkage began to weaken during the 1960s and 1970s when, first, biocenologists finally had to acknowledge that their theory and practice did not cohere; second, they recognized that the theory was out of step with international science; and third, new paradigms and institutions began to challenge their monopoly within Soviet ecological science. These challenges to the ecological community concept, in turn, presented challenges to the raison d'être of the *zapovedniki,* which were amplified by the growth of Soviet tourist demand for scenery. Some environmentalists sought a new scientific theoretical grounding for protected territories to supplant the shaky old one, while others acknowledged their subjective view of the *zapovedniki* as simply sacred space. By the late 1980s the rationales for *zapovedniki* focused more on their roles as protected habitats, buffers for an overindustrialized landscape, aesthetically valuable undisturbed nature, and areas where the flow of life could still go its own way.

Beginning in the late 1960s, philosophers, economists, and political scientists discovered environmental rhetoric—I hesitate to describe it as "advocacy." Although each environmental writer was trying to make a name for himself (or, rarely, herself) and implicitly made claims for his or her discipline's central role in developing environmental theory, collectively these individuals represented a regime-approved and regime-sponsored means of blunting the critical edge of environmental speech. At the same time, the abundance of "environmental" publications in the social sciences represented proof of the regime's good intentions. This appropriation of environmental rhetoric went largely unchallenged by authentic activists because it also served their purposes: it confirmed environmental issues as one of the few zones of relatively free speech in the Soviet Union.

* * *

The evolution of ecological thought in America was neatly summed up by boreal ecologist Hugh Raup, director of the Harvard Experimental Forest in Petersham, Massachusetts:

> Ecological and conservation thought at the turn of the century was nearly all in what might be called closed systems of one kind or another. In all of them some kind of balance or near balance was to be achieved. The geologists had their peneplain; the ecologists visualized a self-perpetuating climax; the soil scientists proposed a thoroughly mature soil profile, which eventually would lose all trace of its geological origin and become a sort of balanced organism in itself. . . . I believe that there is evidence in all of these fields that the systems are open, not closed, and that probably there is no consistent trend toward balance. Rather, in the present state of our knowledge and ability to rationalize, we should think in terms of massive uncertainty, flexibility, and adjustability.[1]

Since 1910 the latter approach had been championed in Russia by Leontii Grigor'evich Ramenskii, but it had made little headway against the holistic, supraorganismic model of the biogeocenosis (ecological community plus abiotic environment) championed by Sukachëv and his many allies.

During the early 1960s things began to change slightly with the appearance of two important articles by V. D. Aleksandrova, which showed a certain sympathy for the continuum notion, but even they fell short of a clean break with organismic holism.[2] Soviet ecologists, particularly plant ecologists, were caught in a bind. On the one hand, they prided themselves on claiming to be an important part of international science. This logic should have pushed them to give up the holistic biogeocenosis for the more Western and relativistic "ecosystem" and for the continuum. On the other hand, they still needed a scientific justification for the *zapovedniki,* and in the absence of any other compelling ecological model clung to that of the biogeocenosis.

The plan the Lavrenko commission (see chapter 11) released in 1957 for future expansion of the network of reserves could have been drafted by G. A. Kozhevnikov himself. At its core was the recommendation that representative *etalony* be chosen on the basis of the geobotanical maps of the USSR that Lavrenko's team at the Botanical Institute had compiled. Once again, the Soviet scientific elite would have its own "archipelago"—islands of natural diversity and research autonomy, embodying the broader values of diversity and autonomy, a kind of "free territory of the intelligentsia," as Stalinist critic Lepeshinskaia once implied. Encouraging results followed endorsements by V. A. Engel'gardt, the new secretary of the Biological Division, and by Academy president Nesmeianov.[3]

However, the specter loomed of another *zapovednik* war like those of the 1920s and early 1930s. Repeating history, the USSR Agriculture Ministry's Glavpriroda, where utilitarians were still ensconced, sought to annex the newly created or restored *zapovedniki* of the Russian Republic's Glavokhota system and of other republican systems to its own, centralized all-Union network. Eerily, the issues had hardly changed since the 1930s, for Glavpriroda's reserves were still pursuing the same income-maximizing goals as before, while the Glavokhota reserves, heirs to the Kozhevnikov tradition, continued to reaffirm both the *etalon* mission of the *zapovedniki* as well as the inviolability of their regime.[4] In this reprise of the debate, ecological questions played a central role.

The *Zapovednik* Question in the 1960s and 1970s

The reiterative nature of the struggle over reserves was amply revealed by the statements of nature protection activists at Glavokhota's conference of *zapovednik* directors held May 22–24, 1963 at the Voronezhskii *zapovednik.* A. M. Krasnitskii, director of the Central Black Earth *zapovednik,* took to task the utilitarian construal of *zapovednik* functions (especially maximizing game) presented by A. G. Bannikov, a kind of quasi-official personage who was generally regarded as working for the secret police. Instead, Krasnitskii unabashedly defended the pure-science nature of research in the reserves.[5] Mikhail Aleksandrovich Zablotskii, the savior of the *zubr,* reminded his coparticipants of how much had been neglected, destroyed, or forgotten as he explicitly invoked the name of V. V. Stanchinskii:

> That which the comrade from the Voronezhskii *zapovednik* just defended here with so much zeal for us [old timers] is not news. In the years before the war, quite formidable scientists worked in our system, among whom the late Prof. V. V. Stanchinskii, doubtless known to many of you, occupied far from last place. . . . Some jokingly say that things go around in a spiral; the circle has closed and now we are again talking about these same interdisciplinary investigations.[6]

Someone else trotted out the memory of Malinovskii, now in retirement, as the symbol of oppression and unhealthy management. What especially rankled the speaker was the "double" expropriation of the *zapovedniki* Malinovskii carried out. Not only was he complicit in the Stalin plan of 1951, but he had the insulting temerity to replace zoologists with foresters as deputy directors for research in the reserves. This was a legacy that cried out to be fully reversed.[7]

Perhaps the most striking defense of the traditional vision of the *zapovednik* during the 1960s was the speech of Glavokhota's director of *zapovedniki,* A. Kondratenko, at the All-Union Conference on *Zapovedniki* on February 12, 1968. Declaring that to give a speech and pretend that the *zapovedniki* were no longer faced with powerful threats and impediments would be "to engage in simple phrase-mongering and empty words," Kondratenko outlined four broad areas in which the "Leninist principles of organization and management of *zapovedniki*" were undercut or trammeled.

First was inviolability. Kondratenko noted that, as a result of the Stalin and Khrushchëv liquidations of 1951 and 1961, "research over many years' duration was interrupted and considerable amounts of state funds were thereby wasted. . . . Consequently, the elimination of *zapovedniki* is not a process linked with the development of our society . . . but merely the thoughtless actions of specific individuals who have misled government and higher political organs."[8] The consequences were sometimes dramatic. After the 1961 elimination of the Bashkirskii *zapovednik,* 1,343 hectares of forest in the most accessible areas along the banks of rivers were cut down. Almost all of the maral, roe deer, and moose were killed off. The Kronotskii reserve fared no better. Almost all of the larch forest around Lake Kronotskoe was chopped down, and geological and topographical expeditions engaged in illegal hunting out of season. The Bogachëv Geological Expedition managed to destroy sixty-four brown bears in two months after the *zapovednik* was decertified in 1961. The Soviet military was not far behind. A military detachment at the Gulf of Ol'ga used a colony of elephant seals as target practice to try out a new gun; the carcasses were just left there, unused. Tourists, unsupervised, engaged in practices that threatened the living and nonliving elements of the landscape.[9]

The second "act of destruction" was Khrushchëv's conversion of some of the most historic and important *zapovedniki* in 1957 to so-called *zapovedno-okhotnich'ia khoziaistva.* "Is there any need to explain the absurdity of this marriage of concepts?" Kondratenko asked. At the very least these areas needed to be renamed.[10]

The third was the regime's failure to enforce existing laws. Kondratenko spoke of the USSR Ministry of Agriculture's "gross violation" of the RSFSR's October 27, 1960 law on nature protection by permitting its *zapovedniki* on RSFSR territory to allocate up to 50 percent of their land to "experiments," some of which entailed significant alteration of natural conditions.[11]

The fourth sin was using the press to promote false and distorted notions of the role of *zapovedniki.* Kondratenko singled out the article "The *Zapovedniki* Need a Single Administration" by the head of the USSR Ministry of Agriculture's *zapovednik* administration and the head of its Party cell,

V. B. Kozlovskii, an agronomist. The article, which ran in 1966 in *Okhota i okhotnich'e khoziaistvo* and the following year in *Nedelia,* in addition to trying to make the case for Ministry of Agriculture suzerainty over all USSR protected territories also promoted the usual ways in which nature within the reserves could be "improved." Kondratenko quoted from a hostile reader response to Kozlovskii as well as from D. L. Armand's *For Us and Our Grandchildren,* which also emphasized the eternal nature of *zapovedniki* as a "closed 'holy of holies.'"[12] In line with this emphasis on inviolability, all experiments in acclimatization had to stop. "It is well known that the most typical way in which the natural balance is destroyed is by introducing nonnative biological species to an area, which often leads to the most unexpected consequences."[13]

If Kondratenko's remarks were sharply phrased, it must be appreciated that as the representative of Glavokhota RSFSR he was not simply echoing the principles of the old-line scientists who dominated his agency's scientific advisory council. He was also defending the historic interests and prerogatives of the Russian Republic against a frequently aggrandizing all-Union center. Less than three years earlier, by means of a decree of the USSR Council of Ministers, the USSR Ministry of Agriculture successfully "raided" seven of Glavokhota RSFSR's *zapovedniki.*[14]

What distinguished these seven reserves was that they were among the most picturesque and the most accessible. Because the USSR Ministry of Agriculture's Glavpriroda had now been designated as the official Soviet "lead agency" for international contacts concerning nature protection and protected territories, there was a concern that the ministry have within its jurisdiction appropriately interesting reserves that it could show to foreign scientists and dignitaries. Not coincidentally, Bannikov's guidebook of the Ministry of Agriculture *zapovedniki,* complete with beguiling color photos, appeared in early 1966.[15] That infuriated "authentic" nature protection activists even more, as their foreign counterparts would be meeting with second- and third-rate scientists and bureaucrats of the Ministry of Agriculture system and would never learn of the existence of their "real" colleagues sequestered in the Glavokhota and other systems.

These and a cavalcade of real abuses committed by and in the USSR Ministry of Agriculture system provided constant reinforcement for the scientific intelligentsia's continuing fixation on the *zapovedniki* as *the* central issue of nature protection well into the Brezhnev era. Trouble started even before the new regime's first anniversary; agriculture minister V. V. Matskevich in October 1965 had issued an order stripping the newly acquired Astrakhanskii and Kavkazskii reserves of *zapovednik* status and converting them into branches of local game management areas, only months after an attempted transfer of 5,000 hectares from the Kavkazskii reserve to local forest authorities had been successfully fought.[16]

Official poaching and other abuses of *zapovedniki* in the Brezhnev era were ubiquitous. In Tadzhikistan's Tigrovaia Balka (where the last tiger's tracks were recorded in 1953), runoff from cotton fields had poisoned the northern portion of the reserve. A hunting lodge had been built on the reserve's territory for the shah of Afghanistan. In the Kzyl-Agach reserve on Azerbaijan's Caspian Sea coast, Marshal Chuikov, the hero of Berlin, led a military-style assault on the wintering waterfowl, scoring a major victory. Several years later, in 1976, Marshal Grechko did Chuikov one better, employing helicopters to obliterate the recalcitrant herons, flamingos, and geese.[17]

One of the worst embarrassments occurred when Minister Matskevich invited Academy of Sciences president Stubbe of the German Democratic Republic to the Voronezh *zapovednik* and then proposed a deer hunt. In deference to the scientific mission of the reserve, Stubbe refused, complicating the diplomatic atmosphere. Matskevich's practices were then held up to sharp criticism at the February 1968 all-Union Conference on Nature Protection, and when the head of the Department of *Zapovedniki* of the USSR Ministry of Agriculture tried to defend his boss, "tens of people started to shout from their seats 'Matskevich is a poacher!' The young naturalists—protégés of P. P. Smolin—made the most noise, for they had been to many of the *zapovedniki* and knew everything firsthand, even to the point of witnessing some of these outrages."[18]

Bannikov and others tried to play a constructive behind-the-scenes role. In August 1966 Bannikov wrote to Matskevich on the eve of the minister's trip to Tadzhikistan about the abuses in the Ramit *zapovednik* and the need to create a *zapovednik* in the Gorno-Badakhshan autonomous *oblast'*.[19] And Glavpriroda began work on a draft law on nature protection for the entire USSR.[20] Yet the prevailing sentiment was that sounded by Glavokhota director Kondratenko: the forces of light (Glavokhota and *nauchnaia obshchestvennost'*) were still locked in battle with the forces of darkness (Glavpriroda and transformism).

One complex of issues that still retained its emotional force was that of acclimatization and the campaign to eliminate large predators. For the scientific intelligentsia, this was an unsurpassed metaphor for the careless, scientifically uninformed, and dangerous transformism of the Soviet regime generally: chaotic capriciousness under the guise of "rational planning." It was difficult to escape the parallels in the realm of social and public policy: the "liquidation" of the "*kulaks* as a class," the deportation of peoples from ancestral lands to places half a continent away, the communal apartment, and other social experiments. Linkage with Lysenko and his ally Pëtr Aleksandrovich Manteifel' only deepened acclimatization's reputation as quackery and "Stalinist science." It was all the more odious that the regime had at various times forced this profane policy into the precincts of the *zapovedniki*.

Khrushchëv's thaw provided the first opportunity for opponents of acclimatization to go public. Writing in MOIP's *Bulletin,* game biologist V. N. Skalon took first crack at the symbolically charged program. "Acclimatization was advanced under the slogan of the transformation of nature," began Skalon, but "it could only be assessed favorably; no criticism of it was tolerated." As a result of that lack of peer review and the consummate haste in which measures were effected, "there was no small number of failures, which cannot be passed over in silence."[21]

Many schemes were hopeless from the first, such as the attempt to improve the Siberian red fox by releasing Canadian foxes. Introduced animals became pests: squirrels in the Crimea, mink in Eastern Siberia (where they preyed on the muskrat, which had also been introduced from North America, but never achieved great enough density to become economically harvestable), and the raccoon dog almost everywhere it was introduced, particularly in central European Russia, where it attacked ground-nesting birds.

Some acclimatization schemes were even fraught with threats to human health. Raccoon dogs carried rabies. Ground squirrels, introduced to the Caucasus, created a variety of epidemiological problems. Although those taken from the Altai were certified to be free of infectious diseases, they quickly became receptive hosts to the endozootics of the Caucasus. When the tufted-eared Altai red squirrel was introduced to Kirgizia, the "planners" were not as careful. It brought with it the flea *Tarsopsylla octodecimdentata,* which carried encephalitis. To top things off, the quality of the fur of the introduced animals declined, in some cases to economic uselessness. One additional consequence of these blunders was a certain resistance to reacclimatization, which is always safe and economically reliable.[22]

There were a few instances in which acclimatization was undertaken for loftier motives. The best example is the acclimatization of the sika deer from the Far East to the Caucasus and other places during World War II, when it was feared that the animal would be driven to extinction should war break out between Japan and the USSR.[23] One or two such examples, however, were hardly enough to outweigh the emotional animus of the scientific intelligentsia. Geptner, in a 1963 article, called once and for all for a renunciation "of that harmful idea that acclimatization may be a basic means of increasing the productivity of game resources."[24] Acclimatization was harmful "not only because it does not hold up, but even more so because it deflects us from serious and thoughtful work to a search for ways of quickly overcoming difficulties and deficiencies without expending a great deal of effort."[25] Skalon weighed in again later in 1963, as did A. A. Nasimovich in 1966 with a major attack on the consequences of the muskrat's acclimatization.[26] In 1965 the Tadzhik Academy of Sciences' Institute of Zoology and Parasitology took the opportunity of Lysenko's downfall to eliminate acclimatization from its research program.[27] Six years later Skalon, in a highly polemical piece, "The Essence of Biotechnics," mounted a spirited attack on the claims of acclimatizers and other "biotechnicians" to speak in the name of science.[28]

One study that struck out at biotechnics and acclimatization simultaneously was Konstantin Pavlovich Filonov's doctoral dissertation on the population dynamics of ungulates in *zapovedniki*.[29] A research scientist in Glavokhota RSFSR, Filonov culled thousands upon thousands of entries on index cards from the *letopisi prirody* (nature logs) kept by each *zapovednik* since the mid-1930s. His aim was to see what effect the policies of provision of salt licks, winter feeding, and other non-natural care of hoofed mammals, combined with campaigns to eliminate wolves and other predators and acclimatization of exotics, had on the ungulates' population dynamics. He found increasingly wide fluctuations: huge increases followed by catastrophic diebacks. As natural predation was eliminated, population densities increased, but so did the percentage of genetically less adaptive individuals. No longer culled from the herd by wolves and bears, these weaker and often sick individuals spread infection throughout the population. In the former Crimean *zapovednik,* deer with six-point antlers constituted 20.5 percent of the herd during the 1920s and only 16.3 percent in the 1950s. More robust nine-pointers, meanwhile, had completely disappeared.[30]

Acclimatized animals, such as the sika deer in the Okskii and Mordovskii reserves, deflected selection pressure from the native moose, becoming another prey species for wolves. In consequence, the moose population was no longer "policed" as efficiently for defective, older, and weak individuals.[31] Ultimately, concluded Filonov, to "undo" the effects of Stalinist "biotechnics," deer and moose herds now had to be thinned on a regular basis. Humans were now condemned, like it or not, to intervene in the life of a no longer pristine nature, but such intervention should always be only a form of damage control.[32]

Once the police power of the Party-state was withdrawn from the arena of biological research and teaching, geneticists and ecologists exerted efforts to decertify Stalin-era "schools" and their practitioners. Controlling scientific credentials was central to the social identity of the scientific intelligentsia and to its norm of scientific autonomy, and scientists lost no time trying to reclaim lost ground.

Paradigms in Motion

Among conservationists, voices of qualification, such as that of G. P. Dement'ev, the new president of the Academy of Science's Conservation Commission, warned that "it was time to renounce the view that *zapovedniki* are a 'higher' form of conservation." Instead, Dement'ev noted that the areas the reserves incorporated were only "conditionally natural" and that the tasks of conservation transcended the preservation of natural areas and their denizens (no matter how worthy that cause).[33] However, his words went largely unheeded by the restorationists.

Nonetheless, a conference called by the Academy's Conservation Commission and those of the republics, meeting at the Zoological Institute in Leningrad on January 25, 1956, revealed the incipient divergence within the ecologist-conservationist camp. Professor V. B. Dubinin, microbiologist and vice president of the Commission, emphasized not *zapovedniki* but resource problems in his keynote address. For the first time there was a vigorous call for the study of the ecological impacts of migrating radioactive compounds, now identified as a serious health threat to humans.[34] Articles on the ecological consequences of pollutants and pesticides also began to appear in the Commission's journal, *Okhrana prirody i zapovednoe delo v SSSR.*

Accompanied by a disturbing photo showing a lifeless stretch of the Kamyshevakha River downstream from a coking plant where unfiltered phenols were discharged, T. E. Nagibina's exposé, one of the first published articles to provide facts and figures on water pollution, cleverly juxtaposed a second photo showing revegetation following a cleanup of the stream, so as to maintain the mood of official optimism.[35] As a sanitary inspector of the USSR Ministry of Public Health, Nagibina approached the problem from the standpoint of human health. Natural conditions of water bodies had been vastly altered, she noted, and the amount of runoff from industry and agriculture had undergone a quantum increase. Pollution threatened the purity of water for drinking and recreational uses, promoted new outbreaks of infectious diseases, and impaired the water bodies' capacity for self-cleansing.[36] She invoked historical precedents: great tsarist-era public health experts such as Erisman and Khlopin had warned of such things, as had Chekhov, who was also a medical doctor. Water pollution was discussed at the 1896 and 1902 Pirogov Society congresses and at the Sewage Congress of 1905, which called for government standards of water quality.[37]

Despite this long-standing awareness plus Soviet-era legislation of 1923 and 1937 and the drinking water standards of 1954, observed Nagibina, the discharge of untreated waste water continued to inflict "great harm on the population and on the economy." The Volga directly received 500,000 cubic meters of untreated wastewater daily, while the Oka basin, which flows into the Volga, received an additional 370,000. Those waters were unusable for drinking or aquatic sports. Some waters were so polluted that they could not be used even by industry. Fishing and agriculture were the major victims.[38]

The main culprits were the oil extracting and refining, pulp and paper, chemical, metallurgical, and some consumer industries. In 1952, 550,000 tons of oil and petroleum products were lost after extraction, 350,000 tons of which polluted river basins through runoff. The concentration of emulsified oil in the Volga near the city of Gor'kii (Nizhnyi) exceeded permissible limits by twenty-five to seventy times. Such pollution also had international implications, added Nagibina, as when flocks of migratory birds died in the Caspian as a result of waters polluted by oil. Rivers and the factories that polluted them were mentioned by name. And while Nagibina pointed to alleged improvement since 1951, "there has still not been the requisite attention paid [to the issue] by the leaders of ministries, agencies, and individual enterprises." Moreover, there were far too few opportunities for public health scientists to test and apply their laboratory findings in the real world; one procedure proposed by scientists in 1951 for filtering petroleum products from waste water had still not been tested under production conditions. Nagibina also criticized the reluctance of industrial ministries to rethink their production processes, which led to an overreliance on technology to detoxify currently generated wastes. Only an all-Union organ for the protection of water bodies with the authority to enforce its decisions could guarantee truly safe production.[39]

That same issue of *Okhrana prirody i zapovednoe delo v SSSR* featured one article about air pollution and another about radioactivity. In much the same vein as Nagibina's piece, the article by M. S. Gol'dberg of the Laboratory of Air Quality in the USSR Academy of Medical Sciences' Institute of General and Public Health focused on public health. However, Gol'dberg also pointed out the special danger of sulfur dioxide to plants, especially trees.[40] A. M. Kuzin and A. A. Peredel'skii of the USSR Academy of Sciences' Institute of Biophysics, researchers in "radiation ecology," provided a history of the emergence of the study of radioisotopes in nature. Recalling the work of Vernadskii, Baranov, Cannon, and others, the authors showed how certain plants functioned as indicator species in areas containing deposits of uranium and other radioactive elements. Those plants that were coadapted to relatively high natural levels of radiation in many cases showed a higher metabolic rate. This was also the case with nitrogen-fixing bacteria exposed to high rates of natural radiation. Consequently, before Hiroshima and Nagasaki, it was believed that radiation was some kind of tonic that could make living things more productive.[41]

Now, with atmospheric testing, the situation had become qualitatively different. Japanese research on contaminated fish conclusively demonstrated the accumulation of isotopes in the internal organs of fish. If the irradiation of plankton was also taken into account, then whole food chains, up to and including humans, were placed in danger. True, some organisms, such as jellyfish and other swimmers, seemed to be largely unaffected, but that was a comparatively minor bright spot in the overall gloom.[42] Most ominous, as was brought out at the 1955 Geneva Conference on the Peaceful Uses of Atomic Energy, were the implications of exposure to high doses of radioactivity for human genetic integrity. Even with the peaceful use of the atom, noted the authors, the problem of disposal of radioactive wastes remained critical.[43]

"Having learned how to use atomic energy," philosophized the authors,

> humanity has still not fully grasped the responsibilities that attend this development, both with respect to our contemporaries and to future generations of people, as well as to nature the whole world over. . . . The expansion of work on the peaceful use of atomic energy demands the development of research in the area of radiation ecology, the training of radiation ecologists, and attracting in the broadest way the attention of scientific public opinion to this problem.[44]

These articles represented a new style of ecology on four counts. First, radiation and pollution were human health issues, to be studied in the tainted earth and waters around nuclear test sites, reactor sites, farms, and factories, remote from the allegedly self-regulating *etalony* of virgin nature. Second, the analytic framework for studying the effects of radiation and pollution was the species population, not the vaguely defined community. Third, this new current of ecological research was pervaded by the optimistic supposition that nature was fully knowable and would eventually be reducible to mathematical description. Finally, adherents of the new ecology believed that each of the new, serious environmental problems confronting society was susceptible of a technical solution, in principle.

This trend was reflected in new institutional arrangements, such as V. A. Kovda's Institute on Soil Science and Stanislav Semënovich Shvarts' Institute for the Ecology of Plants and Animals (established 1955), attached to the Ural Scientific Center of the Academy in Sverdlovsk, which had been fortified by waves of physicists and mathematicians seeking to apply their latest theoretical models to the study of living systems.

S. S. Shvarts, a "new man" of Soviet biology not unlike the geneticist N. P. Dubinin, began developing his critique of the older school as early as the 1950s. A disciple of Nikolai Vladimirovich Timofeev-Resovskii (the "*zubr*" who, exiled to Shvarts' institute, studied ecological aspects of radioactive cycling) and Pavel V. Terent'ev, both of whom strongly championed the population approach and the use of mathematical methods, Shvarts was first drawn to the problem of acclimatization, one of the pet programs of the nature-transformers, although he openly critiqued Manteifel's neo-Lamarckian approach to it from a modern, population-oriented standpoint.[45] Above all Shvarts, traumatized by the arrest in 1937 of his father, an Old Bolshevik, resolved on a course of political acceptability, avoiding even the faintest whiff of dissidence.[46]

After having helped to launch the journal *Ekologiia* (*Ecology*) in 1969, Shvarts began to speak out emphatically on the relationship between ecology as a science and resource development. In a talk at a 1973 special Academy-wide conference on conservation, he first underscored the sharp distinction that needed to be made between professional ecological science and conservation. Owing to a wrongheaded conflation of the two in the public mind, "broad circles of readers began to understand ecology as . . . a science with a social agenda, whose task boiled down to the protection of nature, the amelioration of the microclimate in urban areas, the development of various methods of detoxifying effluents, etc. However, speaking about ecology, it is always essential to emphasize that ecology is a biological discipline with its own . . . specific research methods."[47] Nevertheless, Shvarts did see a central role for ecology in addressing the environmental problems of the day.[48] But to play such a role, ecology needed to be unflinchingly scientific, abandoning all traces of muddy, idealist thinking and values.

One year later, in a talk to Party leaders in the Urals, Shvarts went further, deriding the ecological alarmists. "I am deeply convinced," he declared, "that their assertions are illegitimate." Discussions about the "exhaustion of nature," he continued, "sow doubt about the powers of man. . . . There is a wise aphorism: 'A resource deficit is simply . . . a deficit of knowledge.'"[49] The ultimate goal, he explained, was not some prehuman harmony but the ability "to direct natural processes." "We have no other alternative," he asserted, recommending the development of a general theory of ecological engineering.[50]

Although Shvarts' ecological engineering cannot be equated with I. I. Prezent's voluntaristic call in 1932 for Soviet biologists to become "engineers" in the great transformation of nature, there is at least one common thread: the notion that static natural harmonies do not exist. If, as Shvarts noted, ecology was "a science of the environment," then that environment has become increasingly transformed by humans. Consequently, "the most progressive ecologists see the main task of their science as developing a theory governing the creation of a transformed world." The world "could not remain untransformed," Shvarts declared, adding that such a process of transformation needed to be governed by considerations of human needs.[51] Shvarts radically diverged from his colleagues in considering the framing of economic and developmental strategies as the proper preserve of the political authorities, not of scientists with technocratic aspirations.

A year before his death Shvarts participated in a series of sharp debates with the writer and conservation activist Boris Stepanovich Riabinin, a member of the Central Council of the All-Russian Society for Conservation.[52] Held during the spring of 1975 at the Academy's House of Scholars and the "Ural" Palace of Culture, both in Sverdlovsk, they marked the ultimate development of Shvarts's positions, which provided powerful ecological-scientific justifications for the prodevelopment point of view. At that time Shvarts was perhaps the best-known ecologist in the Soviet Union among the lay public.

Throughout the debates, Shvarts's main argument was that prehuman, "pristine" nature no longer existed. Using the same example offered by Kozhevnikov nearly seventy years earlier, Shvarts noted that almost all of the forests of Western Europe were at least second-growth. However, Shvarts and Kozhevnikov drew diametrically opposite conclusions. Kozhevnikov conjured up the image of German forest plantations as a warning to Russians to preserve what virgin nature remained; for him, the deceptive luxuriance of the human-altered vegetation concealed a less stable, less biologically diverse assemblage of organisms than the community that had been supplanted. Shvarts asserted, on the contrary, that there were no grounds to consider second-growth inferior to original ecosystems. "In general," he noted, "it is not at all easy to determine how a bad or good ecosystem might be defined."[53] The "luxuriant tropical forest," he pointed out, would be choked by industrial effluents in only a few years, while the relatively species-poor taiga was able to withstand such abuse for centuries. The value of a given ecosystem had to be calculated in the context of its value to human society, argued Shvarts, not by some abstract principle of diversity or harmony.[54]

Another aspect of this problem cropped up during the discussion about the place of predators in the modern world. Riabinin quoted from the newspaper article "Nature Has No Stepchildren" to bolster his contention that predators played a necessary role in the economy of nature and should be preserved. He asked Shvarts to comment as an ecologist. Shvarts addressed the fate of the wolf as exemplifying the problem of large predators in the modern world. Through the mid-1950s, the wolf had been hunted down, even in the *zapovedniki*. However, with the triumph of the *etalon* view in the 1960s, the campaigns ceased, and the wolf population surged to over 100,000. Soon wolves once again became an object of public concern. Shvarts distinguished between those few remaining natural areas, such as the tundra, where the wolf still fulfilled a role of sanitary predation, and elsewhere. In the vast, anthropogenetic majority of Russia's modern agricultural landscapes, the wolf needed to be exterminated; there was no going back to the prehuman balance.[55] One practical conclusion that flowed from this was Shvarts's support for active management within nature reserves, which he did not recognize as incorporating self-regulating nature.[56]

In 1979 a roundtable was held in the pages of the main hunting and conservation journal, *Okhota i okhotnich'e khoziaistvo* (*Hunting and Game Management*). The head of the Ministry of Agriculture's Glavpriroda, A. Borodin, repeated Shvarts's argument that *zapovedniki* were only truncated islands of natural systems and, therefore, the ecological argument that wolves were necessary for maintaining those systems' self-regulating properties was spurious.[57] More revealing, however, was the argument of Oleg Kirillovich Gusev, the editor of *Okhota i okhotnich'e khoziaistvo,* who accused the bulk of Soviet biologists of "losing their objectivity" and "idealizing nature" while wolves were destroying thirty million rubles' worth of agricultural stock a year. They were purveying a baseless ecological catechism.[58] Ridiculing the ecologists, Gusev suggested that they had fallen into the teleological fallacy of believing that the wolf was created in order to prey on ungulates, ungulates to eat grass, and "both, in order to testify to the glory of the wise Creator." Their "murky" theory of "natural equilibrium" was the philosophical equivalent of a Divine Plan. Starry-eyed "idealization of nature" was to be contrasted with Gusev's hard-nosed realism: "The crux is this, that with the elimination of predators their place will be taken by other factors of selection, including human beings, whom the entire course of evolution on Earth prepared for a decisive role in the evolution of the biosphere."

This belief in a fated role for humans as the new chiefs of evolution was also sketched out by Shvarts, who, like the Tomsk zoologist-acclimatizer Nikolai Feofanovich Kashchenko (and the Russian revolutionary intelligentsia, many of whom envisioned socialism as a time when humans would become "gods on earth") almost eighty years earlier, proclaimed the end of the wild. All species would come under the management and stewardship of humans. "But this is nothing to fear," Shvarts reassured Riabinin; nature in the future would be better suited to human aspirations and needs, at least according to Shvarts's material understanding of them.[59]

The nub of the matter was a conflict over values. Shvarts dismissed Riabinin's contention that industrialization and urbanization were leading to the "impoverishment of nature" as an emotional reaction not deserving serious consideration. The only relevant understanding of "impoverishment" was in its quantifiable, "professional sense," namely, a lowering of biological productivity. There was no room for aesthetic or emotional criteria. Riabinin stuck to his critique of the urban "rat race," which "cut the heart out of life," warning that "blind faith in science is one of the modern varieties of ignorance."[60] There were absolutely no grounds for technological optimism à la Shvarts; "no," Riabinin warned, "there must not and cannot be easy and quick solutions."[61] For his part, Shvarts declared the "alarmists'" slogan "Back to Nature" not only "reactionary" but also "antiscientific." "Man cannot return to the caves," he intoned.[62]

The Crisis of the Biocenosis

As late as 1967 the great population ecologist and geneticist Timofeev-Resovskii still endorsed the concept of the closed ecological community.[63]

In an article designed to support Soviet involvement in the International Biological Program (IBP), Timofeev-Resovskii stressed the functionalist definition of the biogeocenosis that, inspired by V. I. Vernadskii, he and Sukachëv each had been developing based on observed and measured patterns of closed cycling of nutrients and minerals within circumscribed units of territory. "The biogeocenosis," he stated without qualification,

> is an objectively existing, logically explicable, and irreducible complex, holistic elementary structural unit of the biosphere, which exists in a long-term stasis that could conditionally be defined as a dynamic equilibrium. Biogeocenoses are the elementary building blocks of the biogeochemical activity of the biosphere. . . . Biogeocenology, which was born in the womb of biology, is not a [branch of] biological science, but an independent natural-historical discipline in the same sense that (and to an even greater degree than) Dokuchaev understood genetic soil science to be. The creation of biogeocenology by Sukachëv is one of the very greatest achievements of natural science of our era.[64]

More up-to-date than Sukachëv, Timofeev-Resovskii pinned his hopes on the application of mathematical models that, united with the power of computers and the information derived from natural historical observations (including the use of radioisotopes to trace trophic pathways, which he helped to pioneer in the USSR), would be able to simulate biocenotic systems and identify the conditions under which they were able to maintain "equilibrium." This was particularly important in the assessment of the actual and potential biological productivity of the earth's many biogeographical regions and zones, which was the claimed promise of the IBP.[65] Indeed, like his student Shvarts, Timofeev saw ecology's purchase in the development of its managerial-predictive possibilities, culminating in the eventual realization of Vernadskii's (and Gor'kii's and Stalin's) dream of a planet completely redesigned by human reason (noosphere).[66] Serving as the technical empowering agent of such a transformation would truly clinch for ecology/biocenology the title "queen of the sciences."[67]

Timofeev was a bridge figure in the sense that he was a bona fide field naturalist who also displayed emerging technocratic tendencies. His student, Shvarts, was already much more technocrat than naturalist, and those who came after—Iu. M. Svirezhev, D. O. Logofet, M. I. Budyko, Rem Grigor'evich Khlebopros, V. G. Nesterov, and others—were almost totally absorbed in the theoretical universe of mathematical models.[68]

Of course, ecology's entire past was a story of managerial claims and pretensions; only ecologists' expertise could identify "appropriate land use" or set target quotas for the extraction of living and renewable resources. What was new was a sense that the older, natural history or field research style of investigation was inadequate. Hence the emergence of a new generation of computer whiz kids and modelers, who would restore hope for ecology's technocratic agenda. As everywhere else, however, in the USSR the IBP failed to produce the magic formula for predicting and controlling the multifactoral, nonlinear web of nature. By the 1980s the closed, balanced "biocenosis" began to look less and less tenable.

This crisis of community ecology and of ecology generally was bound up with the crisis the nature protection movement faced in finding workable scientific rationales for the protection of the living world, particularly in light of biocenology's historic service as the scientific justification for the Soviet network of *zapovedniki.* Only in the late 1970s, however, was this linkage consciously addressed, as the crisis of ecological science intersected with a crisis of nature protection strategy in the USSR, especially regarding protected territories.

The Perils of the *Etalon:* A Little Reserve Raises Big Questions

One of the ironies of the history of Soviet nature reserves is that at its moment of resurrection, the ecological *etalon* concept fell into new contradictions. By the end of the 1960s, the ecological program for reserves had swept the field. In 1967, the Ministry of Agriculture finally banned acclimatization in its reserves and had even ceased its raids of Glavokhota *zapovedniki.* Yet, in addition to the serious practical problems of poaching, recreational abuse, and continuing harvesting of resources in the *zapovedniki* (particularly those of the Ministry of Agriculture), conceptual problems also remained.

Critics such as Ramenskii and Shvarts had pointed out ecologists' pervasive ignorance about the most basic problems of their science, beginning with the problem of determining the boundaries of putative integral, natural communities. However, a shift in thinking among ecologists could occur only with the ripening of a number of other developments in their intellectual and social environment. These included the rise of biosphere studies, the emergence of island biogeography theory, the increase of leisure time, a new attitude toward science, and the renewed legitimacy of aesthetic motives for nature protection. By the mid-1970s, leading conservation ecologists began their most daring intellectual journey, one that has not yet ended—the flight from the biocenosis. It led to a radical reconceptualization of the role and nature of protected territories, including *zapovedniki.*

In 1957 the Academy of Sciences' Presidium officially endorsed a plan for the broad expansion of *zapovedniki* throughout the USSR. Developed by a special commission chaired by botanical ecologist Evgenii Mikhailovich Lavrenko,[69] who had been active in conservation biology beginning in the

1920s, the "Long-Range Plan for a Geographical Network of *Zapovedniki* in the USSR" sought to restore or create no fewer than eighty-one new reserves. To justify this ambitious plan three main rationales for *zapovedniki* were advanced:

1. The basic task of *zapovedniki* is the preservation of typical natural landscapes and their constituent elements both for scientific as well as general cultural aims. *Zapovedniki* must be the chief bases for stationary study of natural complexes and their dynamics over a span of many decades and, eventually, centuries. . . . The results of well designed scientific research in *zapovedniki* should, within just a few years, be able to provide highly valuable guidelines for the renewal of natural resources—the restoration of impaired forests and pasturages, of soil fertility, of populations of useful animals and plants.
2. . . . [T]he preservation of [genetic] populations of a large number of species of animals and plants. It is extremely difficult and often even impossible to preserve those species of plants and animals whose numbers are reduced to a few individuals. A good example of this is the European bison, restoration of whose population has proceeded extremely slowly owing to the small number of breeders. Moreover, it is additionally essential to be aware of genetic variability of populations of one and the same species of animal or plant . . . and for that reason we must protect the species in various parts of its overall range of distribution. [The commission here also argued for preservation of biodiversity on practical grounds, because species are ever-new sources of information and resources.]
3. Aside from their scientific importance, *zapovedniki* have enormous general cultural importance. In them not just one but many generations of people may acquaint themselves with age-old parcels of their native nature. *Zapovedniki* must serve as bases for popular tourism.[70]

Although Lavrenko and his colleagues strategically highlighted tourism and mentioned the practical uses of preserving species diversity, the heart of Lavrenko's project was identifying allegedly pristine, representative tracts that could serve as baselines of healthy nature. Increasingly, however, the *etalon* concept began to founder on its own theoretical and practical shortcomings. First, ecology continued to lack a general agreement on the *definition* of the biocenosis (or bio*geo*cenosis, since 1944). Second, none of the numerous competing ecological approaches could satisfactorily resolve nagging problems that were undermining the *etalon* idea. One was that of the so-called downstream effect. Even if one accepted the possibility of encompassing a discrete biocenosis in a nature reserve, there was still no way to isolate such an area from potent, ambient in-migrating factors, such as air- and water-borne pollutants, feral dog packs, fertilizer runoff, and water table drops owing to regional drainage.

The most visible examples of *zapovednik* victims of downstream effect were the Berezina (in Belorussia),[71] the Astrakhan' reserve (in the Volga delta), and the Khopër (in the Black Earth region). Irrigation water taken from the Khopër river together with regional drainage had dried up the pools in the reserve's flood plain, changing its natural character. The most dramatic effect was the further decline of the *vykhukhol'* (desman, or aquatic mole) in one of its last habitats on earth.[72] Another small reserve with a big problem was the Central Black Earth *zapovednik,* which was admittedly too small (4,500 hectares) to be a self-regulating system in any meaningful sense.

Not two years after his plan for a renewed reserve system had been endorsed by the Academy leadership, Evgenii Mikhailovich Lavrenko received a note from L. K. Shaposhnikov, asking him to approve and, if possible, sign on to an extensive set of recommendations for the management of the Central Black Earth reserve prepared by veteran botanical ecologists Genrietta Ivanovna Dokhman and Larisa Vasil'evna Shvergunova.[73] The management plan was a cry for help on the part of the reserve's director, I. N. Iaitskii, and deputy director for research, geobotanist V. N. Golubëv, who wrote to Shaposhnikov that only the authority of the Academy could halt the Kursk *oblispolkom*'s sponsorship of excessive hay-mowing on the territory of the *zapovednik.*[74]

The report began with the initial proposition that the aim of the reserve was to "preserve the natural components characteristic for the central part of the European forest-steppe," including broadleaf forests and meadowlands drying into steppe. However, this undertaking was complicated because the land had at one time or another been subject to intensive human use. Forests were cut and fields were mowed, used as pasture, and even turned over for cropping.

Complicating the issue further, scientists were divided over the very nature of the "primeval" natural condition of the area today identified as forest-steppe. Was today's forest-steppe a consequence of human occupation? Which vegetational components of the contemporary landscape, if any, represented the "primeval" plant community? Such confusion over the baseline conditions of the region implied an equal confusion regarding the management of the reserve lands. If the current biota were in fact the result of human activities and if managers sought to reverse this "impairment of the natural components" of the reserve, then those findings mandated active intervention to restore the natural state and eliminate the consequences of previous human activity.[75] But what, then, would become of the idea of the *zapovednik* as an *existing* baseline of pristine nature? And could other *zapovedniki* have the same problem?

One component of the dispute was the presence of simple, one-storeyed oak groves, which some botanists (V. V. Alëkhin, N. A. Prozorovskii) considered to be of recent vintage. Local botanist G. M. Zozulin, however, disputed that finding, asserting that they were the depauperate remains of complex forests that had been destroyed by human settlement.[76] The reserve officials also agonized over what to do with American maple trees planted when reserves were under greater pressure to acclimatize exotics and over whether to continue with sanitary clearing, pest-control measures, and the collection of windfall.

As for vegetation that the botanists grouped under the general term "meadow-steppe" (*lugovaia step'*), there was an equally sharp dispute among the experts concerning what here was "original" or "natural." Some scientists (again including Alëkhin and Prozorovskii, together with Dokhman) held that the reserve represented a northern type of steppe that had nothing in common with a true meadow. Others (including B. A. Keller) described the vegetation as that of meadow-steppe, a transitional form from meadow to steppe. A third camp, which included Zozulin, M. S. Shalyt, A. P. Shennikov, and Lavrenko himself in his last works, viewed the vegetation as true meadow but with an incursion of steppe vegetation.[77]

Complicating the picture further was the hypothesis that this meadow-steppe seemed to have its origins in centuries-long hay-mowing and pasturing of livestock, which led to an increasingly xerophytic vegetation. "In this sense, the pristine steppe that survived up to the time of its protection as a *zapovednik* may only conditionally be termed a natural vegetation of the central portion of the forest-steppe zone," they wrote.[78] If that were the case, the consequences of a hands-off management regime would be startling and perhaps aesthetically and scientifically disquieting. With the end of hay-mowing the profile of the vegetation would change. A large number of steppe-based species would disappear, and vegetation would become overgrown and characteristic of meadows. If preserving the existing vegetational mix was the goal, asserted Golubëv and Iaitskii, then either large herds of ungulate browsers needed to be introduced or the grass had to be mown. Understandably, recognizing all of the scientific uncertainties even of their own positions, the authors of the plan sought to hedge their bets and recommended retaining a number of unmowed tracts as a permanent experiment.[79]

Whereas in 1959 this conundrum was the subject of discussion in discreetly transmitted memoranda and among a small circle of botanists, the problem of the *etalon* became a very public question for the nature protection community by the early 1980s.

When Aleksei Mikhailovich Krasnitskii, the Central Black Earth reserve's thoughtful late director, first assumed his duties in the late 1960s he believed generally that *zapovedniki* should only incorporate "self-regulating systems." Ruminating about the problem, he posed the key question, first in a series of articles and then, in 1983, in a major monograph: "Even were a steppe *zapovednik* to exist of a dimension one hundred, or even one thousand times as large as the Kursk reserve, what sort of a 'model of nature' would it be?"[80] Since the steppe could easily have been created as a consequence of human economic activity (intensive grazing of livestock), attempting to restore the steppe in the tiny *zapovednik* would ironically amount to the reconstruction, not of a natural *etalon*, but of a previous *anthropogenetic* or human-caused condition.[81] Reacting to this conundrum, Krasnitskii finally tackled the fetish of trying to preserve "pristine" nature head on: "the desire that vegetation in the *zapovednik* have a preagricultural character seems antidialectical," he pronounced. It was an act of intellectual courage.

Instead, Krasnitskii proposed that ecological communities in *zapovedniki* must meet at least two criteria: to be self-regulating communities, and to be maximally insulated from intrusive human factors in the present. With these conditions met, the natural biota would be able to develop spontaneously and the informational value of the system would be saved, he argued. There was no need to make an insupportable fetish of "primeval" prehuman nature.[82] Here, for the first time, was a willingness to accord semi-natural, human-transformed systems the same protection—as "communities" and even *etalony*—that "pristine" biocenoses had enjoyed. This decision opened up a new range of second-growth areas as candidates for status as protected territories—a virtue born of necessity.[83]

Finally, Krasnitskii grappled with the issue of identifying adequate boundaries for the ecological community that the *zapovednik* allegedly encompassed.

Although Krasnitskii began his book with a declaration of faith that "the *zapovednik* is a baseline of nature," thirty pages later he would feel constrained to admit that "anthropogenetic transformations of the environment of our entire planet have attained such dimensions that there hardly remain any biogeocenoses on earth that to one or another degree have not been affected by human activity."[84] Of course, there were vast differences of degree, he noted, proposing a set of "indirect, theoretical criteria for the identification of 'healthy' biogeocenoses," even if they were not "pristine."[85] Taking his scientific cue from Shvarts,[86] Krasnitskii held that such a "healthy" system

1. would not have a great preponderance of phytobiomass over faunal biomass;
2. would have a high level of biomass production and high productivity, with productivity maximally directed toward the increase of biomass;
3. would have a high level of stability, both of the biogeocenosis and of its dominant species, for a broad range of external conditions;

4. should guarantee, through the dynamic equilibrium of the biogeocenosis, homeostasis for the nonliving components of the biogeocenosis: the hydrological regime and the composition of atmospheric gases on its territory;
5. would have a rapid exchange of matter and energy;
6. would have the ability to quickly transform the structure of its community and of its dominant species' populations, and thus to evolve quickly.

For Krasnitskii it did not matter whether the system was located in the distant, uninhabited countryside or in an urban setting. Nevertheless, not even Krasnitskii could part with the dualistic framework of "self-regulating" and "truncated" nature; those areas that would qualify for *zapovedniki,* he repeated, should be those "maximally defended from the influence of human activity" as well as those that in a meaningful sense were "self-regulating."[87]

That view, however, still left open the question of how to determine and certify the presence (or, indeed, existence) of those factors and natural communities. Krasnitskii, attempting to address the linked problem of setting actual boundaries for reserves, saw island biogeography theory as inadequate but flailed around trying to find a suitable alternative.[88] It was obvious that biocenology and the "*zapovednik* cause" (*zapovednoe delo*) were as much theological as scientific issues.

Where to Draw the Line? The Boundary Problem of the Biocenosis

By the early 1980s even diehard defenders of the biocenosis had to acknowledge a crisis in the practical application of their concept to the setting of boundaries for *zapovedniki.* Little scientific progress had been made since the 1920s. The director of planning of new *zapovedniki* of Glavokhota RSFSR's Central Laboratory for Game Management and *Zapovedniki,* Kirill Dmitrievich Zykov, together with coauthor L. D. Alekseeva, explicitly admitted that "up to the present time there has been no theory that could provide either the geographical or the ecological normative bases or practical direction for establishing the area of *zapovedniki.*"[89] Iu. D. Nukhimovskaia of Glavokhota RSFSR, admitting that many existing *zapovedniki* were non-self-regulating, unstable and unrepresentative, proposed a return to a floristic approach; viable *zapovedniki* would include more than 75 percent of the significant species for a given biogeographical zone.[90] Citing her colleague Feliks Robertovich Shtil'mark's estimates, she noted that this approach meant that the minimal area for Arctic zone *zapovedniki* would be one million hectares; those for eastern and western Siberia, 500,000; for the alpine Far East, 300,000; for the Urals, 200,000; for northern European Russia, the southern Urals, and the alpine Caucasus, 100,000; for central European Russia, 50,000; and for the forest-steppe, steppe, and delta zones, 20,000 hectares.[91]

Konstantin Pavlovich Filonov took a more functionalist approach, arguing that the minimal area should assure the "normal functioning" of the biocenosis, but that response begged the question of how to know when an ecological community was "normally functioning." Filonov did take into account the complication of migratory species, arguing that any reserve system needed to protect all essential migratory points and pathways. But he maintained that the outer boundary of the reserve should be set by the range of the most dispersed of the ecological community's members, usually the largest mammals and birds. Examples that proved his point were the Caucasus *zapovednik,* which was not large enough to support its population of brown bears, and the Sikhote-Alinskii *zapovednik,* which was not large enough to prevent the gradual extinction of the Siberian tiger.[92]

What kind of *etalony,* then, were the 145 *zapovedniki* of the USSR (in 1983)? Exceptional for his honesty, A. A. Nasimovich admitted that the boundaries for *zapovedniki* had always been set by trial and error. Indeed, ventured Nasimovich, if Sukachëv were alive today (that is, 1980) he would admit that "pure" *etalony* or biocenoses could not exist in today's biosphere.[93] Scientifically, the *etalon* had no clothes.

The Legacy of Vernadskii

If the concept of biocenoses, those "basic building blocks of the biosphere," was in crisis, the scientific star of the biosphere itself was only rising. From the 1960s on, a new sensitivity to global ecological problems gave wide currency to the ideas of Vladimir Ivanovich Vernadskii, who had pointed to the biosphere as a single system as early as the second and third decades of this century.[94] This trend led to new priorities for conservation. Because, in the words of the late Nikolai Fëdorovich Reimers, one of the most influential of the new ecological-conservation theorists, "one can benefit through reshaping nature in some region [of the biosphere] only by losing in another area,"[95] it was no longer adequate to plan land-use policies on the level of the local biocenosis; a nationwide, even global, perspective was required.

Also on Reimers's mind were questions about how nature works. A creative yet sensible thinker, Reimers found himself increasingly disturbed by the claims made by S. S. Shvarts and partisans of his school about the predictive and technological potency of mathematical ecology. In a series of popularized pamphlets for the Znanie (Knowledge) Society, he began to express serious doubts about ecological engineering. Why, he asked, have environmental disasters occurred? Sometimes, he responded, we are unable to come to grips with the facts at our disposal; sometimes we are not in possession of the facts; sometimes unforeseen circumstances occur; and sometimes circumstances occur that are unpredictable in principle.[96] Given such epistemological limitations, he characterized the recent "fashionable 'prognosis' for the transformation of the biosphere into the technosphere" as folly, a revolutionary conclusion for a Soviet scholar. From an informational standpoint, description of some natural phenomena—let alone their simulation—is impossible. The genetic combinations within one single species, Reimers noted as an example, can range from 10^{50} to 10^{1000} variations.[97] These warnings, plus the wide publicity given to the Meadows (*The Limits to Growth*) and, later, Mesarovich forecasts for the Club of Rome, encouraged conservationists to think on a global scale from the mid-1970s.

Another new ingredient in *zapovednik* affairs had to do with the rise of island biogeography, a branch of ecology. Pioneered by Robert MacArthur and then adapted to conservation problems by Jared Diamond, Michael E. Soulé, and Bruce A. Wilcox, the field focuses not on the dynamics of a putative closed ecological community, but rather on the study of those conditions that affect the viability of populations of individual species living in a particular area, considered as an "island." (Originally, real islands were studied.) For those who were prepared to abandon the concept of a closed ecological community, here seemingly was an empirically oriented conservation program that studied identifiable entities (populations) and whose success could be measured.

Using Jared Diamond's ideas of island biogeography, a group of Belorussian ecologists recommended a "joining up of all the conservation districts having a sufficiently large total area into a single, spatially continuous system preserving the entire diversity of species, populations, and groups" as the most effective way of "controlling balanced development of a natural-anthropogenic complex." Such a complex would also include disturbed lands, since the object was the protection of biodiversity and not the fetishizing of "pristine" nature.[98] One of the most important figures in introducing this perspective in the Soviet Union was Aleksei Vladimirovich Iablokov, who edited the Russian edition of Soulé and Wilcox's *Conservation Biology* and who helped to gain the backing of the influential *Journal of General Biology* (*Zhurnal obshchei biologii*) for that approach.

Sociodemographics also played a role in this story. By the mid-1970s, the highly urbanized population of the USSR was taking to the backroads in increasing numbers in search of scenery and serenity. Whereas the prewar emphasis was on rest homes and sanatoria, modern Soviet vacationers sought more active recreational pursuits.[99] Tourists streamed to *zapovedniki* to see what had frequently been publicized as the "Soviet Yellowstones," and recreational geography became institutionalized.[100]

By the late 1960s the official regime spokesperson for nature protection, Andrei Grigor'evich Bannikov, had broached the issue in a piece in *Priroda* entitled "From *Zapovednik* to Nature Park," in which he noted that the demand for recreation in nature warranted a system of nature parks alongside the inviolable *zapovedniki,* which should remain largely off-limits to tourists.[101] At first, the most committed supporters of nature protection were at most wanly enthusiastic. Some even saw swarms of tourists as human locusts, little less destructive than acclimatizers, flocks of sheep, and belching refineries. However, the Soviet tourist was a potentially powerful ally in the fight to save natural amenities. A new type of protected territory, the national park, could be a boon in the fight to save elements of natural diversity in the USSR.[102]

Along with Bannikov's call for national parks and Krasnitskii's monograph, there appeared a widely discussed book by Reimers and Feliks Robertovich Shtil'mark, *Protected Natural Territories,* published in 1978. Reimers was a biologist affiliated with the Central Mathematical-Economics Institute who had earlier studied the relationship between forest types (by age and species composition) and the population of game animals they could support, while Shtil'mark (see figure 27), trained as a game management specialist, had long worked for the Russian Republic's Glavokhota as one of its key planners of new *zapovedniki.* The salient points, representing more Reimers' global-oriented rethinking than Shtil'mark's sacral view of protected territories, were:

1. The system of protected territories needed to be redefined as an integral, distinct branch of the economy—its stabilizing sector, enabling the rest of the economy to function. Popular perceptions of these territories as unproductive lands, reflected in their zoning status as "nonagricultural lands," needed to be revised. Instead, the USSR should follow the lead of the Kyrgyz SSR, which had established a republican State Land Fund as a special permanent category of land.
2. Principles of siting and determining areas of *zapovedniki* should be revised. Rather than selecting *zapovedniki* according to the old formula of one per biogeographical unit (the *etalon* principle), planners should create *zapovedniki* to provide enough healthy nature in the proper areas so as to ensure no breakdown of the socioecological equilibrium.
3. The socioecological equilibrium, defined as the balance between economic activity and the carrying capacity of the environment that permits a maximum level of production to be sustained, was to be assessed from a broad, nationwide if not global perspective.
4. To accommodate the various needs and levels of conservation,

Figure 27. Feliks Robertovich Shtil'mark (1931–).

> from the protection of rare species to recreation, a new, efficient multifunctional system of protected territories was needed. While appropriate minimum areas of individual reserves dedicated to preserving particular species complexes could be determined with the aid of island biogeography theory, for example, other types of protected territories might have more flexible requirements.[103] These could be regulated so that all of the different protected territories taken together would then be integrated into an overall system of providing for the maintenance of the socioecological equilibrium.[104]

Addressing the constituency of Soviet tourists, Reimers and Shtil'mark warmly greeted the new national park movement, though they emphasized that *natsional'nye parki* should not be established at the expense of or through the conversion of *zapovedniki.*

In 1979, shortly after the appearance of their book, Reimers and Shtil'mark wrote a popular piece for *Priroda i chelovek* (Man and Nature). Although it was titled "Etalony prirody" (Models of Nature) it mentioned little about representative biocenoses. If the old-line biocenology was out of the picture, two themes were salient: diversity and aesthetics. Arguments for diversity were reflected in the protest literature of the "Village Prose School," and especially by the bards of distinctive Siberia and the Far North

(V. Rasputin, V. Astaf'ev, A. V. Skalon, and Shtil'mark himself). And although Shtil'mark and Reimers continued to make scientific arguments for preserving diversity (especially genetic diversity), they struck out on an unabashed, purely literary, nonscientific defense of aesthetic values and diversity (more Shtil'mark than Reimers). In their joint article Shtil'mark resurrected an arresting quotation from the early Russian botanist and conservation leader Valerian Ivanovich Taliev:

> The virgin forest and the unplowed virgin steppe attract the contemporary mature individual not only with the prospect of clean air, wide open spaces, and freedom from the confines of everyday life. They are also sources of experiences of a higher order. They speak to us! . . . [N]ature is not only something outside of us, but it forms together with us an integral whole; we ourselves are only a small unit within the one great organism of nature. To learn how to penetrate to this unity, to feel around oneself the beating of the unbroken pulse of life, means to create a positive foundation for spiritual development, to incorporate into the developing soul a powerful counterweight to the narrow practical "I," and to develop the ability to perceive the world in an artistic and aesthetic way.[105]

Such ideas could not have found an outlet in the days when ecology was dominated by Prezent, and would have been denounced as "idealism" as late as the mid-1960s. Aesthetic argumentation had finally come out of hiding in the USSR.

Enter the Philosophers (and Others): "Environmentalism" as Self-Promotion

Nowhere is the renewed legitimacy of aesthetic rationales for conservation clearer than in the article "The Philosophical Bases of Contemporary Ecology," by the influential philosophers Ivan Timofeevich Frolov, a key Gorbachëv adviser and recent editor of the Party's ideological journal *Kommunist* and later *Pravda,* and Viktor Aleksandrovich Los', of the Soviet Academy of Sciences' Institute of Philosophy.[106]

The aesthetic attraction of nature, asserted the authors, increases in importance as society becomes more and more urbanized. Indeed, they explained, "it would be a mistake to conceive of the biosphere merely as a source of resources or a 'disposer' of wastes."[107] Equally important was the need to reintegrate both aesthetics and values into our way of relating to the world and into our science. Did not Einstein, Bohr, and Heisenberg (no soft humanists or biologists, they, but physicists!) invoke aesthetic criteria in their search for the "best" scientific explanation?[108] Frolov and Los' took pains to debunk a number of "myths" about the society-nature relationship. The first was the myth of the inexhaustibility of nature and nature's capacity to assimilate wastes. Previously, Soviet philosophical writing had stressed (as had

Shvarts) the notion that resources were socially defined and not fixed entities, and that surrogates could be found both for resources and natural processes. The position of Frolov and Los' constituted a reversal of decades of voluntarist thinking.

The second myth they exploded was that of the desirability, or even the possibility, of "man's 'domination' over nature":

> Under the influence of the crisis nature of the developing socioecological situation, man is gradually moving away from the illusion of anthropocentrism and rejecting the traditional hegemonistic relationship to nature. His thinking has ceased to limit itself to notions centering around needs and designs of him and him alone. His activity is acquiring an ever-broader biosphere orientation, and his thinking is drawn to "biocentrism." . . . Biospherocentrism assumes an orientation of human activity and thinking in directions that consider human interests both as subject and as object, as man and nature.[109]

Intriguingly, they recommended a return to Marx's original monistic notion—subsequently elaborated by the great biogeochemist V. I. Vernadskii—that human beings and the environment are parts of a dialectically interactive whole. We act on nature as both subject and object; when we alter our environment, we often create dislocation and dangers for ourselves.[110] This point is crucial because we know that in the Soviet past, owing to a constrictingly narrow definition of the human being, not merely aesthetic but other psychological and, indeed, biological dimensions were disregarded in setting social and economic policies.

Despite their undeniably constructive services of opening up what it was possible to discuss in print and in public, the philosophers (and, we might add, the economists, jurists, systems analysts, human geographers, environmental psychologists, and others) represented a distinct new subgroup within the "environmentalist community" whose impact ultimately must be reckoned in felled trees. Environmentalism comes in a variety of flavors, and for these philosophers and social scientists, it could best be described as a double scoop: professional advancement and maintaining the appearance (not least for themselves) of engagement in relevant, "clean" work. This was particularly poignant given social science's tawdry reputation as the propagandist of Stalinist and post-Stalinist repression.

Joan DeBardeleben has provided a useful portrait of the evolution of those debates.[111] But perhaps more to the point were the innumerable trips to IIASA in Laxenberg-Vienna and the Wenner-Gren Center in Stockholm, the countless conferences and roundtables, and the numberless anthologies whose titles were variations on Nature, Environment, and Society.[112] Especially from the mid-1970s on, there was a torrent of materials from this group touting the promise of the "scientific-technological revolution," systems analysis, and their combined capacity to fix environmental problems.[113]

The major issue was which of the disciplines would provide the conceptual framework for environmental discourse and research.[114] I. P. Gerasimov claimed that role for geography, but this did not go uncontested. The economist P. G. Oldak promoted what he called a "bioeconomics," while Ivan Timofeevich Frolov, editor of *Voprosy filosofii,* together with E. V. Girusov, V. A. Los', and others tried to assert a leading role for philosophy. Championing environmental law were O. S. Kolbasov and V. V. Petrov. Many claimed to follow in the footsteps of Iurii Nikolaevich Kurazhkovskii, who in the late 1950s had advanced the new disciplinary rubric of *prirodopol'zovanie,* or the science of proper land and resource use. Of these, perhaps the most radical was Oldak, who argued that the ultimate criterion of a socioeconomic order's success must be well-being and not the creation of material goods.[115]

One of the most interesting of DeBardeleben's conclusions was that the official "Marxian" categories of analysis and rhetoric precluded any truly critical analysis of the real socioeconomic system in which these social scientists lived. Because, by definition, the Soviet Union was a "socialist" society, observed negative phenomena, by definition, either did not really exist or else represented an aberration. Therefore, researchers could not "expose the structural or socioeconomic causes which lead enterprises and ministry officials to externalize environmental costs."[116] In other words, social scientists could not engage in Marxist analysis of the political economy of their own society.

Occasional insights and the (timid) questioning of the developmental strategies of the regime aside, these economists, philosophers, jurists, and other social scientists must be viewed chiefly as a component of the regime's propaganda apparatus. From Stockholm (1972) to Rio de Janeiro (1992), Soviet rhetoricians made the circuit from UNEP conferences to UNESCO symposia and then off to annual meetings of the International Union for the Conservation of Nature. Nor was the USSR reticent to host such events.[117]

Thus, even as global awareness helped to precipitate a doctrinal crisis for the scientific intelligentsia in its focus on biocenoses and *zapovedniki,* it generated a new scientific justification (island biogeography) for those institutions. It also spawned a new stratum of academics whose social role was to make environmentalism safe for the Soviet regime. For these new men of environmentalism who helped to constitute the public and international face of Soviet concern for the environment, nature protection-as-rhetoric led to picaresque new careers. The Soviet case demonstrates that the social meanings of "environmentalism" are highly variable social constructs, even in the same society. Similar-sounding discourses employing some of the same terms and ostensible referents can have entirely opposite political goals and effects.

CHAPTER EIGHTEEN

Environmental Struggles in the Era of Stagnation

The last half of the Brezhnev regime and the terms of Chernenko and Andropov, Brezhnev's infirm successors, are known by the sobriquet "era of stagnation," not only because the growth rate of the Soviet economy dwindled to zero but also because the system's rulers seem to have been bereft of any positive vision of their society's possibilities for the future. In the ideological vacuum left by a decomposing Marxism, Russian and other nationalisms contended for the loyalties of the educated strata and the general population. One of the few organized forces that continued to uphold a multinational, cosmopolitan, and internationalist vision was the student environmental movement, the *druzhiny*.

Brezhnev wanted to be known as, among other things, the "environmental general secretary." Beginning with the Eighth Five-Year Plan in 1971 he ordered investments in water purification and supply, scrubbers, and other air pollution–control measures.[1] The number and aggregate territory of *zapovedniki* observably grew, without any backsliding "liquidations" as under Brezhnev's two predecessors. International treaties were signed, and the USSR was well represented at Stockholm in 1972. Together with the now hopelessly bureaucratized VOOP, an army of agencies from the USSR Ministry of Agriculture's Glavpriroda to Iurii Izrael''s USSR State Committee on Hydrometeorology, not to mention the increasingly bureaucratized and tame Academy of Sciences, formed the phalanx of the Soviet green image machine. Sadly, naive foreign environmental activists regarded VOOP bureaucrats (see figure 28) as authentic comrades of the same cause, not realizing that their Soviet counterparts culturally had more in common with the Mafia than with themselves.

Newly available archival documents reveal the tight links between VOOP

Figure 28. Sixth Congress of VOOP (1976).

and Soviet foreign policy strategists. As Soviet representatives prepared to attend the Fifteenth General Assembly and technical symposia of the International Union for the Conservation of Nature (IUCN) in Christchurch, New Zealand, in October 1981 they were instructed by the vice president (now president) of VOOP, Ivan Fedotovich Barishpol: "The prime task of the delegation of the [All-Russian] Society for the Protection of Nature is securing adoption at the meetings of the Fifteenth General Assembly . . . of positions and interests of the Soviet side."[2] The delegates themselves were hardly representative of the rank and file: A. M. Borodin, the head of the USSR Ministry of Agriculture's Glavpriroda and vice president of IUCN; V. N. Vinogradov, an academician of VASKhNIL and VOOP's president; and F. Nyymsalu, first deputy minister of forestry of the Estonian SSR.[3] Vinogradov's detailed report upon a similar delegation's return from Gland, Switzerland (IUCN's headquarters), complained about the "increasing influence of the American group, which factually decides all operational questions in the union."[4] Similarly, in April 1981 Barishpol, together with longtime VOOP vice president G. G. Gan and the Presidium secretary met with a consul from the North Korean embassy at his request to coordinate strategy to deny South Korea membership in IUCN. Barishpol gave the North Korean plenipotentiary a generally encouraging answer and then promptly sent a report to the Soviet Ministry of Foreign Affairs Far Eastern Division, evidently seeking further instructions.[5] With twenty-nine million members in the Russian VOOP alone (the Belorussian analog had 3,404,300 in 1984, or 34.5 percent of the republic's total population), in the early 1980s VOOP was a significant player in the Soviet regime's game of image-making.[6]

However, behind the display window the struggle between two visions of development continued. Stalin's signature projects were the Belomorstroi, the Volga-Don and Volga-Moskva River canal systems and the Stalin Plan for the Great Transformation of Nature; Khrushchëv's was the Virgin Lands project. Brezhnev's, which took a decade to unveil, were the Baikal-Amur Mainline (BAM) Railroad and the great river diversion projects. Despite all the environmentalist rhetoric generated by the regime, orders apparently came down in 1975 placing huge categories of environmental data, not to mention criticism of the BAM, and later, the diversion projects, under strict censorship.[7]

At the same times, alternative environmental and social visions persisted in the USSR. Had foreign activists wished to meet their real counterparts, they would have found them not among the officious Soviet representatives to international agencies but in the *druzhiny* and the meetings of the Section on Nature Protection of the Moscow Society of Naturalists and the Moscow branch of the Geographical Society. They would have also needed to seek out the growing literary and scholarly opposition to the regime's megalithic hydroprojects, a discrete movement.

The Further Evolution of the *Druzhiny*

Out of the variety of autonomous groups and initiatives that emerged during the Khrushchëv era, only the *druzhiny* and, to a lesser extent, the KSP (*kluby samodeiatel'noi pesni,* or, roughly, clubs for independent music) managed to preserve their independence and their original character, activist Evgenii Arkad'evich Shvarts noted in 1990. Of course, the *druzhinniki* represented the most self-sacrificing students and those "who had the very highest level of political and general culture," he added. But that answer, although generally true, was inadequate. It seemed to Shvarts that the decisive factor in the survival of the *druzhiny* was their ability to exploit those few outposts of a "law-based society" that were created during the 1950s and early 1960s and that survived the vagaries of Khrushchëv's rule and those of his successors. Foremost among these were the creation of "citizens' inspectorates" in the area of environmental quality.[8]

Alone of all voluntary organizations, VOOP in the 1950s was given the right to organize "citizens' inspectorates," in which presumably duly trained members of the Society would be authorized to inspect for violations of air and water quality, to monitor or even detain poachers, and to write up official complaints to be processed in the courts or through state agencies. Although the officially sponsored "raids" conducted by VOOP were little more than environmental theater, older activists such as V. G. Geptner as well as almost all *druzhina* members kept their official VOOP membership in good order so that they could acquire or retain "citizen inspector" status. Correspondingly, there were always sympathetic local units of VOOP that were willing to credential the students, even if that was not in the interests of the corrupt bureaucrats who now ran the Society. If all the members of a *druzhina* antipoaching unit were credentialed "inspectors," they had the right to detain suspected poachers and bring them up on charges, but the *druzhina* itself did not have the right to credential inspectors. Thus, the *druzhiny* could exist only so long as VOOP—nature protection's false friend and institutional object of the *druzhinniki*'s contempt—existed. Another, more difficult way to become "deputized" as an inspector was through a government agency such as the Main Administration for Small Rivers or Glavokhota; for that reason, former members of student brigades strove to land jobs as inspectors in the various agencies that were the state's resource gatekeepers so that they could coordinate operations with the *druzhiny* and deputize students where necessary.

The *druzhiny* also confronted another problem. According to the law on voluntary societies of 1932 no unofficial organizations were permitted in the USSR; however much they proclaimed their factual independence, the student brigades needed official charters and sponsors. Those sponsors turned out to be the local universities' Komsomol organizations.

Attitudes toward the *druzhiny* on the part of the local and all-Union Komsomols defy simple characterization. Although there was some envy and suspicion of the nature brigades' freewheeling spirit, the Komsomol officials also wanted to see themselves as the kindly intercessors and protectors of these rebellious youths. Thus, in 1972 when Nina Aleksandrovna Gorodetskaia, the chair of the youth section of VOOP's Central Council, complained to the university's Party committee and tried to have the MGU *druzhina* disbanded, the Komsomol interceded and was decisive in saving it.[9]

On the other hand, toward the end of the 1960s the Komsomol leadership decided that it would not abandon the environmental field to the growing *druzhina* movement and in 1968 created its own Council for Nature Protection under its Central Committee.[10] Local Komsomol nature protection groups played constructive roles in a number of places; a notable example is that of the Perm' *oblast'* Komsomol, whose efforts to protect the Chusovaia River in the Western Urals were particularly visible.[11] In the early stages of the construction of BAM, local Komsomols, for example in Amur and Khabarovsk *krais*, were among the leading skeptics and critics of the huge project. Komsomol conferences in Irkutsk in 1975 and again in 1977 provided venues for voicing scientific reservations about BAM as well.[12]

Competition, plus the formal responsibility that the Central Committee of the Komsomol felt it bore for the conduct of the *druzhiny*, however, led to occasional tensions between the two organizations. In late 1976 Evgenii Grigor'evich Lysenko, head of the Komsomol's Council on Nature Protection as well as a deputy chair of the Presidium of VOOP, complained in a speech to VOOP that despite the student brigades' impressive organization of a summer nature-study practicum at Zvenigorod,[13] there was insufficient oversight over these groups by the Komsomol, leading to their taking autonomous decisions and even subverting local Komsomol units. For instance, the Kirov Agricultural Institute, together with the Kirov municipal Komsomol committee, decided to conduct an all-Union seminar for the *druzhiny* without having consulted the Secretariat of the Komsomol Central Committee.[14] Perhaps Lysenko's attention had been drawn to this as a result of the accusations leveled at the *druzhina* movement by Tomsk author N. Laptev, who had charged the brigades with "having broken with the Komsomol line" and with "apoliticism."[15]

Although the Komsomol central bureaucrats' heavy-handedness showed in the literature on nature protection and environmental quality it published,[16] the Komsomol Central Committee did sponsor Viktor Iaroshenko's "Living Water" expedition (described below) and awarded Iaroshenko and Chivilikhin its highest prize for their literary efforts on behalf of nature. Thus, like the Russian Republic, the Komsomol functioned both as patron and co-opter of nature protection activists, reflecting its own status as a core part of the system that nonetheless had pretensions to its own sphere of autonomy.

Highlights of the History of the *Druzhiny*, 1965–1986

One year to the day from that great *druzhina* meeting where Professor Geptner held up the *apparatchik* M. M. Bochkarëv to public ridicule, the Biofak brigade held another conference, this time dedicated to Kedrograd. Over three hundred attended.[17] As Bochkarëv had lost his VOOP presidency but had not yet been removed as chair of the RSFSR Main Forestry Administration, the conference represented not only support for the *kedrogradtsy* but also a continued public struggle against the bureaucrat. Along with a resolution calling for a restoration of the original Kedrograd experiment, which Bochkarëv had cut off, the brigades insisted on the restoration of the Altaiskii *zapovednik* to its previous boundaries—a cause closer to their hearts—and vigorously protested the use of DDT and other chemicals in the cedar forests, particularly around Lake Teletskoe.[18]

The late sixties were golden years for the movement. There were sometimes two conferences in one year, as in 1967, when conferences on both tourism and Baikal were held. One participant recalled:

> Whoever landed in our command in those days was amazed by the unity of this modest collective and by the strength of the friendships that connected

> each member with every other. What also struck newcomers was the attitude toward work as something personally important, done with a feeling of satisfaction. The feelings of pride in belonging to the *druzhina* astonished them as well. . . . We only had a few inspectors but they came almost every Sunday. The excursion was a holiday, it was that blast of a martial trumpet that awakened us from our everyday cares and called us to battle. And many excursions were held in this festive, tense, intrepid mood.[19]

Tartu had boasted the very first such university organization, and Moscow had followed two years later. Brigades were established at Odessa University, the Ul'ianovsk Pedagogical Institute, and the Briansk Technological Institute in 1964–1965, and by 1972 there were thirty-four. With this exceptional growth the MGU Biofak brigade, arguably the social and intellectual center, began to concern itself with "foreign affairs." In late March 1968 a delegation led by former *komandir* Dmitrii Nikolaevich Kavtaradze traveled to Tartu for that group's tenth anniversary. Other bilateral contacts followed, and in September 1972 the various *druzhiny* finally came together in an official first "seminar," held in Moscow. This laid the basis for expanded collective action. By the spring of 1975 the number of brigades had expanded to around forty, with aggregate membership in excess of 2,500, and by the mid-1980s total membership of the 140 or so brigades was estimated at around 5,000.[20]

Soon the brigades began to explore coordinating joint campaigns. The first to be adopted in common was "Vystrel" or Operation Shot, the antipoaching campaign, popularly called "BsB" or *Bor'ba s brakonerstvom* (Struggle with Poaching), adopted at the 1974 Kazan meeting. Under the leadership of Dmitrii Kavtaradze the program now took on a sociological cast, attempting to identify patterns among perpetrators.

Romanticism abounded in the program, and members learned self-defense and practiced lifelike scenarios in their training: indeed, no fewer than six were killed participating in the antipoaching expeditions. On the whole, however, physical showdowns were discouraged; the philosophy was to educate and publicize. If a violator was caught, the *druzhina* sent letters to his workplace. "This was like tossing a pebble into the water; the pebble itself is not terribly big, but the waves it generates can travel a great distance," explained faculty advisor Tikhomirov. The Kazan conference was also noteworthy for stating in a resolution that "above all else it is essential to view the contemporary youth movement for nature protection as a school for citizenship."[21]

Other programs focused on identifying nesting sites of rare and endangered species and other territories in need of protection (Operation Fauna), involving the organization of expeditions staffed by graduate students representing a variety of relevant disciplines. There were programs to design national parks, identify areas for *zakazniki* and *zapovedniki*, and study the environmentally disruptive potential of mass tourism. The movement kept archives as well.

Whereas the student movement through the 1960s was preoccupied with the old agenda of protecting animate nature, by the 1970s it assumed a more assertive posture and began to tackle "sociopolitical problems," in the words of a prominent student leader. It sought the synergistic interaction of "cultural-historical, national, . . . ecological, . . . and socioeconomic issues."[22] One example of this new social planning role was the *druzhina* program "Ecopolis," developed by Dr. Aron Brudnyi, a philosopher with the Kyrgyz Academy of Sciences, and by Kavtaradze, by then head of MGU's Laboratory of Ecology and Conservation (Biological Faculties). Using the town of Pushchino, the Academy of Sciences' Biological Research Center on the Oka River about seventy miles south of Moscow, as an experimental "subject," Brudnyi and Kavtaradze sought to design the new town's services, amenities, and physical features to achieve maximum environmental and aesthetic quality for its residents. Groves were left standing, ecologically sensitive paths wound through forests connecting the town with the accessible, undeveloped riverfront, and conservation educational materials were abundant. By surveying residents for their views the designers attempted to ensure that "Ecopolis" did not simply become two men's vision of an ecological utopia.[23]

Also sponsored by the *druzhiny* was "Operation Cruelty," which began in 1969. A focus on the roots of cruel or sadistic behavior toward animals (again utilizing polling) led both to a broader discussion about "the phenomenon of cruelty" in general and to a widening of the debate beyond the walls of the brigade itself. Perhaps the most significant breakthrough was the convening of a roundtable discussion sponsored by the Conservation Section of MOIP on April 19, 1974, which was subsequently published in the Academy of Science's widely read monthly, *Priroda*. Professor Ksenia Semënova, a child psychiatrist specializing in the rehabilitation of children with cerebral palsy (though at the notorious Serbskii Psychiatric Institute), linked cruelty to animals and to fellow humans as a common failure on the part of many people to develop empathic responses to the pain of others. This, in turn, Semënova attributed to the cruel individual's inability to find any constructive avenues in society for self-affirmation. As a rule, she noted, low academic achievers were overrepresented in this group, which pointed to an implicit socioeconomic pattern; in the Soviet Union, as elsewhere, low academic achievement is highly correlated with poverty and low status.[24]

Urban professionals were not exempt from this behavior, however. Semënova related the case of three young female medical students who were photographed laughing as a just-dissected dog, entrails extruding, regained consciousness from anesthesia. Writing almost at the very moment that Peter

Singer published his controversial *Animal Liberation,* Semënova's piece was perhaps the first in the USSR to call attention to the ethical problems of vivisection and scientific experimentation upon animals.[25] In other words, this outgrowth of the *druzhina* program began to explore the forbidden territory of common behaviors and cultural patterns in Soviet society that were not questioned elsewhere.

Sometimes, programs had indifferent success or even failed. In the early 1980s about thirty young biologists—graduates of Moscow, Gomel, Kazan, and Sverdlovsk schools—began work in five Turkmenian *zapovedniki.* As students they had participated in the summer *druzhina* program "*Zapovedniki*" and tried to continue their *druzhina* program at new places of work. Turkmenia was selected because of its extremely low population density and the biologists' belief that they could still save huge unbroken tracts and the plant and animal species that survived within them. Critics charged that they wanted to turn Turkmenia into a gigantic national park and fence off resources. The mutual incomprehension and lack of acceptance has persisted to the present.[26]

But sometimes the *druzhina* programs had an effect. At the big 1966 conference on Kedrograd the student brigades insisted on the restoration of the Altai *zapovednik.* Was it merely coincidence that soon afterward the Altai *zapovednik* was the first of the reserves axed by Khrushchëv to be reestablished? *Druzhinniki* also took credit for a decree of the USSR State Committee on Forestry in 1967 enhancing protection of the "cedar" stone pine forests.[27]

Ksenia Avilova has penned a personal memoir of her years in and around the *druzhina* movement and in it has compiled some interesting facts and figures. Over twenty-five years, the Moscow University *druzhina* conducted more than 1,300 antipoaching excursions, which resulted in the apprehension of about 4,500 violators; twenty-five expeditions to patrol for illegally cut and sold New Year's trees, which apprehended more than 3,000 violators; thirty conferences, workshops, and seminars; 1,500 lectures on nature protection; seven expeditions to study the effects of mass tourism on natural conditions; more than twenty expeditions to help create national parks; about fifteen expeditions to study the influences of industrial cities and agriculture on pollution; more than ten expeditions to identify habitat sites of rare plants in the Moscow area and to create *zakazniki* for them; more than twenty-five expeditions to identify habitat of rare fauna; six expeditions, in connection with the brigade-sponsored contest "Berkut" (Eagle), to identify nesting sites of rare birds in the Moscow region; and six expeditions to study the influence of poaching on the aquatic life of the Oka River as well as to study the poaching of animals from a sociological perspective. Moreover, the brigades over that time produced plans for more than fifty *zakazniki,* thirty of which were organized by the Moscow *oblast'*; produced plans for ten national parks; and published more than 500 scholarly and popular scientific pieces on nature protection.[28] As the movement celebrated its twenty-fifth anniversary it was at the height of its social prestige. For decades, almost alone it had flown the colors of civic independence; to a great extent, whatever authentic political activism was available to youth during the long years of stagnation was through this movement.

At the stroke of noon on February 3, 1986 the auditorium "P-13" in the Second Humanities Building of Moscow State University was packed to overflowing. With more than 500 in attendance, including more than 100 standees, fire laws were doubtless broken that day. Looking back at twenty-five years of the *druzhina* movement, Professor Vadim Tikhomirov re-emphasized the crowning social achievement of the movement, that "most importantly, the *druzhiny* based their activities on the principle of self-government. They did not subordinate themselves to anyone. They pursued projects prompted by no one other than themselves." In other words, they successfully pursued and defended the subversive right to set their own agenda.

On the question of numbers, Tikhomirov came down on the side of quality over quantity. The movement did not want to become an impersonal bureaucratic outfit like the thirty-six-million-member VOOP, but should remain small and selective. This elitist "caste" mentality also expressed itself in Tikhomirov's understanding of the proper complex of problems with which the *druzhiny* should occupy themselves. As biologists, they should concern themselves for the most part with the protection of living nature, a legitimate concern, and not with "environmental quality," which was already presumably on the slippery slope of reconciliation with industrial civilization.

The All-Russian Convocation of Student *Druzhiny* concluded its last formal session on the morning of February 7, 1986. After lunch sixteen *druzhina* leaders set out for the center of town, where they had an appointment with leading officials of VOOP at the Society's Moscow municipal offices in an old mansion almost opposite the United States embassy on Chaikovskii Street. The *druzhinniki* already had copies of a proposed VOOP charter that would centralize their various organizations into a semi-autonomous, unified all-Russian movement affiliated with VOOP.[29]

In an unambiguous display of power, the five VOOP bigwigs, including vice president German Georgievich Gan, another middle-aged man who was an official in the Ministry of River Transport, an older man who also worked at the Main Administration for Small Rivers and Reservoirs, and two middle-aged women seated themselves at the front of a small auditorium on the raised dais. Seventeen much younger men and women ranged themselves in two rows in the pit.

Although the students were prepared to agree to the terms offered by

VOOP for affiliation in the charter, the meeting quickly went off track. German Georgievich Gan had opened the meeting with a promise that VOOP would indeed respect the students' autonomy. However, it soon became clear that the meaning of "autonomy" was radically different on either side of the dais. Gan had already prepared a list of names for members of the Scientific Council of the proposed united *druzhina* affiliate. It was heavily weighted toward Komsomol officials and ministerial types. Almost unanimously the students rejected the list as an attempt to co-opt their organization and to have bureaucrats dictate policy.

Gan responded that the list was formulated "to facilitate coordination" between the students and government agencies and that, in any case, voluntary organizations existed to assist the state in fulfilling the economic plan (and, implicitly, not to play at being a critical opposition). The students countered with the reasonable argument that coordination could just as well be facilitated by establishing a coordinating committee without including agency types into the policymaking body. Further, they feared a loss of control over their independent "inspectorates" as well as a trivialization of themes.

This fear was confirmed by an offhand remark by the department director of the Main Administration for Small Rivers. Students, he asserted, did not possess sufficient scientific knowledge to involve themselves competently in policy areas, especially in scientific investigations into water quality, land use, and other realms. The *druzhina* representatives displayed a poise and courage remarkable for their years. "And who do you think was responsible for drafting so many existing environmental strategies?" retorted one student, evidently reminding the official of the *druzhiny*'s twenty-five-year track record over a wide spectrum of environmental policies. For three and a half hours the two sides traded barbs. The meeting ended inconclusively, with the students refusing to sign the charter but agreeing to continue negotiations with Gan.

Skirmishing between the *druzhiny* and VOOP continued into the fall when three MGU *druzhina* members, V. Mokievskii, I. Chestin, and the most recent *komandir*, Evgenii Arkad'evich Shvarts, published a devastating exposé of VOOP in *Komsomol'skaia pravda* entitled "You Won't Fool Nature." The subtitle ("Is VOOP Really Interested in Protecting the Environment? It's Hard to Answer 'Yes' to That") cued the reader to the gist of the story. The authors noted sarcastically the odd fact that even as late as just after the Sixth Congress of the Union of Writers of the RSFSR (December 1985), when the entire intellectual community was discussing the river diversions, Baikal, and other environmental problems, "the nature protection society alone, until the last possible moment, had behaved as if the Pechora and the Northern Dvina flowed somewhere in Australia and that Lake Valdai and the hoary Ladoga were located somewhere in the vicinity of, say, Montevideo. . . . As in the past, the work of the approximately 5,000 (!) salaried staff workers in our view boils down to . . . fulfilling the financial plan for dues collection through the mass 'tithing' of workers, students, and pensioners."[30]

To try to penetrate the mountain of self-promotion by VOOP, the *druzhinniki* in the spring of 1985 engaged in a little detective work of their own. In particular, the students sought to learn how effective VOOP's vaunted "raids" on polluters were. In 1985, they discovered, VOOP's several thousand citizen inspectors managed to file just three complaints.

This stagnation was caused by the fact that there were no checks on the leadership, particularly its highest-ranking paid staff, and that membership was fictive in the sense that anyone who paid the thirty-kopeck dues was considered a member. Moreover, many of the members of VOOP's executive organs were high officials in the very ministries and agencies over which the Society was supposed to exercise a kind of citizens' oversight. "Would we allow the winemaker ex officio simultaneously to serve as president of the Temperance Society?" the *druzhina* authors asked.[31]

At the end of the year, in the stately Hall of Columns of the old House of the Nobility, where Bukharin was put on trial in 1938, VOOP held its scheduled national Congress, which now met only every five years. Courageously, *druzhina* member Tat'iana Olegovna Ianitskaia attempted to share some of the students' critical perceptions with the assembled delegates. At first, Ianitskaia summoned the utmost tact to counter the proposal of the Society's president, B. N. Vinogradov, to eliminate the citizens' inspectorates of VOOP and even merge them with the ministerial ones. These were the only officially sanctioned citizen watchdog units in the country, she argued, and they could be made to work more effectively. As an example she pointed to the raids and inspections conducted by the *druzhiny*—all members of which were formally credentialed as "VOOP citizen inspectors." The students were willing to help train other VOOP volunteer inspectors professionally as well. On the issue of protected territories, which had centrally preoccupied the Society for thirty years until its reorganization in 1955, Ianitskaia also asked, "Who else, if not we?"[32]

Next, Ianitskaia tackled the Society's cult of numbers. She was gladdened to hear, for the first time, that the Society decided against any planned increase in membership. But she was disturbed to learn that it had decided to increase the number of student *druzhiny* from their current seventy-odd in the RSFSR to 390 by 1990; this was an approach that even the Komsomol had forsworn. It was not that Ianitskaia did not want to see the *druzhiny* expand, but that she objected to the treatment of free institutions as construction materials or as subjects for labor productivity measurements: "We don't know how many people are transformed into active fighters for nature protection by a series of lectures, seminars, and meetings that they have attended. Those data are not now available to science. And the student *druzhiny* are active fighting units." In a particularly needling aside she also noted the paucity of young people at the Congress—only 7 percent of the delegates were under thirty although 67 percent of the membership were "youths"—and asked: "Can there really be such low activism among youths?"

Finally, she brought up the sorest point: "I believe, comrades, that you are all aware that the Society has become the object of criticism more than once in the press. A particularly serious example was the recent piece in *Komsomolka* [*Komsomol'skaia pravda*]. Serious accusations have been leveled at all of us. It is possible, of course, to try to dispute them, but let us instead find the courage to acknowledge that as a whole the criticism is justified. We have a host of opportunities to rectify the situation, and, in that context, . . . I propose that we evaluate the work of the Society and its Central Council over the past [five-year] period as unsatisfactory." Not surprisingly, Ianitskaia's call went down in defeat. As of September 1996 I. F. Barishpol was still president; the Society has never undergone *perestroika*.

If anything, the *druzhiny*—like their field naturalist forbears—show the utility of viewing even the most activist, democratic groups in Soviet society as continuing expressions of a quite old Russian understanding of social identity: *soslovnost'*, or a corporativist-caste mentality. In a suggestive article I. I. Zhukova described the social milieu of the *druzhiny* as a kind of "our crowd" in nature.[33] You only had to spot the distinctive *druzhina* emblem on a stranger's sweater "to feel confident in going up to that person, knowing that you had come across a like-minded individual, a comrade, someone with whom you could find a common language."[34]

In fact, *druzhina* membership was a lifelong social identity because even though members graduated from the university they continued their ties with the movement as *kuratory*, consultants, mentors, and supporters. The "extremely high degree of support and mutual assistance," according to sociologist Oleg Nikolaevich Ianitskii, was crucial in getting these organizations through the difficult years of the "era of stagnation."[35] It was an esprit de corps modeled after the students' field biologist mentors in the already established nature protection movement, but it especially resembled the old prerevolutionary *studenchestvo*. That six *druzhinniki* were murdered in the course of antipoaching campaigns forged bonds cemented by spilled blood, created a roster of martyrs, and conferred on these young men and women a feeling of significance that was hard to come by in the Soviet Union of those years.

That was not the only movement attribute borrowed from the older generation. The emphasis on the protection of living nature, habitats, rare species, and especially *zapovedniki* grew out of that legacy as well as the fact that the first *druzhina* was established in the Moscow University Biological and Soil Science Faculty (although this agenda would later broaden as the number of *druzhiny* expanded).[36] No one has confirmed the intergenerational link between the old nature protection activists and the brigadiers as well as Sergei Germanovich Mukhachëv, who explained to Ianitskii that although the newly formed *druzhina* at the Kazan Chemical and Technological Institute was formally part of the local branch of VOOP, his group used the formal bureaucratic structures to create an autonomous student organization. "In this way," he said, "we brought about the rebirth of the VOOP that we had in this country during the 1920s."[37]

Another attribute of the movement was its elite quality. The *druzhinniki* were truly the best and the brightest, as Evgenii Shvarts has observed.[38] The brigades' core of support was in the country's major universities, particularly those having biology or geography faculties. The largest, most cosmopolitan universities might have several *druzhiny* in different faculties. On the other hand, of the forty-nine economic higher educational institutions of the former USSR, only three had *druzhiny*. Engineering schools, polytechnical schools, veterinary institutes, and agricultural institutes had similarly low representation.[39] The cleavage here is between a corporate (erudite and partially hereditary) natural science–based intelligentsia and the parvenus in technical schools.

This elite sensibility expressed itself in a number of additional policies, especially the campaigns against poaching and illegal sales of flowers, birds, and Christmas trees, and the student brigades' attempt to limit "ruinous" tourism to natural amenities.[40] The targets of these citizens' "police actions" were mostly lower-class individuals; the white-collar or organized-crime figures who masterminded some of the larger operations and who fenced the ill-gotten state goods were never fingered. *Druzhina* raids often turned into morality plays in which the deficient ethics and *poshlost'* (vulgarity) of the hoi polloi was exposed.[41] The students were neither populists nor thoroughgoing democrats, nor courageous enough to take on the really powerful offenders at the heart of the economic machine. But then, how many in Soviet society were?

The Plan to Reroute the Northern Rivers

One struggle that did challenge a core element of the political economy of the neo-Stalinist state was that against the megalithic river diversions proposed in the 1970s. Perhaps the most far-reaching regime plans for transforming nature during the Brezhnev years were two projects to divert northward-flowing Russian rivers to the south. Robert G. Darst, Jr. discerned that the opposition to the river diversion projects, which were prematurely touted by its sponsors as, collectively, "the project of the century," embraced disparate social visions. Two geographically distinct projects were at issue, each generated by the excess of water use over supply in its respective region. The first project sought to ease the shortage of water for irrigation in the southern portion of the European RSFSR and to stem the perceived fall in the level of the Caspian Sea. To do this, it was proposed to take water from northward-flowing rivers such as the Sukhona, Pechora, and Northern Dvina, as well as from Lakes Lacha, Vozhe, Kubena, and Onega, and pump them southward into the Volga basin.

Even more ambitious was the plan to reverse the flow of the Irtysh River and divert part of the Ob' to the gigantic Sibaral canal, which would stretch for 2,200 kilometers across the Turgai watershed. Here, cotton irrigation had led to withdrawal of almost all the water from the Amu-Dar'ia and Syr-Dar'ia Rivers before they could empty into the Aral Sea. Consequently, the Aral had lost almost half of its volume between 1950 and 1980. Salinity rose to unprecedented levels, and the exposed sands of the desiccated lake bed became a lethal source of pesticides and defoliants, which had precipitated to the lake bottom decades earlier. Winds now blew the toxic sands over fields of vegetables and pastures, spreading desertification, birth defects, and cancer.[42] The water pumped southward was intended to stabilize, if not replenish, the Aral Sea, which had once been one of the most important inland fisheries of the Soviet Union, as well as to meet the growing need for water among Central Asia's fast-growing population.

Exactly when this gargantuan project was first imagined is still clouded in a historical fog. Driven by the fear of a drying-up of the Caspian Sea and after considering a number of proposals, an Academy of Sciences special conference in November 1933 approved a wide-ranging plan for the "reconstruction of the Volga and its basin" in which the following point was included: "To obtain additional flow into the Volga from neighboring river basins by means of a diversion to the Volga of a portion of the waters from the Pechora and Northern Dvina River basins and from Lakes Lacha, Vozhe, and Kubenskoe."[43] Operational design work was assigned to Gidroproekt, S. Ia. Zhuk's hydraulic empire and accessory of the GULAG state-within-a-state, and by the 1950s serious plans were being developed. Khrushchëv in his notorious January 1961 speech to the Central Committee "dug up a six-year-old memo by engineers . . . Zhuk and G. Russo on the feasibility of uniting the Caspian and the Aral to the Arctic Ocean through a series of canals and proposed the idea" to the plenum.[44] As Darst recalls, by the late 1960s the project was already in the hands of design bureaus: "The research and design effort behind the undertaking was huge: over 120 agencies participated in the impact assessment study coordinated by the Institute of Water Problems of the USSR Academy of Sciences, and over a dozen major conferences were held on the subject."[45] Besides making ridiculously low cost estimates (twelve billion rubles), the project's promoters promised that "harvests from newly irrigated lands in Central Asia and Kazakhstan alone would feed an additional 200 million people."[46]

Backers of the projects represented a wide array of bureaucratic interests. Although convict labor had largely been replaced on water projects by free labor during the Khrushchëv years, Zhuk's agency was formally renamed only in October 1965, when it was reincarnated as Minvodkhoz—the USSR Ministry of Land Reclamation and Water Resources. Although eclipsed by the ascending nuclear power industry, Minvodkhoz, with its nearly 70,000 employees, was still a heavy hitter. Its leaders, while never surrendering the monumentalist engineering visions of their past, were able to adapt to new times, learn new rhetorics, and forge new alliances. We have seen how Minvodkhoz posed as a friend of Lake Baikal against the polluting pulp and paper interests. Even more impressively, by the late 1960s under its head E. E. Alekseevskii, the ministry positioned itself as the preeminent champion of a thoroughgoing cleanup of European Russia's waterways and under Brezhnev received significant funding to improve water quality.[47] Between 1966 and 1984 the agency's budget was more than 115 billion rubles. A work force of two million labored on its projects, building more than 700 million kilometers of canals and ditches and draining as well as irrigating vast expanses of land.[48]

Allied with Minvodkhoz were the local Party and government apparatuses of the Central Asian republics, especially Uzbekistan and Turkmenia, which stood to gain most by the diversions. Doubtless tacit support was also provided by the ministries concerned with atomic energy and heavy equipment, both of which would be essential for excavating the enormous canal.[49]

As long as the project seemed to remain in the realm of fantasy, criticism was rare. There were more pressing issues to be concerned about, such as Kedrograd, Baikal, and the fate of the *zapovedniki.* Sergei Pavlovich Zalygin was one of the few to see the potential for huge damage in a system that placed such a premium on monumentality. In a 1961 essay on the literary treatment of Siberia, Zalygin appealed for an end to the ideology of the conquest of nature. "For some reason," he wrote, "our active literary hero, if he comes in contact with nature, does so exclusively according to the principle 'I came, I saw, I conquered.' Otherwise, such a hero is 'not active.'" Meanwhile, there were limits to our ability to change the physical world. "In Siberia the problems of transforming nature are grandiose," he noted, "but the errors made may be equally grandiose. And when people start talking about reversing the waters of the Enisei and the Ob' and diverting them to the Aral Sea, we still have too inadequate an understanding of the consequences of such a transformation. We understand them too poorly, and yet how many dithyrambs have already been sung by our literary brother about that project! And all in the service of grandiosity."[50]

In 1963 one Leningrad professor of the Academy's Arctic and Antarctic Research Institute "pointed to the unforeseeable effects on arctic climate of reducing the flow of fresh water to the Arctic ice pack."[51] In the same year

Zalygin, in one of the more memorable victories of the nature protection movement, organized the defeat of a hydropower dam on the lower reaches of the Ob' River with a memorable series of broadsides he published in *Literaturnaia gazeta*.[52] For the most part, though, public opinion was dormant until 1978, when a propaganda campaign began to promote the project and word leaked out that a thirty-five-meter dam had been erected in Kargopol' and housing built for construction workers, despite an absence of official approval for the project.[53]

Parallel with the development of Minvodkhoz's plans, a group of young scientists, scholars, and journalists had come together as early as 1974 with an interest in examining independently the water resource problems of the country and Minvodkhoz's proposed solutions. Sponsored by the Oleg Poptsov's journal *Sel'skaia molodëzh'* (Rural Youth) and formally credentialed and financed as "The Permanent Ecological Expedition of the Central Committee of the Komsomol *Zhivaia voda* [Living Water]" the group was led by journalist Viktor Afanas'evich Iaroshenko, who set up an energetic program that eventually encompassed thirteen major field expeditions, much in the spirit of the *druzhiny po okhrane prirody.* These included field trips to the Volga, the Caspian, the Russian North, the Pechora, the Sukhona, Lakes Kubena and Onega, the Dnepr, the Pripiat', the Danube, the Amu-Dar'ia, the Aral Sea, the Ob', the Irtysh, Kamchatka, and the Russian Far East.[54] From 1975 the expedition began systematically to investigate the proposed dam sites, canal routes, and other facilities of the river diversion project, and, importantly, to conduct interviews with local residents, scientists, and planners (who, for objectivity's sake, were included in the expedition).[55]

The river diversion scheme was also playing out against another backdrop, the announcement in March 1974 of a massive new campaign to reinvigorate the decaying northern non–Black Earth regions of European Russia. Even within bureaucratic circles this was viewed as throwing good money after bad; an "unseen, quiet bureaucratic war of ranking and positioning [*vedomstvenno-mestnicheskaia voina*] between the North and the South" ensued. The South, with its greater population, soil fertility, and passable roads held a heavy advantage, particularly since northern fields became even less productive after drainage.[56] Soon the non–Black Earth program died on the vine. Hoping to influence the course of events through the Party's high command, the "Living Water" group after its 1976 expedition sent a memorandum to Central Committee secretary for agriculture F. Kulakov warning the Party of the possible negative consequences of river diversions in the North. At the center of their concern was the fate of the unique Kirillo-Belozerskii Monastery, whose walls, murals, and frescoes would be imperiled by a dam and hydropower installation on Siverskii Lake. Kulakov was unimpressed and sent the memo to the USSR State Committee for Science and Technology. "Do you want to mislead the leadership of the country?" the leaders of the expedition were asked. "How can you not understand that the productivity of southern lands is incomparably higher than in the North?!"[57]

The new direction was unmistakable. After the Twenty-fifth Congress of the CPSU in late February 1976, both river diversion projects were written into the "Main Directions for Economic Development for 1976–1980." This was followed by a decree of December 21, 1978 (No. 1048), signed by Brezhnev and Kosygin, which called for the theoretical and economic justifications for the Volga basin diversions to be completed by 1979 and those for Central Asia and Siberia by 1980. Minvodkhoz and its institutes would prepare the reports, while the Academy of Sciences' Institute of Water Problems under Grigorii Vasil'evich Voropaev would provide "scientific justification."[58]

In a newly built high-rise the "All-Union E. Alekseevskii Red Banner of Labor Main Planning, Expeditionary, and Scientific Research Institute on the Problem of the Diversion and Distribution of Northern Waters and of Siberian Rivers" opened in Moscow in 1978, a valedictory accomplishment by Alekseevskii as he retired. Here, battalions of engineers and scribes were at work generating the "scientific" justification for these immense public works. "A full one hundred forty volumes of Technical-Economic Justifications [TEO or *tekhicheski-economicheskiie obosnovaniia*] supported one conclusion: the country would perish if the project were not undertaken," wrote activist Vera Grigor'evna Briusova. Accordingly, censorship was imposed on public statements critical of the diversions. Articles for *Pravda* and even the sympathetic *Nash sovremennik* were suppressed because of objections and real censorship on the part of the USSR State Committee on Hydrometeorology, which was supposedly the lead Soviet agency for environmental monitoring. The committee's head, Iurii A. Izrael', prohibited publication on this theme. For this, plus services at Chernobyl and with regard to legitimizing atomic testing in the Soviet Arctic, Izrael' won a seat on the prestigious *Revkom* (Auditing Committee) of the Party's Central Committee.[59]

Despite this pall of censorship, the "Living Water" expedition managed to smuggle into print in mid-1979 in *Chelovek i priroda* a report on the damage the Volga project would inflict on the Pechora basin, which the expedition visited in 1977, both to nature and to the economy.[60] From Iaroshenko's report of the group's attempt to interview the chief engineer of the Pechora project in the new Gidroproekt office building we may see an important change in the group's sense of its own mission. Defending the right of "public opinion" to be a part of the decision-making process was now as much the issue as evaluating the merits of the project itself.[61]

In 1978 the expedition studied the "Sibaral" variant, traveling to the Amu-Dar'ia from its sources in the high Pamirs to the Aral Sea, and again publishing its findings in *Chelovek i priroda* 7 (1980). Again, the group argued against the diversions, concluding that Central Asian irrigation was already overwatering and thus salinizing the soil. The following year, 1979, the expedition proceeded from the source of the Ob' in Lake Teletskoe in the Altai range down to the great river's delta, again publishing in *Chelovek i priroda,* perhaps the only periodical that escaped the regime of censorship, but one with a national circulation of 100,000.

At this point, the theoretical and economic justifications had to be approved by expert commissions in both Gosplan of the RSFSR and of the USSR. Voropaev was chair of the Gosplan USSR commission. The Twenty-sixth Congress of the CPSU had directed preparatory construction work on the project to commence during the next Five-Year Plan (1981–1985). Everything looked like a sure deal. And then came the bureaucrat's nightmare: a revolt of the experts. The Gosplan USSR expert commission's subgroup on economic planning dismissed the proposal almost out of hand. Courageously, the eight men and women asserted that, owing to its narrow agency-oriented profile of interests, Minvodkhoz was incapable of developing alternative scenarios for improving agriculture in the targeted regions. In the words of Iaroshenko, "This was a reversal—not of rivers but of the situation regarding the public's participation."[62]

Emboldened, a dozen scientists led by geologist and MOIP vice president A. L. Ianshin sent a letter to the country's leaders, "On the Catastrophic Consequences of the Reversal of Part of the Flow of Northern Rivers and the Complex of Measures for Attaining the Food Program of the USSR," in which they demanded the creation of an independent special commission to review the projects.[63] And when a letter by noted "Village Prose" novelist Vasilii Ivanovich Belov, "Will Lakes Vozhe and Lacha Save the Caspian?," calling for a scientific debate on the river diversions, was rejected for publication owing to censorship, he sent it to Paris, where it was published in the *tamizdat* (unapproved foreign publication by Soviet authors).[64] Soon it became well known within the Soviet Union. From Vologda, Belov came to Moscow to pursue the matter, approaching the nationalist historian of the pre-Kievan and Kievan periods, academician B. A. Rybakov, who promised to raise the issue at his well-known Wednesday discussion group on the Volkhonka. When the guests arrived, however, there was a notice that the Wednesday discussion had been canceled, so the group went to Russian art historian V. G. Briusova's apartment nearby. Her home became a clearinghouse for the campaign for the next five years.[65]

"We needed to have the opportunity to discuss these problems among a broad public," wrote Briusova. In December 1981 the antidiversion forces rallied their troops, holding their first real public meeting in the main auditorium of the Union of Artists of the USSR. Chairing the meeting was the president of the Union's Commission for the Preservation of Monuments of Culture and History (VOOPIiK), Dementii Alekseevich Shmarinov, who acted "not because of the office he held but out of a deep, burning conviction, as a true Russian patriot of Russian culture and the Russian land."

Later meetings took place at the House of Artists at 11 Kuznetskii most, in the Central House of Artists on Krymskaia Square, and in a variety of "houses of culture" throughout the city.[66]

At the first expanded session of the preservation of monuments commission, Fotei Iakovlevich Shipunov (of Kedrograd fame), now an avowed Russian nationalist and no longer going by the name Sergei, was among the speakers, who also included Professor S. N. Chernyshev, a geologist, and the economist S. G. Zhukov. Briusova spoke about the threat to the wooden architecture of the north. The meeting, which included writers, artists, and architects, adopted a resolution that the commission prepare a substantive independent assessment of the river diversion project. VOOPIiK obtained for the commission a copy of the official technical-economic justifications. After five months of work the report was readied, and all endangered sites were identified. Like the Belov letter, it was published in *Russkaia mysl'* (Paris) and then republished in Moscow. Altogether the report, which was sent to the Politburo, identified more than 490 threatened historical and architectural sites. In its 120 pages ecological and economic arguments, although not as close to the hearts of the compilers as the cultural ones, were also included to reach the utilitarian minds of the nation's leadership.[67]

This "revolt of the experts" in turn provoked an echo from broader and broader segments of the population. Locals from the threatened city of Kirillov had written to the Central Committee on July 28, 1982 that "Party and civic conscience does not permit us to remain indifferent to the possible destruction of the most magnificent cultural values which, to a significant degree, help to engender a feeling for one's own Motherland."[68] Iurii Efremov and A. A. Kuznetsov, who headed the nature protection section of the Moscow writer's organization, gathered eighty signatures—forty from Party members—on a letter to then RSFSR premier Mikhail S. Solomentsev. When the Moscow writers' leadership protested that eighty was too many, Efremov facetiously suggested perhaps eight to ten, to which the thickheaded leadership, which perhaps had failed to grasp Efremov's sardonic humor, assented. Thus was a letter signed by Zalygin, Belov, Rasputin, Soloukhin, Volkov, Krupin, and others sent to the Russian Republic leadership.[69] The growing piles of letters to all layers of authority, when added to the report of the academicians, created an air of disquiet in the Central Committee. Something had gotten out of hand.

To contain the situation, the Central Committee Agricultural Department convened a special conference. Although the face-off between the academic authors and signatories of the report, on the one hand, and the project planners and their political patrons, on the other, failed to win converts on either side, "the taboo had been broken," in the words of Briusova.

In retrospect, the reluctance of the Central Committee to forcibly silence the critics of the project opened the floodgates for a mass mobilization of the patriotic-minded intelligentsia. While field biologists were conspicuous for their absence in this defense of Russian culture and homeland, in this Party of "memory" were reunited Fotei Shipunov and Vladimir Chivilikhin, joined by the other leading "Village Prose" and patriotic writers: Valentin G. Rasputin, Vasilii I. Belov, Viktor P. Astaf'ev, P. L. Proskurin, and Oleg V. Volkov. Repentant hydrologists included Evgenii Makar'evich Podol'skii and Sergei Pavlovich Zalygin. There were geologists (N. A. Lebedeva, E. M. Pashkin), artists (N. I. Rozov, A. P. Gorskii), economists (Mikhail Iakovlevich Lemeshev, Natalia Petrovna Iurina, L. F. Zelinkina), and the geographer and historical preservationist Iurii Konstantinovich Efremov.[70] The great historian of medieval Russian culture Dmitrii Likhachëv continued to spearhead opposition to the project on the Gosplan USSR Expert Commission.[71] With the help of Iaroshenko and his "Living Water" crew, Sergei Shatalin, working in the All-Union Institute of Systems Analysis, authored two important assessments of the European and Asiatic diversion schemes.[72]

As the bureaucrats lost control, the debate spread to almost every major relevant scientific institution. Brezhnev had died and was succeeded by the infirm Andropov in December 1982. By late 1983 Andropov was mortally ill, and on his death in February 1984 Chernenko assumed the leadership of the Party. The atmosphere of drift lent greater urgency, but it also contributed to a feeling of greater freedom; the old men in the Kremlin no longer seemed capable of enforcing discipline.

When the Gosplan RSFSR Expert Commission formally approved the projects on February 3, 1983, another group of experts sent a memorandum rejecting the commission's conclusions.[73] A month later Briusova daringly sent a letter to Andropov.[74] Two months later, on May 18, 1983, the Presidium of VASKhNIL discussed the economic and agricultural implications of the Asiatic portion of the scheme. Voropaev and the leaders of Minvodkhoz, who were present, were astonished by the "hail" of hostile questions. Even the representative of Kazakhstan, corresponding member S. Mukhamedzhanov, defected. With the exception of Voropaev's Institute of Water Problems, the world of Soviet science was inexorably congealing in a united front against the bureaucrats. Eventually, fifty academicians, twenty-five corresponding members, and five divisions of the Academy of Sciences came around to opposing the project.[75]

The Party and ministerial bureaucrats had the power to legislate the diversion schemes, fund them, and order them constructed. Scientists and experts had no such access to the direct levers of power. However, in James Scott's term, they did possess some "weapons of the weak." One of them was science's residual corporate ability to credential its members, and here they were able to exploit a chink in the bureaucrats' armor. Just as the civil degradation of M. M. Bochkarëv entered the folklore of the scientific intelligentsia as an important moral victory, so the tempestuous and ill-fated doctoral dissertation defense of A. S. Berezner, chief planner of the European portion of the diversion project, became a symbol of social resistance to the Party bureaucracy. As chief planner, Berezner's status as a mere "candidate" (rather than "doctor") of science seemed incommensurate. Using a monograph based on the project's theoretical-economic justification as the scholarly opus to be defended, Berezner prepared to uphold his ideas before the Scholarly Council of the Academy's Institute of Geography, located on the quaint and winding Staromonetnyi pereulok in the old Zamoskvorech'e district. The defense was set for December 4, 1984.[76]

In the Russian and Soviet traditions defenses of scholarly degrees are open to the public, and interested individuals are permitted to submit written evaluations and critiques based on a précis (*avtoreferat*) of the dissertation materials distributed beforehand. Berezner's provoked no fewer than ten negative reviews, an augury of the defense itself. Although Ianshin's evaluation was sharply critical, he tactfully withheld a final judgment as to whether a degree should be conferred. Other evaluations were not as diplomatic. His official opponent, corresponding member of the Academy O. F. Vasil'ev, who had submitted a highly positive appraisal of Berezner's work, was effectively disqualified when, on examination, it became clear that Vasil'ev did not understand the issues involved and probably had not read the work through. The mathematical ecological modeler Iu. M. Svirezhev exposed the deep flaws in Berezner's calculations. As Berezner sank deeper and deeper, the defense had to be extended an unprecedented additional day. Ultimately, the Scholarly Council, though it had sought to avoid trouble, was forced by the weight of academic argument to reject the doctoral defense.[77]

At about this time the unofficial society Pamiat' (Memory) started up its activities. Central among them was agitation against the diversion project. Briusova nostalgically recalled the early days of Pamiat': "At the time," she wrote, "the Chivilikhin-style traditions of healthy patriotism were still alive. Later, it broke into several diverse, frequently mutually hostile groups, whose activities at times took on an extremist cast." In any event, Pamiat''s increasing network of activists helped to arrange meetings in clubs, Houses of Culture, and institutes around the country, including in Moscow, Obninsk, Tula, Novosibirsk, and Irkutsk. Sometimes, as following Briusova's visit and speech in Irkutsk, there was an uproar. In many instances meetings generated a flood of letters to the authorities.[78]

Unprecedented in scale, expense, and general temerity, the river diversion scheme propelled many opponents toward a more vocal extremism. Chivilikhin's own two-volume *Pamiat'*, after all, was based on a belief in the unceasing war between the principles of the "Slavic taiga" and the "Asiatic steppe." Those who would destroy the taiga to water the steppe were traitors to Russian Slavdom. For Rasputin, the diversion scheme was a "conscious act" against the Russian countryside, which he likened in its destructiveness to collectivization.[79] In a letter to Vitalii Vorotnikov, the new RSFSR premier, the writer even threatened to immolate himself on Red Square should the diversion be implemented.

In a letter of April 15, 1985 to Grigorii Vasil'evich Voropaev, Vasilii Belov (then a member of the Vologodskii *obkom* KPSS) denounced the "orientation on the south" and the preference for irrigated agriculture as "antiscientific and against the [Russian] nation [*antinarodnoe*]."[80]

There were repercussions. People were fired, lights went out at meetings, and microphones mysteriously failed. Certain institutions removed themselves from the fray, such as the Ministries of Culture of the USSR and the RSFSR, Moscow State University, the Academy of Artists of the USSR, and even the Central House of Artists on Krymskaia Square, which imposed a ban on evenings with Russian national themes. Briusova was removed from the Academy of Artists' Commission on the Protection of Monuments of History and Culture.[81] The October 1984 Plenum of the Central Committee endorsed USSR premier Nikolai Tikhonov's call for the "completion in the near future of the plans for the diversion of a portion of the flow of Siberian rivers."[82]

Emphysema carried First Secretary Chernenko to his death after barely a year in the Soviet Union's top job, and in March 1985 Mikhail Sergeevich Gorbachëv was formally elevated from acting to *de jure* general secretary of the Party. At first it was unclear whether Gorbachëv was going to take the country in a truly new direction; his initial campaigns, which featured slogans as such *uskorenie* (intensification) and relied heavily on moral exhortation and punitive measures (e.g., the antialcoholism campaign), seemed to cast him as a younger version of the disciplinarian Andropov. Unenthusiastic, both the liberal intelligentsia, including its scientific-environmental wing, and the Russophile intelligentsia and supporters waited skeptically for some sign that Gorbachëv would be different.

In this uncertain atmosphere two highly visible attacks on the diversion project were published. Sergei Zalygin's open letter to Nikolai Fëdorovich Vasil'ev, USSR minister of land reclamation and water resources, which appeared in *Literaturnaia gazeta* on October 2, 1985, was actually a fundamental attack on the vulgarized "Marxist" economics that guided water—indeed, all resource—use in the Soviet Union since the Revolution. Zalygin ridiculed the idea that water had no "cost," for it was this false assumption, he argued, that had led to the squander of such vast quantities of water in Central Asia to begin with. Consequently, the solution did not reside in a monstrously scaled technical scheme but in strict cost accounting and a responsible stewardship and husbanding of existing resources.[83]

Just one month before, Zalygin had succeeded in publishing another important technically argued critique of the diversion plan in the Party's ideological monthly, *Kommunist*.[84] But the article's most memorable point was that scientific public opinion as well as political representatives had a rightful place at the decision-making table:

> Public opinion under our circumstances is an enlightened opinion. It encompasses scientists, engineers, and people who are "abreast of the times," who have passed through the school of civic upbringing and civic activity. And they really do not want terribly much. They want problems to be decided in the open and on a high scientific level, and not just from their technical side. The experience of public opinion in the problems of Baikal and the Lower Ob' Hydropower Station back this up.[85]

As Zalygin repeated in his memorable speech to the Congress of the Union of Writers of the RSFSR in early December 1985, "Technology in its pure form, introduced without taking into account public opinion, is a terrible thing."[86]

Even more striking, as much for where it was published—in *Pravda*—as for its assertion that the project "was from beginning to end in its economic and ecological aspects without justification," was an intrepid piece by two members of the Komsomol's Council of Young Scientists and Scholars that appeared on December 30. The water authorities even demanded an apology from the Council and a disavowal of the article, but the Council did not immediately respond.[87]

Behind the scenes beginning in early 1985 an informal group of scientists, led by Ianshin and including hydrobiologist B. Laskorin and agronomist A. Tikhonov, began work on a letter to Gorbachëv. By July Ianshin had collected twenty signatures from full academicians, and Academy president A. P. Aleksandrov promised to deliver it to the Central Committee, of which he was a member. Strangely, however, the letter was never delivered.[88]

As efforts at the various diversion sites proceeded apace, and with no sign that the new Party boss had shifted from the October 1984 decision, opponents grew more desperate. On August 3, 1985, in a clear breach of protocol, Ianshin and his confederates decided to bypass the Academy president and submitted a new letter, signed as well by V. A. Kovda, Laskorin, Tikhonov, and Ianshin himself, directly to the offices of the Politburo. By late August the "letter of the four" had found its way to the agenda of that body, and Gorbachëv for the first time called for a policy review. Entrusted with that task was Ziia Nuriev, then a deputy chairman of the USSR Council of Ministers. Losing no time, Nuriev called a conference for September 1 after hurried meetings with the authors of the letter.[89]

At his conference Nuriev established nine separate working groups, each of which was to thoroughly study a particular aspect of the European and

Asiatic portions of the project and present their findings. In his choice of chairpersons for these groups, however, Nuriev hewed to the practice of appointing specialists in or around Minvodkhoz. Having already broken the rules, the scientist opponents of the project returned to the Academy of Sciences, where they broke some more by forming their own parallel working groups. Having readied their materials by early January 1986, the scientists' parallel groups delivered them to USSR premier Nikolai Ryzhkov, RSFSR chief Vorotnikov, V. Murakhovskii of Gosagroprom, the new Soviet superministry of agriculture, and A. Nikonov, another Kremlin functionary. Sufficiently troubled by this unprecedented unofficial politicking, Ryzhkov decided to tackle the issue at a special session of the USSR Council of Ministers to be held later.[90]

One of the more piquant moments of the struggle also played a role in reinforcing the praxis of *glasnost'*, which had only recently been elevated to a level with *uskorenie.* A live television broadcast on the river diversions was being aired from Ostankino. All of the speakers had been proponents. To all appearances, at least, it seemed as though the project continued to enjoy official support. But in the closing moments of the show, the script suddenly went awry. A thin, tall, white-haired man—Academy of Sciences vice president A. L. Ianshin—managed to get to the microphone to denounce the river diversions. Nothing like that had ever happened on Soviet television. "It was a real act of daring," wrote geologist Pavel Florenskii and coauthor T. A. Shutova four years later.[91] Of course Ianshin, as a leader of scientific public opinion, had several decades of courageous civic activism under his belt to prepare him for that fateful moment. By spring 1986 the censorship regime of Minvodkhoz and the Hydrometeorology Service was tottering, but the political situation was still unclear and, to a certain extent, unstable.

Meanwhile, the country was on the eve of earthshaking events. The country awoke on the morning of April 27, 1986 to the worst nuclear accident in history. Perhaps things could have played out politically in a completely different way, but as Zhores A. Medvedev argues in his *Legacy of Chernobyl,* true *glasnost',* in a macabre way, was born in the wreckage of reactor block no. 4.

Only in light of Chernobyl can we understand the incredible candor and independence of the speeches at the Eighth Congress of the USSR Union of Writers in late June 1986, an outpouring of criticism of the planners and the bureaucrats for their despoliation of Russia. Iurii Bondarev's speech embodied the anguish and frustration of the literary environmentalist Russian nationalists:

> If we do not stop the destruction of architectural monuments, if we do not stop the violence to the earth and rivers, if there does not take place a moral explosion in science and criticism, then one fine morning, which will be our last . . . , we, with our inexhaustible optimism, will wake up and realize that

> the national culture of great Russia—its spirit, its love for the paternal land, its beauty, its great literature, painting, and philosophy—has been effaced, has disappeared forever, murdered, and we, naked and impoverished, will sit on the ashes, trying to remember the native alphabet . . . and we won't be able to remember, for thought, and feeling, and happiness, and historical memory will have disappeared.[92]

In this atmosphere of upheaval and crisis the Presidium of the USSR Council of Ministers met on July 19. As protocol dictated, minister of land reclamation and water resources Vasil'ev led off with a defense of the entire project. Now, however, Minvodkhoz's opponents were given the opportunity to have their say, if not on a completely equal footing. Ianshin offered the first and crucial rebuttal. An earth scientist, he noted that meteorologists were predicting a period of rising precipitation for northern Russia, thus obviating the need to divert water; besides, the level of the Caspian Sea had been rising for two decades, which many opponents had pointed out previously. As if nature itself sought to buttress Ianshin's case, a violent summer rainstorm began, visible and audible through the windows of the meeting room. Voropaev followed with a complete endorsement of the project, and was followed in turn by Academy president Aleksandrov and by Gurii Marchuk, chair of the USSR State Committee on Science and Technology, both of whom supported the plan with reservations. Gosplan USSR chair N. K. Baibakov also was among the backers. Significantly, though, weighing in against the plan was one of Gorbachëv's most trusted economic advisers, Abel Aganbegian, who questioned the unbelievably low cost estimates in Vasil'ev's presentation. An indication that the planners were in trouble was the conclusion of Vitalii Vorotnikov, who had studied the opponents' critique in detail. "We want to create new seas within the country," he asked, "but where then will we sow our wheat?"[93]

Prime Minister Ryzhkov, now revealing his own opinion on the matter, turned to Gosagroprom minister V. Murakhovskii and asked whether his ministry could find the ninety billion rubles to finance the plan. Of course the minister answered in the negative. To this Ryzhkov responded that the USSR Council of Ministers itself did not have that kind of money either, especially in light of the expenses incurred by the accident at Chernobyl. For that reason, he concluded, the Siberian part of the project needed to be postponed to some time in the next millennium, while the European portion needed to be ended outright, as the scientists had recommended. Ryzhkov even went farther, accusing Aleksandrov and Vasil'ev of disinformation. Neither Vasil'ev nor any of his allies dared to mount a rebuttal to the prime minister, and the decision now went to the Politburo for a final hearing. Gorbachëv supported his prime minister. On August 14, 1986, a joint decree of the Party's Central Committee and the USSR Council of Ministers stopped the further progress of the projects, citing both the objections of "broad circles of public opinion" as well as "the goal of concentrating financial and material resources to enhance the efficiency of water resource use and that of existing reclaimed lands."[94]

For many groups within the Soviet population, even those that did not take active part in the struggle against the diversions, August 14 now stood as a landmark date. Although parts of the project survived—particularly the Volga-Don-2 and Volga-Chograi Canals—another fact far outweighed the defects of the decree. For the first time, the highest Party and state leadership had sided with "public opinion" against the almost united front of bureaucratic empires. For the first time, a "project of the century" was derailed, thanks in good measure to citizen activism. And for the first time, the Soviet intelligentsia and others began to take the regime's commitment to *glasnost'* seriously. This reassessment had the most dramatic consequences for the tenure of Mikhail Sergeevich Gorbachëv.

Three organizations stood out by their official silence on the river diversions. The USSR Academy of Sciences, the Geographical Society of the USSR, and VOOP, all of which—the Geographical Society most recently—had been transformed into the domesticated fiefdoms of Party bureaucrats. Iurii Efremov, who attended both the Geographical Society's congress in Kiev in 1985 and VOOP's in the Hall of Columns of the former Moscow House of the Nobility a year later, noted that "even in hindsight at their congresses [both societies] managed to avoid saying even one word in their official reports about the antidiversion activism of their members." There was "not a single word of approval or encouragement" at the VOOP meeting, despite the fact that some local branches, notably those of Vologda, Leningrad, and the Komi ASSR, figured centrally in the struggle.[95] Traditions of resistance survived, but ironically not in some of the institutions that early nurtured them.

If for field biologists and the *druzhinniki* the Party and state bureaucrats were first and foremost antiscience and anti-intellectual, to many of the activists against the river diversions they were anti-Russian, anti-Volk. For some it was easy to focus on the role of the Jews Iurii Izrael' or Berezner or of the Tadzhik P. A. Polad-Zade; their ethnic identities confirmed the anti-Russian, cosmopolitan rootlessness of these homeland-destroyers. On the other hand, many defenders of Russian culture acted out of their general commitment to culture. Sergei Zalygin and Dmitrii Likhachëv are Russian patriots, but they are also citizens of the world, and they opposed the Party bureaucrats as much for their effect on science and intellectual life as owing to the bureaucrats' threat to Russian culture and Russian rural folk.[96]

In the case of Zalygin various narratives joined to make new combinations. Viewing nature through the eyes of a trained hydrological engineer, he saw the technical problems contingent on the vast nature-transforming schemes. As a sensitive writer who had witnessed (and then written about) the human pain associated with the collectivization of the Siberian peasantry, Zalygin shared the "Village Prose" writers' sensibility that the transformation of nature was an instrument of violence against rural people. As a self-made intellectual who, to a great extent, had absorbed many of the values of the scientific intelligentsia, Zalygin was offended by the Party bosses' rejection of the claims to policy-making roles of credentialed experts and scientists. Yet, he remained a unique voice. Unlike the other "Village Prose" writers, Zalygin never totally embraced a nativist—let alone exclusivist—Russian nationalism. And, unlike the field naturalists, Zalygin was fully aware of the false nature of the dualism "humans and nature" and therefore never fetishized "pristine" nature or considered people to be disruptive and polluting aliens in the otherwise harmonious Eden of aboriginal nature. That is why, for his own reasons, Zalygin could be counted among the supporters of Baikal, *zapovedniki,* and the campaign against the river diversions and was almost alone as an honest broker to whom all factions of Soviet environmentalists could turn.

CHAPTER NINETEEN

Environmental Activism under Gorbachëv

As this book has shown, environmental activism was one sphere of citizen politics tolerated by the Soviet Party-state, perhaps because from the *apparat*'s perspective it looked so little like serious "politics." The various pre-*perestroika* environmental movements had kept alive the essential idea that the Party had no right to a monopoly on decision-making; citizens (even in the narrow, elitist sense of citizen experts embraced by the old field naturalists and the *druzhinniki*) had a right to input on public policies as well. What *perestroika* changed was the cast of actors. No longer did only a relatively few biologists, geographers, writers, and art historians strut the stage of Soviet environmental history. The removal of the worst aspects of the police state brought hundreds of thousands of ordinary working men and women into the streets and squares of the USSR's cities. Where the two older branches of environmentalism were concerned, in one way or another, with sacred space, be it *zapovedniki* or the Russian Northland, the newer actors arrived on the scene with more mundane concerns: a livable environment for themselves and their children.

Russian National Environmentalism

After the stunning victory over the river diversion schemes, for which the nationalist environmental current deserves much credit, its activists continued to keep the spotlight on Russian "national" nature. Particularly visible was a Save the Volga Committee, in which Vasilii Belov and Kedrograd founder Fotei Shipunov played the leading roles. Cooperating with similar committees to save the Don, Ural, and other rivers, notably in Siberia, these groups continued to be attracted by the symbolism of hydropower and canal projects as violent, alien, modernist, technological intrusions that needed to be expunged from Russia. Nationalist environmentalists envisioned the rehabilitation of Russian culture—especially rural culture—and morality, and the restoration of monasteries and churches. Graphic reflections of this vision may be seen in the oversize canvases of the painter Il'ia Glazunov, in which twelfth-century Rus' ships sail untroubled down undefiled rivers.

Economist M. Ia. Lemeshev's Anti-Nuclear Society and the Fund for the Restoration of the Cathedral of Christ the Savior led by writer V. A. Soloukhin were other institutional expressions of this current. Valentin Rasputin was active in the Baikal Fund. Many veterans of the river diversion struggle united around Sergei Zalygin's new organization, Ecology and Peace, which tackled other hydroprojects such as the dam on the Katun' and the Volga-Chograi Canal.

With ideological restrictions removed by the last two years of the decade of the 1980s, some nationalists began to espouse religious mysticism or explicitly chauvinistic positions. For the 1990 anthology *Ecological Alternative,* Fotei Shipunov contributed a philosophical credo in which his developing mysticism was evident. "No one can sense the spirit of nature like a poet-thinker," he wrote. "All of the visible and the unseen world is the creative fruit of this Spirit. . . . [It] creates and organizes the material world like a mechanism and is visible as the biosphere and that which surrounds it in all of its grandeur, brilliance, and beauty, like a Blessed Icon-Frame. And this spiritual organism and its creation . . . are linked to the moral world through its bearers—the individual, peoples, and humanity."[1] One class of people had always been in touch with that spirit—the peasants. "Every peasant," he wrote, "knew that a wisely lived life derived from two sources: (1) from a knowledge of the land and its creatures . . . and (2) from the spiritual attainment arising from the consecration of his land, that is, from faith in something else besides farming alone, besides even his own life—faith in an Absolute Truth reaching down from the eternal, Heavenly world."[2]

During the elections of the spring of 1989 the nationalist environmentalists tried their hand at electoral politics. Briusova complained that although she, Lemeshev, Shipunov, and Iaroshenko all stood for election, they were derailed. "And this stands to reason," she continued, charging a pattern of discrimination on the part of the Electoral Commissions, "insofar as it is now known that patriots of Russia did not make it into the 'Moscow group' of the Congress."[3]

The evolution of some, such as Shipunov before his untimely death, was poignant. As their political role seemed to contract and, after its independence, Russia seemed to adopt a modified pro-Western stance, Shipunov and others became more extreme in their insularity and antimodernism. The embittered Kedrograd founder's last works now included anti-Semitic diatribes reminiscent of the Elders of Zion tracts, having little of the poetic and universalist mysticism of his writings of just two and a half years before. Jews, at the head of a broad anti-Russian coalition that included Westerners, Georgians, and a host of other "others" were driven by the goal of "purposively and sacrilegiously falling on Russia and exacting not only physical but also spiritual vengeance—the ritual murder of that nation!" The Bolsheviks, led by the Jew Bronshtein-Trotskii and others, "turned the land into grotesque fragments [of the former Russian forest] and lifeless deserts, in which humanlike forms wander, casting equally grotesque shadows. . . . The pogrom of the Fatherland in every way included the pogrom of its nature." The end result was that the Russian people "lived in their homeland, but without a homeland, and lived in their Fatherland, but were bereft of a Fatherland."[4]

Although non-Russian movements are outside the scope of this book, it is apposite to note that many environmental activists in the Baltic states, Moldova, and elsewhere underwent a similar evolution. Activists such as Dainis Ivans, who led Latvian resistance to the Daugavpils Dam in the mid-1980s, later became a leader of the nationalist Latvian Popular Front and a supporter of industrial development as a means of strengthening the country's military capability to defend itself against a future Russian invasion.[5]

Democratization and the Crisis of the *Druzhiny*

With the advent of *glasnost'* and the halting democratization of politics from 1986 on, the defensive posture of the nature protection movement became an anachronism. When citizens gained an increasing say in major issues of public concern, the highly symbolic politics of the struggle for *zapovedniki* seemed increasingly abstract and irrelevant. With the legalization of "informal" nongovernmental groups in 1987, the *druzhiny* were no longer the lone knights defending their fragile holdout of civic autonomy against the massed forces of the Party-state bureaucratic machine. *Glasnost'* had destroyed their special status.

The subsequent crisis of the brigades stemmed from the fact that *glasnost'* had unleashed broader social forces that had remained largely quiescent over the previous decades. "Finally, the broad masses of the people rose up," wrote Evgenii Shvarts, "concerned about their own health and the health of their children." But the *druzhiny* as institutions did not always undergo a *perestroika* to keep up with the times. Most, in clutching to their corporativist-elitist social identity, allowed events to pass them by; by the late 1980s the *druzhiny's* glory had faded. The socially most involved students now bypassed the brigades to involve themselves with other causes, sometimes overtly political ones.[6]

By the end of the decade this had led to a drop in membership and a corresponding discussion about whether indeed there was a crisis and, if so, what its causes and dimensions were and how to resolve it. Some vigorous *druzhiny* such as the one in Riazan' managed to adapt to the new times. Galvanizing opposition to an overly hasty plan to build a sewage canal through the floodplain of the Oka River, brigadiers were able to collect 15,000 signatures on a petition opposing that particular route. By June 5, 1988—Earth Day—the protest had spilled onto the streets, Riazan''s first political protest meeting in living memory. A second protest rally followed in July, evidently exhausting the patience of the city fathers, and on September 1 the authorities decertified the *druzhina* for "anti-Soviet speeches." One response of the *druzhina* was to go to court, but the case dragged on for more than two years. Another was to advance the candidacy of former *druzhina komandir* Aleksandr Gavrilov for the USSR Supreme Soviet in the elections of March 1989. Despite the thousands of voters who signed Gavrilov's electoral petition, his candidacy was denied by the Riazan' branch of Gorbachëv's so-called Electoral Commissions. As a crowd of 300 waited peacefully for final word on Gavrilov's candidacy on April 5, 1989 outside the municipal soviet, police with billy clubs moved in, arresting twenty. Gavrilov himself was sentenced to ten days in jail.[7] Ultimately the *druzhinniki* were vindicated as Gavrilov won a sweet victory in the elections for the Russian parliament in 1990. However, Riazan' was more the exception than the rule.

As early as December 1987 at a large *druzhina* conference in Dolgoprudnyi, Moscow *oblast'*, the sociologist Oleg Nikolaevich Ianitskii, who has studied this movement more profoundly than anyone, was able to identify the contours of the movement's debacle. This crisis was not yet apparent to most *druzhinniki* themselves, for the movement had just passed its high-water mark. The representatives of the 110 *druzhiny* had come to Dolgoprudnyi to constitute themselves as a legal, autonomous movement, which did not need to shelter any longer under the legal umbrellas of the Komsomol or local branches of VOOP. VOOP was officially asked to dissolve its Coordinating-Methodological Council for the *druzhina* movement, which the VOOP leadership had established as a curatorial body.[8] Among the problems Ianitskii recognized was that the *druzhiny*, in their emphasis on the protection of wild nature, had projected an image of "rural" irrelevance to the vast mast of the newly politically activated urbanites (ironically, when the *druzhinniki* "went to the country" to pursue their projects, they met with hostility from ruralites as well).[9] Linked with this image of irrelevance were the *druzhinniki*'s reluctance to give sufficient weight to global issues, their reluctance to rethink the political implications of their program and to act and talk in an overtly political way, and their tendency to treat all nature protection issues as scientific problems of ecology to be solved only by biology professionals.[10]

Shvarts added:

> If earlier, the external pressure on the *druzhiny* and their persecution made for a concentration of "the best people" in their ranks, representing a wide spectrum of styles and methods of work (theoreticians, politically oriented, practical people, organizers, etc.), under the new conditions the *druzhiny* began to lose out in favor of sharply "specialized" organizations unmistakably geared toward protests, toward work among the broad public, and toward political forms of activity, on the one hand, and self-supporting ecological organizations, on the other.[11]

By 1990 the former "Green *druzhina*" of Leningrad University had melted away, as had those at the Biology Faculty of Belorussian University, Udmurt University, and a host of others.[12] On the other hand, the *druzhiny* were the nurseries where many of their new competitors were hatched: the Green Party in Leningrad, the Ecological Initiative Movement in Voronezh, the antinuclear movement in Gor'kii (Nizhnyi Novgorod), the ecological-political club Al'ternativa in Kuibyshev (Samara), and a host of others. The most influential organization born from the womb of the *druzhina* movement was the Social-Ecological Union.

To do justice to their sophisticated analyses, Ianitskii's and Shvarts's critiques of the brigades apply from the emergence of *glasnost'* to the end of the Soviet era. The brigades movement had matured politically since its founding, especially from the last years of Brezhnev's rule on. "Consciousness of the need for political actions and for an analysis of the economic interrelationship among various events proceeded quite quickly," wrote Shvarts, "owing to the participation of the *druzhiny* in the struggle to stop various grandiose 'projects of the century,' including the diversion of the northern rivers, the construction of the Danube-Dnepr, Volga-Chograi, and Volga-Don 2 Canals, the Tiumen' gas and chemical complex, and some dangerous projects involving transnational corporations. It is therefore obvious that the process of politicization of the *druzhina* movement began a long time ago."[13] But the most politicized, economically savvy, and broad-minded members of the movement left the brigades for other venues of struggle from 1987 on. Arguably, the best and brightest transferred their efforts to the Social-Ecological Union.

The Social-Ecological Union

Founded on August 6, 1987 during the Third Conference of Former Members of the *Druzhiny,* the Social-Ecological Union (SEU) until its first conference in late December 1988 existed as a movement of *druzhina* veterans. At the 1988 conference, though, held in Moscow's House of Hunters and Fishermen, the SEU managed to attract delegates from eighty-nine cities representing 130 separate organizations in eleven republics.[14] At the December 1988 conference some delegates decided that an overtly environmental political party was needed in addition, and they established the Green Party, based on the ideas of Murray Bookchin, André Gorz, and Ivan Illich, with a founding congress held in Moscow in March 1990. By October 1991 the SEU claimed 4,000 members and sympathizers across seven former Soviet republics.[15]

Reflecting their traditional liberal, internationalist intellectual orientation, SEU leaders quickly forged ties with international groups, including Greenpeace, Friends of the Earth, the Earth Island Institute, the Nature Conservancy, the Natural Resources Defense Council, the Swedish Green Party, the Finnish Green Party, Global 2000 (Austria), the Sacred Earth Network, the Japanese Bird Protection Society, and IUCN.[16] Successes came quickly during those heady days of 1988 and 1989, when free speech was gaining new ground by the day. As early as spring 1988, before its big December congress, the SEU and the *druzhiny* were important members of an environmental coalition that defeated a proposed party and government decree to expand the number of large hydropower stations by ninety during the decade 1990–2000. An unprecedented torrent of letters inundated the Central Committee of the Party, which was forced to rescind the draft decree owing to the almost unanimously negative verdict of public opinion.[17]

Perhaps the most glorious success of the SEU came just months after its official founding when, in nationwide rallies on February 12, 1989 that involved perhaps hundreds of thousands in 100 cities, it was able to collect more than 100,000 signatures against the construction of the Volga-Chograi Canal, a relic of the old river diversion plans. The Moscow rally also marked the first time that a central newspaper, in this case *Vecherniaia Moskva,* published an announcement that an independent, nongovernment demonstration was to take place.[18] Arguably the SEU charted an even more important "first" that day: the first truly nationwide protest in Soviet history. Struck by the power of the public reaction mobilized by the SEU, the government canceled the project and withdrew its earthmoving equipment from Kalmykia the day before the nationwide protest, which proceeded undeterred.[19]

Although many of its continuing struggles, such as that to stop the Katun' hydroelectric station in the Altai mountains—which apparently led to the project's cancellation—and the fight for a number of new *zapovedniki* and national parks, continued traditional themes of opposing development in "pristine" areas, the SEU, unlike the remnants of the *druzhiny,* did in fact move with the times. Local affiliates worked with local communities on issues involving radiation and industrial pollution. Beginning in 1988 the SEU turned its attention to assisting in the fight to shut down the production of a protein-vitamin concentrate for livestock made from oil paraffin. The SEU was particularly helpful in marshaling scientific expertise to buttress citizen efforts; with the organizational efforts by its members in Tomsk, Volgograd, Kremenchug, and Kirishi, the Union put together two scientific-practical conferences on the threat as well as a petition drive. By 1991 the USSR Supreme Soviet adopted a law halting production of the concentrate from that year on.[20]

It would please the leadership of the SEU to believe that it had masterminded this environmental triumph as well. However, in that complicated time other social forces had awakened, and although the links between the *intelligenty* in the SEU and the working mothers and fathers of Tomsk, Volgograd, Kremenchug, and Kirishi cannot be denied, it is not yet possible to allocate credit for the victory. But common sense tells us that the sight of tens of thousands of workers in the streets must have counted for something in the councils of the Central Committee. And it all started with Kirishi.

The Masses Weigh In

Even before the law on meetings and street demonstrations was passed—and even before the founding of the Social-Ecological Union—the mother of all ecological rallies erupted in an almost unknown town of 52,000 on the Volkhov River between Novgorod and Leningrad: Kirishi. In the summer of 1975, after the new biochemical plant had started up, there was an outbreak of a disease never before encountered there: its symptoms included shortness of breath, wheezing, a hoarse cough, and spots on the body. Residents observed that these signs coincided with winds blowing vapors from the direction of the plant, which produced a protein-vitamin concentrate for livestock.

When the plant was built, it transpired, the construction crews cut corners on pollution abatement equipment to complete the job before the official deadline, and the director, Valerii Bykov, was quickly promoted to Moscow as minister of biomedical industries. Seven more such plants were built around the country, including in Volgograd, Novopolotsk, and Kremenchug. Residents of those cities began to suffer from the same condition.

After the initial complaints forced the creation of a commission to look into the situation at Kirishi, the commission's head, USSR deputy minister of public health P. N. Burgasov, the country's chief health inspector, concluded: "Rumors, as always, exaggerate matters. Talk that the discharges from the biochemical plant . . . have increased the number of cases of illness is an exaggeration from fantasyland."[21]

There the matter remained until the spring of 1987. Then calamity struck. In May another outbreak of disease ravaged the town and this time killed eleven children under the age of five. One father who lost a year-old son began a one-person demonstration in the city's main square, holding a photo of the infant bordered in black. Soon he was joined by the other grieving parents and then the whole city. At least twelve thousand turned out on June 1, 1987, Children's Day, with the demand that the factory be closed down. As the episode's chronicler, T. A. Shutova, noted, this was one of the first spontaneous mass street demonstrations in the country in decades. "In the process," she adds, "the residents of Kirishi became citizens."[22]

In the aftermath the citizens of Kirishi organized themselves. Using the infrastructure of the practically defunct local branch of the VOOP so that it could function legally, the citizens' committee added a "Sixth Section" to the society's existing five (pet fish, pine cone crafts, ikebana, and two others). The "Sixth Section," which became a people's shadow government in the city, enrolled thousands. Meetings were organized at schools and factories. Although the factory was closed down the next day, a round-the-clock monitoring of the factory was organized by the residents. One of the leaders of the citizens' revolt and leader of the section, Vladimir Vasil'ev, a mailman, gained national notice when his "mail bomb"—an exposé of the "reprofiling" of the factory—was published in *Komsomol'skaia pravda* on March 15, 1988.[23] When the May Day parade was held in 1989, 15,000 marched under the banner of the "Sixth Section," dwarfing all other contingents.[24]

The easing of international tensions at times created tensions internally in the Soviet Union. In a bid to outdo the United States in dismantling weapons of mass destruction, foreign minister Eduard Shevardnadze announced in January 1989 that the USSR would begin to destroy its stockpiles of chemical weapons unilaterally.[25] Without seeking local consent, the central authorities built a secret facility to destroy the chemical weapons in Chapaevsk, a city of 97,000 in Saratov *oblast'*. The plant—dubbed the "factory of destruction"—had been built in a densely populated area. Once word of its existence got out, local residents were outraged, and a group, Initsiativa, was formed to stop the operations of the plant. Soon 60,000 petition signatures were collected.[26]

Taking on the military and the Foreign Ministry, however, brought environmental protest directly in conflict with "national security." Nevertheless, a coalition of environmental and anarchist organizations from Saratov and Samara provided active support to the local residents. With health and safety issues salient, the struggle attracted tens of thousands of participants in August and September 1989, when rallies culminated in the erection of a tent city of 7,000 that "laid siege" to the factory. Supported additionally by workers' groups from local factories, the "siege" continued for thirty-five days until the central government agreed to convert the plant to a training center.[27]

By 1989 workers as a self-organized group also began to react to issues of occupational and public health. Of particular note were actual or threatened strikes in the automobile factory town of Toliatti, in Ufa, and in the coal fields of the USSR from Karaganda and Kemerovo in Kazakhstan and southern Siberia to the Donets in Ukraine.[28] In the end, not endangered wildlife, besieged *zapovedniki,* or flooded sixteenth-century monasteries lit the fires of righteous civic indignation among the Soviet Union's general working populace but the life-and-death issues of unbreathable air and undrinkable water. In a funny Marxian way, the very biological bases for human survival proved more compelling than the iconographies of the intelligentsia. As Oleg Ianitskii put it, "One way or another ecological protest during the years 1987–1989 in the USSR was the first legal embodiment of *broadly democratic protest and solidarity of citizens as citizens.* This was natural, since the issue was the basic and universal conditions for human survival."[29]

Much more could be written on this fascinating time of environmental activism than is contained in this chapter. But its brevity is indicative of a larger fact: *glasnost'* and democratization, the hour of the traditional environmental movement's greatest triumph, paradoxically set the stage for its marginalization. Against the backdrop of free speech, contested elections, and hundreds of thousands of workers on strike or in the streets, the old-line activists could no longer sustain their social identity as tribunes of the people, intellectual knights fighting an almost isolated struggle against the Party-state *apparat* on the curiously protected terrain of environmental issues. By the early 1990s, as purely economic and political issues edged out even the urgent concerns about public health, as workers were now forced to choose between slow poisoning and unemployment, the fight against pollution did not seem nearly as clear-cut an issue as it had a mere three or four years earlier. For many, putting bread on the table is more urgent than shutting down a factory that causes asthma in a child. Both usually take precedence over fighting for a national park.

An equally profound dynamic also had a hand in remarginalizing environmental activism. By late 1990 powerful brokers within the economic bureaucracies and the Party had recognized that Gorbachëv's reforms, which had been introduced in order to legitimize and save the one-party system and its centralized economy, were leading to exactly the opposite outcome. As a titanic political struggle between two broad alignments—Communists and democrats—seemed to be unfolding, environmental issues were now relegated to a much lower priority. Aleksei Iablokov and Nikolai Nikolaevich Vorontsov, who had each been elected to the Congress of People's Deputies in March 1989 in the curia of seats allotted to the USSR Academy of Sciences, were the two most visible representatives of the old scientific nature protection intelligentsia during late *perestroika.* Elected along with Andrei Dmitrievich Sakharov, they swiftly aligned themselves with the physicist's interregional bloc of democrats. Iablokov's leadership was quickly acknowledged on the Congress's Ecology Committee, where he served as deputy chair, even publicly calling USSR premier Nikolai Ryzhkov an "ecological ignoramus." For Vorontsov fate had another political position in store. In July he was nominated and confirmed as USSR minister of nature protection, making him the first non-Communist in the Soviet cabinet since 1918. Soon Vorontsov and Iablokov both found themselves overwhelmed by the violent and dramatic political events of the day, captured in a photograph by a *Der Spiegel* correspondent showing Nikolai Vorontsov and Boris Yeltsin on top of the tank, together, during the first day of the August 1991 coup.

When Vorontsov (KIuBZ; from the late 1980s, vice president of MOIP) was minister and Iablokov (VOOP Youth Group) was holding forth in the Congress of People's Deputies (he would later become Yeltsin's personal environmental adviser for a time), it seemed that the new day of the scientific intelligentsia had dawned at last. But then the Party pulled back from its reforms, precipitating the crisis of 1990 and 1991 that ultimately brought down Communist monopoly rule and the USSR state structure. For reasons of conscience as well as political exigency, the scientific intelligentsia overcame its caste elitism and supported forces calling for full democracy. With the democrats' program, however, also came a commitment to a "free market."

The country's severe capital shortage meant that what little monies were available for investment would not go into pollution abatement. Exploiting the Russian government's desperate search for capital, foreign concerns were able to move in to extract vast amounts of lumber and other resources. Environmentalists, believing any criticism of Yeltsin and his allies would play into the hands of the Communists, held their tongues. Ultimately, the Communists, as well as Russian nationalists led by Zhirinovskii and Prokhanov and other extremists, were able to repatriate the environmental issue, arguing that Yeltsin and his camp were selling off the country at rock bottom prices.[30] The intelligentsia's environmental activists were disarmed ideologically.

From a social and institutional point of view as well, the new changes were potentially lethal to the scientific intelligentsia. Democratic self-government and the free market under Russian conditions placed institutions of science, learning, and culture in a perilous situation. For their own reasons the Communists had heavily subsidized these sectors for decades. Now, with the cash-starved new order unable and unwilling to subsidize science, culture, and scholarship any longer, we are witness to an unprecedented and potentially cataclysmic meltdown of Russian academic and cultural potential. With the continuing prospect of unimaginable cutbacks in the funding of learning, the scientific intelligentsia is emigrating, finding jobs in business, or living off potatoes grown in out-of-town dachas. The victory of those whom the scientific intelligentsia supported is paradoxically leading to the intelligentsia's virtual eradication, certainly as a privileged social group. Additionally, the elimination of many positions in higher education and science threatens the survival of the scientific intelligentsia as a group with its own traditions and values. For a member of that group it is becoming harder to say what the "lesser of two evils" looks like. I would have liked to end this book with an assurance to the reader that justice and a humane vision of the use of human beings and other living things had triumphed in Russia, promising a national resurrection. There is, however, no "end of history," and some author writing fifty years from now may conclude on an entirely different note. This, at least, is my hope for Russia, the other successor states, and our common home.

Conclusion

Making sense of Russian and Soviet history has never been easy. Often, scholars' understanding of the Soviet system and Soviet society has been influenced by the political climate. During the Cold War, for example, when the Soviet Union was generally viewed as a fomenter of revolutionary challenges to the "Free World" and as a rival locked in a great contest with the United States and its allies for the fate of the globe, many scholars chose themes that they believed would enhance the "Free World's" understanding of top-level Soviet decision-making.

The Cold War also led scholars to pose other questions. Because the overwhelming majority of scholars did not sympathize with the outcome of Russian political developments, they wondered whether Stalin, or indeed Lenin, were really historically inevitable. Many undertook a search for the moment of "original sin." Was it the legacy of serfdom and autocracy, the world war, an accident of poor leadership among alternative contenders for power, the strategic balance of social forces in favor of the Bolsheviks, that party's greater organizational strengths, or other factors that allowed Lenin to come to power in October 1917? Even those few who were in sympathy with the stated ideals of the Bolsheviks wondered whether Lenin's rule inevitably paved the way for Stalin's. For cold warriors and disappointed socialists alike, historical scholarship became a full-scale search for a usable political past. For scholars, the hope of finding in the Soviet past the potential for alternative political and economic development in the future beckoned like a pot of gold. Whoever could assure us that the Lenins, Stalins, Khrushchëvs, and Brezhnevs would not go on forever would bask in the glory and gratitude of the people of the "Free World."[1]

The radicalism of the 1960s changed the image of the Soviet Union, academics' relationship to the Free World establishment's agenda, and consequently the nature of the questions historians asked. Whereas the previous generation had trained its sights on political elites, self-proclaimed "revisionists" began to focus on subaltern groups—workers and peasants. In some cases, they attempted to rehabilitate the "legitimacy" of the Soviet regime by demonstrating mass, especially "working-class," support for it at various times, as if workers' support could in and of itself validate the actions and policies of the regime and demonstrate the essential "socialist," and hence redeeming, nature of the Soviet system as an alternative to capitalism.[2] On this I speak from the inside, because as a graduate student I counted myself one of those new social historians.[3]

In the last decade the wheel has turned again. After Brezhnev, Andropov, and Chernenko, it became difficult if not preposterous to believe that the USSR was in any sense a "revolutionary," let alone progressive, society. More willing now to discount Marxism as a central explanatory element of Soviet regime behavior and social relations and to appreciate powerful similarities between tsarism and Soviet realities, many began to reexamine whether October 1917 constituted such a radical historical break.[4]

The meltdown of Communist regimes in the USSR and Eastern Europe has sent historians and other scholars scurrying to rummage through the past in search of usable seeds from which a new social, economic, and political order might grow (or be grown) in those countries. Particularly in vogue are attempts to find pockets of "civil society" and a democratic resistance to tyranny against the dreary backdrop of seeming conformity and terror. We naively believe that if such hardy seeds could be identified and appropriately nurtured they would eventually outcompete the politically and economically dysfunctional conformist and bureaucratic "weed" vegetation, turning those societies into blooming, democratic, and free-market gardens.

As in the past, some are now eager to map a new social model—"civil society"—onto Russian realities, driven more by political hopes than by a deep knowledge of culture. Consequently, of prepossessing interest lately has been the novelty of public protest, independent and unsupervised cultural communities, and the recent flourishing of nongovernmental public interest organizations, advocacy groups, and political parties.[5] Understandably, the particular prominence of environmental protests during *perestroika* has also drawn scholarly attention to that phenomenon as just such a germ of "civil society."[6]

Setting aside such exceptional events as the crowd of 20,000 Jews at the Moscow Synagogue who greeted Golda Meir on Rosh Hashanah, 1948,[7] the GULAG revolts of 1953–1954, the Novocherkassk riots of 1962, the Tbilisi student demonstrations of the 1950s and 1970s, the Crimean Tatar protests of 1968, and the Estonian students' demonstration of 1980, environmental protests during *perestroika* were the first mass-scale public demonstrations of citizen disaffection in the Soviet Union in recent decades. Protests then expanded up to and including political strikes by organized workers. It was comforting to believe that the suffocating dictatorship of the Communist Party needed only to be removed for Soviet citizens "naturally" to claim their democratic rights in a civil society.

Realities, however, are not so simple. In a major corrective to those who would see in every act of protest or resistance the germ of "civil society," Sheila Fitzpatrick's study of the collective farm peasantry reminds us that people are embedded in social and cultural matrixes that defy easy categorization. They can oppose particular aspects of the system yet assent to others. They can participate in the system out of all kinds of motives—idealism, cynicism, fear, or the pursuit of self-interest—and often a combination of the above. This book attempts to understand environmental activism under Stalin and his successors as just such a complex set of responses by scientists, students, and writers to Soviet conditions. Activists were not do-or-die resisters of the system, on the one hand, nor inoffensive do-gooders, on the other. Rather, they were groups of individuals who used relatively open discursive space to try to carve out independent social and professional identities as best they could within a system that prescribed official models of behavior, ethics, norms, and identity for all.

The activists' single-minded focus on the protection of "pristine" nature by means of the defense and expansion of the network of *zapovedniki* effectively trapped them in a realm of speech concerned with sacred space. It trapped them as well in a delusionary division of nature into "sick" and "healthy," freezing them in a framework of analysis that treated the abstraction "biocenosis" as if it were a real object of nature. They effectively shut themselves off from the more mundane environmental concerns of nonprofessional, ordinary Soviet people as well. To be an active agent in the creation of "civil society" would have meant crossing those discursive and class lines, something the older movement was unable to do. To that extent, it is unlikely that this movement in and of itself could serve as a model for a full-blooded "civil society" based on a deep acceptance of pluralism and a renunciation of elitist claims to leadership.

On the other hand, the activists' story *is* a message of hope, for it tells us that even in the darkest gloom of a terror-ridden society it is possible for individuals to find a way to come together to protect and affirm values and visions radically at odds with those of the rulers. That is the great achievement of scientific public opinion, which for many decades was the only relatively autonomous public opinion in the Soviet Union.

The scientific high intelligentsia tradition in nature protection monopolized the field for many decades. Involvement of a broader public, beginning with other scientists such as in Akademgorodok/Novosibirsk, the literary intelligentsia, and the press only began during the mid-1960s, particularly in connection with the threats to Lake Baikal. Even then, leadership rested with the naturalists, who were viewed as most able to advance expert arguments, speaking with the authority of science.

In the non-Russian republics, especially the Baltics, participation in nature protection embraced broader layers of the population. Two factors contributed to this: first, the Germanic traditions in education, which included highly value-laden attitudes toward the local landscape, and second, the correct perception that industrialization was accompanied by a continuing influx of nonindigenous Slavic migrants—Russians and Ukrainians—whom the Estonians and Latvians saw as swamping their small ethnoses. Nature protection was a benign-sounding argument for keeping factories—and their "foreign" workers—out. That movement, however, largely falls outside the Russian emphasis of this book.

What is common to many of these cases is that nature protection served as a surrogate for politics, as actual political discourse was prohibited and punished. What gave the conservation movement its unique quality was that it was perceived by the regime as a curious trifle, a collection of socially marginal "fool" scientists (*chudaki*) who were not worth the effort of monitoring, detailing, and repressing. By default, the nature protection movement became the only vehicle to express deep feelings of civic concern. For those who breathed the spirit of *obshchestvennost'*—in either its scientific or its civic permutation—environmental activism provided the feeling (and sometimes the fact) that they were tangibly and independently defending the good of the community in the face of a repressive, wasteful, and destructive bureaucratic system.

Inescapably, the presence of sites of civic autonomy like the naturalist societies and VOOP invites us to reexamine our larger understanding of Soviet politics. Was this a unique social phenomenon or will similar movements be uncovered? Have we overestimated the Stalinist regime's abilities to police society or, conversely, underestimated its cleverness in managing potential dissent? We have no definitive answers to the riddle of the regime's failure to snuff out nature protection activism. If we use Fehér, Heller, and Márkus's analysis as a point of departure, perhaps the regime was unable to develop a completely coherent understanding of what constituted real threats to its survival or, conversely, real essentials.[8] Or perhaps the regime thought that it was so strong that it could indulge the survival of one remaining source of social opposition. However, given the regime's history of persecution of even the individual dissent of poets, why should it let thousands of VOOP members continue to meet? Are we looking at a gaping hole of inefficiency in a system of power that had pretensions to total control? Did the regime take nature protection speech at face value and fail to understand the system-related implications of such speech, thereby betraying a shocking dullness, even stupidity? And what about the failure to police "liberal" underlings such as the Russian Republic leadership or that of the USSR Academy of Sciences? How did that integrate into Stalin's ultimate vision of political control? Could such power to give space be effectively limited by Stalin? Was it a failure of political reach? Or do we need to make alterations in our understanding of Stalin's system? I hope the present work has opened up these questions.

* * *

Though not comparative, this study adds to the growing literature on the different national experiences of professionalization, civic activism, and nature protection.[9] Those familiar with the history of the United States will see striking parallels between the Russian scientist activists and Progressive-era conservationists such as Gifford Pinchot or later ecologists such as Victor Shelford, Charles C. Adams, W. C. Allee, and a host of others, who believed that there was one best way to use the environment and that science was the sole institution that could identify that way. In the United States those claims were ultimately rejected or at least severely challenged by business and by voters, whereas in the Soviet Union they were rejected by the ruling Party. Yet, Russian scientists never irrevocably lost hope that enlightened (or just teachable) leaders could come to power in the Party, and consequently hung onto their technocratic claims longer than their American colleagues, perhaps because the lure of access to the Leviathan-state was so powerful, particularly as that Leviathan-state showed no signs of disappearing anytime soon.

Clearly, the case of the Soviet field biologists demonstrates one of the most determined and dogged efforts chronicled anywhere to preserve a traditional professional identity and esprit de corps in the face of adverse and dangerous conditions. Although only one case study, it does point to the hold of such older identities among professionals and to the possibility, even under Stalin's terroristic regime, of reaffirming them.[10]

Yet, even as this study confirms this, it underscores the need to examine each group of professionals on a case-by-case basis. Certainly not all scientists shared the same construal of *nauka*. Unlike field biologists, chemists in the Soviet period and even before became far less attached to basic research ("pure science") as a sacred task. This was understandable. Unlike the field biologists, chemists were valued actors in the push to industrialize the USSR. Consequently, as Nathan Brooks has shown, when the previously independent chemical societies were abolished in 1930–1932 and were replaced by a heavily politically controlled organization under Stalin loyalist A. N. Bakh, chemists were able to trade new lives for old and still feel that they came out ahead; the regime had only recently chartered a "Committee for the Chemization of the Economy" in 1928 and declared chemistry's centrality in the Five-Year Plan. Chemists, Brooks informs us, "enthusiastically threw themselves and their institutes into the task of rapid industrialization."[11]

Just as the relative independent-mindedness of the field biologists was conditioned by their marginal place in the political economy and their desire to study life forms in undisturbed nature with no obligation to generate practical benefits flowing from their research, the political conformity of Soviet chemists, with some few exceptions (N. N. Semënov comes to mind), was heavily determined by their importance in the eyes of the regime and their consequent higher status, and by their vastly greater dependence on that regime for the outfitting of their research facilities, a situation that virtually commanded good behavior. On balance, the promise of status, funding, and research opportunities more than compensated for the loss of professional autonomy, especially in its political dimensions, and of the "pure science" ideal. For field biologists, there really *were* no attractive new lives to trade for old, and so they defended their old professional identities with astonishing persistence. Curiously, that *defense of their identity* also became an important *component* of their professional identity, showing that even as professionals struggled to hang on to prerevolutionary identities, those identities nevertheless evolved. With time, it even became impossible to maintain or reproduce the older model of professional identity, owing to profound changes in education, the workplace, patronage, and science itself.

One noticeable difference between the histories of the nature protection movements of the United States and the Soviet Union is that in the Soviet Union nature protection was used to stake out an independent sphere where activists, whether students, scientists, or writers, could engage in self-initiated civic activity. Through that, they sustained professional and social identities that were also self-generated, in tacit opposition to the behavioral and professional norms set by the Party-state. True, in American history from Henry David Thoreau and John Muir to Earth First! there have been those whose feelings of cultural alienation from what they experienced as an oppressively materialist society led them to fashion socially dissenting identities around nature protection. With the possible exception of African-Americans, other people of color, and gays and lesbians, however, Americans never faced the barriers to the expression of a public social identity that Soviets did, which gives Soviet nature protection a completely different cast.

Moreover, until recently the preeminent social meanings of nature protection in American society have centered around preserving space for leisure, most notably for the middle class. Of the wide spectrum of protected territories in the United States, those enjoying comparable cultural importance to the Russian *zapovedniki* are the national parks. The original impetus for U.S. national parks was to preserve "the spectacular scenery of mountains, chasms, and geological freaks,"[12] or what Alfred Runte, historian of the national parks, called "monumentalism." From their outset, parks in the United States were designed for the delectation of the tourist and for the cultural reassurance of elites. With respect to biota, "the ideal animal was large and stood around in groups in the open—posing nobly in the middle distance against a background of mountain peaks—or entertained tourists with its 'cute' antics," wryly observes Thomas Dunlap. "No one thought of the parks as preserves for all species in a balance dictated by natural forces."[13] Indeed, so great was the emphasis on "display fauna" that broad campaigns of extermination were conducted in the parks to rid them of predators, "varmints," and other pesky life forms that might disrupt the safety or experience of the pleasure-seekers.[14]

If the foundations for "parks" as idealized nature were laid in late eighteenth-century England and Scotland, America brought this idea to its resounding realization. As Karen and Ken Olwig put it, the "baptism in the wilderness" of the "roaming cowboy with his six-shooter and guitar . . . was thought to have created the prototypical free American individual," and it was this "primeval" wilderness that the parks were thought to preserve.[15]

This perception of national parks as "virgin" nature, shared by scientists and laypeople alike, was a self-imposed cultural delusion based on the denial of the land's prior occupancy and transformation by Indians. This irony was noted by the Olwigs, who write:

> The Indians themselves were refused permission to remain in the area and the first Yellowstone tourists reportedly risked having the tranquillity of their sightseeing destroyed by army units clearing the area of Indians. Today, the tourists are no longer disturbed in their reverie of what is predominantly perceived as the "wilderness nature" of Yellowstone, and thus it is easier to maintain the illusion. The Park Service, however, finds it increasingly necessary to preserve the "natural" beauty of the landscape formed by centuries of Indian land use through controlled burning and other methods.[16]

Although American tourists and ecologists/preservationists have different conceptions of the appropriate *use* of national parks, they are united in seeing them as "national museums of our American wilderness," in the words of Stephen Mather.[17] Although American preservationism has frequently been an elite cultural crusade against Babbitry and rampant materialism (including middle-class tourism), whereas automobile-based tourism in the national parks has celebrated the arrival of tourists into the comfortable and materialist middle class—making the national parks into contrastive and contested symbols—both camps unite in viewing the parks as that "pristine" wilderness in which the nation was forged.

The symbolic importance to Russians of protected territories stands in stark contrast. In Russia the Stalinist Party-state was the principal adversary, and for educated citizens, especially the scientific intelligentsia, protected territories took on the aura of a geography of hope, an archipelago of freedom. Like Americans, Soviet nature protection advocates cherished delusions about protected territories: that they encompassed pristine, self-regulating, ecological communities that existed in a healthy equilibrium until the appearance of humanity. However, in the USSR, the stakes were considerably higher; unlike Soviet activists, few Americans risked their lives to assert a symbolic vision of national parks, nor were American environmentalists defending one of the last remaining islands of social autonomy in their country. As the Olwigs conclude: "One society's (or one class's) natural paradise may well be another's weed-overgrown garden; one's wilderness, another's home; one's Manifest Destiny, another's oppression. . . . Parks are no better or worse than the society which produces them."[18] Not just parks, I would add, but nature protection movements as social phenomena.

Nature protection does not exist as a disembodied eternal ideal. Indeed, it can only be understood as a cultural institution functioning in very specific contexts of space, time, and political economy, reflecting ever-changing and constantly contested visions and myths. This should not be read as throwing cold water over the attempts of many both here and abroad to save our planet's dwindling heritage of diversity. Rather, it is a call for those of us who defend that heritage to do so with as much self-awareness as we can muster. I hope that this book contributes to that goal.

GLOSSARY OF TERMS

biocenosis	a closed ecological community, consisting of fauna, flora, and inorganic components
chudak, -i	oddball(s), eccentric(s).
druzhina	lit., a fighting brigade; student organization for nature protection established in 1960; also, volunteer militia created by Khrushchëv in 1958
etalon, -y	ecological baseline(s)
Glavokhota RSFSR	the RSFSR Main Administration for Hunting and *Zapovedniki*
Glavpriroda SSSR	the USSR Main Administration for Hunting and *Zapovedniki* of the USSR Ministry of Agriculture
Gosplan	the State Planning Authority
IGAN	Institute of Geography of the Academy of Sciences
IUPN	International Union for the Protection of Nature (later, International Union for the Conservation of Nature)
Kedrograd	lit., Cedar City; an experimental forestry plantation founded by students of the Leningrad Technical Forestry Academy in the late 1950s, lasting officially to 1976
KIuBZ	the Young Biologists' Circle of the Moscow Zoo
kolkhoz, -y	collective farm(s)
Komsomol	the Young Communist League
kraeved	voluntary society for the study of local lore, crafts, folkways, and nature
krai	a type of province in the USSR; a territory
krai(obl-)ispolkom	provincial government executive committee
kruzhok, kruzhki	circle(s)

leskhoz,-y	Soviet forest plantation(s)
MGB	Ministry of State Security
MGO	Moscow branch of the Geographical Society of the USSR
MGU	Moscow State University
MOIP	Moscow Society of Naturalists
MVD	Ministry of Internal Affairs
nauchnaia obshchestvennost'	scientific public opinion; the self-designation of nature protection activists
nauka	science, with overtones of a sacred calling
NKVD	People's Commissariat of Internal Affairs (including the security police)
oblast'	a Soviet province
obshchestvennost'	public opinion, especially of educated society
RSFSR	the Russian Soviet Federated Socialist Republic; Russia
sovkhoz,-y	state farm(s)
studenchestvo	special student identity, traditionally with radical overtones
SR	Socialist Revolutionary (a non-Marxist party in Russia, fl. 1905–1918)
VOOP	All-Russian Society for the Protection of Nature
VOOPIiK	All-Russian Society for the Preservation of Monuments of Culture and History
VOSSOGZN	All-Russian Society for the Promotion and Protection of Urban Green Plantings
zakaznik	a protected territory established usually for a period of five or ten years, in which all or a part of its natural components may be protected
zapovednik	an inviolable nature reserve dedicated to long-term scientific, especially ecological, study
zubr	European bison; symbol of the old intelligentsia

NOTES

INTRODUCTION

1. For the formative period of this movement through 1933, see Weiner, *Models.*

2. Hosking, "Beginnings of Independent Political Activity," 1–2.

3. Weiner, *Models,* and especially the revised introduction to the Russian-language edition, *Ekologiia v Sovetskom soiuze.* Also Weiner, "Changing Face of Soviet Conservation." Elisabeth S. Clemens provides an illuminating example of how the temperance movement in the United States similarly served as a surrogate avenue for women to engage in politics, and, indeed, to change the way politics was conducted. See Clemens, *People's Lobby.*

4. The NEP (New Economic Policy) was inaugurated in 1921 by V. I. Lenin to help rebuild the Soviet Union's war-ravaged economy. It featured the relegalization of small-scale private enterprise and domestic commerce and was accompanied by a liberalization in the cultural sphere.

5. Zalygin, "Otkroveniia ot nashego imeni."

6. Gerovitch and Struchkov, "Epilogue," 489.

7. As late as 1989, V. V. Dëzhkin wrote in his *V mire zapovednoi prirody,* 158: "However, the main thing for us is our conviction that the question 'How much are *zapovedniki* worth?' is wrongheaded and not germane. *Zapovedniki* are valuable in and of themselves!" The term "geography of hope" is Wallace Stegner's.

8. Pronger, *Arena of Masculinity,* 93–94, sensibly reminds us: "Identity lies in primordial human being; cultural categories are only modes of understanding and action that are variously appropriated and relinquished by human beings. . . . [C]ategories are ways that people think of themselves and others in certain social spheres. These are ways of interpreting oneself in certain situations." Although I use the term *social identity* throughout the book, it is appropriate here to disavow a reified understanding of that term.

9. Yanitsky, *Russian Environmentalism,* 52. One university circle emerged just over a year earlier, independently, in Tartu, Estonia.

10. Interview with Evgenii A. Shvarts, Moscow, May 26, 1996.
11. Yanitsky [Ianitskii], *Russian Environmentalism,* 36.
12. Ibid., 167.
13. Ibid., 61.
14. Hosking, *First Socialist Society,* 408.
15. Huskey, *Russian Lawyers and the Soviet State.*
16. Shlapentokh, *Soviet Intellectuals and Political Power,* 55.
17. TsKhSD f. 17, op. 138, d. 35, listy 26–29.
18. RGAE f. 3, op. 1, d. 335, list 67 rev. Letter of July 23, 1957 to V. A. Varsonof'eva. Upon hearing about Fridman's death, Protopopov, in a letter to Varsonof'eva of February 10, 1959 (RGAE f. 3. op. 1, d. 325, list 40) wrote, "She deserves to be memorialized according to the customs of civic activism."
19. See RGAE f. 3, op. 1, d. 335, listy 64 (postcard of Fridman to Varsonof'eva of April 25, 1956); also letters of July 25, 1958 (listy 75–76 rev.); November 19, 1958 (list 78); late 1958 (undated) (listy 79–81 rev.).
20. Fehér, Heller, and Márkus, *Dictatorship over Needs,* 66.
21. Ibid, 63.
22. Ibid.
23. Yanitsky [Ianitskii], *Russian Environmentalism,* 92.
24. Petro, "'Project of the Century,'" 242.
25. Ibid., 243.
26. Scott, *Domination and the Arts of Resistance.* Hidden transcripts form part of the armamentarium of subordinate groups. See also Scott, *Weapons of the Weak.*
27. O. N. Ianitskii, "Ekologicheskaia politika i ekologicheskoe dvizhenie v Rossii," 19. Speech to the Working Group of eco-sociologists and leaders of ecological organizations, Moscow, May 19–21, 1995. Unpublished typescript, courtesy of O. N. Ianitskii.
28. See, for example, Jancar-Webster, "Eastern Europe and the Former Soviet Union," 210.
29. Yanitsky [Ianitskii], *Russian Environmentalism,* 9.

CHAPTER ONE

1. See Florenskii and Shutova, "Da skroetsia t'ma," 731; on the tradition of *studenchestvo,* see Kassow, *Students, Professors, and the State.*
2. Florenskii and Shutova, "Da skroetsia t'ma," 731. Kedrograd was an experimental forest plantation planned and run by graduates of the Leningrad Technical Forestry Academy, 1959–76. See chapter 15.
3. For an excellent discussion of the specifically Russian elements of professional identity in the late tsarist period, see the editor's introduction and conclusion in Balzer, *Russia's Missing Middle Class.*
4. Goldschmidt, *Human Career.*
5. See introduction and conclusion in Balzer, *Russia's Missing Middle Class.*
6. McClelland, *Autocrats and Academics,* 60. He quotes Kliment A. Timiriazev, a Russian plant physiologist and member of the "generation of the sixties."
7. Parallels to the eighteenth-century Masonic movement are suggestive. The Il-

luminati conceived of themselves as a higher *moral* community precisely because they were intellectually "enlightened," while regarding the defense and promotion of enlightenment as a sacred moral duty. See Koselleck, *Critique and Crisis,* especially chapter 6.

8. McClelland, *Autocrats and Academics,* 65.

9. Ibid., 69–71. See also Weiner, *Models,* 125ff.; Joravsky, *Soviet Marxism and Natural Science,* 153–63; and Graham, *Science in Russia,* 121–23.

10. McClelland, *Autocrats and Academics,* 71.

11. Ibid., 73. I have used McClelland's translation. The quotation is from Mikhal'son, "Rasshirenie i natsional'naia organizatsiia."

12. Quoted from Vernadskii, "Osnovoiu zhizni," 218–19, in Yanitsky, *Russian Environmentalism,* 11.

13. Kassow, *Students, Professors, and the State,* 6.

14. Ibid., 4–6.

15. On the Masons, Koselleck writes: "The Masons have nothing to do with politics directly. . . . [Nonetheless,] their virtue does not cease to be a 'crime'—that is, a threat to the State—until they, not the sovereign, determine what is right and what is wrong. Morality is the presumptive sovereign. Directly nonpolitical, the Mason is indirectly political after all. . . . [This] political secret of the Enlightenment was not to be shrouded just from the outside; as a result of its seemingly non-political beginnings it was concealed from most of the Enlighteners themselves." Koselleck, *Critique and Crisis,* 84–85.

16. Sergei Zalygin endorsed this point in his review of the Russian edition of *Models of Nature.* See Zalygin, "Otkroveniia ot nashego imeni," 215.

17. Notwithstanding the sometimes heroic efforts by Sergei Ivanovich Vavilov and other leaders of the USSR Academy of Sciences to defend the ideal. See Vucinich, *Empire of Knowledge.*

18. See Iaroshevskii, *Repressirovannaia nauka;* Kumanev, *Tragicheskie sud'by;* and Soyfer, *Lysenko.*

19. Kassow, *Students, Professors, and the State,* 8.

20. McClelland, *Autocrats and Academics,* 97; and Kassow, *Students, Professors, and the State, passim.*

21. Hosking, *First Socialist Society,* 403.

22. Zhigulin, *Chërnye kamni.* Zhigulin was one of the few survivors of a group of students in Voronezh who were arrested in 1949 for organizing a political group seeking a return to "Leninist norms."

23. See, for example, Dawson, *Eco-Nationalism.*

CHAPTER TWO

1. Gerovich and Struchkov, "Epilogue," 487.

2. For extensive background on the formative period of Russian nature protection, see Weiner, *Models;* Shtil'mark, *Istoriografiia;* and Boreiko, *Askania-Nova.*

3. See Weiner, *Models,* especially chapters 1 and 6.

4. Ibid., chapter 7.

5. Gor'kii, "Iz stat'i 'O biblioteka Poeta'," 56–57.

6. Weiner, *Models,* chapter 10 and *passim.*

7. Gerovich and Struchkov, "Epilogue," 488–89. On the theme of retreat from Stalinist realities as a journey to Kitezh, see also Turbin, "Kitezhane"; and Anna Akhmatova's 1940 poem, "Kitezhanka."

8. Vorontsov, "Nature Protection," 372ff.

9. Sepp, "Neobkhodimo reorganizovat' nauchnye obshchestva," 20.

10. Until 1933 the Main Administration was subordinated to the Science Sector of the RSFSR People's Commissariat of Education. In September 1933 it was directly subordinated to VTsIK, the All-Russian Central Executive Committee of the RSFSR Congress of Soviets. Later, in 1938 it was transferred to the aegis of the RSFSR Council of People's Commissars (which soon became the RSFSR Council of Ministers).

11. A *zemstvo* is a local, elective organ of self-government with responsibility for education, among other matters.

12. Biographical information is from V. N. Makarov, "Avtobiografiia" (typed and signed), April 19, 1938, and from "Avtobiografiia" (handwritten and signed), October 15, 1950, both from his personal archives. Courtesy of Feliks Robertovich Shtil'mark.

13. Makarov, 1938 "Avtobiografiia" and 1950 "Avtobiografiia." See the 1911 farewell poem from the first through third graders of the Kostroma First Four-Year School, entitled "Proshchal'nyi privet," and the farewell letter, dated September 28, 1913, from fourth-year students of the City School attached to the Moscow Teachers' Institute.

14. Makarov, 1938 "Avtobiografiia" and 1950 "Avtobiografiia."

15. See Weiner, *Models,* chapter 9.

16. See ARAN f. 1593, op. 1, d. 191, listy 1–130. "Stenogramma Vsesoiuznoi faunisticheskoi konferentsii."

17. See Weiner, *Models,* 141ff.

18. Ibid., 135.

19. From V. N. Makarov, letter to the Head of the Scientific Sector of Narkompros RSFSR and to its Party Collective, TsGA f. 404, op. 1, d. 1, listy 9–18. Emphasis in original.

20. Ibid., list 11, quoted from Kol'man, "Sabotazh v nauke," 75.

21. Makarov, letter to the Head of the Scientific Sector, listy 9–lo.

22. Although Oleg V. Khlevniuk, in *In Stalin's Shadow,* 58–60, points to the readiness of some leading factory managers as late as autumn 1933 to stand up to the USSR Prosecutor-General and even to use their private police forces to prevent entry by the state's law enforcement agents, the situations are not exactly comparable.

23. See especially Boreiko's series "Istoriia boli i geroizma"; and Shtil'mark, *Istoriografiia.*

24. The most prominent of these were Dmitrii Evstaf'evich Beling, Evdokiia Grigor'evna Bloshenko, Aleksei Feodos'evich Vangengeim, Pëtr Evgen'evich Vasil'kovskii, Vladimir Georgievich Geptner, Aleksandr Pavlovich Gunali, Iurii Andreevich Isakov, Vladimir Emmanuelovich Martino, Sergei Ivanovich Medvedev, Vasilii Petrovich Nalimov, Mikhail Nikolaevich Poloz, Boris Evgen'evich Raikov, Pëtr Petrovich Smolin, Vladimir Vladimirovich Stanchinskii, Nikolai Mikhailovich Fëdorovskii, Boris Konstantinovich Fortunatov, Khachatur (Khristofor) Georgievich Shaposhnikov, Frants Frantsevich Shillinger, Aleksandr Aleksandrovich Shummer, and Aleksandr

Aloizevich Ianata. From Vladimir Evgen'evich Boreiko, "Razgrom prirodookhrany v SSSR (1929–1939)," in his *Belye piatna istorii,* 161. Boreiko includes a few more names than I have listed here but even his list is surely incomplete.

25. When I first described these events in a previous book, much of my account was inferred from published speeches and journal articles, which were the only sources available then. In the intervening period many more sources about Stanchinskii and Askania-Nova have come to light through the efforts of Ukrainian researcher Vladimir Evgen'evich Boreiko and me, generally confirming the broad flow of events.

26. This question will be taken up in detail in my forthcoming biography of Vladimir Vladimirovich Stanchinskii, in collaboration with Vladimir Evgen'evich Boreiko.

27. Boreiko, *Askania-Nova,* 91. On Stanchinskii and Askania-Nova, see also Weiner, *Models;* Gramma, "Eretiki i buntari"; and Boreiko, *Populiarnyi biografo-bibliograficheskii,* 2:119–27. The last includes a bibliography about Stanchinskii's life and work.

28. Encouragingly, Bukharin's deputy editor on the staff of *Izvestiia,* Kantorovich, paid a visit to the Kavkazskii *zapovednik* and wrote up the trip in the paper. Interview with Andrei Aleksandrovich Nasimovich, April 16, 1980.

29. Letter of F. F. Shillinger, no date cited, ARAN f. 445, op. 4, d. 190, listy 111–12. Cited in Boreiko, "Frants Frantsevich Shillinger," 204–5. Another letter, to the academician N. M. Kulagin and dated November 7, 1937, which contains almost the same language and was probably written at the same time, is cited by Shtil'mark, *Istoriografiia,* 80, in ARAN f. 1674, op. 1, d. 339 (no listy cited).

30. Ol'ga Borisovna Lepeshinskaia, "Dokladnaia zapiska nauchnomu otdelu TsK VKP(b) o rezul'tatakh obsledovaniia zapovednikov v RSFSR. Zametki. Otpusk," nine pages of two-sided typed carbons plus longhand, penciled notes in a small notebook, listy 10–102, in ARAN, f. 1588, op. 1, d. 102, listy 1–102. The author is the same Olga Lepeshinskaia who in the late 1940s and early 1950s became one of the most notorious scientific cranks with her claim that she could produce living beings from albumin.

31. Shtil'mark, *Istoriografiia,* 80. See ARAN f. 445, op. 4, d. 19, listy 76–80.

32. Ibid., list 1 rev.

33. Ibid., listy 2 rev., 4. She implied that in 1922 Shillinger planned the creation of the Crimean reserve to promote a White landing ("Snaipery. Bloshenko, Shelinger [sic]. Ideia o sozdaniia zapovednika na beregu Kryma v 22 godu," list 4).

34. Ibid., list 25. These allegations were totally without evidence and have been denied by Adela Frantsevna Shillinger, the activist's daughter, who still lives in Moscow.

35. Ibid., list 3.

36. Ibid.

37. Ibid., listy 5 and 7 rev.

38. Ibid., list 5.

39. Ibid., list 27.

40. Ibid., list 26 rev.

41. Ibid., list 16 rev. Lepeshinskaia blames Gorbunov and Krylenko for recommending these politically suspect individuals.

42. Ibid., list 28 rev.
43. Ibid., list 7 rev.
44. Ibid., list 8.
45. Ibid., listy 9, 37 rev., and 39.
46. Shtil'mark, *Istoriografiia,* 80.
47. Boreiko, "Frants Frantsevich Shillinger," 206.
48. TsGA f. 404, op. 1, d. 54, list 37.
49. TsGA f. 404 op. 1, d. 1, listy 3–4.
50. Ibid., listy 24–25 rev. "Protokol No. 1 zasedaniia Komissii VOOP po Kavkazu ot 26 fevralia 1937 g."
51. TsGA, f. 404, op. 1, d. 56, list 85.
52. TsGA, f. 404, op. 1, d. 54, list 9.
53. TsGA, f. 404, op. 1, d. 56, list 12.
54. TsGA, f. 404. op. 1, d. 53, listy 50, 13.
55. Ibid., listy 16–17.
56. Ibid., list 25.
57. Ibid., list 26–27.
58. TsGA, f. 404. op. 1, d. 56, list 17.
59. Ibid., listy 28–32.
60. TsGA 404, op. 1, d. 53, list 55.
61. Ibid., list 35 Eventually, VOOP's president Komarov wrote to Soviet premier V. M. Molotov, who then apparently remanded the Division of Nonresidential Housing of the Moscow Soviet to provide premises for the Society. When Molotov's injunction was ignored, Komarov wrote again to the premier. Although the letter is undated, we may surmise it was written in 1938 or 1939. TsGA, f. 404, op. 1, d. 1, list 2.
62. Ibid., list 39.
63. A full financial report was presented by A. G. Giller, who traced the Society's funds from its inception in 1924. In that year, its subsidy from Narkompros amounted to 1,000 rubles, which was increased the following year to 4,000. Despite inflation as well as the growth of the Society, funding levels remained static until 1933, when they were doubled. VTsIK was more generous. In 1935 it contributed 15,000 rubles; in 1936, 20,000; in 1937, 25,000; and 40,000 was promised in 1938. Additionally, the Society received juridical members' dues and contracts worth 30,000 rubles in 1935; 11,000 in 1936; 12,000 in 1937 and 35,000 in 1938. On the outflow side, publications, staff, and operations each consumed about one-third of the budget. Ibid., listy 49–51.
64. TsGA, f. 404, op. 1, d. 52, list 11, in pen, at bottom of typed page.
65. TsGA, f. 404, op. 1, d. 56, list 55.
66. Ibid., list 7.
67. Ibid., list 52.
68. Barishpol and Larina, *U prirody druzei milliony,* 40.
69. Ibid., 44.
70. TsGA, f. 404, op. 1, d. 78, list 1. "Otchet TsS VOOP o rabote obshchestva za 1939 god." 27 listy. Typed. By Susanna N. Fridman.
71. Ibid.
72. Ibid., list 7.

73. See TsGA f. 404, op. 1, d. 1, listy 30–31. Dogovor, August 26, 1938. This agreement between VOOP and the Committee for *Zapovedniki* provided for 4,000 rubles of funding from the Committee and a contribution of 1,157 rubles by the Society. Field research included a desman census in September–October 1938 along the basin of the Tsna, Chelovaia, and Stanovaia Riasa rivers in Tambov and Voronezh provinces.

74. Ibid., list 9.

75. Ibid., listy 11–12.

76. bid., listy 13–17.

77. Ibid., list 4 rev.

78. TsGA f. 404, op. 1, d. 130a, list 3 rev. "Otchet o rabote VOOP za 1938–1947 gg." 8 listy.

79. Ibid., list 7.

80. The environmental consequences of World War II are discussed in Noskov, "Environmental Devastation in the USSR."

81. TsGA, f. 404, op. 1, d. 1, listy 27–28. "Novye zadachi zapovednikov i okhrany," n.d. [c. 1944], typed, carbon.

82. Interview with Nikolai Sergeevich Dorovatovskii, Moscow, January 3, 1980.

83. Ibid.

84. Ibid.

85. Ibid.

86. TsGA RSFSR 404, op. 1, d. 115a, listy 6–7.

87. Medvedev and Nechaeva, "Pamiati V. V. Stanchinskogo," 113.

88. Ramensky, "Basic Regularities of Vegetation Cover," 12.

89. Ramenskii, "Klassifikatsiia zemel' po rastitel'nomu pokrovu," 484.

90. Interview with Iakov Mikhailovich Gall, Leningrad, May 20, 1986. Sukachëv's observation will appear in a new volume of his works now being prepared for publication.

91. A similar argument is made by Robert Hunt Sprinkle in *Profession of Conscience,* who includes in this moral community doctors, molecular geneticists, and other life scientists.

92. Quoted in Struchkov, "Nature Protection as Moral Duty," 428.

93. Shtil'mark, "S drugogo berega," 379.

94. RGAE f. 59, op. 1, d. 96, list 1. Letter of V. N. Makarov to G. P. Dement'ev, August 19, 1934.

CHAPTER THREE

1. TsGA f. 404, op. 1, d. 115, listy 1–3, from *Protokoly zasedanii soveta i prezidiuma ob-va.*

2. Ibid., listy 7, 14–15.

3. Ibid., listy 14–15.

4. Ibid., list 34. The Council took this decision on December 18, 1945.

5. TsGA f. 404, op. 1, d. 120, listy 8, 8 rev., 9.

6. TsGA, f. 404, op. 1, d. 131, list 1. "Otchet o deiatel'nosti VOOP za 1945–1947 gg."

7. Ibid., list 4.
8. Ibid., listy 1, 4.
9. Interview with Mikhail Aleksandrovich Zablotskii, Priioksko-terrasnyi *zapovednik,* April 20, 1991.
10. Browne, "European Bison," 2.
11. Il'inskii, "Belovezhskaia pushcha."
12. Interview with Mikhail Aleksandrovich Zablotskii, Prioksko-Terrasnyi *zapovednik,* April 20, 1991.
13. Ibid.
14. See Bailes, *Technology and Society.*
15. Zablotskii, "Vosstanovlenie zubra v SSSR i za granitsei," 56.
16. On Dubinin, see Soyfer, *Lysenko.*
17. TsGA, f. 404, op. 1, d. 120, list 8.
18. Interview with Mikhail Aleksandrovich Zablotskii, Prioksko-Terrasnyi *zapovednik,* April 20, 1991.
19. *Gosudarstvennaia,* 1:8.
20. Ibid., 7.
21. Interview with Mikhail Aleksandrovich Zablotskii, Prioksko-Terrasnyi *zapovednik,* April 20, 1991.
22. TsGA f. 404, op. 1, d. 99, listy 35 rev., 36.
23. TsGA, f. 259, op. 6, d. 3519, listy 1–6. Emphasis mine.
24. Ibid., listy 11–12.
25. TsGA, f. 259, op. 6, d. 3518, listy 63–66.
26. TsGA, f. 404, op. 1, d. 122, list 23.
27. Ibid., listy 27–31.
28. TsGA f. 404, op. 1, d. 131, list 5 rev.
29. TsGA f. 404, op. 1, d. 122, listy 24–26.
30. Ibid., listy 6–7. The decree was Postanovlenie SM RSFSR No. 642, "Ob okhrane prirody na territorii RSFSR".
31. TsGA f. 404, op. 1, d. 120, listy 17–18.
32. TsGA f. 259, op. 6, d. 5419.416–10, list 17.
33. Ibid.
34. TsGA 404, op. 1, d. 36, listy 6, 6 rev. "V otdel pechati Tsentral'nogo komiteta VKP/b/." No date [late 1946], carbon.
35. Ibid., list 5.
36. Soyfer, *Lysenko,* 227. S. V. Zonn, in an interview conducted in Moscow, August 3, 1995, offered a rather generous and empathetic assessment of Tsitsin. "A very cultured, modest person," Tsitsin had previously worked at the Institute for Agriculture of the South-East (Institut zemledeliia iugo-vostoka) in Saratov in the field of selection and knew N. M. Tulaikov and N. I. Vavilov as colleagues. Accordingly, he took an anti-Lysenkoist position from the start and remained friendly with V. N. Sukachëv. On the other hand, as director of the Academy's main botanical garden, he tried to avoid public controversy, perhaps excessively so. "He found himself in a bind," concluded Zonn. "He was squeezed between his public position of authority and his inner beliefs."
37. RTsKhIDNI f. 17, op. 138, d. 33, list 34.
38. RTsKhIDNI f. 17, op. 138, d. 35, list 2. The report itself, entitled "*K voprosu*

o vetvistykh formakh pshenitsy" [On the Question of Branched Forms of Wheat], occupies listy 3–28. Malenkov came across this report and wrote a note to his aide, Kozlov: "I ask you to read this through and to report back to me" (list 39). However, further archival documents thus far give no clue as to the further or final disposition of this affair.

39. Ibid., list 18.

40. Ibid., list. 19.

41. Ibid., listy 20, 25.

42. Ibid., list 28. Tsitsin ridiculed Lysenko for claiming to be able to guide the "sculpting" (*on tak bukval'no i vyrazhaetsia, "lepit'"*) of the organism, drawing attention to the nonscientific language of the barefoot selectionist. In Russian, the verb *lepit'* is used in connection with pottery and working with clay.

43. Ibid., list 27.

44. Ibid., listy 26–29.

45. Ibid., listy 88–94.

46. Ibid., list 140.

47. Ibid., list 145.

48. Ibid., list 146.

49. Ibid., list 146.

50. Ibid., list 167.

51. Ibid. To the Central Executive Council of the Society were elected forty-one, including such notables as the academicians Vladimir Nikolaevich Sukachëv, L. D. Sheviakov, and Aleksandr Ivanovich Oparin; the head of the Moscow *oblast'* government, P. G. Burylychev; the head of the Crimean *oblast'* government, D. A. Krivoshein; and old-timers Susanna N. Fridman, Sergei N. Ognëv, V. G. Geptner, K. N. Blagosklonov, D. M. Viazhlinskii, R. F. Gekker, P. A. Manteifel', M. P. Rozanov, and A. N. Formozov.

52. TsGA f. 404, op. 1, d. 127, listy 24–25.

53. *Okhrana prirody, sbornik* 1 (1948): 131.

54. TsGA f. 404, op. 1, d. 127, listy 146–47.

55. Ibid., listy 144–45.

56. Makarov, *Okhrana prirody v SSSR,* 11. During the 1940s, beginning even before the war, the journal *Nauka i zhizn'* had begun an episodic series on the *zapovedniki* of the USSR, usually accompanied by photographs and occasionally maps. The geographer and VOOP activist S. M. Preobrazhenskii wrote many of these articles. See especially *Nauka i zhizn'* 4 (1940) on the Pechoro-Ilychskii reserve; 1 (1941) on "Buzulukskii bor"; 1 (1948) on the Voronezhskii reserve; and 3 (1947) for Preobrazhenskii's survey article.

57. See Starr, "Visionary Town Planning"; and Stites, *Revolutionary Dreams.*

58. Leonid Leonov, "V zashchitu Druga," *Izvestiia,* 28 December 1947, reprinted in Leonov, *Sobranie sochinenii,* 10:185–95.

59. TsKhSD f. 17, op. 138, d. 35, list 2.

60. Ibid., list 4.

61. TsGA f. 404, op. 1, d. 136a, listy 37–38.

62. Ibid., list 35.

63. Ibid., listy 63, 63 rev.

64. TsKhSD, f. 5, op. 16, d. 40, listy 184, 184 rev., 185–86.

65. Ibid., list 194.
66. Ibid., listy 183–86.
67. Ibid., listy 192–93 rev.
68. Ibid., listy 198–99.
69. Ibid., listy 200–201.
70. TsGA f. 404, op. 1, d. 136a, listy 136, 146.
71. Ibid., list 126.
72. *Okhrana prirody, sbornik* 3 (1948): 128; 4 (1948): 120–25.
73. TsGA f. 404, op. 1, d. 159, list 14.

CHAPTER FOUR

1. See, for example, "Protokol no. 10 Nauchnogo soveta Voronezhskogo Goszapovednika ot 29 sentiabria 1948 g.," TsGA f. 358, op. 2, d. 612, listy 37–41, at which classical genetics was condemned as "imported science" and all-out assistance pledged to the newly announced Stalin Plan for the Great Transformation of Nature. See also "Vypiska iz protokola zasedaniia uchenogo soveta Darvinskogo Goszapovednika ot 15.IX.1948 g.," ibid., listy 45–46.

2. This is expanded upon in my new work in progress, *"Curiosity for Its Own Sake:" Boris Evgen'evich Raikov and Natural Science Education in the Soviet Union.*

3. See his report, "Itogi raboty attestatsionnoi kommissii Biofaka MGU" [Report of the Certification Commission of the Biological Faculty of Moscow State University], prepared and signed by him as dean not long after the August 1948 session of VASKhNIL. ARAN f. 1593, op. 1, d. 106, listy 1–19. Prezent remarked, apropos of the department of vertebrate zoology, that "to date a reorganization [*perestroika*] of work in accordance with the decisions of the August session . . . is proceeding totally unsatisfactorily. Professors Matveev, Formozov, and Geptner, who suffer from a whole series of the most serious methodological errors, have not subjected those errors to critical reexamination." Formozov was quickly fired.

4. Vorontsov, Afterword to *Stranitsy zhizni,* 325. See also Medvedev, *Rise and Fall of T. D. Lysenko,* 121–30. A third highly readable and personal memoir is that of the late Vladimir Iakovlevich Aleksandrov, *Trudnye gody sovetskoi biologii.*

5. See TsGA f. 259, op. 6, d. 5418, list 30. Shvedchikov was so frequently ill that from late 1947 to December 28, 1949, V. N. Makarov ran the Main Administration as acting director.

6. Ibid.

7. Ibid., listy 33–35.

8. Ibid., listy 31–32. Evidence, including a previous note from Motovilov to Gritsenko, makes it almost certain that this note to Malenkov was written after July 3.

9. Ibid., listy 25–28, especially 27–28. Memo from A. Suchkov, August 13, 1947.

10. Ibid., listy 23–24. Suchkov signed the bottom of Gritsenko's memo and dated it September 6, indicating that Suchkov reported to Gritsenko rather than to Rodionov directly.

11. Ibid., listy 21–22. Note of September 18, 1947 of Rodionov to Malenkov, initialed by Suchkov at the bottom.

12. Ibid., listy 18–20.

13. Ibid., listy 16–17. Dated only October 1947.

14. Ibid., list 17.
15. Ibid., listy 3–4.
16. TsGA f. 358, op. 2, d. 827, listy 35–201.
17. Ibid., list. 189.
18. Ibid., list 189 rev.
19. Ibid., list 182.
20. Ibid.
21. TsGA f. 358, op. 6, d. 5416, listy. 3–5. Cited in Shtil'mark and Geptner, "Tragediia sovetskikh zapovednikov," 100.
22. Ibid.
23. TsGA f. 259, op. 6, d. 5418, list 1.
24. Makarov, "Gosudarstvennye zapovedniki RSFSR," 6.
25. Postanovlenie SM SSSR and TsK KPSS no. 3960 ot 20 oktiabria 1948 g. "O plane polezashchitnykh lesonasazhdenii, vnedreniia pravopol'nykh sevooborotov, stroitel'stva prudov i vodoemov dlia obespecheniia vysokykh i ustoichivykh urozhaev v stepnykh i lesostepnykh raionakh Evropeiskoi chasti SSSR." This had been preceded on May 17, 1948 by a decree of the USSR Council of Ministers, "Ob uporiadochenii pol'zovaniia kolkhoznymi lesami i uluchshenii vedeniia khoziaistva v nikh."
26. TsGA f. 259, op. 6, d. 5411, listy 41–48, which is the report of the chairman of the Bashkir ASSR Council of Ministers to the SM RSFSR, dated December 20, 1948. The report describes the various ASSR- and *raion*-wide meetings of Party, government, farm, and scientific organizations to develop concrete plans for afforestation in Bashkiria. Pressure from the center must have been immense; in a comment on the Bashkir leader's report written eight days later, the RSFSR deputy premier V. Makarov described that republic's fulfillment of the plan as "extremely unsatisfactory" despite the evident flurry of activity.
27. Medvedev, *Rise and Fall of T. D. Lysenko*, 130.
28. Prezent, "Peredelka zhivoi prirody," 8.
29. They were: 1. Saratov–Astrakhan on both sides of the Volga (width, 100 meters; length, 900 kilometers); 2. Penza–Northern Donets (three belts, each sixty meters wide; 300-meter separation from each other; 600 kilometers in length for the parallel belts); 3. Kamyshin–Stalingrad (same as above; length, 170 kilometers); 4. Stalingrad–Cherkessk; 5. Ural River–Caspian Sea; 6. Voronezh–Rostov-na-Donu; 7. Belgorod (Northern Donets)–Don River. See Postanovlenie SM SSSR and TsK KPSS no. 3960 ot 20 oktiabria 1948 g. "O plane polezashchitnykh."
30. Sergei Vladimirovich Zonn, Leonid Fëdorovich Pravdin, and Nikolai Vladimirovich Dylis—all close Sukachëv associates—were fired from their positions as adjunct professors at the Moscow Forestry Institute, while the great soil scientist Aleksei Andreevich Rode lost his job in Briansk. All were labeled "Weismannist-Morganists," or upholders of classical genetics. Interview with Sergei Vladimirovich Zonn, Moscow, August 3, 1995.
31. Ibid.
32. Ibid.
33. Soyfer, *Lysenko*, 168
34. Interview with Sergei Vladimirovich Zonn, Moscow, August 3, 1995.
35. T. D. Lysenko, "Estestvennyi otbor i vnutrividovaia konkurentsiia," 370,

quoted in Shelkovnikov, "K voprosu o 'mal'tuzianstve' v biologii." This was a response to Lysenko from the Sukachëv camp.

36. Anuchin, "Stepnoe lesorazvedenie ne nuzhdaetsia v naukoobraznom uchenii i bigeotsenoze."

37. TsGA f. 259, op. 6, d. 6185, list 4.

38. Ibid., list 1. Appended were a list of talks.

39. Ibid., list 5.

40. Ibid., listy 25 rev. and 26.

41. TsGA f. 259, op. 6, d. 6188, listy 5–9.

42. Ibid., list 5.

43. Ibid.

44. Ibid., list 6.

45. Ibid., list 7.

46. TsGA f. 259, op. 6, d. 7150, list 58. See TsGA f. 358, op. 2, d. 827, list 193.

47. TsGA f. 259, op. 6, d. 7150.415–6, papka 2, list 60. "Spravka o rabote Glavnogo Upravleniia po zapovednikam pri SM RSFSR," Po porucheniiu Sel'khozotdela TsK VKP(b) i SM RSFSR, conducted by G. S. Svetlakov, starshii pomoshchnik zaveduiushchego sekreteriatom SM RSFSR, December 28, 1949.

48. Ibid., list 61.

49. Ibid., list 62.

50. TsGA f. 259, op. 6, d. 7150, listy 63–100.

51. Postanovlenie SM RSFSR no. 1093 ot 28.XII.1949.

52. On Malinovskii, see Grigor'ev, "Aleksandr Vasil'evich Malinovskii," 29; and "A. V. Malinovskii" [obituary], 28. His most complete biography was compiled in January 1951 for V. N. Merkulov, USSR Minister of State Control. See GARF, f. 8300, op. 24, d. 1786, list 32.

53. GARF, f. 8300, op. 24, d. 1786, list 32.

54. Ibid.

55. See TsGA f. 259, op. 6, d. 7148, list 23. Chaired by M. N. Shcherbakov of the RSFSR Ministry of State Control, its other members were A. I. Korshunov of the RSFSR Ministry of Finances and I. A. Grigor'ev from the RSFSR Ministry of Forestry.

56. TsGA f. 259, op. 6, d. 7147, list 55.

57. Ibid. The subtext was that the RSFSR should not enact any future budgets or plans for the reserve system or the Main Administration until the Party's intentions became clear.

58. Conversation with Andrei Aleksandrovich Nasimovich, Moscow, April 18, 1980.

59. TsGA f. 259, op. 6, d. 7148, listy 21–23 rev., and f. 259, op. 6, d. 7147, list 22. The latter document is the "Protokol sobraniia aktiva sotrudnikov Glavnogo Upravleniia po zapovednikam pri SM RSFSR, 10-ogo ianvaria 1950 g. po voprosu oznakomleniia s priemo-sdatochnym aktom, v sviazi s naznacheniem novogo Nachal'nika . . . i s vyvodami i predlozheniiami Pravitel'stvennoi Komissii." Twenty-one persons attended the meeting, which was addressed by M. N. Shcherbakov, chair of the Government Commission, and by Malinovskii himself.

60. TsGA f. 259, op. 6, d. 7147, 22 rev. They sought inclusion in the provisions of the Decree of the USSR Council of Ministers of March 6, 1946 no. 514, which

only applied to scientific institutes, from which category the *zapovedniki* were technically excluded.

61. TsGA f. 358, op. 2, d. 836, list 17, cited in Shtil'mark and Geptner, "Tragediia sovetskikh zapovednikov," 102.

62. Shtil'mark and Geptner, "Tragediia sovetskikh zapovednikov," 102.

63. TsGA f. 358, op. 2, d. 838, list 55. Dated December 8, 1949.

64. Ibid., list 54.

65. Ibid. listy 53–52.

66. TsGA f. 358, op. 2, d. 838, listy 76 rev., 76, 75 rev., 75. The territory represented about 10 percent of the total of the southern island of Novaia Zemlia.

67. RGAE f. 8512, op. 1, d. 411. A note of July 25, 1950 from G. Orlov, minister of the paper and woodworking industry of the USSR to Georgii M. Malenkov complained that the ministry was experiencing a severe shortage of specialists despite a decree of September 3, 1948 obligating Gosplan USSR and the Ministry of Higher Education to make 75 percent of all graduating students of forestry academies and institutes available to the ministry.

68. For the draft, with marginal notes that may be Stalin's own, see RGAE f. 8512, op. 1, d. 438.

69. RGAE f. 8512, op. 1, d. 438, listy 130–39.

70. Ibid., esp. list 137.

71. TsGA f. 358, op. 2, d. 838, listy 238–45. The typewriter used to produce an appendix (listy 243–45), signed by Malinovskii, seems to be the same used to type the draft of the proposed All-Union Decree. Additionally, they were found in the same set of documents.

72. Ibid., list 242.

73. TsGA f. 358, op. 2, d. 836, listy 3–4, quoted in Shtil'mark and Geptner, "Tragediia sovetskikh zapovednikov."

74. TsGA f. 358, op. 2, d. 836, listy 242–41.

75. Ibid., list 241.

76. Decrees of SNK SSSR of March 6, 1946, no. 514, and SM SSSR of August 28, 1947, no. 3020.

77. TsGA f. 358, op. 2, d. 836, list 240.

78. It is possible that the so-called Presidential Archives, containing much sensitive material from the former Central Committee archives, may hold the key to this question, if Stalin or his Politburo or Secretariat colleagues became involved in this issue personally at this stage.

79. The materials are from February 20, 1950.

80. TsGA f. 358, op. 2, d. 836, list 239.

81. Ibid., listy 238–39.

CHAPTER FIVE

1. TsGA f. 358, op. 2, d. 838, list 133. Makarov was asked to prepare a review and revision of the statute on the Main Administration and on *zapovedniki* by February 15, 1950, which presupposes that he was at least given guidelines if not shown the whole texts of the draft law and reorganization plan.

2. TsGA f. 358, op. 2, d. 827, listy 73, 129.

3. TsGA f. 358, op. 2, d. 838, listy 133–37. The last page is dated and signed by Malinovskii. The date appears to be February 20, 1950, but the figure for the day is not completely clear.

4. TsGA f. 358, op. 2, d. 838, listy 134, 135, 137. We have no documentation of Malinovskii's appearance at the ministry, but we do know that on May 24 he was present at a session of the Bureau of the RSFSR Council of Ministers chaired by Chernousov.

5. The reserve was founded by decree of the SM RSFSR no. 489 of May 18, 1948. The USSR Rasporiazhenie no. 7611-r was dated June 16, 1948. Cited TsGA f. 259, op. 6, d. 7152.415–9, list 1. The April 28, 1950 decree was no. 1789.

6. TsGA f. 259, op. 6, d. 7152.415–9, list 2.

7. Ibid., list 1.

8. Ibid. In pen at the bottom of his letter it was noted that the question was revisited by the USSR Council of Ministers on June 29, 1950 in decree no. 14194-r.

9. TsGA f. 358, op. 2, d. 838, listy 283 and 283 rev.

10. TsGA f. 358, op. 2, d. 838, listy 229–30.

11. "Proekt. Tovarishchu Saburovu, zam. preds. SM SSSR, preds. Gosplanu SSSR. Iul' 1950," 2–3. Five pages, carbon. From the personal papers of A. N. Formozov, courtesy F. R. Shtil'mark.

12. Ibid., 4–5.

13. GARF f. 8300, op. 24, d. 1704, listy 18–19, cited in Boreiko, "Eto ne dolzhno," 8.

14. Ibid. He has found such a request in the Ukrainian republican archives, TsGAVO Ukrainy f. 2, op. 2, d. 4516, list 16.

15. As per "Poruchenie Prezidiuma SM SSSR ot 30-ogo iulia 1950 g." See RGAE f. 4372, op. 50, d. 17(2), list 103.

16. The RSFSR Ministry of State Control was a Union-republic ministry. In light of its importance in matters of state security—it was a major investigative organ—it seems likely that the republican-level ministries were at least partially controlled by the Union ministry, which was represented by "its" people on the local level.

17. TsGA f. 358, op. 2, d. 838, list 305.

18. TsGA f. 404, op. 1, d. 182a, list 104.

19. Ibid., listy 87–90.

20. Ibid.

21. Ibid., list 102.

22. Ibid., list 100. Apparently, Makarov was ailing during this time and these letters were signed by Dement'ev and Kuznetsov.

23. TsGA f. 259, op. 6, d. 7148, list 20.

24. TsGA f. 259, op. 6, d. 7152, list 7.

25. TsGA f. 358, op. 2, d. 838, list 266.

26. Ibid., listy 339, 339 rev.

27. Ibid., list 340.

28. Boreiko, "Eto ne dolzhno," 9.

29. RGAE f. 4372, op. 50, d. 17(2), listy 103, 111. Evidently another copy of Saburov's report exists in the archives; Boreiko, "Eto ne dolzhno," 9, provides the following location as well: GARF f. 8300, op. 24, d. 1786, listy 90–99.

30. The figure of 1,864 is from RGAE f. 4372, op. 50, d. 17(2), list 114.

31. Ibid. Saburov officially presented his findings to the USSR Council of Ministers in the draft decree "Ob uluchshenii raboty zapovednikov SSSR" on November 18, 1950, no. 10024 of Gosplan USSR. Three days later the academician Ivan P. Bardin, a leader of technology development and a vice president of the Academy of Sciences, sought to take advantage of the shaky position of the reserve system to claim the Il'menskii *zapovednik* for the Academy of Sciences. Under Bardin's patronage, the reserve would presumably have focused exclusively on mineralogical studies. Tellingly, Saburov wrote to Bardin on December 4 rejecting his proposal and arguing that the Main Administration was better suited to protect all of the components of the reserve, including the fauna and flora. See RGAE 4372/50/17(2) l. 170. We do know that Saburov and Kaganovich opposed Khrushchëv's Virgin Lands program and, in the words of T. A. Iurkin, who accused them at the December 1958 Central Committee Plenum, "tried to call science to their aid." See Conquest, *Power and Policy in the U.S.S.R.*, 237. This raises the intriguing question of Saburov's ties to the scientific intelligentsia.

32. Boreiko was able to find some of the paper trail in the state archives of Ukraine. See TsGAVO Ukrainy f. 2, op. 2, d. 4516, list 16, cited in Boreiko, "Eto ne dolzhno," 9.

33. Boreiko, "Eto ne dolzhno," 9.

34. Ibid.

35. TsGA f. 358, op. 2, d. 838, list 389.

36. Boreiko, "Eto ne dolzhno," 9. The order to the RSFSR Ministry of State Control came down a day or two later, as it was only on November 27 that this ministry in turn issued instructions to its operatives in order no. 307. See TsGA f. 339, op. 1, d. 4987, list 4.

37. Ibid., listy 2–3.

38. Ibid.

39. RSFSR minister of state control N. Vasil'ev's order no. 307 of November 27, 1950, a response to decree no. 1161 of the USSR Ministry of State Control of November 24, 1950, is found in TsGA f. 339, op. 1, d. 4987, list 1. Vasil'ev gave a deadline of December 2 for the completion of the investigation of the Main *zapovednik* Administration and of December 12 for that of the individual reserves.

40. TsGA f. 339, op. 1, d. 4987, list 4.

41. Ibid., list 6.

42. TsGA f. 339, op. 1, d. 4987, list 12. Kornilov's article appeared in the May 20, 1950 number of *Moskovskaia pravda*.

43. TsGA f. 358, op. 2, d. 991, listy 5–8.

44. Ironically, both Nasimovich and Dul'keit were supporters of the "Mendelist-Morganist-Weismannist" position, as classical genetics was derogatively characterized by the Lysenkoist camp. However, Nasimovich did not actively lie in his assessment of Dul'keit's work. There was no *evident* allusion to Mendelian genetics in Dul'keit's zoological studies.

45. Boreiko, "Eto ne dolzhno," 9. Boreiko also includes Trofim D. Lysenko among those present at the meeting, but in the archival document "Spisok uchastnikov soveshchaniia u Ministra Gos. Kontrolia SSSR tov. Merkulova V. N. po voprosu o rabote gos. zapovednikov," GARF f. 8300, op. 24, d. 1786, list 9, Lysenko's name is crossed out in pencil.

46. MOIP Archives, d. 54, list 63. Cited in Shtil'mark and Geptner, "Tragediia sovetskikh zapovednikov," 109–10.

47. TsGA f. 358, op. 2, d. 882, listy 3–4. "Stenogramma zasedaniia Nauchnogo soveta pri Glavnom Upravlenii po zapovednikam pri Sovete Ministrov RSFSR ot 25-ogo dekabria 1950 g."

48. Ibid., list 4.

49. Ibid., list 8.

50. Ibid., list 6.

51. Ibid., list 7.

52. Ibid., list 10.

53. Conversation with Mikhail Aleksandrovich Zablotskii, to whom the remark was made. Priroksko-Terrasnyi *zapovednik,* April 20, 1991. ("Malinovskii—eto zloi genii zapovednikov.")

54. TsGA f. 358, op. 2, d. 882, list 43.

55. Ibid., listy 43–44.

56. Ibid., list 44.

57. Ibid., list 45.

58. Ibid., list 46.

CHAPTER SIX

1. Boreiko, "Eto ne dolzhno," 9. He cites TsGAVO Ukrainy f. 539, op. 1, d. 2401, list 10.

2. Ibid.

3. GARF f. 8300, op. 24, d. 1786, listy 1–8. A note in pen on the archival copy, dated January 4, reads: "One copy of these notes has been used for further processing and was [then] destroyed."

4. RTsKhIDNI f. 17, op. 138, d. 329, list 159.

5. GARF f. 8300, op. 24, d. 1786, list 2.

6. Ibid., list 1.

7. Ibid., list 2.

8. Ibid.

9. Ibid., list 4.

10. Ibid., listy 4–6.

11. Ibid., list 7.

12. Ibid., list 8.

13. Ibid., list 19.

14. Ibid., listy 19–20.

15. RTsKhIDNI f. 17, op. 138, d. 329, listy 157–58.

16. Boreiko, *Istoriia zapovednogo dela v Ukraine,* 123–25.

17. RTsKhIDNI f. 17, op. 138, d. 329, list 159.

18. TsGA f. 259, op. 6, d. 8666, listy 11–13. My own handwritten notes list the edinitsa khraneniia (delo) as d. 8664, which makes sense in light of the related document cited in note 19 and that delo 8664 concerns the period of May 1951, whereas delo 8666 contains materials from August and September; however, my official xeroxed copy of the document, ordered later, was identified by the archival staff as d. 8666.

19. TsGA f. 259, op. 6, d. 8664.415–1, listy 2–3.

20. Boreiko, *Istoriia zapovednogo dela v Ukraine,* 125. Boreiko (132n. 26) provides the following archival citation: GARF f. 544ss, op. 59, d. 7878, listy 1–229.

21. Ibid., list 3.

22. TsGA f. 358, op. 2, d. 1036, list 3.

23. Ibid., list 6.

24. Ibid., list 7.

25. Ibid., listy 106–7.

26. TsGA f. 358, op. 2, d. 991, list 1.

27. Letter to M. M. Bessonov, February 14, 1951. TsGA f. 358, op. 2, d. 991, list 4.

28. The letter-writer put "Tianshanskii" in quotes, presumably to call attention to the origin of the second part of Semënov's family name: his grandfather, the explorer and legislator Pëtr Petrovich Semënov, was granted this second name by Tsar Nikolai II in recognition of his exploration of the Tien Shan Mountains and his service to the empire.

29. TsGA f. 358, op. 2, d. 991, list 48.

30. Ibid.

31. Ibid., list 55. Oleg Izmailovich Semënov-tian-shanskii on August 25 had sent Bykov, the bookkeeper and auditor of the Main Administration, a detailed letter refuting the charges. Only five birds (one black cock, one capercaille, two willow ptarmigans, and one duck, all males) were taken in May and June and 277 fish, all previously studied in the lab before being cooked, for the entire first six months of 1951. Ibid., listy 49–50. Reserve director Chernenko sent an even more forceful response, asking for the identification of those who claim to have "gone hungry," those allegedly fired without cause, etc. Ibid., listy 51–51 rev.

32. MOIP Archive, d. 1858, list 13. Dated June 1, 1951.

33. It is difficult to say whether the senders were aware that their protest was technically misdirected, as Merkulov and the Kremlin high command were wielding the hatchet, not Chernousov.

34. TsGA f. 259, op. 6, d. 8666, listy 40–41.

35. Ibid., list 47.

36. GARF f. 8300, op. 24, d. 1786, listy 156–57.

37. Ibid.

38. TsGA f. 259, op. 6, d. 8666, list 46. Dated August 18, 1951 no. 0303/441.

39. Ibid.

40. Ibid., listy 70–72. Entitled "Spravka o predlozheniiakh krai(obl)ispolkomov i Sovetov Ministrov avtonomnykh respublik RSFSR po voprosu ispol'zovaniia zemel' uprazniaemykh zapovednikov." There is no date but it was initialed on August 29, 1951, so it was circulating prior to that date. A second document, "Spravka o dopolnitel'nykh predlozheniiakh po zapovednikam," was dated August 20, 1951 to Merkulov from N. Fetisov and is located in GARF f. 8300, op. 24, d. 7486, listy 33–34. This second document records the responses of the Union republics.

41. Ibid. For a fascinating look at the subtle correspondence between Merkulov and the Ukrainian Party leader Mel'nikov on this, see Boreiko, *Istoriia zapovednogo dela v Ukraine,* 124–25.

42. TsGA f. 358, op. 3, d. 581, listy 121–23.

43. Boreiko, *Istoriia zapovednogo dela v Ukraine,* 125–26. He cites GARF f. 544ss, op. 59, d.7878, listy 1–229.

44. Conversation between Vladimir Boreiko and Oleg Kirillovich Gusev, per Vladimir Boreiko, April 20, 1991. Gusev was a close friend of Malinovskii's and is editor in chief of *Okhota i okhtnich'e khoziaistvo.*

45. For the complete text of the decree see Boreiko, "Eto ne dolzhno," 9–11.

46. Boreiko, *Istoriia zapovednogo dela v Ukraine,* 128, makes this point.

47. RTsKhIDNI f. 17, op. 138, d. 329, listy 107–8. Titov had apparently been approached by the acting head of the Krymskoe Upravlenie lesnym khoziaistvom, Potapenko, who had begun making complaints against the *zapovednik* in June.

48. Ibid., list 113.

49. Ibid., listy 211–14. Malenkov had asked the Central Committee Agricultural Department to keep current on this matter, and Malinovskii had to send copies of all correspondence connected with it. Hence its present location in the Party archives.

50. TsGA f. 358, op. 2, d. 991, list 115. Letter of October 31, 1951. The firings were to take effect November 1.

51. See Boreiko, *Istoriia zapovednogo dela v Ukraine,* 128.

52. "O raspredelenii zemel' zapovednikov, vydelennykh kolkhozam i sovkhozam," Postanovl. no. 4164 of the USSR Council of Ministers, October 29, 1951.

53. Boreiko, *Istoriia zapovednogo dela v Ukraine,* 128.

54. Boreiko, "Eto ne dolzhno," 11.

55. ARAN f. 1674, op. 1, d. 240, listy 23 and 23 rev. Dated November 11, 1951, Moscow, in response to a telegram sent to Protopopov from Puzanov.

56. Ibid., list 23 rev.

57. Postanovlenie SM SSSR no. 4139, Polozhenie o Glavnom upravlenii po zapovednikam pri SM SSSR and Polozhenie o gosudarstvennykh zapovednikakh. For the text, see Glavnoe upravlenie po zapovednikam pri SM SSSR, *Polozhenie o Glavnom upravlenii po zapovednikam pri SM SSSR; Polozhenie o gosudarstvennykh zapovednikakh SSSR; Polozhenie ob okhrane gosudarstvennykh zapovednikakh SSSR.*.

58. My xerox is of copy no. 800.

59. RGAE f. 9466, op. 5, d. 376, listy 1–9.

60. GARF f. 8300, op. 24, d. 1847, listy 1–8. Beria's instructions are on listy 1–2.

61. Ibid., listy 2–8.

62. Ibid., listy 6–7.

63. Ibid., list 19.

64. Ibid.

65. Interview with Aleksandr Leonidovich Ianshin, Moscow, July 24, 1992.

66. Interviews with Andrei Aleksandrovich Nasimovich, Moscow, April 16 and 18, 1980.

67. When I published *Models of Nature* this was the only "insider" explanation I had, and I was unable to assess it critically.

68. Shtil'mark and Geptner [Heptner], "Tragediia sovetskikh zapovednikov," 101.

69. Ibid., 112.

70. TsGA f. 259, op. 6, d. 8666, list 69. Handwritten note from Malinovskii to

RSFSR deputy premier Arsenii Mikhailovich Safronov. Marked "Soglasen, A.S." by Safronov, signaling the RSFSR's concurrence with the ultimate disposition of the territory formerly held by its reserves.

71. RTsKhIDNI, f. 17, op. 138, d. 466, listy 114, 115. Malenkov's note is dated December 6, 1952.

72. Ibid., listy 123–24.

73. Ibid., listy 131–34.

74. Ibid., list 148.

75. Ibid., list 149.

76. Ibid., listy 168–72.

77. Boreiko, "Eto ne dolzhno," 9, provides these examples in support of the kind of reassessment of Malinovskii that I am attempting here.

78. Conversation with Andrei Aleksandrovich Nasimovich, Moscow, April 18, 1980.

79. Shtil'mark and Geptner [Heptner], "Tragediia sovetskikh zapovednikov," 111.

CHAPTER SEVEN

1. TsGA f. 404, op. 1, d. 136a, list 51. Dated July 24, 1948. This was a first draft and had hand-written corrections by Zaretskii, who signed it at the bottom.

2. Interview with Konstantin Mikhailovich Efron, August 2, 1995.

3. Ibid.

4. TsGA f. 259, op. 6, d. 5419.416–10. "Vypiska iz protokolov no. 16 zasedaniia Biuro SM RSFSR ot 2 aprelia 1948 g.," list 34; "no. 17 ot 9 aprelia 1948 g.," list 32.

5. TsGA f. 339, op. 1, d. 4734, list 1a. This was an order issued by N. Vasil'ev, the minister, to the Group for Cultural and Educational Institutions. The investigation was to be completed in twelve days; a senior controller and a controller were assigned to this.

6. TsGA f. 339, op. 1, d. 4737, list 7.

7. Ibid., see listy 8–12.

8. Ibid., list 12.

9. Ibid., list 13. The text of the report to N. M. Vasil'ev, minister of state control of the RSFSR, is contained on listy 1–6.

10. Ibid., list 14.

11. Ibid., list 15.

12. Ibid., listy 15–16.

13. Ibid., list 19.

14. TsGA f. 404, op. 1, d. 175, listy 42–43. "Protokol zasedaniia Prezidiuma TsS VOOP no. 9 za 2 avgusta 1950 g." Ibid., list 51. "Protokol zasedaniia Prezidiuma TsS VOOP no. 13."

15. TsGA f. 259, op. 6, d. 8667, listy 148–49.

16. Figures for VOOP's membership are found in TsGA f. 259, op. 6, d. 8667, list 168.

17. TsGA f. 404, op. 1, d. 182a, listy 33–77.

18. TsGA f. 404, op. 1, d. 174, list 4. "Stenograma zasedaniia plenuma TsS VOOP ot 20-ogo aprelia 1950 g."

19. Ibid., list 5.
20. Ibid., listy 71–72.
21. Ibid., list 73.
22. Ibid.
23. Those two governments then appealed to the RSFSR Council of Ministers to help in their appeal to the State Staffing Commission, but those appeals had not yet been acted upon. TsGA f. 404, op. 1, d. 182a, list 73.
24. Ibid., listy 73–74.
25. TsGA f. 404, op. 1, d. 183a, list 2.
26. Ibid., list 1.
27. Ibid., list 17.
28. Ibid., list 13.
29. See ibid., list 19.
30. Ibid., list 18.
31. Ibid., list 31. The IUPN was created in July 1947 and was based in Switzerland. Gams was active in the 1949 Lake Success technical conference of the IUPN. See Boardman, *International Organization and the Conservation of Nature,* 51.
32. TsGA f. 404, op. 1, d. 183a, list 31. A copy of the completed application may be found in TsGA f. 404, op. 1, d. 205, listy 2–3. Significantly, at the place where the questionnaire from the International Union for the Protection of Nature asked whether the applying organization was of a governmental, semi-governmental, or private (*chastnyi*) character, the VOOP respondents typed and underlined *chastnyi* and added: "a voluntary association of citizens" (*dobrovol'noe ob"edinenie grazhdan*), list 3.
33. TsGA f. 404, op. 1, d. 175, listy 47–50.
34. TsGA f. 404, op. 1, d. 194, list 7. Undated. Dated in pen February 12, 1951, probably by aides in the RSFSR Council of Ministers.
35. See TsGA f. 404, op. 1, d. 182a, list 2. Joint letter of Makarov and the academician N. A. Maksimov, president of the *orgburo* of VOSSOGZN, of January 13, 1950 to Georgii Maksimilianovich Malenkov, proposing a merger of the two societies.
36. TsGA f. 259, op. 6, d. 8667.418–12, papka 1, list 105.
37. The draft letter to the Central Committee is in ibid., list 106, and the follow-up note is on listy 111–12.
38. TsGA f. 259, op. 6, d. 8667, list 145.
39. Ibid.
40. Ibid.
41. Ibid., list 146.
42. Ibid.
43. Ibid.
44. TsGA f. 259, op. 6, d. 8668, list 138. Dated August 27, 1951.
45. Ibid., listy 138–39.
46. Ibid.
47. Ibid.
48. Ibid., list 140.
49. Ibid., list 101.
50. Ibid.
51. Ibid., list 102.

52. Ibid., list 103. "Protokol no. 7 ot 21-ogo marta 1951 g."
53. Ibid., list 105.
54. Ibid., list 134.
55. Ibid.
56. Ibid.
57. TsGA f. 259, op. 6, d. 8667.418–12, papka 1, list 90. Letter of S. V. Kuznetsov, secretary of the Party cell of the central staff of VOOP, and P. V. Ostashevskii, deputy secretary, to Arsenii Mikhailovich Safronov, June 14, 1951.
58. Ibid. "Apolitical" meant unwilling to endorse or promote the regime's political campaigns.
59. Ibid., listy 90–91.
60. Ibid.
61. Ibid., list 92.
62. Ibid., list 93.
63. Ibid., list 94.
64. Makarov Papers, courtesy of F. R. Shtil'mark. "Rezolutsiia," dated 1951. "The resolution of a closed meeting of the Party organization of the Main Administration . . . on the question of criticism of the article by comrade V. N. Makarov 'Zapovedniki Sovetskogo Soiuza' in the book *Zapovedniki SSSR*, vol.1."
65. Makarov Papers, courtesy of F. R. Shtil'mark. On Main Administration letterhead. Copy. Dated September 29, 1951, letter no. 74. Signed as correct by the head of the Secretariat, L. Surina.
66. Makarov Papers, "Rezolutsiia."
67. TsGA f. 259, op. 6, d. 8667, listy 10–28. "Spravka o rabote VOOP." June 11, 1951.
68. Ibid., list 19.
69. Ibid., listy 19–20.
70. Ibid., listy 21–22.
71. Ibid., listy 25, 28.
72. TsGA f. 404, op. 1, d. 192, list 167. "Vypiska Protokola no. 14 zasedaniia Prezidiuma TsS VOOP ot 28-ogo iulia 1951 g."
73. Ibid.
74. Ibid.
75. TsGA f. 259, op. 6, d. 8667, list 4.
76. Ibid., listy 167–72. "Protokol zasedaniia no. 14 ot 28-ogo iulia 1951 g."
77. Ibid., list 68. "Vypiska iz Protokola no. 33 Biuro SM RSFSR ot 27-ogo iulia 1951 g."
78. TsGA f. 259, op. 6, d. 8668.418–12, papka 2, list 71.
79. Ibid., list 75.
80. Ibid., list 76.
81. Ibid., list 77.
82. Ibid., listy 77–82.
83. Ibid., list 92.
84. Ibid., listy 93–94.
85. Ibid., list 95.
86. Ibid., listy 95–96.

87. Ibid., list 96.
88. Ibid., list 97.
89. Ibid., listy 105–6.
90. Ibid., list 106.
91. Ibid., list 108.
92. Ibid., list 69.
93. Ibid., list 280.
94. TsGA f. 404, op. 1, d. 192, listy 271–73. S. Kuznetsov, "O sostoianii izdatel'skogo portfelia." Report to the Presidium of VOOP, from "Protokol no. 18 zasedaniia prezidiuma TsS VOOP ot 3-ego oktiabria 1951 g."
95. Ibid., listy 280–82.
96. Ibid., listy 282–84.
97. Ibid., list 65. "Vypiska iz Protokola no. 45 Biuro SM RSFSR ot 5-ogo sentiabria 1951 g." VOOP was the fifth item on the agenda for that day.
98. Ibid., list 66. TsK VKP(b) Malenkovu, G. M. ot Chernousova. No date. Carbon. It is not known whether this was sent, judging from the preserved document.
99. Ibid., listy 66–67.
100. Ibid., list 53. "Vypiska iz protokola no. 54 Biuro SM RSFSR."
101. Ibid., listy 54, 48. Memorandum of Kostoglodov to Safronov of October 18, 1951.
102. Ibid., list 46. "Vypiska iz protokola no. 58 ot 26-ogo oktiabria 1951 g."
103. Postanovlenie no. 1359 SM RSFSR, October 31, 1951. "O nezakonnom i bezkhoziaistvennom raskhodovanii denezhnykh sredstv Vserossiiskim Obshchestvom Okhrany Prirody."
104. TsGA f. 404, op. 1, d. 192, list 292.
105. Ibid., listy 306–11.
106. Ibid.
107. Ibid.
108. Ibid., list 9.
109. Ibid., list 10.
110. Ibid.
111. Ibid.
112. Ibid., list 11.
113. Ibid.
114. Ibid., list 12.
115. Ibid., list 13.

CHAPTER EIGHT

1. TsGA f. 404, op. 1, d. 201a, listy 15–16. "Protokol no. 1 Tsentral'nogo soveta VOOP ot 24-ogo ianvaria 1952 g."
2. Ibid., list 18.
3. Ibid., list 20.
4. Ibid.
5. Ibid., listy 20–21.
6. Ibid., list 23.

7. Ibid., list 24.
8. Ibid., list 25.
9. Ibid.
10. Ibid., listy 34–35.
11. TsGA f. 404, op. 1, d. 203, listy 23–24.
12. TsGA f. 259, op. 6, d. 8668.418–12, papka 2, listy 3–6. Report of Koz'iakov to Bessonov, January 29, 1952. Quotation is from list 6. Koz'iakov relied for his figures on the resolution and accompanying report sent by VOOP following its Central Council meeting on January 24. See TsGA f. 404, op. 1, d. 203, listy 25–27.
13. TsGA f. 259, op. 6, d. 8668.418–12, papka 2, list 6.
14. RTsKhIDNI, f. 17, op. 138, d. 466, listy 1–3.
15. Ibid.
16. Ibid., list 4.
17. Ibid., list 8.
18. Ibid.
19. Ibid., listy 8–9.
20. Ibid., list 10.
21. Ibid., listy 11–19.
22. Ibid., listy 20.
23. TsGA f. 404, op. 1, d. 201, list 91.
24. Ibid., list 83.
25. Ibid.
26. Ibid., list 96. "Vypiska iz protokola no. 10 ot 1-ogo iulia 1952 g."
27. TsGA f. 259, op. 6, d. 8668.418–12, papka 2, listy 106–7. Memo of Koz'iakov to A. M. Safronov of June 20, 1952.
28. Ibid., list 106.
29. Ibid., list 107.
30. TsGA f. 259, op. 7, d. 1894, list 7.
31. TsGA f. 404, op. 1, d. 204, list 67.
32. Ibid., list 62. For detailed branch-by-branch assessments, see RGAE f. 3, op. 1, d. 147, listy 91–111. "Spravka o rezul'tatakh obsledovaniia Vserossiiskogo Obshchestva okhrany prirody."
33. Ibid., list 63. See also TsGA f. 404, op. 1, d. 204, listy 26–45. "Spravka o rezul'tatakh obsledovaniia VOOP." This is the full report.
34. Ibid., listy 63–66.
35. Ibid., listy 47–50.
36. Ibid., list 50.
37. TsGA f. 259, op. 42, d. 8530, list 17. Letter of D. Korolëv, deputy minister of trade of the RSFSR, to the RSFSR Council of Ministers, September 17, 1952.
38. TsGA f. 259, op. 7, d. 1894.074–63, papka 1, list 30.
39. Ibid., and TsGA f. 404, op. 1, d. 201, listy 108–9.
40. TsGA f. 404, op. 1, d. 201, list 110.
41. Ibid.
42. Ibid., list 112.
43. Ibid., list 114. See remarks of Krivoshapov.
44. Ibid., list 111.

45. Ibid., list 115.

46. Ibid.

47. Ibid., list 117.

48. Ibid., list 119. In her speech she apparently used the word *unichtozhena* (destroyed). However, the stenogram was later corrected to read *sokrashchena* (reduced in size, truncated).

49. Ibid.

50. Ibid., list 121.

51. TsGA f. 259, op. 7, d. 1894, listy 22–25. "Spravka G. A. Avetisiana po povodu zaiavleniia S. V. Kuznetsova," (to RSFSR Council of Ministers), January 5, 1953.

52. TsGA f. 404, op. 1, d. 201, list 121.

53. Chernousov's daughter, Galina Borisovna Chernousova, kindly spoke with me about her father in Moscow on June 4, 1996. After his abrupt dismissal, she related, he was clearly worried that he would be arrested. He remained uncomfortably without an alternative appointment until February 8, 1953.

54. TsGA f. 259, op. 7, d. 1894.074–63, papka 1, listy 28–29. Memo of Koz'iakov to Maslov, December 20, 1952. A letter of Lebedev to Koz'iakov, c. November 1952, also contains a good deal of information about the history of the Green Plantings Society. His evaluation was that the death of its first chair, academician N. A. Maksimovich, was fatal to the organization. See also list 130.

55. TsGA f. 259, op. 7, d. 1895, listy 98–100. Memo of V. Maslov to A. M. Puzanov, January 29, 1953.

56. TsGA f. 259, op. 7, d. 1894.074–63, papka 11, listy 28–29.

57. TsGA f. 259, op. 7, d. 1895, listy 98–100. Memo of V. Maslov to A. M. Puzanov, January 29, 1953.

58. Ibid., list 64. "Vypiska iz Protokola no. 23 zasedaniia Biuro SM RSFSR ot 4-ogo marta 1953 g."

59. TsGA f. 259, op. 7, d. 1894, listy 16–17. His list included Iosif Ivanovich Chadroshvili, deputy head of the Main Administration for Shelter Belts of the USSR Council of Ministers and head of the Forest Protection Section of VOOP; Pëtr Aleksandrovich Manteifel'; Pavel Stepanovich Melekhov, head of the USSR Ministry of Forestry's Section of Forests of Special Importance; Vasilii Vasil'evich Gusev, deputy minister of municipal services of the RSFSR; Aleksandr Vasil'evich Mel'nikov, deputy minister of forestry of the RSFSR; Ivan Fëdorovich Lotsmanov, deputy chair of the Moscow Soviet; and Aleksandr Nikolaevich Volkov, head of the *Oblast'* Station for the Protection of Green Plantings.

60. See the letter to A. M. Puzanov from Avetisian, Varsonof'eva, Geptner, Protopopov, Krivoshapov, Chernenko, and Chaianov of November 13, 1952. TsGA f. 259, op. 7, d. 1894, papka 1, listy 13–14. See also "Dokladnaia zapiska tovarishchu A. M. Puzanovu" of November 12, 1952, TsGA f. 404, op. 1, d. 204, listy 85–87.

61. TsGA f. 259, op. 7, d. 1895, list 61. Note from Maslov to the Sel'sko-khoziaistvennyi otdel, TsK KPSS (Agricultural Department of the Central Committee), of April 30, 1953.

62. TsGA f. 259, op. 7, d. 1894, listy 31–36. See esp. listy 31–32.

63. Ibid. listy 35–36.

64. TsGA f. 404, op. 1, d. 212, listy 13–14. "Tov. V. A. Maslovu ot G. Avetisiana, July 17, 1953."

65. TsGA f. 259, op. 7, d. 1894, listy 74–75. Memorandum of March 27, 1953 from V. Gusev, deputy minister of municipal services of the RSFSR to V. A. Maslov. Maslov asked Gusev to canvass a set of *oblasts* about the merger of the two societies. Of fifteen *oblast'* leaderships surveyed, all but one thought the merger should be approved. However, the *oblast'* leaders were split on the question of allowing the new voluntary society to maintain a permanent staff.

66. TsGA f. 259, op. 7, d. 1895, listy 96–97.

67. TsGA f. 259, op. 7, d. 1894.074–63, papka 1, list 103.

68. TsGA f. 259, op. 7, d. 1895, listy 43–44. "V. Galitskii Vasiliiu Maksimovichu Shakhanovu, nachal'niku sel'sko-khoziaistvennogo otdela pri SM RSFSR, August 25, 1953." The *Krokodil* piece, titled "O pchëlakh i osakh," was by K. Eliseev. How best to rebut the claims of this piece was the chief subject of VOOP presidium meeting of February 20, 1953. See TsGA 404, op. 1, d. 211, listy 7–17. The denunciation of Avetisian as an anti-Lysenkoist appeared in *Za sotsialisticheskoe zemledelie* for October 1948.

69. TsGA f. 259, op. 7, d. 1895, list 11. "Vypiska iz Protokola no. 68 Biuro SM RSFSR ot 15-ogo iulia 1953 g."

70. TsGA f. 259, op. 7, d. 1895.074–63, papka 2, list 1. "Ob ob"edinenii VOOP i Vserossiiskogo obshchestva sodeistviia stroitel'stvu i okhrane gorodskikh zelenykh nasazhdenii bo Vserossiiskoe obshchestvo sodeistviia okhrane prirody i ozeleneniia naselennykh punktov." The organizational bureau of the new society was given two months to devise a charter and was permitted to call a congress for December 1953 to elect permanent officers. Avetisian's note of September 4, 1953 may be found at TsGA f. 404, op. 1, d. 212, list 16.

71. TsGA f. 404, op. 1, d. 211, listy 8–10.

72. Ibid., list 39. "Vypiska iz Protokola no. 6 zasedaniia prezidiuma VOOP ot 2-ogo iunia 1953 g."

73. Ibid., list 49. "Vypiska iz Protokola no. 8 ot 15-ogo sentiabria 1953 g."

74. RGAE f. 3, op. 1, d. 335, listy 40–43. Letter of June 30, 1953 from Susanna N. Fridman to Vera A. Varsonof'eva. (list 40 rev.)

75. Ibid., listy 41 rev., 42.

76. TsGA f. 404, op. 1, d. 212, listy 17–18. Letter from Avetisian to Maslov of October 12, 1953. The other members were: V. A. Varsonof'eva, V. I. Ivanov, I. S. Krivoshapov, G. P. Motovilov, F. N. Petrov, V. V. Prokof'ev, A. P. Protopopov, and I. O. Chernenko.

77. Ibid., listy 19–19 rev. Letter of G. I. Lebedev to the Presidium of the Organizing Committee of the All-Russian Society for the Promotion of the Protection of Nature and the Greening of Population Centers, November 19, 1953.

78. On the official lawsuit brought against Galitskii by the new Presidium, see TsGA f. 404, op. 1, d. 215, listy 22–40.

CHAPTER NINE

1. See RGAE f. 3, op. 1, d. 148, listy 43–44. "Poiasnenie k prikazu t. Avetisiana no. 34 ot 16.X.54 g.," a letter by I. Chernenko of October 21, 1954 to the Organizing Committee of VOOP announcing his resignation as scholarly secretary in protest

against Avetisian's overly authoritarian governance of the Society. In a letter of August 5, 1954 from Susanna N. Fridman to Vera A. Varsonof'eva, Fridman described Avetisian as a "typical speculator using science to make his career" (*tipichnyi spekuliant ot nauki*). See RGAE f. 3, op. 1, d. 335, list 53 rev.

2. See TsGA f. 404, op. 1, d. 226, list 9. "Protokol no. 1 zasedaniia Orgkomiteta ot 20-ogo iunia 1955 g."

3. Ibid., list 11.

4. RGAE f. 3, op. 1, d. 148, listy 82–83.

5. Ibid., list 83. I have been unable to locate the personal papers of Susanna Fridman or any trace of the history that she had been working on.

6. *Materialy Pervogo,* 5–6.

7. Ibid., 82.

8. Ibid., 82–83.

9. TsGA f. 404, op. 1, d. 253, list 88.

10. See *Materialy Pervogo,* 89–91, for a list of those elected to the Central Council.

11. TsGA f. 404, op. 1, d. 229, listy 1–5. "Protokol no. 1 zasedaniia Ts. Soveta ot 19 avgusta 1955 g."

12. TsGA f. 404, op. 1, d. 253, listy 281–96. Letter of Lakoshchënkov to V. M. Molotov, February 6, 1957.

13. Profiting from official or state-chartered positions, although technically illegal, was tolerated as a bureaucratic perquisite.

14. Iukhno, of course, asked all the foresters of the reserve to keep silent. The Moscow *oblast' prokuror* (prosecutor) was ready to jail Iukhno, but he was saved by Eliseev, who named him director of the Astrakhanskii *zapovednik,* far outside the jurisdiction of the Moscow *oblast'* justice machinery. Iukhno then proceeded to cause a scandal in Astrakhan', and Eliseev named him acting director of the Barguzinskii *zapovednik* on Lake Baikal, where he was supposed to wage a war against poaching. However, he was stashing the confiscated pelts in a trunk and was ultimately caught at it. Eliseev then appointed him the main hunting inspector of Ul'ianovskaia *oblast'.* There, Iukhno made a poster showing animals that could be hunted in season. Unfortunately, he included the marmot, which was protected. This blunder motivated A. N. Formozov to write a satirical piece in the *Literaturnaia gazeta* "Gosokhotinspektor—brakoner" (The State Hunting Inspector Is a—Poacher!). Per interview with M. A. Zablotskii, Prioksko-Terrasnyi *zapovednik,* April 20, 1991.

15. TsGA f. 404, op. 1, d. 253, listy 289, 292.

16. Ibid., list 278.

17. Ibid., list 292.

18. Ibid., list 281.

19. Ibid., list 259.

20. Ibid., listy 274–78. F. K. Alëkhin also added his signature to the letter, although it was written by Lakoshchënkov.

21. Ibid., list 274–75.

22. Ibid., list 276.

23. Ibid., list 133.

24. TsGA f. 404, op. 1, d. 237, list 208.

25. TsGA f. 404, op. 1, d. 227, list 15.

26. TsGA f. 404, op. 1, d. 252, list 112.

27. TsGA f. 404, op. 1, d. 253, list 135.

28. RGAE f. 3, op. 1, d. 335, list 80. Letter from Fridman to Varsonof'eva, no date (late 1958).

29. Ibid., listy 80 rev., 81.

30. TsGA f. 404, op. 1, d. 277, listy 27–28.

31. TsGA 404, op. 1, d. 281, listy 1–4. A similar request was repeated on August 4, 1959, in a letter from Vice President V. Nikitin to chairman of the RSFSR Supreme Soviet N. N. Organov. Organov asked Ageev of the RSFSR Council of Ministers staff to discuss the matter with a representative of the Central Committee's Bureau for the RSFSR, Semënov, who finally responded that the Central Committee would not support the request because the Academy of Sciences' Commission on Nature Protection was already representing the Soviet Union in the International Union for the Conservation of Nature and Natural Resources. Although the Central Committee may have been reluctant to pay out an additional $250 in membership dues for VOOP, another important reason is that the Party was probably wary of having more than one institution represent the Soviet Union's position, which was supposed to be unitary, in the international organization. See TsGA f. 259, op. 42, d. 3823, listy 11–15. (Two documents entitled "Spravka.")

32. TsGA f. 404, op. 1, d. 274, list 26.

33. Ibid., list 13.

34. TsGA f. 404, op. 1, d. 252, listy 245–48, for example.

35. VOOP, *Materialy vtorogo,* 11–12. Outgoing president Motovilov admitted that the Presidium was tardy in preparing for the Congress, which should have taken place in 1958. As of December 1, 1959, membership within the sixty-seven branches consisted of 355,000 adults and 561,000 members in youth sections. There were 4,362 organizations that had joined as "juridical members," thereby committing all their employees to VOOP membership.

36. Ibid., 30.

37. Ibid., 32.

38. Ibid., 38–39.

39. Ibid., 49.

40. Ibid., 59. The original draft of her talk is in RGAE f. 3, op. 1, d. 24, listy 74–82.

41. VOOP, *Materialy vtorogo,* 60.

42. Ibid.

43. Ibid., 71

44. Ibid., 82.

45. Ibid., 82–83.

46. Ibid., 63–66, 78–79, 85.

47. RGAE f. 3, op. 1, d. 148, list 54. It is not known how this ballot remained in her hands. Either she ultimately did not vote or she obtained a duplicate for her records.

48. F. N. Mikhailov, "Okhrana vodoëmov ot zagriazneniia," 3–6.

49. Bosse and Iablokov, *Okhrana prirody i eë znachenie dlia nashei strany.*

CHAPTER TEN

1. Malinovskii, "Zapovedniki sovetskogo soiuza."

2. TsGA f. 404, op. 1, d. 216, list 8. These were the remarks of the system's deputy director, Korol'kov, at the May 1954 conference on *zapovedniki.*

3. Ibid., list 9.

4. Ibid.

5. See my discussion of Veitsman in *Models,* 225–26. For the ideas of Arkhipov and Boitsov, see Arkhipov, "Instruktsii dlia organizatsii zapovednogo khoziaistva"; and Boitsov, "O sostoianii i perspektivnom plane."

6. Filonov, "Dinamika chislennosti," 158–59.

7. Ibid., 162–63.

8. Ibid., 163.

9. See the two-volume *Akklimatizatsiia okhotnich'e-promyslovykh zverei i ptits v SSSR;* and *Akklimatizatsiia zhivotnykh v SSSR.*

10. Filonov, "Dinamika chislennosti," 161.

11. Ibid., 167.

12. Oborin, "Troitskoe uchebno-opytnoe lesnoe khoziaistvo Permskogo," esp. 120–21.

13. K. P. Filonov, in his major study of the consequences of nature transformation in the *zapovedniki,* "Dinamika chislennosti," does not make such a categorical distinction between the Makarov and Malinovskii eras. See his table 37, p. 183, for example.

14. Postanovlenie AN SSSR no. 169 ot 28-ogo marta 1952 g.

15. Interview with academician Aleksandr Leonidovich Ianshin, Moscow, July 24, 1992.

16. Ibid.

17. Ibid.

18. Makarov Papers, "Prilozhenie k Postanovleniiu Prezidiuma AN SSSR No. 169 ot 28 marta 1952 g."

19. Interview with Sergei Vladimirovich Zonn, Moscow, August 3, 1995.

20. RGAE f. 544, op. 1, d. 2, list 1.

21. Ibid., list 15. "Protokol rasshirennogo zasedaniia Biuro ot 9-ogo iulia 1952 g."

22. RGAE f. 544, op. 1, d. 3, listy 43–44.

23. Ibid., list 44.

24. RGAE f. 544, op. 1, d. 7, list 1. "Zakliuchenie po utochënnomu proektu plana nauchno-issledovatel'skikh rabot, nauchnykh, i nauchno-tekhnicheskikh meropriiatii i planu vnedreniia zakliuchënnykh rabot po gosudarstvennym zapovednikam na 1953 god." This document is not dated, but it probably was generated in the last week of February or in early March 1953. It is a response to the Proekt plana. . . . Glavnogo upravleniia po zapovednikam pri SM SSSR No. 40/315 of February 16, 1953.

25. Ibid., list 6.

26. RGAE f. 544, op. 1, d. 6, list 11. From "Stenogramma zasedaniia Komissii po zapovednikam AN SSSR ot 4-ogo marta 1953 g."

27. Ibid.

28. These were divided as follows: Academy of Sciences of the USSR: Il'menskii,

Suputinskii, Kedrovaia Pad', and Kivach; Academy of Sciences of the Ukrainian SSR: Streletskaia steppe, Khomutovskaia steppe, Mikhailovskaia tselina, Veselye Boboben'ki, Dendropark Trostianets, Ustinovskii dendropark, and Kamennye mogily; Academy of Sciences of the Tadzhik SSR: Tigrovaia balka; Academy of Sciences of the Turkmen SSR: Repetekskii; Academy of Sciences of the Lithuanian SSR: Zhivuntas.

29. RTsKhIDNI f. 17, op. 138, d. 466, list 64. May 17, 1952.

30. Ibid.

31. Ibid., list 66.

32. Makarov Papers. V. N. Makarov, "Doklad," 6. Undated (written after March 15, 1953). Presented at the conference of representatives of the Academy's *zapovedniki* in Moscow, April 20–22, 1953.

33. Ibid., 15.

34. Ibid., 19.

35. Ibid., 26.

36. "Proekt Polozheniia o zapovednikakh Akademii nauk SSSR i akademii nauk soiuznykh respublik," 2 pp. Carbon, typed. Courtesy of F. R. Shtil'mark.

37. On MOIP, see especially Varsonof'eva, *Moskovskoe obshchestvo;* and Lipshits, *Moskovskoe obshchestvo ispytatelie prirody.*

38. Its membership, however, rose noticeably in the postwar period, from 717 in 1948 to 974 full and 87 corresponding members on January 1, 1951. A surprisingly large number (189) were women. The membership was overwhelmingly non-Party (825). The membership was particularly distinguished, and included 80 members or corresponding members of the USSR Academy of Sciences or its republican affiliates and 100 laureates of the Stalin Prize. The Society's staff also increased from 7 in 1945 to 29 in 1950, caring for a library of more than 100,000 volumes and an active publication schedule, among other things. See MOIP Archives, d. 1792, listy 17–18.

39. Interview with Aleksandr Leonidovich Ianshin, Moscow, July 24, 1992.

40. See Vorontsov, "Nature Protection."

41. Yanitsky [Ianitskii], *Russian Environmentalism,* 19.

42. Ibid.

43. MOIP Archives, d. 1792, list 13. "Protokol zasedaniia [Prezidiuma] no. 2, 12 aprelia 1950 g." Originally the Society had received its space on the old campus in exchange for an urban mansion, willed to it in the early nineteenth century, which it passed on to the university, which built an astronomical observatory at the site. The official address of MOIP in 1950 was 9 Mokhovaia Street.

44. Ibid., list 19.

45. The literature on the Lysenko affair is vast. See the bibliography in Soyfer, *Lysenko.*

46. Vorontsov, afterword to *Stranitsy zhizni,* 325.

47. Ibid.

48. Ibid.

49. Ibid.

50. MOIP Archives, d. 1790, list 5.

51. MOIP Archives, d. 1792, list 56. A later meeting of the new Council settled the contest for the vice president slots by secret ballot as well. Varsonof'eva was in first place with twenty-nine ayes and only two nays.

52. See Soyfer, *Lysenko,* chapter 13. On Sukachëv's role, see especially V. Ia. Aleksandrov, *Trudnye gody sovetskoi biologii.*

53. On this see Soyfer, *Lysenko,* 225–42; and Joravsky, *Lysenko Affair,* 157–59.

54. ARAN f. 1557, op. 2, d. 141, list 1. Letter of N. N. Vorontsov to V. N. Sukachëv, June 5, 1955.

55. TsGA f. 404, op. 1, d. 216, listy 1–2. Stenogramma soveshchaniia "O sostoianii i perspektivakh razvitiia zapovednogo dela v SSSR, ot 12-ogo maia 1954 g.," vol. 1 (of 2).

56. Ibid., list 3.

57. Ibid., list 4.

58. Ibid.

59. Ibid., listy 4–5.

60. Ibid., list 5.

61. Ibid., list 6.

62. Ibid., listy 6, 10.

63. Ibid., listy 27–28.

64. Ibid., list 35.

65. Ibid., list 38.

66. Ibid., list 52.

67. Ibid., listy 52–55.

68. Ibid., list 54.

69. Ibid.

70. Ibid., list 38.

71. Ibid., list 56.

72. Ibid., listy 57–58.

73. Much of this information is from Formozov, *Aleksandr Nikolaevich Formozov,* esp. 111–31.

74. Ibid., 116.

75. Ibid., 116–17.

76. TsGA f. 404, op. 1, d. 216, listy 64–65.

77. Ibid., list 65.

78. Ibid.

79. Ibid., list 66.

80. Ibid., listy 66–67.

81. Ibid.

82. Ibid., list 68.

83. Ibid., list 69.

84. Ibid.

85. Ibid., list 70.

86. Ibid., listy 70–71.

87. Ibid., list 71.

88. Ibid., list 72.

89. Ibid.

90. Ibid., list 73.

91. Ibid., list 74.

92. Ibid., listy 74–75.

93. Ibid., list 76.
94. Ibid.
95. Ibid., list 80.
96. Ibid., list 82.
97. Ibid.
98. Ibid., listy 82–83.
99. Ibid., list 83.
100. Ibid., list 84. A note in the text stated: "V zale shum i smekh."
101. Ibid.
102. MOIP Archives, d. 55, list 1. Another stenogram of the session, separately typed and differing in some minor details and locutions, is located in TsGA f. 404, op. 1, d. 217, listy 1–48.
103. MOIP Archives, d. 55, list 2.
104. Ibid., listy 8–8a; TsGA f. 404, op. 1, d. 217, listy 7–8.
105. MOIP Archives, d. 55, list 9.
106. Ibid., listy 10–11.
107. Ibid., list 11.
108. Ibid.
109. Ibid., list 12, and TsGA f. 404, op. 1, d. 217, list 11.
110. TsGA f. 404, op. 1, d. 217, list 11.
111. Ibid.; MOIP Archives, d. 55, listy 12–13.
112. See Malinovskii, "Zapovedniki sovetskogo soiuza." Cited on TsGA f. 404, op. 1, d. 217, list 14.
113. TsGA f. 404, op. 1, d. 217, listy 17–18.
114. Ibid., list 30.
115. Ibid., listy 30–31.
116. Ibid., list 33.
117. Ibid., list 34.
118. Ibid., list 35. Evidence for his pessimism was based not only on his observations but on personal experience. For example, his Commission for the Protection and Restoration of the European Bison of the Academy of Science's Biological Division had identified a number of *zapovedniki* where bison could be introduced or reintroduced, noted Zablotskii, but their findings were completely ignored by Malinovskii in 1951, and none of the areas identified were spared.
119. Ibid.
120. Ibid.
121. Ibid., list 36.
122. Ibid., listy 41–42.
123. Ibid., list 43.
124. Ibid., listy 45–46.
125. Ibid., list 46.
126. Ibid., list 48.
127. MOIP Archives, d. 55, list 67.
128. Ibid., list 68.
129. Ibid., list 69.
130. Ibid., list 73.

131. Ibid., list 74.
132. Ibid., list 75.

CHAPTER ELEVEN

1. TsGA f. 259, op. 7, d. 5736, listy 61–62.
2. Ibid., listy 58–60.
3. Ibid., list 57. "Vypiska iz Protokola no. 103 zasedaniia Biuro SM RSFSR ot 18-ogo dekabria 1954 g."
4. Ibid., list 40.
5. Ibid., listy 39, 18, 27, and 37.
6. Ibid., listy 1–8. Postanovlenie SM RSFSR No. 1004 ot 9-ogo avgusta 1955 g. "Ob obrazovanii Glavnogo Upravleniia okhotnich'ego khoziaistva i zapovednikov pri SM RSFSR."
7. TsGA f. 259, op. 7, d. 8335, list 75. From Glavnoe upravlenie okhotnich'ego khoziaistva i zapovednikov pri Sovete ministrov RSFSR, *Otchet o deiatel' nosti za 1955 god,* listy 74–101.
8. Ibid., listy 82, 84.
9. Ibid. Citizens' inspectors were among Khrushchëv's innovations designed to transfer part of the burden of administration from the Party-state to carefully controlled citizens' groups. Eventually, the legal right to become credentialed citizens' inspectors for game and nature protection was restricted to members of VOOP.
10. "Stenogramma Vsesoiuznogo soveshchaniia direktorov zapovednikov i okhotovedcheskikh khoziaistv," 27–28 ianvaria 1956 g. TsGA f. 358, op. 3, d.160, list 1.
11. Ibid., list 7.
12. Ibid., listy 8–9.
13. Ibid., listy 9–10.
14. Ibid., listy 19–22.
15. Ibid., list 23.
16. Ibid., list 25.
17. TsGA 358, op. 3, d. 581, list 181.
18. Ibid., listy 181–82.
19. Ibid., list 183.
20. Ibid., list 189.
21. Ibid., listy 37–39.
22. Ibid., list 121.
23. Ibid., list 198. This use of *zubr* is almost identical to D. Granin's use of the term in his biography of N. V. Timofeev-Resovskii, *Zubr.* On *zubry* see especially Gusev, "V zashchitu zubra i zubrov."
24. Ibid.
25. TsGA f. 358, op. 3, d. 581, listy 145–57. Many examples were drawn from the United States.
26. Ibid., listy 199–200.
27. Ibid., list 202.
28. Ibid., listy 94–95.
29. TsGA f. 259, op. 7, d. 7021, listy 39–44. See especially listy 41–42. Report (*spravka*) of N. Krutorogov to the RSFSR Council of Ministers, April 4, 1956.

30. Ibid., list 42.

31. Ibid.

32. Ibid.

33. TsGA f. 404, op. 1, d. 237, listy 139–40.

34. The 1956 figure is in TsGA 259, op. 7, d. 7021, list 41. The 1960 figure is in TsGA f. 259, op. 42, d. 5807, list 39. *Otchet o deiatel'nosti Glavnogo upravleniia okhotnich'ego khoziaistva i zapovednikov pri SM RSFSR za 1959 god.*

35. TsGA f. 259, op. 42, d. 5807, list 39.

36. In 1959 two academicians, thirty-two doctors of science, and eighty-five candidates of science as well as twenty-seven foreign scientists visited the reserves, chiefly to conduct research. Ibid., list 42. The report also informs us that in 1959 alone ten fascicles of scientific proceedings of the *zapovedniki* were published, which included 113 separate articles by seventy-nine authors. Additionally, 125 other articles resulting from research conducted in the reserves were published in such journals as *Priroda, Okhota i okhotnich'e khoziaistvo, Botanicheskii zhurnal, Zoologicheskii zhurnal,* and other publications.

37. TsGA f. 404, op. 1, d. 247, listy 13–14 rev.

38. TsGA f. 259, op. 6, d. 8330, listy 15–31. See especially list 25. Letter of N. Masterov to V. A. Karlov, May 18, 1957.

39. *Okhota i okhotnich'e khoziaistvo* 4 (1957): 63.

40. Bel'skii, "Zapovedniki i okhotnich'e khoziaistvo."

41. Interview with Iurii Konstantinovich Efremov, Moscow, August 4, 1995. Postanovlenie Prezidiuma Akademii nauk SSSR ot 11 marta 1955 g. no. 106, "O reorganizatsii Komissii po zapovednikam v Komissiiu po okhrane prirody." See "Khronika."

42. Ibid., 129. The only contradictory note sounded by the decree was the transfer of the reorganized committee from a direct attachment to the Presidium to affiliation with the Academy's Division of Biology.

43. Ibid.

44. Ibid., 129–31.

45. RGAE f. 544, op. 1, d. 141, listy 6–8.

46. See Dement'ev, "Deiatel'nost'"; and the transcript of his remarks in Protokol koordinirovannogo soveshchaniia Komissii po okhrane prirody AN SSSR s predstavilteliami komissii po okhrane prirody AN soiuznykh respublik, ministerstv, i vedomstv v Moskve, 6 aprelia 1956 g. RGAE, f. 544, op. 1, d. 30, listy 26–27.

47. Dement'ev, "Deiatel'nost'," 4.

48. Ibid.

49. Ibid., 5.

50. Ibid., 11.

51. Ibid., 6–7.

52. Ibid., 7.

53. *Nauchnye s"ezdy, konferentsii i soveshchaniia v SSSR,* 2:118.

54. Dement'ev, "Deiatel'nost'," 9.

55. Ibid., 9–10.

56. RGAE f. 544, op. 1, d. 30, list 26.

57. These included the Institute for Nature Protection of the Polish Academy of Sciences, the Commission for Nature Protection of the Romanian Academy of Sciences, the Society for Nature Protection of the Bulgarian Academy of Sciences, the

Yugoslav Institute for Nature Protection, the Hungarian Nature Protection Council, the Administration for the Protection of Nature of the Czechoslovakian Ministry of Culture, and even the Scientific Institute of Albania.

58. Dement'ev, "Deiatel'nost'," 10. The visit was that of a delegation of the Académie Française, of which IUPN's president, Roger Heym, was a member.

59. Ibid., 12–13.

60. Ibid., 14–15.

61. Dubinin, "Soveshchanie po okhrane prirody SSSR," 1436–38. It was almost certainly Dubinin who raised the issue, since it occupies a full third of his overall report on the conference.

62. Borisov, "V Komissii po okhrane prirody Akademii nauk SSSR," 145–48.

63. Ibid., 147.

64. Ibid.

65. Shaposhnikov, "Deiatel'nost' Komissii," 106–7.

66. Ibid., 108.

67. Ibid., 109.

68. Ibid., 113–14.

69. It is reproduced in Kabachnik, *Aleksandr Nikolaevich Nesmeianov,* 238–42.

70. TsGA f. 259, op. 42, d. 1513, listy 64–65.

71. RGAE f. 544, op. 1, d. 141, listy 22–24.

72. Ibid., list 25.

73. Ibid., list 27.

74. Ibid., list 28.

75. Ibid., listy 32–37.

CHAPTER TWELVE

1. Interview with Konstantin Mikhailovich Efron, Moscow, August 2, 1995.

2. I have taken many ideas here from Iurii Konstantinovich Efremov, whom I interviewed in Moscow on August 4, 1995.

3. Interview with Iurii Konstantinovich Efremov, Moscow, August 4, 1995.

4. Ibid.

5. Letter of Iurii Andreevich Zhdanov to the author, July 28, 1995.

6. Efremov, *Nauchno-obshchestvennoe,* 12.

7. Ibid.

8. Personal papers of Iu. K. Efremov, Vystuplenie Iu. K. Efremova, Soveshchanie u Glavnogo redaktora "Pravdy" P. A. Satiukova, June 5, 1957, list 2.

9. Ibid.

10. Ibid., list 3.

11. Ibid., listy 3–4.

12. Ibid., list 5.

13. Ibid., list 6.

14. Ibid.

15. Ibid., list 7.

16. Ibid., list 9.

17. Ibid.

18. Interview with Iurii Konstantinovich Efremov, Moscow, August 4, 1995.

19. The text of the law was published in *Pravda* on October 28, 1960. I have drawn largely from Philip R. Pryde's translation, contained in appendix 3 of his *Conservation in the Soviet Union,* 180–83.

20. See *Geograficheskoe obshchestvo.* The figure is for 1967.

21. Sobolev and Syroechkovskii, "Voprosy okhrany prirodnykh resursov strany," 142–43. For a list of twelve of those talks, see *Voprosy geografii* 48 (1960): 303–4.

22. *Geograficheskoe obshchestvo,* 52–66.

23. Ibid., 61.

24. Ibid., 77.

25. See MOIP Archives, d. 63, "Stenogramma sovmestnogo zasedaniia Komissii po okhrane prirody MOIP i Mammalogicheskoi sektsii Vserossiiskogo obshchestva sodeistviia okhrany prirody i ozeleneniia naselennykh punktov, 15 fevralia 1955 g." Geptner noted at the meeting, "Naturally, nature protection is of concern to very broad circles of public opinion. That is one of the reasons why the Society of Naturalists created a Commission for Nature Protection and so that the society, which has always held issues of nature protection close to its heart, could have some sort of organizational means to facilitate public discussion and to bring its influence to bear on the problem" (listy 1–2). The section was created in 1954 under the chairmanship of F. N. Petrov.

26. ARAN St. Pbg., f. 996, op. 5, d. 372, listy 1–2. In this letter of December 29, 1956, for example, Protopopov wrote to corresponding member Evgenii Mikhailovich Lavrenko, head of the Academy Commission's group for developing a prospective plan for new *zapovedniki,* asking him to serve as the keynote speaker.

27. *MOIP: Otchet o rabote obshchestva,* 34.

28. *Trudy soveshchaniia po voprosam okhrany redkikh i ischezaiushchikh vidov rasteniii i zhivotnykh i unikal'nykh geologicheskikh ob"ektov (25–30 marta 1957 g.)* (Moscow, 1959), 14. Unpublished, hand-bound proof, located in the MOIP Archives. There is also a *Stenogramma soveshchaniia po redkim, tsennym i ischezaiushchim vidam rastenii i zhivotnykh, po unikal'nym geologicheskim ob"ektam* in the MOIP Archives as well, d. 110 (the title of the conference varies with the editor, stenographer, etc.).

29. *Trudy soveshchaniia,* 18.

30. Ibid., 19.

31. Ibid.

32. Ibid., 26.

33. Ibid.

34. Ibid., 26–27.

35. Ibid., 27–28.

36. Ibid., 28–29. In an article Peredel'skii coauthored with A. M. Kuzin, "Okhrana prirody i nekotorye voprosy radioaktivno-ekologicheskikh sviazei," there is a copious bibliography of both Soviet and foreign works. However, N. V. Timofeev-Resovskii, the premier population geneticist, is not mentioned. That is thoroughly understandable, however, because Timofeev-Resovskii was still in bad political odor and was also working in a secret nuclear-related facility in the southern Urals.

37. MOIP Archives, d. 110. *Stenogramma soveshchaniia po redkim, tsennym i ischezaiushchim vidam rastenii i zhivotnykh, po unikal'nym geologicheskim ob"ektam,* zasedanie ob"edinennoi sessii, 29 marta 1957 g., vecher, list 14.

38. Ibid., list 19.
39. Ibid.
40. Ibid., list 20.
41. Ibid., list 21.
42. Ibid., list 26.
43. Ibid., list 27.
44. Ibid., list 34.
45. Ibid., list 35.
46. In the archive of V. A. Varsonof'eva there is an earlier typed draft of the resolutions with the editing that significantly softened that text. For instance, the sentence beginning "Nepopravimoe zlo prinosit blizorukoe planirovanie" (Nearsighted planning causes irreparable harm) was edited to "Nepopravimoe zlo prinosit v otdel'nykh sluchaiakh blizorukoe" (Nearsighted planning in some instances causes irreparable harm). See RGAE f. 3, op. 1, d. 148, listy 59–70. The actual resolutions were published in *Voprosy geografii* 48 (1960): 294–97.
47. Interview with Konstantin Mikhailovich Efron, Moscow, August 2, 1995.
48. Ibid. Today the pages are too brittle to reproduce by xerox, raising the possibility that, without conservation, the censor may yet win in the end. Nevertheless, the story of the conference and of the fate of its proceedings both continue to live as inspirational stories among the scientific intelligentsia.
49. Bosse, Gekker, Geptner, Kabanov, Nasimovich, Nikol'skii, Formozov, and Chernenko, "V zashchitu zapovednikov."
50. This is the figure given in *MOIP: Otchet o rabote obshchestva,* 34.
51. TsGA f. 358, op. 5, d. 39, listy 3–4. "Stenogramma Vsesoiuznogo soveshchaniia po zapovednikam, 17–20 marta 1958 goda." The proceedings of the conference are divided between two archival *dela,* d. 39 and d. 45. Sadly, some of the proceedings were apparently lost or never saved.
52. Ibid., listy 4–5.
53. Ibid., list 84.
54. Ibid., listy 90–91.
55. Ibid., list 91.
56. *Voprosy geografii* 48 (1960): 297–301.
57. TsGA f. 259, op. 42, d. 3813, listy 16–18.
58. *Nauchnye s"ezdy, konferentsii, i soveshchaniia,* 186–88. The year of sputnik was also the year of the All-Union Botanical Congress, the Far Eastern Inter-Provincial Conference on Problems of Nature Protection, and the First Transcaucasus Conference on the Protection of Nature. In the following year in June the First All-Union Conference of Commissions for the Protection of Nature of the Academies of Sciences of the USSR and of the Union Republics was held in Tbilisi, Georgia, and there were regional conferences in the Urals and in Siberia. Nor did the pace slacken in 1959, with a regional conference on the Lower Volga and Northern Caspian, a discussion on Baikal organized by MGO, an All-Union Conference on Waterfowl Protection, the Second All-Union Conference of Academy of Sciences Commissions for the Protection of Nature in Vilnius, Lithuania, and a second conference in the Urals. In 1960 there was a regional conference in Bashkiria, one on the Crimea, and the Third All-Union Conference of Academy of Sciences Commissions in Dushanbe,

Tadzhikistan. Additionally, there were botanical, zoological, and geographical congresses where these issues were front and center.

59. *MOIP: Otchet o rabote obshchestva,* 9.

60. Varsonof'eva, *Moskovskoe obshchestvo,* 11, 40, 41. They were called "les élèves de la société."

61. Ibid., 41.

62. Ibid., 42.

63. Gershkovich, Razorënova, and Maksimov, "Pëtr Petrovich Smolin."

64. Conversation with Elena Alekseevna Liapunova, Moscow, August 12, 1995.

65. Letter of Varvara Ivanovna Osmolovskaia to author, n.d. (1992).

66. Manteifel', "Vospominaniia," 34–35.

67. Interview with Elena Alekseevna Liapunova, Moscow, August 12, 1995.

68. Il'ina, "Poslednye gody," 39.

69. Ibid.

70. Smolin, "Organizatsiia i rabota ," 151–52.

71. Interview with Elena Alekseevna Liapunova, Moscow, August 12, 1995. In their memoir of those years, "Da skroetsia t'ma," Pavel V. Florenskii and T. A. Shutova confirmed this experience, with the following description of KIuBZ: "In the children's *kruzhok* a collective of like-minded souls came together on the basis of democratic structures, self-government, and continuity between the generations. Here were forged principles of morality, traditions of friendship, consciousness of a unity with nature and of the need for an eternal dialogue with nature. The free young naturalist's life was a living alternative to the dry, bureaucratized school and the withered Pioneer and Komsomol organizations. Having been ourselves in our childhood and youth members of that outspoken young community, we feel to this day the freshness of the oaths we then took of eternal loyalty to our friendships and to nature" (731).

72. Interview with Elena Alekseevna Liapunova, Moscow, August 12, 1995.

73. Ibid.

74. Ibid.

75. Tikhomirov, "Istoriia i deiatel'nost'," 12.

76. Interview with Konstantin Mikhailovich Efron, Moscow, July 5, 1991.

CHAPTER THIRTEEN

1. Most of those executed for "economic crimes" were Jewish. Khrushchëv cynically sought to deflect public anger against corruption from the Party-state elite as a whole to a particular ethnic group. See Juviler, *Revolutionary Law,* 84.

2. Boreiko, *"Tsarskie okhoty,"* 21.

3. Quoted ibid., 21–22.

4. Ibid., 22.

5. Ibid. The decree "Ob assignovanii sredstv na vosstanovlenie sluzhebnykh i podsobnykh pomeshchenii v gosudarstvennykh zakaznikakh U(kr)SSR" may be found in TsGAVO Ukrainy f. 2, op. 7, d. 2099, listy 140–42.

6. Boreiko, *"Tsarskie okhoty,"* 23–24.

7. [Khrushchëv], *Khrushchev Remembers,* 600.

8. Ibid., 384.

9. Boreiko, *Istoriia zapovednogo dela,* 140.

10. Ibid., 141. Boreiko here cites TsGAVO Ukrainy f. 2, op. 9, d. 2898, listy 76–79.

11. TsKhSD f. 5, op. 45, d. 85, listy 24–28.

12. Ibid.

13. TsKhSD f. 5, op. 45, d. 140, listy 151–52. Letter of October 10, 1956. This was the letter Burdin described to the 1957 conference on rare and endangered species.

14. Ibid., list 153.

15. Ibid., list 154.

16. TsKhSD, f. 5, op. 45, d. 161, listy 34–37. "Dokladnaia zapiska Pervomu sekretariu Khrushchëvu, 24 aprelia 1957 g." Signed by F. N. Petrov as chairman of the conference.

17. V. E. Boreiko and I attempted to interview V. Matskevich, then USSR minister of agriculture, but he refused to be interviewed.

18. TsGA f. 259, op. 42, d. 8527, listy 28–29.

19. In 1958 the uniquely beautiful Church of the Odigitriia in Pskov was saved, barely, by the strong intervention of the RSFSR Ministry of Culture. See TsGA f. 259, op. 42, d. 688, listy 28–29.

20. *Plenum Tsentral'nogo komiteta Kommunisticheskoi,* 602.

21. Ibid., 602–3.

22. Boreiko, *Istoriia zapovednogo dela,* 150.

23. Ibid., 151. See also Boreiko, "Vtoroi razgrom zapovednikov," 12–15. See also TsGA f. 259, op. 42, d. 7532, listy 25.

24. Ibid.

25. TsGA f. 259, op. 42, d. 7528, listy 44–45. "Reshenie rasshirennogo zasedaniia biuro Komissii po okhrane prirody Akademii nauk SSSR ot 30-ogo ianvaria 1961 g." Copy. Places for signatures by Dement'ev and L. K. Shaposhnikov. Copy verified by K. L. Segal'.

26. Ibid., listy 44–45.

27. Ibid., list 49.

28. TsGA f. 259, op. 42, d. 7532, list 51–55. Letter of T. A. Aleksandrova et al. of February 9, 1961 to V. P. Zotov and to the RSFSR Council of Ministers. See also list 56 of this archival dossier for the minutes of the Kuibyshev Pedagogical Institute's Agrobiology Faculty, which resolved on February 7, 1961 to protest the proposed elimination of the Zhiguli (Kuibyshevskii, Tsentral'no-Volzhskii) *zapovednik,* which resolution was attached to the letter.

29. RGAE f. 544, op.1, d. 99, listy 36–37. Letter of V. V. Krinitskii to G. P. Dement'ev, February 13, 1961.

30. TsGA f. 259, op. 42, d. 7531, list 9–10. A letter from the of the Altai *kraiispolkom* to the RSFSR Council of Ministers of February 6, 1961 urged liquidation of the Altaiskii *zapovednik* and its conversion to a revived *Teletskii leskhoz* (forest plantation) under the control of Glavleskhoz RSFSR with provision for logging up to 300,000 cubic meters of wood per year. Some tourism could also be accommodated, the letter suggested. Another interesting case revolved around the Tsentral'no-Lesnoi *zapovednik,* which the RSFSR Council of Ministers restored on February 23,

1960 with the direct support of Glavokhota and of Nesmeianov in the Academy over the opposition of the Kalininskaia *oblast' sovnarkhoz*. After Khrushchëv's January Plenum speech, however, the *sovnarkhoz* tried to block the transfer of some woodlands from local *leskhozy* to the *zapovednik* by referring to the premier's talk. Premier Dmitrii Polianskii of the RSFSR apparently held firm and apparently was supported by Zotov in Gosplan of the USSR; the reserve survived the Khrushchëv "liquidation." See TsGA f. 358, op. 3, d. 1911, listy 1–79, esp. 4–5, 30, 33–34, 78–79.

31. TsGA f. 259, op. 42, d. 7528, list 42. See also RGAE f. 544, op. 1, d. 99, list 45.

32. MOIP Archives, d. 2045, list 50, 53.

33. MOIP Archives, d. 2044, list 4.

34. MOIP Archives, d. 2041, listy 2–5.

35. TsGA f. 259, op. 42, d. 7528, listy 47–49. Letter of V. F. Riabov et al. of January 31, 1961 to V. P. Zotov and A. S. Bukharov.

36. MOIP Archives, d. 2041, list 3.

37. Ibid., listy 3–4.

38. Ibid., list 5.

39. TsGA f. 259, op. 42, d. 7528, list 42. *Spravka* (summary report) of aide A. Ageev, n.d. but reviewed and returned to him on February 28 with the instruction to contact those who had written letters of protest and assure them that "the question of putting the *zapovednik* system in order and the elimination of superfluous elements connected with [that system] will be reviewed, taking into account the necessity to preserve the most valuable monuments of nature, rare animals, birds, and plants, as well as to create the essential conditions for study of natural phenomena using the *zapovedniki* as bases."

40. TsGA f. 259, op. 42, d. 7531, listy 2, 3, and 28–35. These are a summary report (*spravka*) by RSFSR Council of Ministers aide A. Ageev of January 25, 1961 to that body's Presidium to guide discussion at the February 11 meeting (listy 2–3). A letter from deputy premier of the RSFSR A. S. Bukharov (listy 28–29), followed by an addendum (listy 30–35) prepared and signed by N. Eliseev, details the specific reserves to be eliminated or reduced in size with their areas in hectares before and after.

41. Interview with Konstantin Mikhailovich Efron, Moscow, August 2, 1995.

42. Interview with Iurii Konstantinovich Efremov, Moscow, August 4, 1995.

43. Ibid.

44. Ibid.

45. See Pryde, *Conservation in the Soviet Union*, appendix 7, part 2, 206–8, for a list of reserves abolished or merged in 1961 (although he errs in his inclusion of the Crimean, which had already been transformed in 1957). See also TsGA f. 259, op. 42, d. 7532, list 21 for Zotov's preliminary list of cuts versus those already set by Eliseev. That list was further amended in the spring, however. The final decree, signed by Kosygin, was issued by the USSR Council of Ministers on June 10, 1961, no. 521, "Ob uporiadochenii seti gosudarstvennykh zapovednikov i okhotnich'ikh khoziaistv." Each republic then subsequently issued its own decree. For the RSFSR, see Postanovlenie SM RSFSR no. 841 ot 28-ogo iunia 1961 g., TsGA f. 259, op. 42, d. 7532, listy 1, 1 rev., 2, 2 rev., and 3 for the reserves eliminated or transformed in that republic. The Ukraine's Postanovlenie no. 118 was issued on July 22, 1961.

46. TsGA f. 259, op. 42, d. 7532, list 26. From letter of V. Zotov of April 20, 1961,

no. 1035 of Gosplan USSR (listy 25–29). At the top, dated April 25, 1961, there is a note from Aleksei Kosygin that reads: "Examine this at the Presidium of the Council of Ministers." This letter was also routed to D. S. Polianskii, whose name is stamped at the very top.

47. Ibid., list 26. Zotov even named E. M. Lavrenko's *Geografiia rastenii* (*Plant Geography*), V. V. Alëkhin's *Rastitel'nyi pokrov SSSR* (*Vegetational Cover of the USSR*), N. V. Pavlov's *Botanicheskaia geografiia SSSR* (*Botanical Geography of the USSR*), A. A. Rode's *Pochvennaia vlaga* (Soil Moisture), and the three-volume *Ptitsy Sovetskogo soiuza* (*Birds of the Soviet Union*). These were solid studies, and some had even become classics.

48. Boreiko, *Istoriia zapovednogo dela,* 151. On Pidoplichko and Pogrebniak, see Boreiko, *Populiarnyi biografo-bibliograficheskii,* 2:27–34. For the actual transcript of the meeting see TsGAVO Ukrainy f. 2, op. 9, d. 9121, list 12–19.

49. Boreiko, *Istoriia zapovednogo dela,* 152.

50. Ibid., 152–53.

51. Ibid., 153.

52. MOIP Archives, d. 2042, list 96. Copy, dated May 16, 1961.

53. Ibid.

54. Ibid., list 97.

55. Ibid.

56. Ibid., listy 97–98.

57. MOIP Archives, d. 2042, listy 93–94. Letter of A. N. Formozov to the Presidium of MOIP and to N. V. Eliseev, head of Glavokhota RSFSR, copy, n.d., list 93–95.

58. Ibid., list 95.

59. MOIP Archives, d. 2042, listy 17–20. Letter of F. N. Petrov and V. A. Varsonof'eva to the Otdel nauki (Science Department) and Otdel Pechati (Press Department) of the Central Committee of the CPSU, May 16, 1961. Copy. The quotation is from list 19.

60. Ibid., list 20.

61. For the letter of Sukachëv et al., see MOIP Archives, d. 2042, list 23. For that of Lavrent'ev, see RGAE f. 544, op. 1, d. 99, listy 54–55.

62. RGAE f. 533, op. 1, d. 141, list 76. Letter of G. Orlov, deputy chairman of Gosplan SSSR, to the USSR Council of Ministers of December 21, 1961. Kosygin reviewed it on December 29 and sent it on to the Council's Komissiia po tekushchim delam (Commission on Current Business), per note at top.

63. Ibid., listy 91–97. "Prilozhenie k prikazu po Gosplanu SSSR ot 19-ogo noiabria 1962 g. no. 266."

64. Ibid., listy 104–5. Letter of L. K. Shaposhnikov to Gosplan USSR Chairman V. È. Dymshits, November 10, 1962.

65. ARAN St. Pbg. f. 996, op. 4, d. 35, listy 271, 271 rev., 272. Letter of L. K. Shaposhnikov to E. M. Lavrenko, December 24, 1961.

66. Ibid., list 284. Note of Lavrenko to the Gosplan Commission of April 17, 1962.

67. RGAE f. 544, op. 1, d. 141, listy 125–28, with attached diagrams and organizational flow charts (listy 130–31).

68. Ibid., listy 160–64. Letter of V. S. Prokrovskii, deputy scholarly secretary of

the Commission for Nature Protection and head of its laboratory, to the Presidium of the USSR Council of Ministers and to Foreign Minister Andrei A. Gromyko. Dated September 16, 1963.

69. The Council of Mutual Economic Assistance, headquartered in Moscow, had been established in 1949 to unite the Communist-run nations' economies.

70. RGAE f. 544, op. 1, d. 141, list 164.

71. Ibid., listy 165–66. The letter is dated September 18, 1963. Next to Petrov's name is a note that he had been a Party member since 1896. The quotation is from list 165. He is probably referring to the A. Starker Leopold Commission, which published its findings as *Future Environments of North America,* following a conference it organized.

72. Ibid., list 166.

73. Ibid., list 167.

74. RGAE f. 3, op. 1, d. 261, listy 46–47.

75. Ibid., listy 49–51.

CHAPTER FOURTEEN

1. Moskovskii gosudarstvennyi universitet, Molodëzhnyi sovet po okhrane prirody, *Organizatsiia molodezhnogo,* 7.

2. Ksenia V. Avilova, Memoir-history of the Moscow University Biofak *druzhina po okhrane prirody,* [5–6]. Untitled, typed (xeroxed), unpaginated. Courtesy of the author.

3. Tikhomirov, "Istoriia i deiatel'nost'," 14.

4. Hough and Fainsod, *How the Soviet Union Is Governed,* 226–27, 302. On the volunteer auxiliary police, see also Juviler, *Revolutionary Law,* chapter 4.

5. Decree "On the Participation of the Working People in the Maintenance of Public Order." See Juviler, *Revolutionary Law,* 78–79.

6. I am grateful to Loren R. Graham for urging me to explore this connection.

7. Tikhomirov, "Istoriia i deiatel'nost'," 12.

8. Avilova, Memoir-history, [8].

9. Ibid., [12–13].

10. E. Smantser mentioned these antecedents to the MGU *druzhina* in a speech commemorating the latter's tenth anniversary. See Sotsio-Ekologicheskii Soiuz, *Tridtsat' let dvizheniia,* 5.

11. Tikhomirov, "Istoriia i deiatel'nost'," 13–14. Kharitonov and Avilova, in "Okhrana prirody," 107–8, mention that, of the early generation of *druzhinniki,* Smantser, V. Maksimov, A. Agadzhanian, and V. Tsvetkov had been in KIuBZ; M. Gribov, Sergei Ivanov, M. Cherniakhovskii, and N. Chernyshev had been in the VOOP youth group; and A. Giliarov, B. Goncharov, A. Tishkov, and P. Tomkovich had been in the MOIP one. The authors note that this list could have been extended considerably.

12. Avilova, Memoir-history, [6].

13. Sotsio-Ekologicheskii Soiuz, *Tridtsat' let dvizheniia, 1960–1987,* 4. The textbook they coauthored is Blagosklonov, Inozemtsev, and Tikhomirov, *Okhrana prirody.*

14. Avilova, Memoir-history, [6].

15. For violating the rules of conduct on raids members could incur suspension from the antipoaching raids, revocation of their civilian inspector certificates, or, in the worst cases, expulsion from the *druzhina.*

16. Sotsio-Ekologicheskii Soiuz, *Tridtsat' let Druzhiny, 1960–1987,* 5.

17. Ibid.

18. Tikhomirov, "Istoriia i deiatel'nost'," 14.

19. Ibid., 15.

20. Ibid.

21. Sotsio-Ekologicheskii Soiuz, *Tridtsat' let Druzhiny, 1960–1987,* 6.

22. Ibid., 6–7.

23. Tikhomirov, "Istoriia i deiatel'nost'," 15. As a gesture of gratitude, the *druzhina* elected Naumov to honorary membership.

24. Yanitsky [Ianitskii], *Russian Environmentalism,* 20.

25. Ibid., 66. The Komsomol also feared the *druzhina's* potential to attract students to its banner; by stepping in as its official sponsor, the Komsomol created an aura of liberalism for itself and thereby became more attractive, or at least less unattractive, to those more independent-minded students.

26. Fitzpatrick, *Education and Social Mobility in the Soviet Union.*

27. Chivilikhin, "Shumi," 8–9.

28. Ibid., 9.

29. Ibid., 10.

30. Ibid., 11.

31. Ibid., 17–20.

32. Krylov, "Slovo o sibirskom kedre." Krylov, a biologist, was chairman of the Siberian Branch of the Academy of Sciences' Commission on the Protection of Nature.

33. Ibid., 155–56.

34. See Weiner, *Models of Nature, passim.*

35. Parfënov, "Kedrovye lesa."

36. Ibid.

37. Kiriasov, "Skazanie o Kedrograde," 2–3.

38. Interview with Vitalii Feodos'evich Parfënov, Moscow, August 21, 1995.

39. Chivilikhin, "Shumi," 20.

40. Interview with Vitalii Feodos'evich Parfënov, Moscow, August 21, 1995.

41. Ibid. Chivilikhin, "Shumi," 21, had another take on the episode: "The students did not really grasp the essence of the matter and childishly raised a hue and cry. Sergei convened a meeting of the Komsomol in the professor's defense and, like the other youths, spouted all sorts of nonsense out of his sense of outrage."

42. Chivilikhin, "Shumi," 22.

43. Ibid., 23.

44. Interview with Vitalii Feodos'evich Parfënov, Moscow, August 21, 1995.

45. Ibid.

46. Ibid.

47. TsKhDMO f. 1, op. 9, d. 458, "Predlozhenii Altaiskogo kraikoma VLKSM ob organizatsii v Gorno-Altaisoi oblasti komsomolskogo kompleksnogo leskhoza," listy 1–9.

48. Kovalevskii, "Bol'she molchat' nel'zia!" Parfёnov feels that "he was put up to this, without a doubt." Interview, Moscow, August 21, 1995.

49. Parfёnov, "Nash otvet." That was followed in the same issue by a rebuttal by Kovalevskii.

50. Chivilikhin, "Shumi," 49. Of the forested area of the *krai* (4,497,000 ha.), cedars occupied 23 percent, or 1,025,000 hectares. Annual logging per 1,000 hectares was 280,000–300,000 cubic meters of wood, 88 percent of which was allegedly mature or overmature (140–250 years old). The forestry academy graduates asked for 85,000 hectares, of which 71,000 were forested, as well as for financial coverage of the first year's expenses.

51. Ibid., 33.

52. Yanitsky [Ianitskii], *Russian Environmentalism,* 177.

53. Ibid. Cherkasova mistakenly places Armand's book in the period of the "thaw." In reality, his book appeared at the end of Khrushchёv's rule, in 1964.

54. Chivilikhin, "Shumi," 51–52.

55. Chivilikhin, "Mesiats v Kedrograde," 67.

56. Ibid., 57.

57. Sotsio-Ekologicheskii Soiuz. *Tridtsat' let dvizheniia, 1960–1987,* 15.

58. Interview with Vitalii Feodos'evich Parfёnov, Moscow, August 21, 1995.

59. Ibid.

60. Ibid.

61. Parfёnov, "Kedrovye lesa," 3–4.

62. Ibid., 3.

63. Chivilikhin, "Taiga shumit," 78.

64. Parfёnov, "Kedrovye lesa," 3.

65. Ibid., 4.

66. Ibid.

67. Ibid.

68. Ibid.

69. Chivilikhin, "Taiga shumit," 74–82.

70. Ibid.

71. Chivilikhin, "Piatiletie Kedrograda," 1, 3.

72. Ibid.

73. Ibid.

74. Ibid.

75. Ibid.

76. Parfёnov, "Kedrogradu byt'!" 2.

77. Ibid.

78. Ibid.

79. Chivilikhin, "Piatiletie Kedrograda."

80. Ibid.

81. Ibid.

82. Ibid.

83. Interview with Vitalii Feodos'evich Parfёnov, Moscow, August 25, 1995.

84. Ibid. Years later, in 1984, the journal *Priroda i chelovek,* later *Svet,* was founded

with his participation. Bochkarëv took up the pen in 1965 to defend his policies. Bochkarëv, "Problemy kedrovoi taigi," 2.

85. Chivilikhin, "Piatiletie Kedrograda."

86. Chivilikhin, "V mire," 102–9.

87. Kazarkin, "Pamiat'," 156–57.

88. Quoted, ibid.

89. Literary influences on Chivilikhin are discussed in "Beseda v *Nashem sovremennike*," 401.

90. Chivilikhin, "Uroki Leonova," 256.

91. Chivilikhin, "Mesiats v Kedrograde," 70.

92. Quoted, Kazarkin, "Pamiat'," 158.

93. Ibid., 157.

94. Ibid., 158.

95. Ibid., 162.

96. Ibid., 159–62. Of course, this is not the first time in Russian social thought that the ancestors of the Slavs (or Europeans) were sought for in Siberia; Nikolai Rërikh, the artist and seeker, did so in the 1920s and Gumilëv's entire opus was dedicated to this very set of problems. Forest versus steppe was an eternal, epic cultural conflict played out on the vast Eurasian stage.

97. Chivilikhin, "V mire," 108–9.

98. Ibid.

99. Ibid.

100. Ibid.

101. "Vsë prodolzhaetsia v nas," esp. 247.

102. Ibid.

103. Chivilikhin, "Shvedskie ostanovki," 576.

104. Chivilikhin, *Pamiat'*.

105. "Pamiat'—kategoriia nravstvennaia" (Iz interv'iu gazete *Sotsialisticheskaia industriia*), in Chivilikhin, *Zhit' glavnym*, 235.

106. Chivilikhin, "Svetloe oko Sibiri."

107. Interview with Vitalii Feodos'evich Parfënov, Moscow, August 21, 1995.

108. Chivilikhin, "V mire," 108–9.

109. Chivilikhin, "Mesiats v Kedrograde," 39.

110. Chivilikhin, "Shvedskie ostanovki," 579.

111. Ibid., 576.

112. "I khram, i masterskaia!" (Iz besedy za 'Kruglym stolom' v redaktsii *Literaturnoi gazety*), in Chivilikhin, *Zhit' glavnym*, 232–33.

113. Ibid.

114. Although John B. Dunlop, in his perceptive and influential study *The Faces of Contemporary Russian Nationalism*, noted that preservationism was a "relatively 'safe' way of contesting the direction being taken by the regime and encouraged by the ruling ideology" (64) without overtly revealing the Russian nationalist sentiment beneath, he saw this nationalism as simply a reaction to modernization. My analysis points to a process of initial support for the regime followed by disillusionment and subsequent evolution of a nationalist ideology.

CHAPTER FIFTEEN

1. TsGA f. 259, op. 42, d. 8530, listy 18–26.
2. Ibid., list 27. Memo of A. Ageev [to premier of RSFSR] dated August 4, 1962.
3. TsGA f. 404, op. 1, d. 791, listy 79–136, esp. list 92.
4. Ibid., list 92.
5. The *zubr,* or European bison (*Bison bonasus*), a highly endangered cousin of the American bison, was commonly used from the 1920s on to represent intellectuals, especially scientists, of the old prerevolutionary type.
6. TsGA f. 404, op. 1, d. 791, listy 94–95.
7. Ibid., list 123. Copy of report of September 16, 1964, "Prezidiumu VOOP," from N. Katin, acting senior fishing inspector of Riazan' *oblast'.*
8. Ibid., list 99.
9. Ibid., list 101. Report of N. Katin to Professor V. G. Geptner, September 1, 1964.
10. A copy of the feulliton was preserved in VOOP archives concerning the incident. Ibid., list 88.
11. Ibid., listy 107–9. V Prezidium Ts. Soveta VOOP. February 1965 (date cut off on xerox).
12. See Blagosklonov's letter of January 24, 1965, ibid., list 111, 111 rev., as well as the letter from Vladislav Petrovich Tydman, February 19, 1965, listy 102–3, as examples of critical correspondence.
13. Ibid., listy 79–87, esp. 86–87. Postanovlenie [Prezidiuma TsS VOOP] ot 24 fevralia 1965 goda No. 16 "O stat'e v zhurnale 'Krokodil' No. 1 za 1965 god 'Vot kakie karasi.'"
14. Ibid., list 130. Copy of letter of Geptner to Katin of September 5, 1964.
15. Ibid., listy 134–36.
16. TsGA f. 404, op. 1, d. 790, listy 17–20. "Protokol No. 1 zasedaniia Tsentral'nogo soveta VOOP."
17. Interview with Vitalii Feodos'evich Parfënov, Moscow, August 21, 1995.
18. Ibid.
19. Ibid.
20. Parfënov, "Kedrovye lesa," 4.
21. Ibid.
22. Interview with Vitalii Feodos'evich Parfënov, Moscow, August 21, 1995.

CHAPTER 16

1. Struchkov, "Nature Protection as Moral Duty," 426–27.
2. See Weiner, *Models,* 183.
3. Pryde, *Environmental Management,* 84.
4. Taurin, *Daleko v strane Irkutskoi,* 117.
5. Josephson, *New Atlantis Revisited.* See chapter 5, "Siberian Scientists and the Engineers of Nature."
6. TsKhSD f. 5, op. 37, d. 27, listy 28–31. Report to the Bureau of the Central Committee of the CPSU for the RSFSR from N. Kaz'min, director of the Department

for Sciences, Schools, and Culture for the Bureau, and N. Dikarev, *instruktor.* The report was dated September 3, 1958. There were two other representatives of the Central Committee at the conferences in addition.

7. Ibid., list 93.

8. Ibid., list 31.

9. Taurin, *Daleko v strane Irkutskoi,* 118.

10. Ibid.

11. "V zashchitu Baikala," *Literaturnaia gazeta,* 21 October 1958, reprinted in *Slovo v zashchitu Baikala,* 13–16.

12. See RGALI f. 634, op. 4, d. 2116, listy 150 and 189, for examples of letters supporting "V zashchitu Baikala."

13. Taurin, *Daleko v strane Irkutskoi,* 124.

14. MOIP Archives, d. 110. Stenogramma soveshchaniia po redkim . . . , Zasedaniia ob"edinnënykh sektsii, Evening, March 29, 1957, list 16. Remarks by L. K. Shaposhnikov. See also RGALI f. 634, op. 4, dela 1507, 1838, 2116, etc., which are "Otzyvy chitatelei o probleme okhrany okruzhaiushchei sredy" (Readers' letters on environmental problems) for 1956–58.

15. Skalon, *Okhraniaite prirodu!*

16. Taurin, *Daleko v strane Irkutskoi,* 126.

17. Ibid., 126–31.

18. Taurin, "Baikal dolzhen byt' zapovednikom," reprinted in *Slovo v zashchitu Baikala,* 17–20.

19. Dement'ev et al., "V zashchitu Baikala!" reprinted in *Slovo v zashchitu Baikala,* 20–21.

20. Löwenhardt, *Decision Making in Soviet Politics,* 71.

21. Ibid., 73.

22. Josephson, *New Atlantis Revisited,* chapter five, 19.

23. Archives of *Literaturnaia gazeta* at the editorial offices. Letter of G. L. Pospelov, February 8, 1963, unpublished. Filed by year and date in folders.

24. Pospelov, "Razmyshleniia o sud'be Baikala," 160–64.

25. Ibid., 164.

26. Chivilikhin, "Svetloe oko Sibiri."

27. MOIP Archive, d. 2050, list 74. Letter of August 7, 1963. Carbon.

28. "Snova o Baikale," 171–74.

29. Ibid., 171.

30. Ibid., 173–74. Twenty-one such organizations and conferences were listed.

31. Ibid., 174.

32. Ibid.

33. Ibid.

34. Merkulov, "Trevoga o Baikale," 4.

35. Volkov, "Tuman nad Baikalom," reprinted in *Slovo v zashchitu Baikala,* 45–52.

36. Ibid., 50–52.

37. Sobolev, Sartakov, Taurin, et al., "V zashchitu Baikala!" reprinted in *Slovo v zashchitu Baikala,* 52–53.

38. Orlov, "Snova o Baikale," reprinted in *Slovo v zashchitu Baikala,* 57–59.

39. Ibid., 59.
40. Volkov, "Tuman ne resseialsia," reprinted in *Slovo v zashchitu Baikala*, 62–69.
41. Trofimuk, "Tsena vedomstvennogo upriamstva," reprinted in *Slovo v zashchitu Baikala*, 70–77. Quotation, 76–77.
42. MOIP Archive, d. 2050, list 8.
43. MOIP Archive, d. 2062, list 2. Letter of January 11, 1966 from MOIP to minister of agriculture V. V. Matskevich. Draft carbon. Originally typed January 8, 1966 and amended three days later.
44. Ibid., listy 2, 3. The words used were *bezzakoniia* and *proizvol*.
45. Ibid., list 3.
46. Ibid., listy 3–4.
47. Ibid., list 5.
48. Konstantinov et al., "Baikal zhdёt," reprinted in *Slovo v zashchitu Baikala*, 78–83. Quotation, 80.
49. Ibid., 80.
50. Ibid., 82.
51. Ibid., 82–83.
52. MOIP Archive, d. 2062, list 163.
53. MOIP Archives, d. 2063. Draft letter of July 5, 1966 to Brezhnev from Varsonof'eva, Prozorovskii, and Professor V. I. Pelevin, deputy chair of the Section on Nature Protection of the Academy of Sciences' House of Scholars.
54. MOIP Archive, d. 2063. Reshenie sobraniia uchёnykh, spetsialistov i predstavitelei sovetskoi obshchestvennosti, sostoiavshegosia v Dome uchёnykh AN SSSR 13 iunia 1966 g. 4 pp.
55. MOIP Archive, d. 2063. Protokol no. 5 zasedaniia Prezidiuma Soveta Moskovskogo obshchestva ispytatelei prirody ot 16-ogo iunia 1966 g. On 3 pp. No "listy" show up on these xeroxed pages.
56. Ibid. The resolutions are contained in the same *delo*.
57. MOIP Archive, d. 2063, Letter of September 20, 1966 of Sukachёv et al. to Viktor Borisovich Sochava. Carbon copy. 4 pp.
58. Ibid.
59. Ibid.
60. "Tol'ko dokumenty" (Only documents), *Komsomol'skaia pravda*, 13 May 13 1966 (no. 134), reprinted in *Slovo v zashchitu Baikala*, 83–90. Quotation, 83–84.
61. Ibid., 89.
62. Ibid., 90.
63. "Vystuplenie Akademika P. L. Kapitsy na sovmestnom zasedanii Kollegii Gosplana SSSR, Kollegii Goskomiteta SSSR po nauke i tekhniki, i Prezidiuma Akademii nauk SSSR," 22 iunia 1966 g. First published in *Khimiia i zhizn'* 7 (1987), reprinted in *Slovo v zashchitu Baikala*, 90–95. Quotation, 91.
64. Ibid., 92.
65. Ibid., 95.
66. Goldman, *Spoils of Progress;* Gustafson, *Reform in the Soviet Union;* Jancar-Webster, *Environmental Management;* Kelley, "Environmental Policy-Making in the USSR"; Löwenhardt, *Decision-Making in Soviet Politics;* Pryde, *Conservation in the Soviet Union;* Komarov, *Unichtozhenie prirody v Sovetskom soiuze.*
67. Pryde, *Environmental Management*, 86–87.

CHAPTER SEVENTEEN

1. Quoted in Worster, *Nature's Economy,* 249.

2. Aleksandrova, "Rastitel'noe soobshchestvo," and "Problema razvitiia v geobotanike." For a highly useful bibliography of her works and those of her colleagues, see Levina, *Geobotanika v Botanicheskom institute.*

3. "Postanovlenie Biuro otdeleniia biologicheskikh nauk Akademii nauk SSSR ot 18 iunia 1957 g., Protokol no. 21, P. 1, 'O ratsional'noi seti zapovednikov SSSR,'" in "Khronika" (1958), 112–13.

4. See, for instance, Zharkov, *Prosteishie nabliudeniia v prirode;* or Stepanov, "Zapovednoe delo v Kazakhstane," 4–6.

5. TsGa f. 358, op. 3, d. 3404, list 90. "Stenogramma soveshchaniia direktorov gosudarstvennykh zapovednikov i ikh zamestitelei po nauchnoi rabote s uchastiem storonnykh nauchnykh uchrezhdenii v Voronezhskom gosudarstvennom zapovednike."

6. Ibid., list 142.

7. TsGA f. 358, op. 3, d. 3405, list 147.

8. TsGAf. 358, op. 3, d. 5082, listy 21–22.

9. Ibid., listy 22–23.

10. Ibid., list 23.

11. Ibid., listy 23–24.

12. Ibid., listy 25–26.

13. Ibid., list 26.

14. The seven *zapovedniki* were: Astrakhanskii, Voronezhskii, Kavkazskii, Okskii, Prioksko-Terrasnyi, Sikhote-Alinskii, and Tsentral'no-Lesnoi. The decree was Postanovlenie SM SSSR No. 390 ot 18 maia 1965 g. Glavokhota acceded with a corresponding decree on August 7, 1965. See also TsGA f. 358, op. 3, d. 3876, listy 1–2: "Akt peredachi gosudarstvennykh zapovednikov Glavokhota RSFSR v vedenie Ministerstva sel'skogo khoziaistva SSSR."

15. Bannikov, *Po zapovednikam SSSR.*

16. Boreiko, *"Tsarskie okhoty,"* 24. A. A. Nasimovich Archives, Institute of Geography of the Russian Academy of Sciences, uncatalogued. Letter of Director I. Tsepliaev of the Kavkazskii Gosudarstvennyi *zapovednik* to Andrei Aleksandrovich Nasimovich of March 23, 1965. The letter warmly thanks Nasimovich for his help in fighting the transfer.

17. Readers are referred to Komarov, *Destruction of Nature in the Soviet Union,* chapter 6, as well as to virtually any issue of *Okhota i okhotnich'e khoziaistvo.* See also Volkov, "Tropa brakonera," 11; and Shtil'mark, *Istoriografiia rossiiskikh zapovednikov.*

18. Nasimovich Archives, Institute of Geography of the Russian Academy of Sciences, uncatalogued. A. A. Nasimovich's letter to the editor of *Komsomol'skaia pravda,* September 13, 1968, p. 1 of 3.

19. RGAE f. 7486, op. 33, d. 60, list 103–6. Note of August 23, 1966.

20. Ibid., listy 40–54. "O merakh po uluchsheniiu okhrany prirody v SSSR. Proekt."

21. V. N. Skalon, review of *Trudy Voronezhskogo gosudarstvennogo zapovednika,* no. 5 (Voronezh, 1954), in *Biulleten' MOIP* 63, no. 1 (1958): 120–22.

22. Ibid. On the Altai red squirrel, see also Grebënok, "Parazity belki," 340–42.

The reacclimatization of the European bison, the beaver, and the sable were among the most successful rescue operations of twentieth-century nature protection. Beavers, 3,340 of which were reacclimatized to former habitats via the *zapovedniki,* had attained an overall population of 100,000 by the early 1980s. From 1946 to 1977 the sable population increased by 600 percent and stood at more than 650,000 by the early 1980s as well. At one 1975 auction, 45,500 pelts sold for $3,420,000. See *Geograficheskoe razmeshchenie zapovednikov,* 4.

23. Interview with Andrei Aleksandrovich Nasimovich, Moscow, April 18, 1980.

24. Geptner, "Kakovy zhe puti obogashcheniia fauny." Quoted in *Akklimatizatsiia okhotnich'e-promyslovykh,* 1:4.

25. Ibid., 5.

26. *Akklimatizatsiia okhotnich'e-promyslovykh,* 1:5; Nasimovich, "Ekologicheskie posledstviia vkliucheniia novogo vida," 1593–98.

27. *Akklimatizatsiia okhotnich'e-promyslovykh,* 1:5.

28. Skalon, "Sushchnost' biotekhnii."

29. Filonov, "Dinamika chislennosti."

30. Ibid., 88.

31. Ibid., 89.

32. Ibid., 203–4 and passim.

33. Dement'ev, "Deiatel'nost'," 11.

34. "Soveshchanie po okhrane prirody."

35. Nagibina, "Zadachi sanitarnoi okhrany vodoemov."

36. Ibid., 36.

37. Ibid., 36–37.

38. Ibid., 38.

39. Ibid., 39–46.

40. Gol'dberg, "Chistota atmosfernogo vozdukha i ego okhrana." Another article about air pollution appeared four years later in the same journal: Voroshilov and Nedotko, "Ispol'zovanie mineral'nogo topliva."

41. Kuzin and Peredel'skii, "Okhrana prirody," esp. 65–70.

42. Ibid., 72–73.

43. Ibid., 75.

44. Ibid., 76.

45. S. S. Shvarts, "Voprosy akklimatizatsii mlekopitaiushchikh na Urale," 10. He noted that acclimatization always referred to the successful formation of a *population* working from the existing genotypes of the introduced specimens, and not from the falsely held power of the environment to work directed, adaptive hereditary changes in the introduced individuals.

46. One Soviet informant, a close collaborator of Timofeev-Resovskii's, described Shvarts as "a classic yes-man."

47. Quoted in Bol'shakov, *Ekologicheskie osnovy okhrany prirody,* 5.

48. Ibid., 3–4.

49. S. S. Shvarts, *Tekhnicheskii progress i okhrana prirody,* 1, 12–13.

50. Ibid., 15.

51. Ibid., 14.

52. *Dialog o prirode.*

53. Ibid., 127. It is significant that Shvarts uses the modern term *ecosystem* rather than the older and more platonic *biocenosis.*

54. Ibid.

55. Ibid., 37–38.

56. Ibid., 49.

57. Borodin, "Usilit' bor'bu s volkami," 4.

58. Gusev, "Protiv idealizatsii prirody," 26.

59. *Dialog o prirode,* 43.

60. Ibid., 125; 24.

61. Ibid., 22.

62. Ibid., 125.

63. Timofeev-Resovskii and Tiuriukanov, "Biogeotsenologiia i pochvovedenie."

64. Ibid., 109.

65. Ibid., 110–11.

66. Ibid., 114–15.

67. We see a similar approach in Giliarov, Vinberg, and Chernov, "Ekologiia": "Precisely ecology has been called upon to create the scientific basis for the rational exploitation of resources [and] prognostication of the changes in nature cause by human activity" (5).

68. See Nesterov, *Voprosy upravleniia prirodoi;* Svirezhev and Logofet, *Ustoichevost' biologicheskikh sistem;* Bazykin, *Matematicheskaia biofizika vzaimodeistvuiushchikh populiatsii;* Budyko, *Global'naia ekologiia;* Gorban' and Khlebopros, *Demon Darvina.*

69. The deputy chair of the commission was S. V. Kirikov. Other members were: G. E. Burdin, P. I. Valeskaln, A. G. Voronov, V. G. Geptner, I. P. Gerasimov, N. E. Kabanov, and A. N. Formozov. The commission formally was named the USSR Academy of Sciences' Commission on Nature Protection.

70. Lavrenko, Geptner, Kirikov, and Formozov, "Perspektivnyi plan," 4–5.

71. See Nasimovich, "V Berezinskom zapovednike," 111–13. Upstream water users siphoned off water for irrigation of tributaries of the Berezina River, causing downstream desiccation and fall of the water tables, drying up the marsh, and affecting the habitat of fauna.

72. Interview with A. A. Nasimovich, Moscow, April 18, 1980. In 1946–47 there were 2,500 desman in the Khopër reserve but by 1975 there were only 500. An additional problem was that acclimatized axis deer and boar destroyed the rodents' burrows when they came to drink at the stream's edge. Acclimatized muskrat competed with the desman for food and burrows as well. The situation called into question the "natural" state of the territory protected within the boundaries of the *zapovednik.*

73. ARAN—St. Pbg. f. 996, op. 4, d. 35, list 202. Shaposhnikov's handwritten note to Lavrenko was included on a typed letter from I. N. Iaitskii, director of the reserve, of March 24, 1959, to Shaposhnikov and the USSR Academy of Sciences Commission on Nature Protection, which supervised the scientific research and management strategies of the reserve. Attached to the letter on listy 203–15 was the proposed management plan itself.

74. Ibid., list 202. The plan, entitled "Rezhim Tsentral'no-Chernozëmnogo Goszapovednika," was dated January 22, 1959, on list 203–15.

75. Ibid., list 203.
76. Ibid., list 204.
77. Ibid., list 208.
78. Ibid.
79. Ibid., list 209.
80. Krasnitskii, "Lesokhoziaistvennye meropriiatiia"; Krasnitskii and Dyrenkov, "Sravnitel'naia otsenka lugovykh"; Dyrenkov and Krasnitskii, "Osnovnye funktsii zapovednykh territorii"; and Krasnitskii, *Problemy.*
81. See discussion of this in Rashek, Vasil'ev, and Chumakova, "Okhrana soobshchestv." Also see Krasnitskii, *Problemy,* 111.
82. Krasnitskii, *Problemy,* 112.
83. See Gilyarov, "Agrocenology."
84. Krasnitskii, *Problemy,* 58, 89.
85. Ibid., 112.
86. Shvarts, "Teoreticheskie osnovy," cited in Krasnitskii, *Problemy,* 112.
87. Krasnitskii, *Problemy,* 112.
88. Ibid., 144–47.
89. Alekseeva and Zykov, "Differentsiatsiia razmerov zapovednikov," 38–39.
90. Nukhimovskaia, "Biologicheskie i geograficheskie," 23.
91. Ibid.
92. Interview with Konstantin Pavlovich Filonov, Moscow, March 6, 1986.
93. Interview with Andrei Aleksandrovich Nasimovich, Moscow, April 18, 1980.
94. Fedorenko and Reimers, "Nature Conservation," 87.
95. Ibid.
96. Reimers, "Bez prava na oshibku," 17.
97. Ibid., 29.
98. Sushchenyz, Parfënov, Vynayev, and Rykovsky, "Scientific Principles," 121–24.
99. See Kotliarov, *Geografiia otdykha i turizma,* 29–30; and Chizhova, *Rekreatsionnye nagruzki v zonakh otdykha.*
100. The Academy of Sciences' Institute of Geography, Department of Biogeography has a section led by N. A. Kazanskaia that studies the ecological effect of recreation on landscapes.
101. Bannikov, "Ot zapovednika do prirodnogo parka."
102. On "natsional'nye parki," see Nikolaevskii, *Natsional'nye parki;* and Filippovskii, *Natsional'nyi park.*
103. Reimers and Shtil'mark, *Osobo,* 179.
104. Ibid., 161 and *passim.* An example of a land-use zoning plan based on these principles may be seen in Iagomagi and Pallok, "Funktsional'noe zonirovanie i sel'skoe khoziaistvo."
105. Reimers and Shtil'mark, "Etalony prirody," 15–16.
106. Frolov and Los', "Filosofskie osnovaniia sovremennoi ekologii."
107. Ibid., 16.
108. Ibid., 21.
109. Ibid., 15.
110. Ibid., 22.
111. DeBardeleben, *Environment and Marxism-Leninism.*

112. As a sampler, see Ursul, *Philosophy and the Ecological Problems of Civilisation; Obshchestvo i prirodnaia sreda;* and Fëdorov, *Ekologicheskii krizis.* Readers are referred to DeBardeleben's bibliography and particularly to Pidzhakov, *Sovetskaia ekologicheskaia politika,* esp. 142–45, nn. 15–30.

113. DeBardeleben, *Environment and Marxism-Leninism,* 55.

114. Ibid., 115.

115. Ibid., 191, 214.

116. Ibid., 268.

117. See Alimov, Sukhov, Nikolaev, and Khodanovich, *Problemy ekologii i sovremennost'.*

CHAPTER EIGHTEEN

1. See Gustafson, *Reform in the Soviet Union.* For legislation of this period see also Pidzhakov, *Sovetskaia ekologicheskaia politika.*

2. TsGA f. 404, op. 1, d. 2912, listy 153–54. Dated October 5, 1981.

3. Ibid., listy 155–57.

4. Ibid., list 25. The document is V. N. Vinogradov, "Otchët o rezul'tatakh komandirovaniia v sostave sovetskoi delegatsii predsedatelia Prezidiuma TsS VOOP chlena Ispolnitel'nogo Soveta MSOP na VIII zasedanie Mezhdunarodnogo soiuza okhrany prirody i prirodnykh resursov (MSOP)." Undated (but written shortly after November 15, 1980, when Vinogradov returned from seven days in Switzerland).

5. TsGA f. 404, op. 1, d. 2912, listy 107–8. Evidently sent April 29, 1981.

6. *Belorusskoe obshchestvo okhrany prirody.*

7. Komarov, *Destruction of Nature,* 16.

8. Botvich, Zagrishiev, and Shvarts, "Iz opyta raboty," 19–20. Shvarts is the author of chapter 3, "Dvizhenie druzhin po okhrane prirody."

9. Interview with Dmitrii Nikolaevich Kavtaradze, Moscow, March 23, 1991.

10. Postanovlenie TsK VLKSM, "Ob uchastii komsomol'skikh organizatsii, komsomol'tsev i molodëzhi v okhrane prirody, ratsional'nom ispol'zovanii i vostanovlenii eë resursov," April 29, 1968, in Bulgakov, "Uchastie komsomol'skikh organizatsii," 119.

11. TsKhDMO f. 1, op. 36, d. 288, listy 44–45. Letter of A. Chubukov, secretary of the Perm' VLKSM (Komsomol), to F. Oborin, deputy president of the Perm' *oblast'* Council of VOOP.

12. TsKhDMO f. 1, op. 36, d. 423a, listy 1–229, "Stenogramma naucho-prakticheskoi konferentsii po okhrane prirody i ratsional'nomu ispol'zovaniiu prirodnykh resursov v zone stroitel'stva BAM'a, 19–20 ianvaria 1977 g.," and f. 1, op. 36, d. 345, "O zonal'nom soveshchanii komsomol'skikh rabotnikov, aktivistov po okhrane prirody, i spetsialistov po voprosu 'Okhrana prirody i ratsional'noe ispol'zovanie prirodnykh resursov v zone BAM'a,' 26–28 ogo iunia 1975, Irkutsk." See especially the speech of V. N. Skalon, list 6.

13. Faculty adviser K. N. Blagosklonov was the leading spirit of that program, which was started by the MGU *druzhina* in 1975. It was the first summer school for the training of citizen inspectors.

14. TsKhDMO f. 1, op. 36, d. 378, listy 83–151. Speech of Evgenii Grigor'evich Lysenko, "Zadachi komitetov komsomola i organizatsii VOOP po privlecheniiu

molodëzhi k aktivnomu uchastiiu v okhrane prirody i ratsional'nom ispol'zovanii prirodnykh resursov v svete trobovanii XXV s"ezda KPSS." See esp. listy 95–96 and 148.

15. "Iz istorii studencheskogo dvizheniia za okhranu prirody," 2. Typed, xeroxed, n.d., no author.

16. See, for example, Lysenko and Terent'ev, *Komsomol i okhrana prirody.*

17. Ksenia V. Avilova, Memoir-history of the Moscow University Biofak *druzhina po okhrane prirody,* [4]. Untitled, typed (xeroxed), unpaginated.

18. Sotsio-Ekologicheskii Soiuz, *Tridtsat' let dvizheniia,* 17.

19. Ibid., 24.

20. Ibid., 70 (for 1975). Figures for 1985 from Tikhomirov speech, February 3, 1986, Moscow.

21. TsKhDMO f. 1, op. 36, d. 279, list 41. Sergei Germanovich Mukhachëv and Iurii Sergeevich Kotov, "Spravka o Kazan'skoi druzhine po okhrane prirody, 8 sentiabria 1976 g." See also Sotsio-Ekologicheskii Soiuz, *Tridtsat' let dvizheniia,* 42–43.

22. Moskovskii gosudarstvennyi universitet, Molodëzhnyi sovet po okhrane prirody, *Organizatsiia,* 99.

23. On "Ecopolis," see Brudny and Kavtaradze, "'Ecopolis.'"

24. "Fenomenologiia zhestokosti," 89–90.

25. Ibid., 101. See also Singer, *Animal Liberation.*

26. Florenskii and Shutova, "Da skroetsia t'ma," 736. On the present-day status of this conflict, see recent issues of the *Eco-Stan News,* edited by Eric Sievers.

27. Blagosklonov, "Komsomol'skaia druzhina po okhrane prirody," 95.

28. Ksenia V. Avilova, Memoir-history, [4].

29. I witnessed the events described.

30. Mokievskii, Chestin, and Shvarts, "Prirodu ne obmanesh'."

31. Ibid. N. G. Ovsiannikov, VOOP's president from 1965 to 1971, was the RSFSR minister of land reclamation and water resources. N. F. Vasil'ev, the USSR minister of land reclamation and water resources from 1979 to 1988, was a member of the Society's Central Council. In 1980, the president of the Kazakh society was chairman of that republic's Supreme Soviet. Other notables on the Central Council included the deputy head of the Agriculture Section of the Kazakh party Central Committee, the first deputy minister of energy and electrification of Kazakhstan, the deputy ministers of finance, fishing, forestry, agriculture, education, culture, and public health, the deputy chair of Gosplan, and a secretary of the Komsomol. This example could be duplicated for the Russian, Kyrgyz, and other republics' societies. E. A. Shvarts, in "O politicheskikh pravakh samodeiatel'nykh organizatsii," 91, notes that in almost every case the *oblast'* branch president of VOOP is the deputy chairman of the *oblispolkom.*

32. Xerox of speech of Ianitskaia, personal copy of author.

33. Zhukova, "Svoi liudi u prirody," 15–23.

34. Ibid., 15.

35. Cited ibid.

36. I am elaborating on a basic point made by Zhukova in ibid., 16.

37. Yanitsky [Ianitskii], *Russian Environmentalism,* 197.

38. Botvich, Zagrishiev, and Shvarts, "Iz opyta raboty samodeiatel'nykh," 19.

39. Ibid., 33–36.

40. On tourism, see Zhukova, "Svoi liudi u prirody," 16–17.

41. Yet even these initial inclinations could be modified. One excellent example of such social learning was the Riazan' *druzhina,* whose active enlistment of tourists as allies ultimately led to the creation of the Meshcherskii National Park in 1991. Ibid., 19.

42. Darst, "Environmentalism in the USSR," 226–27.

43. Iaroshenko, "U istokov," 62. See also Akademiia nauk SSSR, *Trudy noiabrskoi sessii 1933 goda,* 222–26.

44. Petro, "'Project of the Century,'" 236.

45. Darst, "Environmentalism in the USSR," 227.

46. Petro, "'Project of the Century,'" 237. This is a claim of I. Gerardi, chief engineer of the project. For articles defending diversion, see *Izvestiia AN SSSR, seriia geograficheskaia,* no. 2 (1977) and no. 2 (1981).

47. See Gustafson, *Reform in Soviet Politics.*

48. Iaroshenko, "U istokov," 63.

49. Ibid., 238.

50. Zalygin, "Pisatel' i Sibir'" (1961), reprinted in Zalygin, *Kritika i publitsistikia,* 280.

51. Quoted in Petro, "' Project of the Century,'" 238.

52. See Zalygin, "Lesa, zemli, vody," originally published in *Literaturnaia gazeta,* 26 June 1962, reprinted in Zalygin, *Pozitsiia,* 9–16. See also his follow-up articles in *Literaturnaia gazeta:* "Lesa, zemli, vody i vedomstvo," published January 26, 1963, reprinted in Zalygin, *Pozitsiia,* 16–21; and "Delo narodnoe, a ne vedomstvennoe!" published August 1, 1963, reprinted in ibid., 24–30.

53. Zalygin, *Pozitsiia,* 662.

54. Iaroshenko, "U istokov," 64.

55. Ibid., 64–65.

56. Ibid., 65.

57. Ibid., 67, 72.

58. Ibid., 68–69.

59. Briusova, "'Antirekam,'" 660–61.

60. See "Pechora sverkhu vniz" in *Chelovek i priroda* 5 (1979), cited in Iaroshenko, "U istokov," 70.

61. Iaroshenko, "U istokov," 72–73.

62. Ibid., 75.

63. Ibid. 75, 79.

64. Belov, "Spasut li Kaspii Vozhe i Lacha?" 10.

65. Briusova, "'Antirekam,'" 663.

66. Ibid.

67. Ibid., 663–78. The report "Materialy k probleme perebroski chasti stoka severnykh rek na iug" (Sostavlennyi po porucheniiu Komissii po okhrane pamiatnikov istorii i kul'tury pri Soiuze khudoznikov SSSR i RSFSR, [Moscow, 1982]) is provided on 664–78, radically abridged and without notes.

68. Iaroshenko, "U istokov," 78.

69. Efremov, *Nauchno-obshchestvennoe,* 16–17.

70. Briusova, "'Antirekam,'" 678–79.

71. Iaroshenko, "U istokov," 79.
72. Ibid.
73. Briusova, "'Antirekam,'" 680.
74. Letter of V. G. Briusova to Iu. V. Andropov, March 14, 1983, published in *Grani* 133 (1984): 262–63.
75. Iaroshenko, "U istokov," 80; Briusova, "'Antirekam,'" 687. In December 1983 Leonid Leonov, Dmitrii Likhachëv, B. V. Raushenbakh, and I. V. Petrianov-Sokolov sent a letter to Andropov as well. See *Grani* 133 (1984): 263–66.
76. The official publication that served as the subject of defense was Berezner, *Territorial'noe pereraspredelenie.*
77. Briusova, "'Antirekam,'" 680.
78. Ibid., 679. See Tenditnik, "Mnogo shuma."
79. Briusova, "'Antirekam,'" 679, from excerpted letter of V. Rasputin of April 14, 1983.
80. Ibid., 681.
81. Ibid., 685.
82. Ianshin and Melua, *Uroki ekologicheskikh proschëtov,* 12.
83. Zalygin, "Vodnoe khoziaistvo bez stoimosti . . . vody?" reprinted in Zalygin, *Pozitsiia,* 51–55.
84. Zalygin, "Proekt," 63–73.
85. Ibid., reprinted in Zalygin, *Pozitsiia,* 80.
86. Sergei Zalygin, Speech, "S"ezd pisatelei RSFSR," *Literaturnaia gazeta,* 18 December 1985, 7, quoted in Petro, "'Project of the Century,'" 244.
87. See Iaroshenko, "U istokov," 81.
88. Ianshin and Melua, *Uroki ekologicheskikh proschetov,* 12.
89. Ibid., 17.
90. Ibid., 18.
91. Florenskii and Shutova, "Da skroetsia t'ma," 737.
92. *Literaturnaia gazeta,* 2 July 1986, quoted in Darst, "Environmentalism in the USSR," 242.
93. Ianshin and Melua, *Uroki ekologicheskikh proschëtov,* 19.
94. Cited in ibid., 19. The decree was entitled "O prekrashchenii rabot po pereboske chasti stoka severnykh i sibirskikh rek." Ryzhkov's accusations against Vasil'ev and Aleksandrov are in Briusova, "'Antirekam,'" 688.
95. Efremov, *Nauchno-obshchestvennoe,* 19.
96. This point is made by both Darst and Petro.

CHAPTER NINETEEN

1. Shipunov, "Biosfernaia etika," 447, 453.
2. Ibid., 453.
3. Briusova, "'Antirekam,'" 689.
4. Shipunov, *Istina Velikoi Rossii,* 54, 88.
5. On Moldova, see V. Mikhailov, "Natsional'noe dvizhenie"; on the Baltics, see Lieven, *Baltic Revolution,* esp. 220. Meeting with Dainis Ivans, Riga, Latvia, May 7, 1991.
6. Botvich, Zagrishiev, and Shvarts, "Iz opyta raboty," 22.

7. Zhukova, "Svoi liudi," 21–23.

8. Rezoliutsiia Vsesoiuznoi konferentsii dvizheniia druzhin po okhrane prirody, December 11, 1987, in Sotsio-Ekologicheskii Soiuz, *Tridtsat' let dvizheniia,* 191–201.

9. Dudenko, "Obshchestvennoe prirodookhrannoe dvizhenie," 6–14.

10. Ianitskii, "Problemy razvitiia studencheskogo prirodookhrannogo dvizheniia," also cited in Botvich, Zagrishiev, and Shvarts, "Iz opyta raboty," 23.

11. Botvich, Zagrishiev, and Shvarts, "Iz opyta raboty," 24.

12. Ibid.

13. Ibid., 27.

14. Sotsio-Ekologicheskii Soiuz, *Sotsio-ekologicheskii soiuz,* 1.

15. Sotsio-Ekologicheskii Soiuz, *Tridtsat' let dvizheniia,* 241–43. For some additional sources on the history of the SEU, see Sotsial'no-Ekologicheskii Soiuz, *Informatsionno-metodicheskoe pis'mo;* a number of xeroxed or rexographed *Informatsionnye pis'ma* from 1988 and 1989; and the episodic booklets *Vsia nasha zhizn'* (All Our Life).

16. Sotsio-Ekologicheskii Soiuz, *Sotsio-ekologicheskii soiuz,* 4. This is just a partial list. Interestingly, the SEU and the Natural Resources Defense Council collaborated in 1991 to stop one of Armand Hammer's companies from building a polyvinylchloride factory in the Ukrainian city of Kalusha in Ivanovo-Frankovsk *oblast'* (ibid., 5).

17. Ibid., 4.

18. Sotsio-Ekologicheskii Soiuz, *Tridtsat' let dvizheniia,* 244. For more on the Volga-Chograi action, see Evgenii Simonov's description in Sotsial'no-Ekologicheskii Soiuz, *Vyzhivëm vmeste,* 77–85.

19. Sotsio-Ekologicheskii Soiuz, *Sotsio-ekologicheskii soiuz,* 4.

20. Ibid.

21. Shutova, "Kirishskii sindrom," 693–95.

22. Ibid., 695.

23. Razin, "'Bomba pochtal'ona Vasil'eva."

24. Shutova, "Kirishskii sindrom," 693–98.

25. Peterson, *Troubled Lands,* 201.

26. Ibid.

27. Fomichëv, "Zelënye," 240–41; Peterson, *Troubled Lands,* 201.

28. Peterson, *Troubled Lands,* 211–15.

29. O. N. Ianitskii, "Ekologicheskaia politika i ekologicheskoe dvizhenie v Rossii," 19. Speech to the Working Group of Eco-sociologists and Leaders of Ecological Organizations, Moscow, May 19–21, 1995. Xeroxed personal copy.

30. See, for example, "'Daiu ustanovku: unichtozhaite gosudarstvo!'" *Zavtra* 18, no. 23 (March 1994), which, among other things, also targets me as an abettor of the spoliation of Russia. A critic of the behavior of foreign corporations in Russia, I deny that accusation vigorously.

CONCLUSION

1. Stephen F. Cohen's *Bukharin and the Bolshevik Revolution* and his subsequent works seem to hold out the prospect of a democratic socialist or gradualist tradition within Bolshevism as a viable alternative to the repressive, hypercentralized Stalinist

model that has dominated Soviet life since the 1930s. Moshe Lewin, meanwhile, found the seeds of hope for a more democratized Soviet future in postwar demographic and occupational trends. The predominance of cities over the countryside as where most Russians live, combined with generally higher levels of education, promised the withering away of a social base for authoritarian rule, according to Lewin. See Lewin, *Gorbachev Phenomenon.* Among prominent historians of the older generation who were sympathetic either to the Bolsheviks or to revolutionary or socialist goals of some kind and who searched for moments of "original sin," we might mention the work of Isaac Deutscher, Robert V. Daniels, Paul Avrich, Israel Getzler, and perhaps Oliver Radkey. For a particularly hard-hitting critique of Deutscher, see Labedz, *Use and Abuse of Sovietology.*

2. For a critique of this trend, see Burbank, "Controversies over Stalinism."

3. I still treasure a letter from Stephen F. Cohen critiquing my uncritical "leftist" views as they were set out in a seminar paper in 1975. When I finally published a reworked version of that paper sixteen years later, it no longer hinted that collectivization could be justified as the "legitimate" reaction by the "proletariat" against peasant migrants to the cities who were outcompeting them for jobs. Rather, it simply tried to analyze workers' attitudes as a sociological problem. See Weiner, "'*Razmychka?*'"

4. For this view, see Nicholas Timasheff's old but still suggestive *Great Retreat,* as well as Nina Tumarkin's *Lenin Lives!*

5. See Darst, "Environmentalism in the USSR"; Petro, "'Project of the Century'"; Goldman, *Spoils of Progress;* Peterson, *Troubled Lands;* Pryde, *Environmental Management;* Feschbach and Friendly, *Ecocide in the USSR;* Green, *Ecology and Perestroika;* French, *Green Revolutions;* Jancar-Webster, *Environmental Management;* Singleton, *Environmental Problems;* Stewart, *Soviet Environment;* and Ziegler, *Environmental Policy.*

6. Hosking, Aves, and Duncan, *Road to Post-Communism;* and Sedaitis and Butterfield, *Perestroika from Below.* For a good discussion of the Masonic lodge as the germ of "civil society" in eighteenth-century France, see Koselleck, *Critique and Crisis,* especially chapters 5 and 6.

7. Kostyrchenko, *Out of the Red Shadows,* 104. In *V plenu u krasnogo faraona,* 114, Kostyrchenko gives the figure as between 15,000 and 20,000.

8. See Fehér, Heller, and Márkus, *Dictatorship over Needs.* Gábor Tamás Rittersporn, in his *Stalinist Simplifications and Soviet Complications,* 1–55, argues that the regime was aware of some problems that seemed to threaten its legitimacy but could take no consistent measures to remedy those problems without creating equally serious new problems for its stability. Hence, it was always engaged in damage control, trying to limit "negative" trends before they imperiled supreme power, but limiting those corrective measures before the corrective measures themselves became severe liabilities. The "permanent purge" then acquires greater rationality as a kind of political juggling act.

9. See Jarausch, *Unfree Professions,* for an interesting analysis of German professional identities and their relationships with the various German regimes during the period studied. Jarausch also supplies an excellent bibliography to the national and comparative literatures on the professions. On German nature protection, see Dominick, *Environmental Movement in Germany;* Groening and Wolschke-Buhlmahn, "Politics, Planning, and the Protection of Nature"; and Bramwell, *Ecology in the*

Twentieth Century. On colonial or imperial nature protection, see Grove, *Green Imperialism;* and Osborne, *Nature, the Exotic, and the Science of French Colonialism.* One of the few works to examine the social meaning of environmentalism in the United States is Samuel P. Hays's classic, *Conservation and the Gospel of Efficiency.*

10. Although there were as yet scant case studies such as this one available, historian of Soviet professions Harley Balzer several years ago presciently asserted this view in his conclusion, in Balzer, *Russia's Missing Middle Class,* 295.

11. Brooks, "Chemistry in War." Cited from draft manuscript, 26–27.

12. Dunlap, "Wildlife, Science, and the National Parks," 188.

13. Ibid., 188–89.

14. Ibid., *passim.*

15. Olwig and Olwig, "Underdevelopment and the Development of 'Natural' Park Ideology," 17–18.

16. Ibid., 21.

17. Quoted in Schmitt, *Back to Nature,* 156.

18. Olwig and Olwig, "Underdevelopment and the Development of 'Natural' Park Ideology," 53.

BIBLIOGRAPHY

ARCHIVAL SOURCES

ARAN (Archives of the Russian Academy of Sciences), Moscow. Formerly AAN (Archives of the Academy of Sciences of the USSR), Moscow.

ARAN—St. Pbg. (Archives of the Russian Academy of Sciences, St. Petersburg Branch), St. Petersburg. Formerly AAN—LO (Archives of the Academy of Sciences of the USSR, Leningrad Branch).

Archives of Films and Recordings (Arkhiv kinofonodokumentov), Krasnogorsk, Moscow *oblast'*.

Archives of *Literaturnaia Gazeta,* Moscow.

Efremov, Iurii (Georgii) Konstantinovich, Papers, Moscow.

GARF (State Archive of the Russian Federation), Moscow. Formerly TsGAOR (Central State Archive of the October Revolution and Socialist Construction of the USSR).

Makarov, Vasilii Nikitich, Papers. Access courtesy of Feliks Robertovich Shtil'mark, Moscow.

MOIP (Archive of the Moscow Society of Naturalists), Moscow.

Nasimovich, Andrei Aleksandrovich, Papers, Institute of Geography of the Russian Academy of Sciences, Moscow.

RGAE (Russian State Archive of the Economy), Moscow. Formerly TsGANKh (Central State Archive of the National Economy of the USSR).

RGALI (Russian State Archive for Literature and Art), Moscow. Formerly TsGALI (Central USSR State Archive for Literature and Art).

RTsKhIDNI (Russian Center for the Preservation and Study of Documents of Contemporary History), Archive of the Central Committee of the CPSU, Moscow.

TsGA [RSFSR] (Central State Archive of the RSFSR), now part of GARF, Moscow.

TsGAVO Ukrainy (Central State Archive of Ukraine), Kiev. Formerly TsGAOR UkrSSR (Central State Archive of the October Revolution and Socialist Construction of the Ukrainian Republic).

TsKhDMO (Center for the Storage of Documents on Youth Organizations), Moscow. Formerly VLKSM Archives (Archive of the All-Union Lenin Communist Union of Youth).

TsKhSD (Center for the Storage of Contemporary Documentation), Archive of the former Institute of Marxism-Leninism under the Central Committee of the CPSU, Moscow. Formerly TsPA (Central Party Archive).

PUBLISHED SOURCES

The following abbreviations are used in this List of Published Sources:

MOIP Moskovskoe obshchestvo ispytatelei prirody. Otdel biologicheskii
OPZD *Okhrana prirody i zapovednoe delo v SSSR*

"A. V. Malinovskii" [obituary]. *Okhota i okhotnich'e khoziaistvo* 12 (1981): 28.

Abramov, Lev Solomonovich. *Opisanie prirody nashei strany: Razvitie fiziko-geograficheskikh kharakteristik.* Moscow: Mysl', 1972.

Akademiia nauk SSSR. *Trudy noiabrskoi sessii 1933 goda: Problemy Volgo-Kaspiia.* Leningrad: Izdatel'stvo AN SSSR, 1934.

Akademiia nauk SSSR. Institut geografii. *Opyt raboty i zadachi zapovednikov SSSR.* Series: Problemy konstruktivnoi geografii. Ed. A. A. Nasimovich and Iu. A. Isakov. Moscow: Nauka, 1979.

Akademiia nauk SSSR. Materialy k biobibliografii uchenykh. *Nikolai Vasil'evich Tsitsin.* Foreword by P. I. Lapin. Comp. I. G. Bebikh and N. M. Anserova. Series: Biologicheskie nauki: Botanika, no. 12. Moscow: Nauka, 1988.

Akademiia nauk SSSR. Nauchnyi sovet po problemam biosfery and Institut mezhdunarodnogo rabochego dvizheniia. *Sotsial'nye aspekty ekologicheskikh problem.* Series: Sovremennye problemy biosfery. Moscow: Nauka, 1982.

Akademiia nauk SSSR. Sektsiia khimiko-tekhnologicheskikh i biologicheskikh nauk. Sovetskii komitet po programme IuNESKO "Chelovek i biosfera." *Ekologicheskaia propaganda v SSSR,* ed. D. M. Gvishiani. Moscow: Nauka, 1984.

Akklimatizatsiia okhotnich'e-promyslovykh zverei i ptits v SSSR. 2 vols. Kirov: Volgo-Viatskoe knizhnoe izdatel'stvo, Kirovskoe otd., 1973–74.

Akklimatizatsiia zhivotnykh v SSSR: Materialy konferentsii po akklimatizatsii zhivotnykh v SSSR (10–15 ogo maia 1963 g. v g. Frunze). Alma-ata: Izdatel'stvo AN Kaz SSR, 1963.

Aleksandrov, Daniil Aleksandrovich. "Istoricheskaia antropologiia nauki v Rossii." *Voprosy istorii estestvoznaniia i tekhniki* 4 (1994): 3–22.

Aleksandrov, V. Ia. *Trudnye gody sovetskoi biologii. Zapiski sovremennika.* St. Petersburg: Nauka—St. Petersburg Branch, 1992.

Aleksandrova, V. D. "Problema razvitiia v geobotanike." *Biulleten' MOIP* 67, no. 2 (1962): 86–107.

———. "Rastitel'noe soobshchestvo v svete nekotorykh printsipov kibernetiki." *Biulleten' MOIP* 66, no. 3 (1961): 101–12.

Alekseeva, L. V., and K. D. Zykov. "Differentsiatsiia razmerov zapovednikov." In *Sotsial'no-ekonomicheskie i ekologicheskie aspekty sovershenstvovaniia deiatel'nosti zapovednikov, Sbornik nauchnykh trudov,* 38–39. Moscow, TsNIL Glavokhota, 1985.

Alimov, A. A., V. N. Sukhov, A. V. Nikolaev, and V. I. Khodanovich. *Problemy ekologii i sovremennost'*. Leningrad: Leningradskaia Lesotekhnicheskaia Akademiia, 1991.

Anuchin, N. P. "Stepnoe lesorazvedenie ne nuzhdaetsia v naukoobraznom uchenii i biogeotsenoze." *Les i step'* 9 (1952): 32–39.

Arkhipov, S. S. "Instruktsii dlia organizatsii zapovednogo khoziaistva." *Nauchno-metodicheskie zapiski* 2 (1939): 51–90.

Artomonov, V. I. "'Daiu ustanovku: unichtozhaite gosudarstvo!'" *Zavtra* 18, no. 23 (March 1994): 4.

Bailes, Kendall E. *Science and Culture in an Age of Revolutions: V. I. Vernadsky and His Scientific School, 1863–1945*. Bloomington: Indiana University Press, 1990.

———. *Technology and Society under Lenin and Stalin*. Princeton: Princeton University Press, 1978.

Balzer, Harley D., ed. *Russia's Missing Middle Class: The Professions in Russian History*. Armonk, N.Y.: M. E. Sharpe, 1996.

Bannikov, A. G. "Ot zapovednika do prirodnogo parka." *Priroda* 4 (1968): 89–97.

———. *Po zapovednikam SSSR*. Moscow: Mysl', 1966.

Barishpol, Ivan Fedotovich, and Vera Gennad'evna Larina. *U prirody druzei milliony*. Moscow: Lesnaia promyshlennost', 1984.

Bazykin, A. D. *Matematicheskaia biofizika vzaimodeistvuiushchikh populiatsii*. Moscow: Nauka, 1985.

Belaya, Galina, comp. *In Search of Harmony*. Moscow: Progress, 1989.

Belorusskoe obshchestvo okhrany prirody: Kratkaia spravka. Minsk: Uradzhai, 1984.

Belov, V. I. "Spasut li Kaspii Vozhe i Lacha?" *Russkaia mysl'*, 15 July 1982, 10.

Bel'skii, P. "Zapovedniki i okhotnich'e khoziaistvo." *Okhota i okhotnich'e khoziaistvo* 10 (1957): 27–30.

Berezner, A. S. *Territorial'noe pereraspredelenie rechnogo stoka Evropeiskoi chasti RSFSR*. Moscow: Gidrometeoizdat, 1985.

"Beseda v *Nashem sovremennike*." In *Svetloe oko*, 399–406. Moscow: Sovremennik, 1980.

Blagosklonov, Konstantin Nikolaevich. "Iz istorii iunnatskogo dvizheniia." *Biulleten' MOIP* 85, no. 1 (1980): 123–28.

———. "Komsomol'skaia druzhina po okhrane prirody." In *Okhrana prirody*, comp. V. I. Pelevin, 91–97. Posobie dlia uchitelei, no. 3. Moscow: Prosveshchenie, 1971.

Blagosklonov, K. N., A. A. Inozemtsev, and V. N. Tikhomirov. *Okhrana prirody*. Moscow: Vysshaia shkola, 1967.

Boardman, Robert. *International Organization and the Conservation of Nature*. Bloomington: Indiana University Press, 1981.

Bochkarëv, E. M. "Problemy kedrovoi taigi." *Lesnaia promyshlennost'*, 2 September 1965.

Boitsov, L. V. "O sostoianii i perspektivnom plane nauchno-issledovatel'skoi raboty goszapovednikov na tret'e piatiletie (1938–1942) po razdelu biotekhniki i akklimatizatsii." *Nauchno-metodicheskie zapiski* 1 (1938): 56–59.

Bol'shakov, Vladimir Nikolaevich. *Ekologicheskie osnovy okhrany prirody (V pomoshch' lektoru)*. Moscow: Znanie RSFSR, 1981.

Boreiko, V[ladimir] E[vgen'evich]. *Askania-Nova: Tiazhkie versty istorii (1826–1993)*. Series: Istoriia okhrany prirody, no. 1. Kiev: Kievskii ekologo-kul'turnyi tsentr, 1994.

———. *Belye piatna istorii prirodookhrany: SSSR, Rossiia, Ukraina.* 2 vols. Series: Istoriia okhrany prirody, nos. 6–7. Kiev: Kievskii ekologo-kul'turnyi tsentr, 1996.

———. "Eto ne dolzhno povtorit'sia." *Okhota i okhotnich'e khoziaistvo* 9 (1991): 8–11.

———. "Frants Frantsevich Shillinger." In *Populiarnyi biografo-bibliograficheskii slovar'-spravochnik deiatelei zapovednogo dela i okhrany prirody Ukrainy, tsarskoi Rossii i SSSR (1860–1960),* 2:204–5. Kiev: Kievskii ekologo-kul'turnyi tsentr and Tsentr okhrany dikoi prirody SoES, 1995.

———. Series: "Istoriia boli i geroizma: Skorbnyi spisok deiatelei okhrany prirody i zapovednogo dela SSSR, repressirovannykh v 20–50-e gody." *Okhota i okhotnich'e khoziaistvo* 1 and 4 (1995) and other issues (continuing).

———. *Istoriia okhrany prirody Ukrainy (X vek–1980 g.).* 2 vols. Series: Istoriia okhrany prirody, nos. 10–11. Kiev: Kievskii ekologo-kul'turnyi tsentr, 1997.

———. *Istoriia zapovednogo dela v Ukraine.* Series: Istoriia okhrany prirody, no. 2. Kiev: Kievskii ekologo-kul'turnyi tsentr, 1995.

———. *Ocherki o pionerakh okhrany prirody.* 2 vols. Series: Istoriia okhrany prirody, nos. 8–9. Kiev: Kievskii ekologo-kul'turnyi tsentr, 1996.

———. *Populiarnyi biografo-bibliograficheskii slovar'-spravochnik deiatelei zapovednogo dela i okhrany prirody Ukrainy, tsarskoi Rossii i SSSR (1860–1960).* 2 vols. Series: Istoriia okhrany prirody, nos. 4–5. Kiev: Kievskii ekologo-kul'turnyi tsentr and Tsentr okhrany dikoi prirody SoES, 1995.

———. *"Tsarskie okhoty": Ot Vladimira Monomakha do Vladimira Shcherbitskogo.* Series: Istoriia okhrany prirody, no. 3. Kiev: Kievskii ekologo-kul'turnyi tsentr, 1995.

———. "Vtoroi razgrom zapovednikov." *Okhota i okhotnich'e khoziaistvo* 1 (1993): 12–15.

Boreiko, V. E., and V. Berlin, comps. *Iz literaturnogo nalslediia Andreia Petrovicha Semënova-tian-shanskogo (1866–1942): Proza, Stikhotvoreniia, Epigrammy.* Series: Istoriia okhrany prirody, no. 8. Kiev: Kievskii ekologo-kul'turnyi tsentr, Tsentr okhrany dikoi prirody SoES, and Apatitskii sovet VOOP, 1996.

Borisov, V. A. "V Komissii po okhrane prirody Akademii nauk SSSR." OPZD 2 (1956): 145–48.

Borodin, A. "Usilit' bor'bu s volkami." *Okhota i okhotnich'e khoziaistvo* 7 (1979): 4–5.

Bosse, G., R. Gekker, V. Geptner, N. Kabanov, A. Nasimovich, G. Nikol'skii, A. Formozov, and I. Chernenko. "V zaschitu zapovednikov." *Izvestiia,* 6 April 1957, 3.

Bosse, G. G., and A. V. Iablokov. *Okhrana prirody i ee znachenie dlia nashei strany (Materialy k lektsii).* Moscow: VOOP, 1958.

Botvich, E. L., M. Zagrishiev, and E. A. Shvarts. "Iz opyta raboty samodeiatel'nykh ob"edinenii studentov." In Nauchno-Issledovatel'skii Institut Problem Vysshei Shkoly [Moscow], *Sistema vospitaniia v vysshei shkole: Obzornaia informatsiia* 5 (1990): 19–20.

Bramwell, Anna. *Ecology in the Twentieth Century.* New Haven: Yale University Press, 1989.

Briusova, V. G. "'Antirekam'—Narodnoe NET!" In *Ekologicheskaia al'ternativa: Istoki bedy, znaki bedy, ekologicheskaia al'ternativa,* ed. M. Ia. Lemeshev and A. V. Iablokov, 660–89. Moscow: Progress, 1990.

Brooks, Nathan. "Chemistry in War, Revolution, and Upheaval: Russia and the Soviet Union, 1900–1929." *Centaurus* (November 1997).

Browne, Malcolm W. "European Bison, Once Near Extinction, Now a Big Herd in Polish-Soviet Forest." *New York Times,* 29 December 1975.

Brudny, Aron, and Dmitri Kavtaradze. "'Ecopolis': Prognosis Turned into Project." *Social Sciences* 2 (1984): 183–93.

Budyko, M. I. *Global'naia ekologiia.* Moscow: Mysl', 1977.

Bulgakov, L. N. "Uchastie komsomol'skikh organizatsii v okhrane i vosproizvodstve prirodnykh resursov." In *Molodëzh' v okhrane prirody.* Saratov, 1972.

Burbank, Jane R. "Controversies over Stalinism: Searching for a Soviet Society." *Politics and Society* 19, no. 3 (Sept. 1991): 325–40.

Chesnokov, Nikolai Ivanovich. *Dikie zhivotnye meniaiut adresa.* Moscow: Mysl', 1989.

Chivilikhin, Vladimir. *Izbrannoe: V dvukh tomakh.* 2 vols. Moscow: Molodaia gvardiia, 1978.

———. "Mesiats v Kedrograde." In *Svetloe oko,* 8–77. Moscow: Sovremennik, 1980.

———. *Pamiat' (Roman-esse).* 2 vols. Moscow: Khudozhestvennaia literatura, 1988.

———. "Piatiletie Kedrograda." *Literaturnaia gazeta,* 28 January 1965.

———. *Po gorodam i vesiam: Puteshestviia v prirodu.* Moscow: Sovremennik, 1976.

———. "Shumi, taiga, shumi!" In *Zhit' glavnym,* 8–74. Moscow: Molodaia gvardiia, 1988.

———. "Shvedskie ostanovki." In *Izbrannoe: V dvukh tomakh,* 2:451–589. Moscow: Molodaia gvardiia, 1978.

———. *Svetloe oko.* Moscow: Sovremennik, 1980.

———. "Svetloe oko Sibiri." In *Svetloe oko,* 78–144. Moscow: Sovremennik, 1980.

———. "Taiga shumit." In *Zhit' glavnym,* 74–82. Moscow: Molodaia gvardiia, 1988.

———. "Uroki Leonova." In *O Leonove,* comp. V. Chivilikhin, 248–71. Moscow: Sovremennik, 1979.

———. "V mire bol'shikh zabot" (Iz rechi na XV s"ezde VLKSM). In *Zhit' glavnym,* 102–9. Moscow: Molodaia gvardiia, 1988.

———. *Zhit' glavnym.* Moscow: Molodaia gvardiia, 1988.

Chizhova, V. P. *Rekreatsionnye nagruzki v zonakh otdykha.* Moscow: Lesnaia promyshlennost', 1977.

Clemens, Elisabeth S. *The People's Lobby: Organizational Innovation and the Rise of Interest Group Politics in the United States, 1890–1925.* Chicago: University of Chicago Press, 1997.

Cohen, Stephen F. *Bukharin and the Bolshevik Revolution: A Political Biography, 1888–1938.* New York: A. A. Knopf, 1973.

Conquest, Robert. *Power and Policy in the U.S.S.R.: The Struggle for Stalin's Succession, 1945–1960.* New York: Harper and Row, 1961.

Croker, Robert A. *Pioneer Ecologist: The Life and Work of Victor Ernest Shelford, 1877–1968.* Washington, D.C.: Smithsonian Institution Press, 1991.

Darst, Robert G., Jr. "Environmentalism in the USSR: The Opposition to the River Diversion Projects." *Soviet Economy* 4, no. 3 (1988): 223–51.

Dawson, Jane I. *Eco-Nationalism: Anti-Nuclear Activism and National Identity in Russia, Lithuania, and Ukraine.* Durham, N.C.: Duke University Press, 1996.

DeBardeleben, Joan. *The Environment and Marxism-Leninism: The Soviet and East German Experience.* Boulder, Colo.: Westview, 1985.

Dement'ev, G. P. "Deiatel'nost' Komissii po okhrane prirody AN SSSR za pervyi god ee sushchestvovaniia" (abridged presentation to the Plenary session). OPZD 2 (1956): 3–15.

Dement'ev, G. P., et al. "V zashchitu Baikala!" *Literaturnaia gazeta,* 7 April 1959.

Dëzhkin, V. V. *V mire zapovednoi prirody.* Moscow: Sovetskaia Rossiia, 1989.

Dialog o prirode. Sverdlovsk: Sredne-ural'skoe knizhnoe izdatel'stvo, 1977.

Dialog o Sibiri. Publitsisticheskii sbornik. Razgovor o problemakh ekologii, nravstvennosti, ekonomiki, sotsial'noi zhizni sibiriakov. Irkutsk: Vostochno-Sibirskoe knizhnoe izdatel'stvo, 1988.

Dokhman, Genrietta Isaakovna. *Istoriia geobotaniki v Rossii.* Moscow: Nauka, 1973.

Dokuchaev, Nikolai Petrovich. *Srednee zveno. Raionnyi (gorodskoi) sovet Obshchestva. Praktika raboty.* Gor'kii [Nizhnyi Novgorod]: VOOP and Volgo-Viatskoe knizhnoe izdatel'stvo, 1987.

Dombrovskii, Iurii. *Khranitel' drevnosti.* Moscow: Izvestiia, 1991.

Dominick, Raymond H. *The Environmental Movement in Germany: Prophets and Pioneers, 1871–1971.* Bloomington: Indiana University Press, 1992.

Dubinin, V. B. "Soveshchanie po okhrane prirody SSSR." *Zoologicheskii zhurnal* 35, no. 9 (1956): 1436–38.

Dudenko, N. B. "Obshchestvennoe prirodookhrannoe dvizhenie v sel'skoi mestnosti i rol' narodnykh traditsii (na primere Ukrainskoi SSR)." In *Ekologiia, demokratiia, molodëzh',* ed. O. N. Ianitskii, 6–14. Moscow: Filosofskoe obshchestvo SSSR, Moskovskoe otdelenie, Molodëzhnaia sektsiia, 1990.

Dunlap, Thomas R. "Wildlife, Science, and the National Parks, 1920–1940." *Pacific Historical Review* 59, no. 2 (May 1990): 187–202.

Dunlop, John B. *The Faces of Contemporary Russian Nationalism.* Princeton: Princeton University Press, 1983.

Dyrenkov, S. A., and A. M. Krasnitskii. "Osnovnye funktsii zapovednykh territorii i ikh otrazhenie v rezhime okhrany lesnykh ekosistem." *Biulleten' MOIP* 87, no. 6 (1982): 105–14.

Efremov, Iu. K. *Nauchno-obshchestvennoe dvizhenie soprotivleniia ukhudzhaiushchim preobrazovaniiam prirody.* Moscow: Moskovskii tsentr Geograficheskogo obshchestva RF, 1992.

Eilart, Jaan. *Man, Ecosystems, and Culture.* Tallinn: Perioodika, 1976.

———. "Pervyi studencheskii kruzhok po okhrane prirody v SSSR." *Voprosy ukhoda za landshaftom i prirodookhranitel'nogo prosveshcheniia v Estonskoi SSR.* Tartu: Tartu Riikliku Ülikool, 1978.

Ekologicheskaia al'ternativa: Istoki bedy, znaki bedy, ekologicheskaia al'ternativa. Ed. M. Ia. Lemeshev and A. V. Iablokov. Series: Perestroika, demokratiia, sotsializm. Moscow: Progress, 1990.

Ekologicheskaia propaganda v SSSR. Moscow: Nauka, 1984.

Eliseev, K. "O pchëlakh i osakh." *Krokodil* 16 (10 June 1952): 13.

Fedorenko, N., and N. Reimers. "Nature Conservation—Growing Proximity of Economic and Ecological Goals." In *Man and the Biosphere.* Moscow: Nauka, 1984.

Fëdorov, E. K. *Ekologicheskii krizis i sotsial'nyi progress.* Moscow: Gidrometeoizdat, 1977.

Fehér, Ferenc, Agnes Heller, and György Márkus. *Dictatorship over Needs: An Analysis of Soviet Societies.* Oxford: Basil Blackwell, 1983.

"Fenomenologiia zhestokosti." *Priroda* 1 (1975): 89–107.

Feschbach, Murray, and Alfred Friendly Jr. *Ecocide in the USSR.* New York: Basic Books, 1992.

Filippovskii, N., ed. *Natsional'nyi park: Problema sozdaniia. Sbornik.* Moscow: Znanie, 1979.

Filonov, Konstantin Pavlovich. "Dinamika chislennosti kopytnykh zhivotnykh i zapovednost'" [Thesis for degree of Doctor of Science]. In *Okhotovedenie,* 1–232. Moscow: TsNIL Glavokhota RSFSR and Lesnaia promyshlennost', 1977.

Fitzpatrick, Sheila. *Education and Social Mobility in the Soviet Union, 1921–1934.* Cambridge, Eng.: Cambridge University Press, 1979.

Fitzpatrick, Sheila. *Stalin's Peasants: Resistance and Survival in the Russian Village after Collectivization.* New York: Oxford University Press, 1994.

Florenskii, P. V., and T. A. Shutova. "Da skroetsia t'ma . . ." In *Ekologicheskaia al'ternativa: Istoki bedy, znaki bedy, ekologicheskaia al'ternativa,* ed. M. Ia. Lemeshev and A. V. Iablokov, 727–43. Moscow: Progress, 1990.

Fomichëv, S. R. "Zelënye: Vzgliad iznutri." *POLIS (Politicheskie issledovaniia)* 1–2 (1992): 238–45.

Formozov, A. A. *Aleksandr Nikolaevich Formozov (1899–1973).* Moscow: Nauka, 1980.

French, Hilary F. *Green Revolutions: Environmental Reconstruction in Eastern Europe and the Soviet Union.* Worldwatch Paper no. 99. Washington, D.C.: Worldwatch Institute, 1990.

Fridman, S. "O russkom lese." *Literaturnaia gazeta,* 23 March 1954.

Frolov, I. T., and V. A. Los'. "Filosofskie osnovaniia sovremenoi ekologii." In *Ekologicheskaia propaganda v SSSR,* 5–26. Moscow: Nauka, 1984.

Geograficheskoe obshchestvo SSSR, 1917–1967. Comp. I. B. Kostrits and D. M. Pinkhenson. Moscow: Mysl', 1968.

Geograficheskoe razmeshchenie zapovednikov v RSFSR i organizatsiia ikh deiatel'nosti. Sbornik. Moscow: TsNIL Glavokhota RSFSR, 1981.

Geptner, V. G. "Kakovy zhe puti obogashcheniia fauny." *Okhota i okhotnich'e khoziaistvo* 2 (1963): 21–26.

Gerovitch, Vyacheslav, and Anton Struchkov. "Epilogue: Russian Reflections." *Journal of the History of Biology* 25, no. 3 (Fall 1992): 487–95.

Gershkovich, N. A., A. P. Razorënova, and A. A. Maksimov. "Pëtr Petrovich Smolin." *Biulleten' MOIP* 81, no. 5 (1976): 120–25.

Giliarov [Ghiliarov], M. S., G. G. Vinberg, and Iu. M. Chernov. "Ekologiia—Zadachi i perspektivy." *Priroda* 5 (1977): 3–11.

Gilyarov, M. "Agrocenology—An Important Field of Modern Biogeocenology." In *Man and the Biosphere,* 18–25. Moscow: Nauka, 1984.

Glavnoe upravlenie okhotnich'ego khoziaistva i zapovednikov pri Sovete ministrov RSFSR. Tsentral'naia nauchno-issledovatel'skaia laboratoriia. *Aktual'nye voprosy zapovednogo dela. Sbornik nauchnykh trudov.* Moscow: TsNIL Glavokhota, 1988.

———. *Sotsial'no-ekonomicheskie i ekologicheskie aspekty sovershenstvovaniia deiatel'nosti zapovednikov. Sbornik nauchnykh trudov.* Moscow: TsNIL Glavokhota, 1985.

Glavnoe upravlenie po zapovednikam pri SM SSSR. *Polozhenie o Glavnom upravlenii po zapovednikam pri SM SSSR; Polozhenie o gosudarstvennykh zapovednikakh SSSR; Polozhenie ob okhrane gosudarstvennykh zapovednikakh SSSR.* Moscow, 1952.

Gol'dberg, M. S. "Chistota atmosfernogo vozdukha i ego okhrana." *OPZD* 1 (1956): 47–64.

Goldman, Marshall I. *The Spoils of Progress: Environmental Pollution in the Soviet Union.* Cambridge, Mass.: MIT Press, 1972.

Goldschmidt, Walter. *The Human Career: The Self in the Symbolic World.* Cambridge, Eng.: Basil Blackwell, 1990.

Golley, Frank Benjamin. *A History of the Ecosystem Concept in Ecology: More Than the Sum of the Parts.* New Haven: Yale University Press, 1993.

Gorban', A. N., and R. G. Khlebopros. *Demon Darvina: Ideia optimal'nosti i estestvennyi otbor.* Moscow: Nauka, 1988.

Gor'kii, Maksim. "Iz stat'i 'O biblioteka Poeta' . . . ." In *Gor'kii i nauka: stat'i, pis'ma i vospominaniia,* 56–57. Moscow: Nauka, 1964.

Gosudarstvennaia plemennaia kniga zubrov i bizonov (chistokrovnykh, chistoporodnykh i gibridnykh). Moscow: Ministerstvo sel'skogo khoziaistva SSSR, 1956.

Graham, Loren R. *The Ghost of the Executed Engineer: Technology and the Fall of the Soviet Union.* Cambridge, Mass.: Harvard University Press, 1993.

———. *Science in Russia and the Soviet Union: A Short History.* New York: Cambridge University Press, 1993.

Gramma, V. N. "Eretiki i buntari: O pervykh sovetskikh uchënykh-prirodookhraniteliakh." In *Zavtra budet pozdno,* ed. G. G. Kostak, S. V. Razmetaev, and S. A. Taglin, 6–61. Khar'kov: Prapor, 1990.

Granin, Daniil Aleksandrovich. *Zubr.* First serialized in *Novy mir* 1 and 2 (1987).

Grebënok, R. V. "Parazity belki, akklimatizirovannoi v Kirgizii." In *Akklimatizatsii zhivotnykh v SSSR. Materialy konferentsii po akklimatizatsii zhivotnykh v SSSR (10–15 maia 1963 g. v g. Frunze),* 340–42. Alma-Ata: Izdatel'stvo AN Kaz SSR, 1963.

Green, Eric. *Ecology and Perestroika.* Washington, D.C.: American Committee on U.S.-Soviet Relations, 1990.

Grigor'ev, I. "A. V. Malinovskii" [obituary]. *Okhota i okhotnich'e khoziaistvo* 12 (1981): 28.

———. "Aleksandr Vasil'evich Malinovskii." *Okhota i okhotnich'e khoziaistvo* 8 (1980): 29.

Groening, Gert, and Joachim Wolschke-Buhlmahn. "Politics, Planning, and the Protection of Nature: Political Abuse of Early Ecological Ideas in Germany, 1933–1945." *Planning Perspectives* 2 (1987): 127–48.

Grove, Richard H. *Green Imperialism: Colonial Expansion, Tropical Island Edens, and the Origins of Environmentalism, 1600–1860.* Cambridge, Eng.: Cambridge University Press, 1995.

Gusev, O[leg]. "Protiv idealizatsii prirody." *Okhota i okhotnich'ego khoziaistvo* 11 (1978): 25–27.

———. "V zashchitu zubra i zubrov." In *Nikolai Vladimirovich Timofeev-Resovskii: Ocherki, vospominaniia, materialy,* ed. N. N. Vorontsov, 371–75. Moscow: Nauka, 1993.

Gustafson, Thane. *Reform in the Soviet Union: Lessons of Recent Politics on Land and Water.* Cambridge, Eng.: Cambridge University Press, 1981.

Hays, Samuel P. *Conservation and the Gospel of Efficiency: The Progressive Conservation Movement, 1890–1920.* Cambridge, Mass.: Harvard University Press, 1959.

Hicks, Barbara. *Environmental Politics in Poland: A Social Movement between Regime and Opposition.* New York: Columbia University Press, 1996.

Hosking, Geoffrey A. "The Beginnings of Independent Political Activity." In *The Road to Post-Communism: Independent Political Movements in the Soviet Union, 1985–1991*, by Geoffrey A. Hosking, Jonathan Aves, and Peter J. S. Duncan, 1–2. London and New York: Pinter Publishers/St. Martin's Press, 1992.

———. *The First Socialist Society: A History of the Soviet Union from Within*. 2d ed. Cambridge, Mass.: Harvard University Press, 1992.

Hosking, Geoffrey A., Jonathan Aves, and Peter J. S. Duncan. *The Road to Post-Communism: Independent Political Movements in the Soviet Union, 1985–1991*. London and New York: Pinter Publishers/St. Martin's Press, 1992.

Hough, Jerry F., and Merle Fainsod. *How the Soviet Union Is Governed*. Cambridge, Mass.: Harvard University Press, 1979.

Huskey, Eugene. *Russian Lawyers and the Soviet State: The Origins and Development of the Soviet Bar, 1917–1939*. Princeton: Princeton University Press, 1986.

Iagomagi, Iu. E., and V. I. Pallok, "Funktsional'noe zonirovanie i sel'skoe khoziaistvo." *Sel'skoe khoziaistvo i okhrana prirody, Nauchnye trudy po okhrane prirody* [Tartu] 8 (1985): 36–41.

Ianitskii, Oleg N. *Ekologicheskoe dvizhenie v Rossii. Kriticheskii analiz*. Moscow: Rossiiskaia Akademiia Nauk. Institut sotsiologii, 1996.

———. "Problemy razvitiia studencheskogo prirodookhrannogo dvizheniia: tochka zreniia sotsiologa." In *Ekologiia, demokratiia, molodëzh'*, ed O. N. Ianitskii, 106–17. Moscow: Filosofskoe obshchestvo SSSR, Moskovskoe otdelenie, Molodëzhnaia sektsiia, 1990.

———. *Sotsial'nye dvizheniia. 100 interv'iu s liderami*. Moscow: Moskovskii rabochii, 1991.

Ianitskii, O. N., ed. *Ekologiia, demokratiia, molodezh'. Sbornik*. Moscow: Filosofskoe obshchestvo SSSR, Moskovskoe otdelenie, Molodezhnaia sektsiia, 1990.

Ianshin, A. L., and A. I. Melua. *Uroki ekologicheskikh proschetov*. Moscow: Mysl', 1991.

Iaroshenko, Viktor. *Ekspeditsiia "Zhivaia Voda."* Moscow: Molodaia gvardiia, 1989.

———. "U istokov." *Chelovek i priroda* 10 (1987): 18–84.

Iaroshevskii, M. G., ed., *Repressirovannaia nauka*. Leningrad: Nauka—Leningradskoe otdelenie, 1991.

Il'ina, E. D. "Poslednye gody." *Okhota i okhotnich'e khoziaistvo* 7 (1982): 39.

Il'inskii, A. P. "Belovezhskaia pushcha i perspektivy razvërtyvaniia v nei nauchno-issledovaltel'skoi raboty." *Sovetskaia botanika* 3 (1941): 3–11.

Issledovaniia v oblasti zapovednogo dela: Sbornik nauchnykh trudov. Moscow: Minsel'khoz SSSR, 1984.

Jancar-Webster, Barbara. "Eastern Europe and the Former Soviet Union." In *Environmental Politics in the International Arena: Movements, Parties, Organizations, and Policy*, ed. Sheldon Kamieniecki, chapter 10. Albany: State University of New York Press, 1993.

———. *Environmental Management in the Soviet Union and Yugoslavia: Structure and Regulation in Federal Communist States*. Durham, N.C.: Duke University Press, 1987.

Jarausch, Konrad H. *The Unfree Professions: German Lawyers, Teachers, and Engineers, 1900–1950*. New York: Oxford University Press, 1990.

Joravsky, David. *The Lysenko Affair*. Cambridge, Mass.: Harvard University Press, 1970.

———. *Soviet Marxism and Natural Science, 1917–1932.* New York: Columbia University Press, 1961.

Josephson, Paul R. *New Atlantis Revisited: Akademgorodok, the Siberian City of Science.* Princeton: Princeton University Press, 1996.

Juviler, Peter H. *Revolutionary Law and Order: Politics and Social Change in the USSR.* New York: Free Press, 1976.

Kabachnik, M. I., ed. *Aleksandr Nikolaevich Nesmeianov: Uchenyi i chelovek.* Series: Uchënye SSSR. Ochernki, vospominaniia, materialy. Moscow: Nauka, 1988.

Kassow, Samuel D. *Students, Professors, and the State in Tsarist Russia.* Berkeley and Los Angeles: University of California Press, 1989.

Kazarkin, A. "Pamiat'—Ekologiia kul'tury: (k 60-letiiu V. Chivilikhina)." *Sibirskie ogni* 3 (1988): 156–62.

Kelley, Donald R. "Environmental Policy-Making in the USSR: The Role of Industrial and Environmental Interest Groups." *Soviet Studies* 28 (October 1976): 570–89.

Kharitonov, N. P., and K. V. Avilova. "Okhrana prirody i rabota iunatskikh kruzhkov." In *Studenchestvo i okhrana prirody: Materialy konferentsii, posviashchennoi 20-letiiu Druzhiny biologicheskogo fakul'teta MGU po okhrane prirody, 11–12 dekabria 1980 g.*, ed. K. V. Avilova, 107–9. Moscow: Izdatel'stvo Moskovskogo universiteta, 1982.

Khlevniuk, Oleg V. *In Stalin's Shadow: The Career of "Sergo" Ordzhonikidze.* Trans. David J. Nordlander. Armonk, N.Y.: M. E. Sharpe, 1995.

"Khronika." *OPZD* 1 (1956): 129–31.

"Khronika." *OPZD* 3 (1958): 112–13.

[Khrushchëv, N. S.] *Khrushchev Remembers.* With an introduction, commentary, and notes by Edward Crankshaw. Trans. and ed. Strobe Talbott. Boston: Little, Brown and Co., 1970.

Kiriasov, Viktor. "Skazanie o Kedrograde." *Lesnaia gazeta,* 12 January 1995, 2–3.

Kol'man, Arnosht. "Sabotazh v nauke." *Bolshevik* 2 (1931): 75.

Komarov, Boris [Vol'fson, Ze'ev]. *Unichtozhenie prirody v sovetskom soiuze.* Frankfurt on Main: Possev-Verlag, 1978. Translated as *The Destruction of Nature in the Soviet Union.* Armonk, N.Y.: M. E. Sharpe, 1980.

Konstantinov, B. P., et al. "Baikal zhdët" (Pis'mo v redaktsiiu). With editorial commentary. *Komsomol'skaia pravda,* 11 May 1966.

Korzhikhina, Tat'iana Petrovna. *Sovetskoe gosudarstvo i ego uchrezhdeniia, noiabr' 1917 g.—dekabr' 1991 g.* Moscow: Rossiiskii gosudarstvennyi gumanitarnyi universitet, 1994.

Koselleck, Reinhart. *Critique and Crisis: Enlightenment and the Pathogenesis of Modern Society.* Cambridge, Mass.: MIT Press, 1988.

Kostyrchenko, Gennadi. *Out of the Red Shadows: Anti-Semitism in Stalin's Russia.* Amherst, NY: Prometheus Books, 1995.

———. *V plenu u krasnogo faraona.* Moscow: Mezhdunarodnye otnosheniia, 1994.

Kotliarov, E. A. *Geografiia otdykha i turizma.* Moscow: Mysl', 1978.

Kovalevskii, A. D. "Bol'she molchat' nel'zia!" *Nash sovremennik* 3 (1960): 206–12.

Krasnitskii, A. M. "Lesokhoziaistvennye meropriiatiia i ikh mesto v zapovednom dele (o rubkakh lesa v zapovednikakh)." *Biulleten' MOIP* 80, no. 2 (1975): 18–29.

———. *Problemy zapovednogo dela.* Moscow: Lesnaia promyshlennost', 1983.

Krasnitskii, A. M., and S. A. Dyrenkov. "Sravnitel'naia otsenka lugovykh i stepnykh ekosistem, formiruiushchikhsia pri kosimom i nekosimom rezhimakh zapovednoi okhrany." *Biulleten' MOIP* 87, no. 4 (1982): 102–10, with response by A. A. Nasimovich and T. A. Rabotnov, 110–11.

Krasnopol'skii, A. V., ed. *Otechestvennye geografi (1917–1992). Biobibliograficheskii spravochnik.* 3 vols. St. Petersburg: Rossiiskaia akademiia nauk and Russkoe geograficheskoe obshchestvo, 1993.

Krylov, G. V. "Slovo o sibirskom kedre." *Sibirskie ogni* 5 (1964): 154–65.

Kublitskii, G. *Chtoby priblizit' vek griadushchii.* Moscow: Detgiz, 1982.

Kumanev, V. A., ed. *Tragicheskie sud'by: Repressirovannye uchënye Akademii nauk SSSR, Sbornik statei.* Comp. I. G. Aref'eva. Moscow: Nauka, 1995.

Kurazhkovskii, Iu. N. *Zapovednoe delo v SSSR.* Rostov/Don: Izdatel'stvo Rostovskogo universiteta, 1977.

Kuzin, A. M., and A. A. Peredel'skii. "Okhrana prirody i nekotorye voprosy radioaktivno-ekologicheskikh sviazei." *OPZD* 1 (1956): 65–78.

Labedz, Leopold. *The Use and Abuse of Sovietology.* New Brunswick, N.J.: Transaction, 1987.

Lavrenko, E. M., V. G. Geptner, S. V. Kirikov, and A. N. Formozov. "Perspektivnyi plan geograficheskoi seti zapovednikov SSSR (proekt)." *OPZD* 3 (1958): 3–92.

Leonov, Leonid. *Sobranie sochinenii.* Moscow: Khudozhestvennaia literatura, 1972.

———. "V zashchitu Druga." *Izvestiia,* 28 December 1947.

Levina, F. Ia., comp. *Geobotanika v Botanicheskom institute im. V. L. Komarova AN SSSR.* 2 vols. Leningrad: Nauka—Leningradskoe otdelenie, 1971 and 1978.

Lewin, Moshe. *The Gorbachev Phenomenon: A Historical Interpretation.* Berkeley and Los Angeles: University of California Press, 1988.

Lieven, Anatol. *The Baltic Revolution: Estonia, Latvia, Lithuania, and the Path to Independence.* New Haven: Yale University Press, 1993.

Lipshits, S. Iu. *Moskovskoe obshchestvo ispytatelie prirody za 135 let ego sushchestvovaniia, 1805–1940.* Moscow: MOIP, 1940.

Löwenhardt, John. *Decision Making in Soviet Politics.* New York: St. Martin's Press, 1981.

Lysenko, E. G., and I. Terent'ev. *Komsomol i okhrana prirody.* Series: Biblioteka komsomol'skogo aktivista. Moscow: Molodaia gvardiia, 1978.

Lysenko, T. D. "Estestvennyi otbor i vnutrividovaia konkurentsiia." *Izbrannye sochineniia.* Moscow: Moskovskii rabochii, 1953.

Makarov, V. N. "Gosudarstvennye zapovedniki RSFSR k 30-letiiu velikoi oktiabrskoi sotsialisticheskoi revoliutsii." *Nauchno-metodicheskie zapiski Glavnogo upravleniia po zapovednikam pri SM RSFSR* 10 (1948): 3–8.

———. *Okhrana prirody v SSSR.* Moscow: Goskul'tprosvetizdat, 1947.

———. *Zapovedniki SSSR.* 2 vols. Moscow: Gosudarstvennoe izdatel'stvo geograficheskoi literatury, 1951.

Malinovskii, A. V. "Zapovedniki sovetskogo soiuza." *Dostizheniia nauki i peredovogo opyta v sel'skom khoziaistve* 7 (1953): 72–77.

Man and the Biosphere. Moscow: Nauka, 1984.

Manteifel', B. P. "Vospominaniia o P. A. Manteifele." *Okhota i okhotnich'e khoziaistvo* 7 (1982): 34–35.

McClelland, James C. *Autocrats and Academics: Education, Culture, and Society in Tsarist Russia.* Chicago: University of Chicago Press, 1979.

Matveev, E. "Vot kakie karasi!" *Krokodil* 1 (January 10, 1965).

Medvedev, S. I., and N. T. Nechaeva. "Pamiati V. V. Stanchinskogo." *Biulleten' MOIP* 82, no. 6 (1977): 109–17.

Medvedev, Zhores A. *Nuclear Disaster in the Urals.* Trans. George Saunders. New York: Random House, 1979.

———. *The Rise and Fall of T. D. Lysenko.* Garden City, N.Y.: Doubleday, 1971.

Merkulov, A. "Trevoga o Baikale." *Pravda,* 28 February 1965.

Mikhailov, F. N. "Okhrana vodoemov ot zagriazneniia—vsenarodnoe delo." *Okhrana prirody i ozelenenie* 6 (1960): 3–6.

Mikhailov, V. "Natsional'noe dvizhenie (Moldavskii variant)." *Polis* 4 (1992): 85–93.

Mikhel'son, V. A. "Rasshirenie i natsional'naia organizatsiia nauchnykh issledovanii v Rossii." *Priroda* 5–6 (1916), cols. 696–98.

Ministerstvo sel'skogo khoziaistva SSSR. Glavnoe upravlenie lesnogo khoziaistva. Upravlenie po zapovednikam i okhotnich'emu khoziaistvu. *Gosudartstvennaia plemennaia kniga zubrov i bizonov (chistokrovnykh, chistoporodnykh i gibridnykh).* Vol. 1. Moscow: Izdatel'stvo Ministerstva sel'skogo khoziaistva SSSR, 1956.

MOIP. *Otchet o rabote obshchestva v techenie 1969–71 gg.* Ed. and comp. K. M. Efron. Moscow: Nauka, 1972.

———. *Otchet o rabote obshchestva v techenie 1972–74 gg.* Ed. and comp. K. M. Efron. Moscow: Nauka, 1975.

———. *Otchet o rabote obshchestva v techenie 1984–86 gg.* Comp. K. M. Efron. Moscow: MOIP, 1988.

———. *Otchet o rabote obshchestva za 1957–1959.* Comp. K. M. Efron. Moscow: Nauka, 1960.

Mokievskii, V., I. Chestin, and E. Shvarts. "Prirodu ne obmanesh'." *Komsomol'skaia pravda,* 5 September 1986.

Moskovskii gosudarstvennyi universitet. Molodëzhnyi sovet po okhrane prirody. *Organizatsiia molodëzhnogo dvizheniia po okhrane okruzhaiushchei sredy i ratsional'nomu prirodopol'zovaniiu.* Ed. I. I. Rusin. Moscow: Izdatel'stvo Moskovskogo universiteta, 1988.

Nagibina, T. E. "Zadachi sanitarnoi okhrany vodoëmov." *OPZD* 1 (1956): 36–46.

Nasimovich, A. A. "Ekologicheskie posledstviia vkliucheniia novogo vida v materikovye biotsenozy (ondatra v Evrazii)." *Zoologicheskii zhurnal* 45, no. 11 (1966): 1593–98.

———."V Berezinskom zapovednike." *Biulleten' MOIP* 84, no. 3 (1979): 111–13.

Nauchnye s"ezdy, konferentsii i soveshchaniia v SSSR, 1954–1960: Bibliograficheskii ukazatel'. Moscow: Nauka, 1966.

Nauchnyi tsentr biologicheskikh issledovanii AN SSSR. Moskovskii gosudarstvennyi universitet. Molodëzhnyi sovet po okhrane prirody, Druzhina Biofaka MGU po okhrane prirody, Laboratoriia ekologii i okhrany prirody kafedry vysshikh rastenii Biologicheskogo fakul'teta. Rabochaia gruppa po obrazovaniiu i podgotovke spetsialistov sovetskogo komiteta po programme IuNESKO "Chelovek i biosfera." *Napravleniia i metody raboty po programme "FAUNA."* Pushchino: Nauchnyi tsentr biologicheskikh issledovanii AN SSSR, 1983.

Nesterov, V. G. *Voprosy upravleniia prirodoi.* Moscow: Lesnaia promyshlennost', 1981.

Nikolaevskii, A. G. *Natsional'nye parki.* Moscow: Agropromizdat, 1985.

Noskov, Yu. G. "Environmental Devastation in the USSR during the Second World War." *Environmental Management in the USSR* 4 (1985): 87–95.

Novikov, Georgii Aleksandrovich. *Ocherk istorii ekologii zhivotnykh.* Leningrad: Nauka, Leningradskoe otdelenie, 1980.

Nukhimovskaia, Iu. D."Biologicheskie i geograficheskie predposylki optimizatsii territorii zapovednikov." In *Geograficheskie razmeshchenie zapovednikov v RSFSR i organizatsii ikh deiatel'nosti. Sbornik,* ed. N. K. Noskova, 23–59. Moscow, TsNIL Glavokhota, 1981.

O Leonove. Comp. V. Chivilikhin. Moscow: Sovremennik, 1979.

Oborin, A. I. "Troitskoe uchebno-opytnoe lesnoe khoziaistvo Permskogo gosudarstvennogo universiteta." *Okhrana prirody na Urale* 1 (1960): 111–15.

Obshchestvo i prirodnaia sreda: Sbornik. Moscow: Znanie, 1980.

Olwig, Karen Fog, and Kenneth Olwig. "Underdevelopment and the Development of 'Natural' Park Ideology." *Antipode* 11, no. 2 (1979): 16–25.

Orlov, G. M. "Snova o Baikale" (pis'mo v redaktsiiu). *Literaturnaia gazeta,* 10 April 1965.

Osborne, Michael A. *Nature, the Exotic, and the Science of French Colonialism.* Bloomington: Indiana University Press, 1994.

Parfёnov, V. F. "Kedrogradu byt'!" *Komsomol'skaia pravda,* 19 February 1963.

———. "Kedrovye lesa: Vchera, segodnia, zavtra." *Vestnik Federal'noi sluzhby lesnogo khoziaistva Rossii* 11 (July 7, 1994): 3–4.

———. "Nash otvet A. D. Kovalevskomu." *Nash sovremennik* 5 (1960): 217–20.

Peredel'skii, A. A., and A. M. Kuzin "Okhrana prirody i nekotorye voprosy radioaktivno-ekologicheskikh sviazei." OPZD 1 (1956): 65–78.

Peterson, D[emosthenes] J[ames]. *Troubled Lands: The Legacy of Soviet Environmental Destruction.* Boulder, Colo., San Francisco, and Oxford: Westview Press and RAND, 1993.

Petro, Nicolai N. "'The Project of the Century': A Case Study of Russian National Dissent." *Studies in Comparative Communism* 20, nos. 3–4 (1987): 235–52.

Pidzhakov, A. Iu. *Sovetskaia ekologicheskaia politika 1970-kh—nachala 1990-kh godov.* St. Petersburg: Izdatel'stvo Sankt-Peterburgskogo universiteta ekonomiki i finansov, 1994.

Plenum Tsentral'nogo komiteta Kommunisticheskoi partii sovetskogo soiuza, 10–18 ianvaria 1961 g., Stenograficheskii otchet. Moscow: Gosizdpolitlit, 1961.

Pospelov, Gennadii. "Razmyshleniia o sud'be Baikala." *Sibirskie ogni* 6 (1963): 154–64.

Preobrazhenskii, S. M. "Gosudarstvennye zapovedniki SSSR." *Nauka i zhizn'* 3 (1947): 31–34.

Prezent, I. I. "Peredelka zhivoi prirody." *Ogonёk* 10 (1949): 8.

Prirodookhrannoe obrazovanie v universitetakh. Uchebnoe posobie. Ed. V. E. Sokolov. Moscow: Izdatel'stvo Moskovskogo universiteta, 1985.

Prirodookhrannoe vospitanie i obrazovanie. Moscow: Izdatel'stvo Moskovskogo universiteta, 1983.

Pronger, Brian. *The Arena of Masculinity: Sports, Homosexuality, and the Meaning of Sex.* New York: St. Martin's Press, 1990.

Pryde, Philip R. *Conservation in the Soviet Union.* Cambridge, Eng.: Cambridge University Press, 1972.

———. *Environmental Management in the Soviet Union.* Cambridge, Eng.: Cambridge University Press, 1991.

Ramenskii, L. G. "Klassifikatsiia zemel' po rastitel'nomu pokrovu." *Problemy botaniki* 1 (1950): 484–512.

Ramensky, L. G. "Basic Regularities of Vegetation Cover and Their Study (on the Basis of Geobotanic Researches in Voronezh Province)." Trans. John L. Brooks. *Bulletin of the Ecological Society of America* 64, no. 1 (March 1983). Orig. publ. in *Vestnik opytnogo dela Sredne-chernozemnoi oblasti* (1924): 37–73.

Rashek, V. L., N. G. Vasil'ev, and A. V. Chumakova. "Okhrana soobshchestv v zapovednikakh." In *Issledovaniia v oblasti zapovednogo dela: Sbornik nauchnykh trudov,* 3–21. Moscow: Minsel'khoz SSSR, 1984.

Razin, S. "'Bomba pochtal'ona Vasil'eva." *Komsomol'skaia pravda,* 15 March 1988.

Reimers, Nikolai Fëdorovich. "Bez prava na oshibku." *Chelovek i priroda* 10 (1980): 13–64.

Reimers, N. F., and F. R. Shtil'mark. "Etalony prirody." *Chelovek i priroda* 3 (1979): 7–63.

———. *Osobo okhraniaemye prirodnye territorii.* Moscow: Mysl', 1978.

Rittersporn, Gábor Tamás. *Stalinist Simplifications and Soviet Complications: Social Tensions and Political Conflicts in the USSR, 1933–1953.* Chur, Switzerland: Harwood Academic Publishers, 1991.

Rossiiskaia akademiia nauk. Materialy k biobibliografii uchënykh. *Vadim Nikolaevich Tikhomirov.* Foreword by V. S. Novikov, I. A. Gubanov, and T. A. Rabotnov. Comp. I. A. Gubanov and R. I. Kuz'menko. Series: Biologicheskie nauki: Botanika, no. 13. Moscow: Nauka, 1994.

Schmitt, Peter. *Back to Nature: The Arcadian Myth in Urban America.* New York: Oxford University Press, 1969.

Scott, James C. *Domination and the Arts of Resistance: Hidden Transcripts.* New Haven: Yale University Press, 1990.

———. *Weapons of the Weak: Everyday Forms of Peasant Resistance.* New Haven: Yale University Press, 1985.

Sedaitis, Judith B., and Jim Butterfield, eds. *Perestroika from Below: Social Movements in the Soviet Union.* Boulder, Colo.: Westview Press, 1991.

Sepp, E. K. "Neobkhodimo reorganizovat' nauchnye obshchestva." *VARNITsO* 1 (1930): 20–22.

Shaposhnikov, L. K. "Deiatel'nost' Komissii po okhrane prirody Akademii nauk SSSR v 1956–1957 gg." OPZD 3 (1958): 105–11.

Shelkovnikov, S. S. "K voprosu o 'mal'tuzianstve' v biologii." *Biulleten' MOIP* 59, no. 3 (1954): 89–108.

Shipunov, F. Ia. "Biosfernaia etika." In *Ekologicheskaia al'ternativa: istoki bedy, znaki bedy, ekologicheskaia al'ternativa,* ed. M. Ia. Lemeshev and A. V. Iablokov, 447–54. Moscow: Progress, 1990.

———. *Istina Velikoi Rossii.* Moscow: Molodaia gvardiia, 1992.

Shlapentokh, Vladimir. *Soviet Intellectuals and Political Power: The Post-Stalin Era.* Princeton: Princeton University Press, 1990.

Shtil'mark, Feliks Robertovich. *Istoriografiia rossiiskikh zapovednikov (1895–1995)*. Moscow: LOGATA, 1996.

———. "S drugogo berega" [Afterword to the Russian edition of *Models of Nature*]. In *Ekologiia v sovetskoi rossii. Arkhipelag svobody: Zapovedniki i okhrana prirody*, by D. Vainer [Uiner], trans. E. P. Kriukova, ed. F. R. Shtil'mark, 375–84. Moscow: Progress Publishers, 1991.

Shtil'mark, F. R., and M. V. Geptner [Heptner]. "Tragediia sovetskikh zapovednikov (k 40-letiiu 'reorganizatsii' zapovednoi sistemy v SSSR)." *Biulleten' MOIP* 98, no. 2 (1993): 97–113.

Shutova, T. A. "Kirishskii sindrom (Khronika chetyrëkh let perestroiki)." In *Ekologicheskaia al'ternativa: Istoki bedy, znaki bedy, ekologicheskaia al'ternativa*, ed. M. Ia. Lemeshev and A. V. Iablokov, 693–95. Moscow: Progress, 1990.

Shvarts, E. A. "O politicheskikh pravakh samodeiatel'nykh organizatsii." In *Ekologiia, demokratiia, molodëzh'*, ed. O. N. Ianitskii, 85–96. Moscow: Filosofskoe obshchestvo SSSR, Moskovskoe otdelenie, Molodëzhnaia sektsiia, 1990.

Shvarts, S. S. *Tekhnicheskii progress i okhrana prirody: Lektsiia*. Sverdlovsk: Sverlovskii gorkom KPSS and Sverdlovskaia oblastnaia organizatsiia obshchestva "Znanie," 1974.

———. "Teoreticheskie osnovy global'nogo ekologicheskogo prognozirovaniia." In *Vsestoronnyi analiz okruzhaiiushchei prirodnoi sredy*, 181–91. Leningrad: Gidrometeoizdat, 1976.

———. "Voprosy akklimatizatsii mlekopitaiushchikh na Urale." *Trudy instituta biologii Ural'skogo filiala Akademii nauk SSSR* 18 (1959): 3–22.

Singer, Peter. *Animal Liberation*. New York: Random House, 1975.

Singleton, Fred, ed. *Environmental Problems in the Soviet Union and Eastern Europe*. Boulder, Colo.: Lynne Rienner, 1987.

Skalon, V. N. *Okhraniaite prirodu!* Irkutsk: Irkutskoe knizhnoe izdatel'stvo, 1958.

———. "Sushchnost' biotekhnii." *Biologicheskie nauki* (Kazakhskii gosudarstvennyi universitet) 1 (1971): 165–75.

Slovo v zashchitu Baikala: Materialy diskussii. Comp. B. F. Lapin. Irkutsk: Vostochno-Sibirskoe knizhnoe izdatel'stvo, 1987.

Smolin, P. P. "Organizatsiia i rabota tsentral'nogo kruzhka iunykh biologov iunosheskoi sektsii VOOP." *Okhrana prirody* 14 (1951): 151–52.

"Snova o Baikale." *Oktiabr'* 10 (1963): 171–74.

Sobolev, Leonid, Sergei Sartakov, Frants Taurin, et al. "V zashchitu Baikala!" *Literaturnaia gazeta*, 18 March 1965.

Sobolev, L. N., and E. E. Syroechkovskii. "Voprosy okhrany prirodnykh resursov strany." *Izvestiia AN SSSR. Seriia geograficheskaia* 6 (1956): 142–43.

Sokolov, V. V. *Ocherki istorii ekologicheskoi politiki Rossii*. St. Petersburg: Izdatel'stvo Sankt-Peterburgskogo universiteta ekonomiki i finansov, 1994.

Sotsial'no-Ekologicheskii Soiuz. *Informatsionno-metodicheskoe pis'mo*. Moscow: Sotsial'no-Ekologicheskii Soiuz, 1990.

———. *Vyzhivëm vmeste/Surviving Together, sbornik materialov*. Moscow: "Put'," 1991.

Sotsial'no-ekonomicheskie i ekologicheskie aspekty sovershenstvovaniia deiatel'nosti zapovednikov: Sbornik nauchnykh trudov. Moscow: TsNIL Glavokhota RSFSR, 1985.

Sotsio-Ekologicheskii Soiuz. *Sotsio-ekologicheskii soiuz: Istoriia i real'nost'. Sbornik materialov*. Moscow: SoES, 1994.

———. *Tridtsat' let dvizheniia: Neformal'noe prirodookhrannoe molodëzhnoe dvizhenie v SSSR. Fakty i dokumenty.* Ed. and comp. Sviatoslav Zabelin and Sergei Germanovich Mukhachev. Kazan': Terra, 1993.

———. *Tridtsat' let dvizheniia: Neformal'noe prirodookhrannoe molodëzhnoe dvizhenie v SSSR. Fakty i dokumenty, 1960–1987.* Ed. and comp. Sviatoslav Zabelin. Moscow: Put', 1990.

Soyfer, Valery N. *Lysenko and the Tragedy of Soviet Science.* Trans. Leo Gruliow and Rebecca Gruliow. New Brunswick, N.J.: Rutgers University Press, 1994.

Sprinkle, Robert Hunt. *Profession of Conscience: The Making and Meaning of Life-Sciences Liberalism.* Princeton: Princeton University Press, 1994.

Starr, S. Frederick. "Visionary Town Planning during the Cultural Revolution." In *Cultural Revolution in Russia, 1928–1931,* ed. Sheila Fitzpatrick, 207–40. Bloomington: Indiana University Press, 1978.

Stepanov, V. A. "Zapovednoe delo v Kazakhstane." In *Zapovedniki Kazakhstana: Ocherki,* 4–6. 2d ed. Alma-ata: Kazakhskoe gosudarstvennoe izdatel'stvo and Upravlenie okhotnich'em khoziaistvom pri Minsel'khoze Kaz SSR, 1963.

Stewart, John Massey, ed. *The Soviet Environment: Problems, Policies, and Politics.* Cambridge, Eng.: Cambridge University Press, 1992.

Stites, Richard. *Revolutionary Dreams: Utopian Vision and Experimental Life in the Russian Revolution.* New York: Oxford University Press, 1989.

Struchkov, Anton Yu. "Nature Protection as Moral Duty: The Ethical Trend in the Russian Conservation Movement." *Journal of the History of Biology* 25, no. 3 (Fall 1992): 413–28.

Sushchenyz, L. M., V. I. Parfënov, G. V. Vynayev, and G. F. Rykovsky. "Scientific Principles of Designing a System of Nature Reserves in Byelorussian SSR." In *Conservation, Science, and Society: First International Biosphere Reserve Conference, Minsk, September 26–October 2, 1983,* 1:121–24. Paris: UNESCO-UNEP, 1984.

Svirezhev, Iu. M., and D. O. Logofet. *Ustoichevost' biologicheskikh sistem.* Moscow: Nauka, 1978.

Taurin, Frants. "Baikal dolzhen byt' zapovednikom." *Literaturnaia gazeta,* 10 February 1959.

———. *Daleko v strane Irkutskoi: Sibirskoe povestvovanie.* Moscow: Sovetskii pisatel', 1986.

Tenditnik, N. "Mnogo shuma—iz-za chego?" *Izvestiia,* 15 August 1985.

Tikhomirov, Vadim Nikolaevich. "Istoriia i deiatel'nost' druzhiny biologicheskogo fakul'teta MGU po okhrane prirody." In *Studenchestvo i okhrana prirody: Materialy konferentsii, posviashchennoi 20-letiiu Druzhiny biologicheskogo fakul'teta MGU po okhrane prirody, 11–12 dekabria 1980 g.,* ed. K. V. Avilova, 12–23. Moscow: Izdatel'stvo Moskovskogo universiteta, 1982.

Timasheff, Nicholas. *The Great Retreat: The Growth and Decline of Communism in Russia.* New York: Dutton, 1946.

Timofeev-Resovskii, N. V., and A. N. Tiuriukanov. "Biogeotsenologiia i pochvovedenie." *Biulleten' MOIP* 72, no. 2 (1967): 106–17.

Tobey, Ronald C. *Saving the Prairies: The Life Cycle of the Founding School of American Plant Ecology, 1895–1955.* Berkeley and Los Angeles: University of California Press, 1981.

Trofimuk A. "Tsena vedomstvennogo upriamstva: otvet ministru SSSR G. M. Orlovu." *Literaturnaia gazeta,* 15 April 1965.

Tumarkin, Nina. *Lenin Lives! The Lenin Cult in Russia.* Cambridge, Mass.: Harvard University Press, 1983.

Turbin, Vladimir Nikolaevich. "Kitezhane: Iz zapisok russkogo intelligenta." In *Pogruzhenie v triasinu (anatomiia zastoia), sbornik,* 346–70. Moscow: Progress, 1991.

Ursul, A. D., ed. *Philosophy and the Ecological Problems of Civilisation.* Trans. H. Campbell Creighton. Moscow: Progress, 1983.

Varsonof'eva, V. A. *Moskovskoe obshchestvo ispytatelei prirody i ego znachenie v razvitii otechestvennoi nauki.* Moscow: MOIP, 1955.

Vernadskii, V. I. "Osnovoiu zhizni—Iskanie istiny." *Novyi mir* 3 (1988): 218–19.

Volkov, Oleg. "Tropa brakonera," *Literaturnaia gazeta,* 6 February 1980.

———. "Tuman nad Baikalom." *Literaturnaia gazeta,* 6 February 1965.

———. "Tuman ne resseialsia." *Literaturnaia gazeta,* 13 April 1965.

VOOP. *Materialy Pervogo s"ezda Vserossiiskogo obshchestva sodeistviia okhrane prirody i ozeleneniiu naselënnykh punktov, prokhodivshego 15–17ogo avgusta 1955 g. v Moskve.* Moscow: VOOP, 1956.

———. *Materialy vtorogo s"ezda Vserossiiskogo obshchestva okhrany prirody i ozeleneniia naselënnykh punktov, prokhodivshego 15–18 dekabria 1959 goda v gorode Moskve.* Moscow: VOOP, 1960.

Vorontsov, Nikolai Nikolaevich. Afterword to *Stranitsy zhizni: Istoriia odnogo issledovaniia,* by M. M. Zavadovskii, 316–35. Moscow: Izdatel'stvo Moskovskogo universiteta, 1991.

———. "Nature Protection and Government in the USSR." *Journal of the History of Biology* 25, no. 3 (Fall 1992): 369–83.

Voroshilov Iu. I., and P. A. Nedotko. "Ispol'zovanie mineral'nogo topliva i izmenenie prirodnoi sredy." *OPZD* 6 (1960): 5–14.

"Vsë prodolzhaetsia v nas" [Interview with correspondent V. Shurygin of *Na boevom postu*]. In *Zhit' glavnym,* 246–49. Moscow: Molodaia gvardiia, 1988.

Vserossiiskoe ordena trudovogo krasnogo znameni obshchestvo okhrany prirody. *Sbornik rukovodiashchikh dokumentov, primeniaemykh v organizatsiiakh VOOP.* Moscow: VOOP, 1988.

Vsesoiuznoe mineralogicheskoe obshchestvo. Vsesoiuznyi nauchno-issledovatel'skii geologicheskii institut. Vsesoiuznoe istoriko-prosvetitel'skoe obshchestvo "Memorial." Sankt-Peterburgskii gornyi institut. *Repressirovannye geologi. (Biograficheskie materialy).* St. Petersburg: Vsesoiuznoe mineralogicheskoe obshchestvo, Vsesoiuznyi nauchno-issledovatel'skii geologicheskii institut, Vsesoiuznoe istoriko-prosvetitel'skoe obshchestvo "Memorial," Sankt-Peterburgskii gornyi institut, 1992.

Vsesoiuznyi institut nauchno-tekhnicheskoi informatsii po sel'skomu khoziaistvu. *Okhrana prirody i zapovednoe delo v SSSR za 50 let. Obzor literatury.* Fasc. 14 (85). Moscow: Ministerstvo sel'skogo khoziaistva SSSR and VINITI, 1967.

Vucinich, Alexander. *Empire of Knowledge: The Academy of Sciences of the USSR (1917–1970).* Berkeley and Los Angeles: University of California Press, 1984.

Weiner, Douglas R. "The Changing Face of Soviet Conservation." In *The Ends of the Earth,* ed. Donald Worster, 252–73. Cambridge, Eng.: Cambridge University Press, 1989.

———. "Demythologizing Environmentalism." *Journal of the History of Biology* 25, no. 3 (Fall 1992): 385–411.

———. "Environmental Problems under Gorbachev." In *The Gorbachev Encyclopedia,* ed. Joseph L. Wieczynski, 133–51. Salt Lake City: Charles Schlacks, Jr., 1993.

———. "Man of Plastic: Gor'kii's Vision of Humans in Nature." *Soviet and Post-Soviet Review* 22, no. 1 (Spring 1995): 65–88.

———. *Models of Nature: Ecology Conservation and Cultural Revolution in Soviet Russia.* Bloomington: Indiana University Press, 1988. A slightly revised Russian edition was published as *Ekologiia v sovetskom soiuze. Arkhipelag svobody: Zapovedniki i okhrana prirody.* Trans. Elena Kriukova, with an afterword by Feliks Robertovich Shtil'mark. Moscow: Progress, 1991.

———. "Prometheus Rechained: Social Sources of Change in Soviet Ecological Thought." In *Science and the Soviet Social Order,* ed. Loren R. Graham, 70–93. Cambridge, Mass.: Harvard University Press, 1990.

———. "'*Razmychka?*' Urban Unemployment and Peasant In-migration as Sources of Social Conflict." In *Russia in the Era of NEP: Explorations in Soviet Society and Culture,* ed. Sheila Fitzpatrick, Alexander Rabinowitch, and Richard Stites, 144–55. Bloomington: Indiana University Press, 1991.

———. "Three Men in a Boat: The All-Russian Society for the Protection of Nature (VOOP) in the Early 1960s." *Soviet and Post-Soviet Review* 20, nos. 2–3 (1993): 195–212.

Worster, Donald. *Nature's Economy: A History of Ecological Ideas.* 2d ed. New York: Cambridge University Press, 1994.

Yanitsky, Oleg. *Russian Environmentalism: Leading Figures, Facts, and Opinions.* Moscow: Mezhdunarodnyje Otnoshenija, 1993.

Zabelin, I. M. *Ocherki istorii geograficheskoi mysli v SSSR, 1917–1945.* Moscow: Nauka, 1989.

Zablotskii, M. A. "Vosstanovlenie zubra v SSSR i za granitsei." OPZD 4 (1960): 56.

Zalygin, Sergei. *Kritika i publitsistika.* Moscow: Sovremennik, 1987.

———. "Otkroveniia v nashem imeni." *Novyi mir* 10 (1992): 214–16.

———. *Pozitsiia.* Moscow: Sovetskaia Rossiia, 1988.

———. "Proekt: Nauchnaia obosnovannost' i otvetstvennost'. Razdum'ia pisatelia." *Kommunist* 13 (September 1985): 63–73.

———. "Vodnoe khoziaistvo bez stoimosti . . . vody?" *Literaturnaia gazeta,* 2 October 1985.

Zavadovskii, M. M. *Stranitsy zhizni: Istoriia odnogo issledovaniia.* Moscow: Izdatel'stvo Moskovskogo universiteta, 1991.

Zharkov, I. V. *Prosteishie nabliudeniia v prirode (posobie dlia nabliudatelei zapovednikov).* 2d ed. Moscow: Minsel'khoz SSSR, 1956.

Zhigulin, Anatolii. *Chërnye kamni: Avtobiograficheskaia povest'.* Moscow: Moskovskii rabochii, 1989.

Zhukova, I. I. "Svoi liudi u prirody." In *Ekologiia, demokratiia, molodëzh',* ed. O. N. Ianitskii, 15–23. Moscow: Filosofskoe obshchestvo SSSR, Moskovskoe otdelenie, Molodëzhnaia sektsiia, 1990.

Ziegler, Charles E. *Environmental Policy in the USSR.* Amherst: University of Massachusetts Press, 1987.

Zonn, Sergei Vladimirovich. *Vladimir Nikolaevich Sukachëv, 1880–1967*. Moscow: Nauka, 1987.

Zubkova, Elena Iur'evna. *Obshchestvo i reformy, 1945–1964*. Afterword by P. V. Volobuev. Series: "Pervaia monografiia" of the Assotsiatsiia issledovatelei rossiiskogo obshchestva XX veka. Moscow: Izdatel'skii tsentr "Rossiia molodaia," 1993.

INTERVIEWS

Borodulina, Tat'iana Leonidovna. Interview with author. Moscow, August 9, 1995.

Chernousova, Galina Borisovna. Interview with author. Moscow, June 4, 1996.

Dorovatovskii, Nikolai Sergeevich. Interview with author. Moscow, January 3, 1980.

Efremov, Iurii (Georgii) Konstantinovich. Interview with author. Moscow, August 4, 1995.

Efron, Konstantin Mikhailovich. Interviews with author. Moscow, July 5, 1991 and August 2, 1995.

Filonov, Konstantin Pavlovich. Interview with author. Moscow, March 6, 1986.

Gall, Iakov Mikhailovich. Interview with author. Leningrad, May 20, 1986.

Ianitskii, Oleg Nikolaevich. Interview with author. Moscow, August 20, 1995.

Ianshin, Aleksandr Leonidovich. Interview with author. Moscow, July 24, 1992.

Kavtaradze, Dmitrii Nikolaevich. Interviews with author. Moscow, March 23 and July 11, 1991.

Liapunova, Elena Alekseevna. Interview with author. Moscow, August 12, 1995.

Masing, Viktor and Linda Poots. Interview with author. Tartu, Estonia, July 25, 1995.

Nasimovich, Andrei Aleksandrovich. Interviews with author. Moscow, April 16 and April 18, 1980.

Parfënov, Vitalii Feodos'evich. Interview with author. Moscow, August 21, 1995.

Podol'skii, Evgenii Makar'evich. Interview with author. Moscow, July 9, 1991.

Shvarts, Evgenii Arkad'evich. Interview with author. Moscow, May 26, 1996.

Zablotskii, Mikhail Aleksandrovich. Interview with author. Prioksko-Terrasnyi zapovednik, April 20, 1991.

Zonn, Sergei Vladimirovich. Interview with author. Moscow, August 3, 1995.

INDEX

Italicized numbers refer to pages with illustrations.

Compositor: Prestige Typography
Text: 10/12 Baskerville
Display: Baskerville

www.ingramcontent.com/pod-product-compliance
Lightning Source LLC
Chambersburg PA
CBHW030632270326
41929CB00007B/46

9780520232136